Dairy 4.0

Revolutionizing the Dairy Sector Through Industry 4.0 Technologies

The Fourth Industrial Revolution (Industry 4.0) in the Food Sector: Revolutionizing Agri-Food and Seafood through Smart Technologies

Dairy 4.0

Revolutionizing the Dairy Sector Through Industry 4.0 Technologies

Edited by

Abderrahmane Aït-Kaddour

VetAgro Sup, Departement of Sustainable Foods, Management and Innovation/Clermont Auvergne, INRAE, VetAgro Sup, Cheese Research Unit, Lempdes, France

Abdo Hassoun

Faculty of Agricultural Engineering, University of Aleppo, Aleppo, Syria; Sustainable AgriFoodtech Innovation & Research (SAFIR), Arras, France

Series Editor

Abdo Hassoun

ELSEVIER

ACADEMIC PRESS

An imprint of Elsevier

Academic Press is an imprint of Elsevier
125 London Wall, London EC2Y 5AS, United Kingdom
50 Hampshire Street, 5th Floor, Cambridge, MA 02139, United States

Notices

ISBN: 978-0-443-44730-3

For Information on all Academic Press publications
visit our website at https://www.elsevier.com/books-and-journals

Publisher: Mara Conner
Acquisitions Editor: Vaishali Singh
Editorial Project Manager: Sadhna Rajan
Production Project Manager: Fahmida Sultana
Cover Designer: Mark Rogers

Typeset by MPS Limited, Chennai, India

Dedication

To the pioneers of food innovation and to the next generation of researchers, technologists, and changemakers, who dare to reimagine how food is produced, processed, shared, and sustained. May your curiosity spark discovery, your creativity inspire transformation, and your commitment help build food systems that are smarter, more resilient, more inclusive, and more sustainable for generations to come.

Contents

Part III
Current development and future landscape of dairy technology

List of contributors

Qurratu Aini Alya Adzkia Department of Chemistry, Universitas Padjadjaran, Bandung, Indonesia, France

Abderrahmane Aït-Kaddour VetAgro Sup, Departement of Sustainable Foods, Management and Innovation/Clermont-Auvergne, INRAE, VetAgro Sup, Cheese Research Unit, Lempdes, France

Moutaz Alhamada Istom, Higher School of International Agro-Development, Angers, France; Department of Animal Husbandry, Faculty of Agronomy, Al-Furat University, Deir ez-Zur, Syria

Sadia Aslam School of Food and Agricultural Sciences, University of Management and Technology, Lahore, Pakistan

Olfa Ben Moussa Research Laboratory: Technological Innovation and Food Security LR22-AGR01, Higher Institute of Food Industries (ESIAT), University of Carthage, Tunis, Tunisia

Sofiane Boudalia Département d'Écologie et Génie de l'Environnement, Université 8 Mai 1945, Guelma, Algeria; Laboratoire de Biologie, Eau et Environnement, Université 8 Mai 1945 Guelma, Algeria

Oumayma Boukria University Mohammed IV Polytechnic (UM6P), African Sustainable Agriculture Research Institute (ASARI), College of Agriculture and Environmental Sciences (CAES), Ben Guerir, Morocco

Mouna Boulares Research Laboratory: Technological Innovation and Food Security LR22-AGR01, Higher Institute of Food Industries (ESIAT), University of Carthage, Tunis, Tunisia

Alexandre Cabagnols Department of Humanities, INP Clermont-Auvergne, Polytech Clermont-Ferrand, CleRMa (Clermont Recherche Management—Research in Sustainable Management), Clermont-Ferrand, France

Yana Cahyana Department of Food Technology, Universitas Padjadjaran, Bandung, Indonesia, France

Benoit Cournoyer Laboratoire d'Ecologie Microbienne, Université Claude Bernard Lyon 1, INRAE, VetAgro Sup, Marcy l'étoile, France

Sami Fattouch Department of Biological Engineering and Chemistry, National Institute of Applied Sciences and Technology (INSAT), Laboratory Ecochimie (LR21ES02), University of Carthage, Tunis, Tunisia

Geneviève Gagne UMRF, Université Clermont Auvergne, INRAE, VetAgro Sup, Aurillac, France

Wessam Galia Laboratoire d'Ecologie Microbienne, Université Claude Bernard Lyon 1, INRAE, VetAgro Sup, Marcy l'étoile, France

Zakia Gueboudji Department of Microbiology and Biochemistry, Faculty of Natural and Life Sciences, University of Batna 2 Mustapha Ben Boulaid, Batna, Algeria

Zouheyr Hadri Department of Agricultural sciences, Faculty of Sciences and Technology, University of Relizane, Relizane, Algeria

Ahmed Hamad Food Hygiene and Control Department, Faculty of Veterinary Medicine, Benha University, Benha, Egypt

Abdo Hassoun Sustainable AgriFoodtech Innovation & Research (SAFIR), Arras, France; Faculty of Agricultural Engineering, University of Aleppo, Aleppo, Syria

Samuel Imathiu Department of Food Science and Technology, Jomo Kenyatta University of Agriculture and Technology, Nairobi, Kenya

Anum Ishaq School of Food and Agricultural Sciences, University of Management and Technology, Lahore, Pakistan

Rivan Satria Jaluwardhana Department of Food Technology, Universitas Padjadjaran, Bandung, Indonesia, France

Peter Kahenya Department of Food Science and Technology, Jomo Kenyatta University of Agriculture and Technology, Nairobi, Kenya

Imane Kouadri Department of Process Engineering, Laboratory of Industrial Analysis and Materials Engineering, University of May 8 1945 Guelma, Guelma, Algeria

Atef Lakoud Research Laboratory: Technological Innovation and Food Security LR22-AGR01, Higher Institute of Food Industries (ESIAT), University of Carthage, Tunis, Tunisia

Mohd Nizam Lani Faculty of Food Science and Agrotechnology, Universiti Malaysia Terengganu, Kuala Nerus, Terengganu, Malaysia

Françoise Leriche UMRF, Université Clermont Auvergne, INRAE, VetAgro Sup, Aurillac, France

Yassine Louzzani Department of Sustainable Food Systems, Management and Innovation, VetAgro Sup, CleRMa (Clermont Recherche Management—Research in Sustainable Management), Lempdes, France

Imen Mahmoudi Research Laboratory: Technological Innovation and Food Security LR22-AGR01, Higher Institute of Food Industries (ESIAT), University of Carthage, Tunis, Tunisia

Cécile Moussard UMRF, Université Clermont Auvergne, INRAE, VetAgro Sup, Aurillac, France

Atiek Rostika Noviyanti Department of Chemistry, Universitas Padjadjaran, Bandung, Indonesia, France

Isaac Osuga Department of Animal Sciences, Jomo Kenyatta University of Agriculture and Technology, Nairobi, Kenya

Eddy Owaga Institute of Food Bioresources Technology, Dedan Kimathi University of Technology, Nyeri, Kenya

Olfa Rebai Department of Biological Engineering and Chemistry, National Institute of Applied Sciences and Technology (INSAT), Laboratory Ecochimie (LR21ES02), University of Carthage, Tunis, Tunisia

Jasur Safarov Tashkent State Technical University named after Islam Karimov, Tashkent, Uzbekistan

Amna Sahar National Institute of Food Science and Technology, University of Agriculture, Faisalabad, Pakistan

Shakhnoza Sultanova Tashkent State Technical University named after Islam Karimov, Tashkent, Uzbekistan; Tashkent City Khakimiyat, Tashkent, Uzbekistan

Ines Tarchi National School of Engineers of Sfax (ENIS), Laboratory of Electrochemistry and Environment, University of Sfax, Sfax, Tunisia

Hassen Touati Département d'Écologie et Génie de l'Environnement, Université 8 Mai 1945 Guelma, Guelma, Algeria; Laboratoire de Biologie, Eau et Environnement, Université 8 Mai 1945 Guelma, Guelma, Algeria

Hafiz Ubaid ur Rahman School of Food and Agricultural Sciences, University of Management and Technology, Lahore, Pakistan; Department of Animal Sciences, Purdue University, West Lafayette, IN, United States

Caroline Wybraniec Laboratoire d'Ecologie Microbienne, Université Claude Bernard Lyon 1, INRAE, VetAgro Sup, Marcy l'étoile, France

Yostina Atef Zaky Food Hygiene and Control Department, Faculty of Veterinary Medicine, Benha University, Benha, Egypt

About the editors

Dr. Abderrahmane Aït-Kaddour earned his engineering degree in food science and technology in 2003 (Graduate School of Chemistry and Physics of Bordeaux—formerly Institut of Science and Technology of Bordeaux, France), master's degree in food engineering in 2004 (University of Science in Montpellier, France), PhD in food engineering in 2007 (Institut Agro Montpellier—formerly Montpellier SupAgro until 2022, France), and completed his postdoctoral fellowship in food packaging in 2008 (INRAE Reims, France). He is currently a professor in food science and technology at VetAgro Sup Agronomic Campus Institut (Lempdes, France). Dr. Aït Kaddour is affiliated with the Department of Sustainable Foods, Management and Innovation for his teaching activities and with the Research Unit UMR sur le Fromage (specialized on cheese and dairy products) for his research activities.

His research work lies at the intersection of chemometric modeling, spectroscopy, multispectral/hyperspectral imaging, and food chemistry for food quality control and authentication. He authored more than 100 peer-reviewed publications, in high-impact factor journals, related to milk and cheese and, to a lesser extent, meat, fish, and cereal products. He has edited several books and contributed to several books' chapters on innovative food processing, analytical approaches to food preservation, characterization, and quality assessment. Prof. Aït Kaddour has been actively involved in numerous regional, national, and international research projects, fostering strong collaborations between academia and industry.

Beyond his research activities, he serves as an adjunct professor and/or visiting scholar at several international universities (Morocco, Uzbekistan, and Indonesia), where he teaches courses and supervises graduate students in food science, spectroscopy, multivariate data analysis, and advanced food processing. Renowned for his multidisciplinary outlook and scientific leadership, Prof. Aït Kaddour continues to advance the field of food science through a blend of rigorous analytics, innovation, and cross-border collaboration.

Dr. Abdo Hassoun (PhD in engineering of biological functions, Lille University of Science and Technology, France) is a multidisciplinary researcher specializing in food science, technology, innovation, and sustainability. Recognized among the world's top 2% of scientists, he has made substantial contributions to advancing research and innovation in these fields over the past decade. He has also recently been affiliated with the University of Aleppo, Syria, and is recognized as the most cited researcher from Syria.

Dr. Hassoun earned his bachelor's degree in agricultural engineering in 2007 from the University of Aleppo, Syria; obtained a master's degree in 2011 from Blaise Pascal University, France; and completed his PhD in 2016 at Lille University of Science and Technology, France. Following his doctorate, he joined the Norwegian Food Research Institute (Nofima) as a postdoctoral researcher, where he worked from 2017 to 2020.

In 2021, he founded Sustainable AgriFoodtech Innovation & Research (SAFIR), a nonprofit research organization that collaborates with academic and industrial partners worldwide to address food sustainability challenges and promote healthier and more affordable food systems. He also served at Université du Littoral Côte d'Opale (ULCO), France, during 2022 and 2023.

Currently, Dr. Hassoun is the chief executive officer of SAFIR and works as an independent researcher and scientific consultant. His research spans emerging food trends, Industry 4.0 and 5.0 technologies, food loss and waste reduction, innovative processing and analytical methods, and broader sustainability issues. With more than 150 peer-reviewed articles and reviews published in leading international journals, along with numerous book chapters and edited volumes, Dr. Hassoun continues to be a leading voice in advancing sustainable and intelligent food systems. Through his books and scientific contributions, he remains strongly committed to connecting science with society and to shaping a more sustainable future for food systems worldwide.

Preface

This volume is part of a book series edited by Dr. Abdo Hassoun entitled *The Fourth Industrial Revolution (Industry 4.0) in the Food Sector*. The series provides a comprehensive exploration of how Fourth Industrial Revolution technologies, commonly referred to as Industry 4.0, are transforming diverse sectors of the food and agricultural industries. Comprising multiple volumes, each dedicated to a specific food sector or category, the series highlights the transformative impact of advanced digital, physical, and biological technologies on food production, processing, and distribution. The volumes address major themes such as sustainability, quality enhancement, safety, traceability, and waste management across sectors including seafood, dairy, meat, olive farming, aquaponics, aquaculture, beekeeping, horticulture, grain production, and baking, among others. Key topics include automation, digitalization, precision farming, predictive maintenance, and the integration of artificial intelligence, the Internet of Things, smart sensors, robotics, blockchain, 3D printing, nanotechnology, and biotechnological advances.

The series also examines barriers to the adoption of Industry 4.0 technologies, current opportunities, and future perspectives, while providing valuable insights, practical applications, and case studies for researchers, practitioners, industry professionals, policymakers, educators, and students. This pioneering book series aims to bridge knowledge gaps, foster innovation and continuous improvement in the food and agricultural industries, and highlight how current developments in Industry 4.0 are paving the way for future advancements toward Industry 5.0.

In this context, the present volume dedicated to Dairy 4.0 focuses on the profound transformations occurring within the dairy sector in the era of Industry 4.0. Traditionally regarded as a cornerstone of the agri-food industry, the dairy sector is now at a critical crossroads, facing complex challenges related to sustainability, resource efficiency, market volatility, and evolving consumer expectations. These challenges, combined with rapid technological advancements, are driving the urgent need for transformation toward more intelligent, resilient, and sustainable dairy systems.

This volume is structured into three main parts, each addressing key dimensions of this transition.

Part I: Introduction and core technologies lays the foundation by exploring the current state of the dairy sector and the forces shaping its evolution. It begins by examining the major challenges and drivers pushing the sector toward transformation, followed by a comprehensive overview of the Dairy 4.0 concept, its enabling technologies, and key application areas. Emphasis is placed on how Industry 4.0 technologies contribute to advancing sustainability within dairy systems, integrating environmental and social considerations.

Part II: Enhancing dairy operations using Industry 4.0 technologies focuses on the practical implementation of these technologies across the dairy value chain. It highlights how digitalization and smart systems can enhance product quality and safety while strengthening traceability and transparency. In addition, this part addresses critical operational aspects such as waste management and the emergence of Packaging 4.0, demonstrating how innovative solutions can reduce environmental impact while improving efficiency and product integrity.

Part III: Current development and future landscape of dairy technology provides a forward-looking perspective on the evolution of the dairy sector. It explores cutting-edge developments such as precision dairy farming and smart, data-driven processing systems, which enable real-time decision-making and optimized resource use. This part also examines the differentiated socio-economic impacts of Dairy 4.0, particularly in biologically variable production systems, highlighting both opportunities and inequalities. Furthermore, it discusses the key barriers to the adoption of Industry 4.0 technologies, including economic, technical, and organizational challenges, while outlining current opportunities and future perspectives. The volume concludes by addressing the transition from Dairy 4.0 to Dairy 5.0, emphasizing a shift toward more human-centered, sustainable, and ethically grounded innovation.

By bringing together contributions from leading international experts, this volume offers a comprehensive and multidisciplinary perspective on Dairy 4.0. It aims to bridge knowledge gaps, support informed decision-making, and foster innovation across the dairy sector. Ultimately, this work seeks to contribute to the development of a smarter, more sustainable, and resilient dairy industry, aligned with the broader evolution toward Industry 5.0.

Part I

Introduction and core technologies

Chapter 1

The dairy sector at a crossroads: challenges, drivers, and the need for transformation

Ahmed Hamad[1], Yostina Atef Zaky[1], Abderrahmane Aït-Kaddour[2], and Abdo Hassoun[3,4]
[1]*Food Hygiene and Control Department, Faculty of Veterinary Medicine, Benha University, Benha, Egypt,* [2]*VetAgro Sup, Departement of Sustainable Foods, Management and Innovation/ Clermont-Auvergne, INRAE, VetAgro Sup, Cheese Research Unit, Lempdes, France,* [3]*Sustainable AgriFoodtech Innovation & Research (SAFIR), Arras, France,* [4]*Faculty of Agricultural Engineering, University of Aleppo, Aleppo, Syria*

1.1 Introduction

The dairy industry is a major part of global food systems, contributing significantly to food security, economic activity, and rural livelihoods. Dairy milk is produced across all regions of the world and supplies essential nutrients that are present prominently in human diets; milk's contribution to global nutrient supply comprises key macro- and micronutrients, making it an important component of food and nutrition security (Hamad and Awad, 2026). Additionally, the dairy sector creates millions of jobs and sustains rural economic structures, with a substantial portion of the labor force in many countries connected directly or indirectly to dairy production and processing (Hill, 2024). However, biological complexity, market fluctuations, longer value chains and growing societal and environmental expectations are affecting the modern dairy industry. At the foundation of dairy economics are ongoing and transformative changes at the socioecological level, as reflected in structural changes of where milking and consumption take place; longer agendas of production for particular markets; greater understanding about environmental concerns, including but not limited to dietary impacts on environment; and increasing market knowledge about alternative dairy product formulations (Bojovic and McGregor, 2023).

Traditional management and organizational systems are struggling to deal with accelerating uncertainty, variability, and systemic risk across dairy value

Dairy 4.0. DOI: https://doi.org/10.1016/B978-0-443-44730-3.00003-1

chains. Market disruptions and climate variability are problems. Traditional operational approaches are increasingly inadequate for addressing these challenges. In addition, dairy systems must contend with highly fragmented supply chains and growing consumer demand for healthier products, greater transparency regarding product origin, production practices, and sustainability performance. Some of the issues are related to markets that are not connected, technology that is not used in a consistent way and small farmers not being part of the official supply chain. These things make it even harder to deal with the problems we are facing with market disruptions and climate variability and supply chain issues (Hamad and Awad, 2026).

The arrival of Industry 4.0 has revolutionized multiple industrial sectors through the integration of digital technologies such as artificial intelligence (AI), the Internet of Things (IoT), big data analytics, automation, and cyber-physical systems. Manufacturing, logistics, healthcare, and energy systems have undergone profound transformations toward data-driven, intelligent, and interconnected operations (Hassoun, 2024, 2025). The dairy sector is increasingly influenced by these digital paradigms, creating new opportunities for precision farming, smart processing, predictive quality control, enhanced traceability, and sustainability optimization across the entire value chain.

In order to lay the groundwork for the ensuing debate of Dairy 4. 0, this chapter evaluates the dairy industry's structural characteristics, key challenges, and drivers of change. In doing so, it brings together research from current literature on global dairy systems and food systems transformation, underlining the need for comprehensive, technology-driven solutions to growing challenges in productivity, sustainability, welfare, quality, welfare, and supply chain resilience. These insights set the stage for a systematic exploration of Dairy 4.0 technologies and their potential to address complex dairy system challenges.

1.2 Current state of the dairy sector: a global perspective

The dairy industry plays an important role in the global food system, providing a variety of dairy products (i.e., cheeses, yogurts, cream) that are consumed by a lot of people worldwide. Dairy farming develops a daily source of income, generating employment opportunities as well as in transportation, services sectors, milk processing, and the agricultural supply. However, the dairy industry is increasingly facing structural constraints and intense competition, particularly in saturated markets, highlighting the need for farms to conduct comprehensive economic sustainability assessments that go beyond simple production monitoring.

1.2.1 Importance of dairy worldwide

Globally, dairy farming is predominant in many developing economies, where it helps smallholder farmers and provides an economic safety net (Felis et al.,

2025). The dairy industry is an essential part of the global food system (Shamsuddoha et al., 2023), providing a lot of dairy products that are consumed by millions of people worldwide. The dairy industry stands out due to its ability in transforming crops that are not suitable for human consumption into high-quality food product. Through a symbiotic interaction with rumen bacteria, dairy cows convert indigestible plant resources into milk, which provides important nutrients for human livelihood. Distiller grains and soybean meal are examples of by-products of human food processing that are used as supplements for dairy cattle, improving nutrient utilization and reducing wastage. Furthermore, dairy animals' digestive systems aid to the decomposition process, producing high-quality fertilizer that enhances the soil. The dairy sector has distinct characteristics that distinguish it from other agricultural industries. First, milk is composed of around 87% water, making it a commodity with significant bulk that is produced every day of the year. Secondly, milk has a short shelf life and is perishable (Fiorillo and Amico, 2024) because it is an optimal medium for bacterial growth. Third, dairy production is highly sensitive to animal health, feed quality, climatic conditions, and management practices, making it inherently variable and complex to manage.

Dairy production is highly valued in low- and middle-income countries, and demand for milk will continue to rise significantly over the next few decades (Adesogan and Dahl, 2020). Dairy is a worldwide food system, with over six billion people consuming milk and dairy products, making it one of the most globally dispersed agricultural commodities. Its importance extends beyond nutrition, strongly integrated in rural identities, cultural traditions, and local economies throughout a wide range of societies, from pastoralist communities in East Africa to highly industrialized systems in Northern Europe. The sector includes a wide range of production patterns, ranging from subsistence farming with one or two animals to highly profitable operations milking thousands of cows, creating a complex and diverse global landscape. This variety demands context-specific innovations and policies

1.2.2 Role in food security and nutrition

Milk is one of the most complete and highly nutritious food that provides high-quality proteins and bioavailable amino acids. It is also an essential source of minerals like calcium, a variety of vitamins (B2, A, B5, B12), and a vital source of lysine (Smith et al., 2022). It is also a source of fats and carbohydrates. Beyond basic nutrition, milk contains numerous bioactive compounds such as phospholipids, fatty acids including omega-3 and omega-6 polyunsaturated fatty acids, conjugated linoleic acid (CLA), gangliosides, and bioactive peptides. These compounds are associated with multiple physiological benefits, including the prevention and management of chronic diseases such as cardiovascular disease, diabetes, obesity, and certain cancers.

These bioactive components reinforce the role of milk not just as a food, but as a functional dietary component contributing to improved health outcomes (Punia et al., 2020; Verardo et al., 2017). This comprehensive nutrient profile makes milk critical for supporting growth, development, and overall nutritional well-being across different age groups (Park and Haenlein, 2013). Milk and dairy product consumption supports food security by supplying nutrient-rich, accessible food that can be incorporated into diets globally. It is especially important for vulnerable groups such as children, pregnant and lactating women, and the elderly, who have higher nutritional needs. The presence of essential micronutrients such as calcium, vitamin D, riboflavin, and vitamin B12 in milk addresses common nutritional deficiencies and supports bone health, muscle function, and overall metabolism. This makes milk an efficient tool for combating malnutrition as part of food-based nutrition programs, especially in developing countries (Park and Haenlein, 2013).

Moreover, dairy products deliver nutritional advantages beyond their nutrient content. They provide probiotics and prebiotics, which promote gut health and affect systemic immunity positively. Milk fat globule membranes contain bioactive molecules that have been linked to antiinflammatory effects and reduction in risk factors for cardiovascular diseases. Such multifaceted nutritional and health benefits underscore milk's contribution to food security by enhancing not only food availability but also food quality and safety (Sanjulián et al., 2025; Tunick and Van Hekken, 2015). Despite the rise of plant-based milk alternatives, dairy milk remains unparalleled in its nutrient composition and functional roles in human nutrition. Plant-based substitutes often do not provide equivalent quantities and bioavailability of key nutrients naturally found in dairy milk. This difference is critical in food security contexts where milk is a primary source of key nutrients and bioactive components necessary for optimal health (Beckett et al., 2024). In conclusion, milk supports food security by ensuring a reliable source of nutrient-dense food essential for human health and development. It contributes to nutritional adequacy, provides bioactive health-promoting compounds, and plays a role in combating malnutrition and chronic disease risk, thus enhancing the quality and stability of food systems worldwide (Park and Haenlein, 2013; Kliem and Givens, 2011).

1.2.3 Production trends

Global milk production has consistently increased in recent decades, due to population growth, economic development, and increasing consumer demand, particularly in low- and middle-income countries. Breeding, feeding, and health management practices result in increased production and livestock population. Advancements in dairy processing, cold chain logistics and quality control are advancing efficiency, stability and safety of food. At the same time, sustainability measures and programs, regulations and government schemes are driving responsible sourcing, animal welfare, as well as environmental footprint reduction. The global dairy market is undergoing continuous transformation as the industry seeks to reconcile tradition with innovation, advancing toward more sustainable, health-conscious, and consumer-driven models.

The dairy products industry is experiencing several notable trends driven by consumer health awareness, technological advances, and sustainability concerns. Functional and fortified dairy products are a key development area. There is increasing demand for yogurt and other dairy foods enhanced with probiotics, prebiotics, and bioactive compounds to promote health beyond basic nutrition. Functional yogurts, including probiotic, prebiotic, synbiotic, and lactose-free varieties, are widely studied and marketed for their health benefits such as improving gut health, supporting immunity, and alleviating disease symptoms (Sarıtaş et al., 2024). The demand for lactose-free and low-lactose dairy products has risen sharply due to global lactose intolerance prevalence and heightened health awareness. Advances in lactose hydrolysis technologies have improved the sensory qualities and nutritional profiles of lactose-free dairy, rapidly expanding this segment, which is now among the fastest growing in the dairy sector (Li et al., 2023). Value-added and premium dairy products focus on innovation in processing, flavor, and nutritional enhancements. These products often incorporate functional ingredients or novel fermentation processes to improve taste, texture, and health properties, catering to consumers willing to pay more for enhanced quality and benefits (Sarıtaş et al., 2024).

Digitalization and smart dairy supply chain management are transforming the industry by improving production efficiency, quality control, and traceability. Technologies such as IoT, data analytics, and blockchain enable real-time monitoring of milk quality, animal health, and supply logistics, enhancing transparency and sustainability across the dairy supply chain (Hamad and Awad, 2026). Additionally, plant-based milk alternatives are emerging strongly as complements or substitutes to traditional dairy products. These alternatives cater to lactose-intolerant, vegan, or sustainability-focused consumers and are rapidly evolving through improved fermentation technologies to enhance nutritional completeness, sensory properties, and bioactive potential, although they still generally differ nutritionally from bovine milk (Erem and Kilic-Akyilmaz, 2024; Hidalgo-Fuentes et al., 2024).

Dairy is one of the agricultural industries with above-average direct greenhouse gas (GHG) emissions. For several years, the International Dairy Federation (IDF) has been responsible for reporting dairy emissions. To assess the global dairy sector, from farming to processing, to have a robust benchmark for calculating carbon footprints for precompetitive discussions and to make continued progress in both researching and applying new technologies to effectively reduce the sector's GHG emissions, IDF created a common carbon footprint approach for dairy (OECD-FAO Agricultural

Outlook 2025–2034, 2025). Overall, the dairy product industry is innovating rapidly in functional food development, lactose-free options, premium product creation, and adopting sustainable and digital approaches to meet changing consumer demands and global sustainability goals (Li et al., 2023; Sarıtaş et al., 2024).

1.2.4 Consumption patterns

Dairy consumption is especially significant in controlling malnutrition, particularly in developing countries (Felis et al., 2025). Dairy consumption patterns differ greatly across regions, reflecting cultural preferences, economic conditions, dietary habits, and demographic trends. Globally, dairy products remain an important part of people's diets, with fresh milk, yogurt, cheese, and butter broadly consumed by people of different ages. According to recent market reports, dairy products could become one of the fastest-growing food categories internationally, likely due to rising population, urbanization, and changing dietary preferences toward nutrient-rich foods. Strong demand in India and Pakistan is promoting worldwide dairy consumption. Although milk is an extremely perishable product that must be handled quickly after collection, most milk is utilized in the form of fresh dairy products, particularly fermented and pasteurized varieties. The global consumption of fresh dairy products is expected to rise over the next decade as a result of stronger demand development in India and Pakistan, which is explained by income, urbanization and population expansion. World per capita consumption of fresh dairy products is expected to grow by 1.0% per year over the next decade, mostly driven by higher income per capita income (OECD-FAO Agricultural Outlook 2025–2034, 2025).

Milk intake per capita (in terms of milk solids) varies significantly across countries, owing to differences in income and regional preferences. The most significant enhancements are anticipated in India and Pakistan, where milk solids consumption is predicted to rise to 30 kg and 46 kg per capita, respectively. China's average fresh dairy consumption per capita is substantially lower than that of the European Union and North America, although a great increase is expected over the next decade. In low- and lower middle-income countries the majority of the production is consumed in the form of fresh dairy products (OECD-FAO Agricultural Outlook 2025–2034, 2025).

1.3 Structural characteristics of the dairy sector

1.3.1 Biological specificity and inherent variability of milk production

Milk production exhibits biological specificity and inherent variability influenced by multiple genetic, physiological, and environmental factors. It is a highly dynamic biological process regulated by the differentiation and activity of mammary secretory cells, which modulate milk quantity and composition throughout lactation. Models describe milk yield changes as outcomes of secretory cell differentiation, cell death, and secretion rate per cell. Among milk components, fat secretion rates display the greatest variability, while lactose and water secretion rates remain relatively constant during lactation, reflecting tightly regulated biosynthetic pathways (Pollott, 2004). Genetic variability plays a significant role in affecting milk production traits. Genome-wide analyses have identified numerous candidate genes influencing milk yield, fat, and protein content, which have pleiotropic effects on milk quantity and composition as well as health traits like mastitis resistance. Genetic polymorphisms exhibit not only additive but also dominance and parent-of-origin effects, underscoring the complexity of genetic regulation in milk production. Such genetic determinants underlie breed-specific milk production potential and adaptability to environmental conditions (Ayalew et al., 2024; Cai et al., 2020; Kuehn et al., 2007).

Environmental factors also induce variability in milk yield and composition. Seasonal changes and maternal condition influence milk nutrient profiles, as shown in wild mammals like eastern gray kangaroos, where milk lipid and protein concentrations vary with season, maternal size, and offspring sex. These adjustments optimize offspring survival under fluctuating resource availability, highlighting adaptive maternal investment strategies through milk (Quesnel et al., 2017). Moreover, milk harbors a microbiome shaped by evolutionary, ecological, and environmental factors, exhibiting variability across mammalian species. Dietary intake directly and indirectly influences milk microbiome composition, which in turn affects the immunological development and health of offspring. This microbial variability exemplifies the biological complexity and adaptive nature of milk beyond its nutritional components (Keady et al., 2023).

In sum, milk production's biological specificity is marked by distinct cellular and molecular processes regulating component synthesis, while inherent variability arises from complex interactions between genetic makeup, environment, physiology, and maternal investment strategies. Understanding these factors is critical for improving dairy production efficiency, animal health, and offspring survival (Keady et al., 2023; Kuehn et al., 2007; Pollott, 2004; Quesnel et al., 2017).

1.3.2 Natural variation of milk composition and yield

The milk's composition is influenced by various factors, involving physiology (age, stage of lactation, health status), genetics (breed), feeding, and environmental conditions (season, housing conditions), among others (Hamad and Awad, 2026). Improvement in milk yield can be obtained through genetic selection, as well as development in cow nutrition and management. Nutritional adjustment in the composition and proportions of individual components can efficiently result in a rise in milk yield and alterations in its composition. One of the major issues influencing milk yield is poor udder health, especially due to mastitis (Bednarski and Kupczyński, 2024). Age and parity significantly impact milk yield in dairy animals. Age at first calving is a critical factor; cows that calve between about 22 and 26 months tend to have higher first lactation yields, greater lifetime milk production, and longer productive lifespans compared to those calving later. Delayed first calving beyond 28 months typically results in lower milk yields in the first lactation and overall reduced lifetime productivity, along with shortened longevity and increased culling rates (Haworth et al., 2008).

Parity also influences milk production, generally showing an increase in milk yield with advancing parity up to approximately the fifth parity, after which yields may decline. For example, milk yield tends to increase with parity due to physiological maturation of the mammary gland but is often accompanied by greater body condition loss, slower recovery postlactation, and heightened risk of health disorders such as retained placenta, metabolic disorders, and reproductive failures in higher parity cows (Lee and Kim, 2006; McNitt and Lukefahr, 1990). Milk composition and quality also vary with parity and stage of lactation. With increased parity and days in milk, somatic cell counts and markers of udder health problems like β-hydroxybutyrate (BHBA), a ketone body marker, tend to increase, potentially lowering milk quality (Sabek et al., 2021; Yoon et al., 2004). Studies have shown that milk yield and components such as fat, protein, total solids, and lactose are influenced by these factors, with milk fat content generally increasing in mid and late lactation compared to early lactation, though parity may have less effect on certain milk quality parameters (Washaya & Cheruwa, 2025).

With increasing parity and days in milk, somatic cell count (SCC), a key indicator of udder health and milk quality, tends to rise. Elevated SCC correlates with indicators of udder infection or inflammation, which can degrade milk quality by altering its composition. For instance, milk from cows with high SCC often shows increased total protein and whey protein, but a decline in components like lactose and casein (Auldist et al., 1995; Nudda et al., 2003). Additionally, somatic cell counts increase with parity and lactation stage in goats and cows, reflecting a potential decline in udder health as lactation progresses and cows age (Salomone-Caballero et al., 2024). BHBA increases with increasing days in milk and parity, associating with metabolic stress and udder health problems. Elevated BHBA levels in milk are linked with subclinical ketosis, which can negatively affect milk fat and protein content, typically increasing fat concentration and reducing protein concentration. This metabolic state corresponds with decreased milk quality and increased susceptibility to diseases such as mastitis and other udder infections (Guliński, 2021; Sabek et al., 2021). Genetic studies have indicated variability in BHBA levels depending on parity and lactation stage, highlighting its role in monitoring both animal health and milk quality (Lee et al., 2016).

The relationship between rising SCC and milk quality deterioration is supported by evidence showing that high SCC is associated with reduced milk yield and altered milk components including protein and lactose. This effect tends to be more pronounced with higher parity and prolonged lactation, which coincide with increased SCC and BHBA, marking worsening udder health and metabolic stress (Cinar et al., 2015; Sabek et al., 2021). Furthermore, changes in milking practices during late lactation, such as reduced milking frequency, can exacerbate increases in SCC and related changes in milk composition, further contributing to reductions in milk quality (Kelly et al., 1998).

1.3.3 Nutrition

Animal nutrition profoundly impacts milk production, composition, and quality in dairy animals. Feed intake and energy balance are critical determinants; negative energy balance due to feed restriction reduces milk yield and alters milk composition by decreasing lactose and protein content while increasing fat concentration. These nutritional deficits affect animal metabolism, hormonal regulation, and mammary gland activity, influencing milk synthesis and cell turnover (Leduc et al., 2021). Feed composition also shapes milk's fatty acid profile and nutritional value. Dietary supplementation with vegetable oils, seeds, microalgae, and phytogenic additives enriches milk with healthy fatty acids such as omega-3, α-linolenic acid, and conjugated linoleic acid (CLA), which possess anticarcinogenic and health-promoting properties. Modulating dairy rations with agroindustrial by-products rich in bioactive polyphenols can similarly improve milk fatty acid profile without compromising milk yield, supporting economic and environmental sustainability (Correddu et al., 2023; Kholif and Olafadehan, 2022).

Additionally, nutritional strategies are essential to mitigate heat stress, which negatively impacts feed intake, milk yield, protein content, and overall animal health. Nutritional interventions including dietary fats, fibers, minerals, vitamins, microbial additives, and plant extracts have been shown to alleviate some adverse effects of thermal stress on dairy cows, helping maintain production and milk quality under challenging environmental conditions (Min et al., 2019). Good grazing and pasture management enhance milk production sustainability and nutritive value. Optimizing pasture feeding through supplementation and advanced technologies can improve feed efficiency, milk output per cow, and fatty acid composition, contributing to both animal welfare and milk quality that is perceived as superior by consumers (Wilkinson et al., 2020). Functional additives such as phytoactive compounds have shown benefits for animal health and milk quality by modulating rumen fermentation, enhancing nutrient digestibility, reducing somatic cell counts, and increasing beneficial unsaturated fatty acids in milk, thus improving both productivity and milk safety (de Vitt et al., 2024). Water intake, a critical dietary component, influences feed consumption, metabolism, and milk production. Adequate and quality water supply is vital for maintaining milk yield and the physiological processes underlying lactation (Golher et al., 2021).

In summary, optimal nutrition, including energy balance, feed composition, strategic supplementation, and water provision, is fundamental to maximizing milk yield, improving milk composition and quality, and ensuring health and welfare of dairy animals, especially under environmental stresses like heat (de Vitt et al., 2024; Kholif and Olafadehan, 2022; Leduc et al., 2021; Min et al., 2019; Golher et al., 2021; Wilkinson et al., 2020).

1.3.4 Season

During the colder months (spring and winter), the highest daily milk yield, with elevated milk fat and protein contents, was recorded, while summer had the lowest values, illustrating the negative impact of heat stress on production results. These findings clearly demonstrate the necessity for a seasonally adapted method to herd management, with an emphasis on optimizing feeding, enhancing microclimatic conditions, and improved use of precision dairy farming technologies (Gavran et al., 2025). Seasonal variations significantly influence milk production and composition. Several studies highlight that milk yield and key components such as fat, protein, lactose, solids-not-fat (SNF), and pH levels are highest in winter, accompanied by the lowest SCC and total bacterial counts, reflecting better milk quality during this season. Conversely, adverse conditions in warmer months tend to reduce milk yield and alter composition negatively (Bokharaeian et al., 2023; Bondan et al., 2018).

For example, an extensive study across dairy farms showed peak milk yield, fat, and protein in January and March, with the highest milk quality indicators observed during the colder months. Such changes are linked to ambient temperature and humidity fluctuations that affect cow physiology and feed availability (Bokharaeian et al., 2023). The interaction between breed physiology and seasonal conditions further shapes milk quality, with breeds like Brown Swiss and Holstein exhibiting significant seasonal differences in milk protein, lactose, fat content, and total solids depending on rainy or dry seasons. These findings suggest environmental factors seasonally modulate milk constituents in a breed-dependent manner (Estremadoyro et al., 2024). In pasture-based dairy systems, seasonal weather impacts grass growth, leading to variations in diet quality that influence milk composition such as crude protein, casein, and solids content. Extreme weather conditions may reduce pasture quality, thereby altering milk nutritional profile and processing characteristics. Forecasting milk composition based on seasonal patterns can assist processors in optimizing production logistics (Hayes et al., 2023).

Overall, season-associated environmental factors including temperature, humidity, feed availability, and breed interact intricately to affect both milk yield and quality, emphasizing the need for season-specific dairy management strategies to maintain production efficiency and milk safety (Bokharaeian et al., 2023; Bondan et al., 2018; Estremadoyro et al., 2024; Hayes et al., 2023).

1.3.5 Stage of lactation

Macronutrients such as fat, protein, total solids, and casein improve throughout lactation regardless of whether a pasture or concentrate-based regime is utilized, whereas lactose concentrations fall and micronutrients such as calcium change with lactation, analogous to casein content, due to storage in casein micelles. So, these findings highlight the essential impact of the lactation stage on both the macro- and micronutrients composition of bovine milk and illustrate its critical role in milk processing and the production of downstream dairy products (Connolly et al., 2023).

The stage of lactation exerts a profound impact on milk production and composition in dairy animals. Milk yield typically peaks in early lactation, often within the first 6 weeks, where the mammary gland is most responsive to stimuli such as frequent milking or offspring suckling. This early lactation period is critical because frequent milking during this window can permanently increase the mammary gland's milk production capacity, reflecting the "use it or lose it" principle in lactation management (Wall and McFadden, 2008). As lactation progresses to mid and late stages, milk yield generally decreases due to physiological changes in the mammary gland, including reductions in secretory cell numbers and activity. Late lactation is often characterized by mammary gland regression and lower milk volumetric output, although milk fat concentration may rise during this phase. Reduced milking frequency in late lactation has been shown to increase somatic cell counts and alter milk composition negatively, potentially affecting milk quality (Bondan et al., 2018; Kelly et al., 1998).

Milk composition changes with stage of lactation; protein and lactose concentrations are generally higher during peak lactation periods, correlating with higher milk yields, while fat proportions often increase in late lactation. Stages of lactation also influence molecular profiles in milk, such as dynamic changes in microRNAs that may regulate milk synthesis and composition in different sheep breeds, indicating biological complexity underlying the lactation cycle (Bondan et al., 2018).

Other factors tied to lactation stage include SCC, which tends to rise as lactation progresses, negatively impacting milk quality. Nutritional and physiological stresses during different lactation phases also affect milk yield and composition, underscoring the need for stage-specific management strategies to optimize dairy production (Cinar et al., 2015; Leduc et al., 2021). In summary, effective management across lactation stages, from exploiting early lactation's responsiveness to navigating late lactation's challenges, is vital to maximize milk yield, maintain desirable milk composition, and ensure dairy product quality (Bondan et al., 2018; Kelly et al., 1998; Wall and McFadden, 2008).

1.3.6 Genetics

The genetic parameters are useful for explaining the selection approach, expecting the direct and associated response to selection, selecting the breeding strategy to be used for future development, and evaluating genetic gain (Wodajo Tirfie, 2023). Understanding the genetic alterations underlying desired phenotypes is the aim of today's animal producers. It has been well-established that the production of milk can be improved through genetic marker-assisted selection. Various methods, including mapping of quantitative trait loci, genome-wide association study (GWAS), RNA-sequencing, whole-genome sequencing, and candidate gene analysis, have been applied to screen out the causal genes or their mutations associated with milk production characteristics (Ma et al., 2021).

Animal genetics has a significant impact on milk production and composition, influencing the efficiency and quality of dairy production. Milk production traits are moderately heritable and controlled by numerous genes, making genetics a key focus for enhancing dairy cattle productivity. Identification of genetic markers such as single nucleotide polymorphisms through GWAS, RNA sequencing, and other genomic tools enables more accurate selection and faster genetic progress by shortening generation intervals (Ma et al., 2021). Specific candidate genes have been associated with milk production traits, including genes involved in fatty acid metabolism, mammary gland development, and milk protein synthesis. For instance, in Barka cattle adapted to semiarid environments, genes like ACAA1, P4HTM, and SLC4A4 play fundamental roles in milk production and composition traits, underscoring the genetic basis for breed-specific productivity and adaptability (Ayalew et al., 2024).

Genomic studies in dairy sheep have identified key loci and candidate genes (e.g., PPARGC1A, LYPLA1, LEP, MYH9) linked to milk yield, fat, and protein percentages, contributing to genetic improvement programs in less favorable environments (Mohammadi et al., 2022). Similarly, in Holstein cattle, polymorphisms in genes such as PPARGC1A and OPN correlate significantly with milk fat content, protein yield, and overall milk production, offering practical targets for marker-assisted selection (Pasandideh et al., 2015). Furthermore, genotype-by-environment interactions (G × E) emphasize the importance of selecting animals with genetic profiles that ensure consistent production and health across diverse environmental conditions, including heat stress and disease challenges. Genomic approaches allow for the integration of environmental adaptability traits with milk production genetics to optimize breed performance and welfare under variable conditions (Habimana et al., 2023; Silva Neto et al., 2025).

Beyond production, genetics also influence fertility traits linked to milk yield. While intensive selection for increased milk output can challenge reproductive success, well-structured breeding programs leveraging genetic variation can achieve gains in both traits without severe trade-offs (Berry et al., 2016). In summary, animal genetics profoundly shape milk yield, composition, adaptability, and reproductive traits. Advances in genomic technologies enhance the precision and speed of improvement, enabling sustainable dairy production tailored to specific environments and breeds (Ayalew et al., 2024; Ma et al., 2021; Mohammadi et al., 2022; Silva Neto et al., 2025).

1.3.7 Health status

One of the biggest topics influencing milk yield is poor udder health, especially due to mastitis. Better mastitis management can lead to improved milk quality and enhance more responsible use of important antimicrobial agents on dairy farms (Bednarski and Kupczyński, 2024). Any other disturbance in the health of dairy cows causes a reduction in milk yield and its quality. Furthermore, we do not consider only adult cattle, as calf diseases can affect subsequent productivity, fertility, health, and longevity (Franceschini et al., 2025). Additionally, the milk somatic cell count, which is an indicator of udder health in cattle, and the differential somatic cell count have been developed as an attempt to promote the accuracy of mastitis detection (Magro et al., 2023).

Animal health critically affects milk yield, milk quality, and overall dairy production efficiency. Udder health, particularly prevention and management of mastitis, is pivotal; mastitis reduces milk quality and yield due to infection and inflammation, and increasing heat stress exacerbates risks by imposing physiological strain on dairy cows, lowering their productivity and milk safety. Emerging technologies like infrared thermography and biosensors aid early detection and management of udder health issues, improving both animal welfare and milk quality (Neculai-Valeanu and Ariton, 2022). General animal health constraints influence milk production via metabolic and physiological pathways. For instance, negative energy balance during early lactation and limitations in nutrient availability impede milk synthesis and can compromise animal well-being. Only healthy animals can fully express their genetic potential for high milk yields, emphasizing the interdependence of health and productivity (Gross, 2022).

Common health problems like lameness cause significant drops in milk yield, impairing feed intake and fertility. Lameness is linked with increased economic losses and animal welfare concerns, necessitating focused prevention and management on farms to sustain productivity (Garvey, 2022). Moreover, the use of antibiotics in mastitis treatment is critical but raises concerns regarding antimicrobial resistance, necessitating balanced strategies that safeguard animal health while maintaining food safety. Evidence-based therapeutic approaches and selective treatments can reduce antibiotic usage while ensuring udder health (Krömker and Leimbach, 2017). Nutritional and management strategies aimed at alleviating stress—such as heat stress and transition period challenges—are essential for maintaining health and welfare, which in turn uphold milk production and reproductive efficiency. Functional feed additives, including essential oils, minerals, and probiotics, have shown potential in supporting immune function and antioxidant capacity in dairy ruminants (Habimana et al., 2023; Kotsampasi et al., 2024). The COVID-19 pandemic demonstrated how disruptions in feed availability, veterinary access, and overall animal care can reduce milk production substantially. Reduced feed intake and limited treatments correlated with lowered milk yields, highlighting the sensitivity of dairy production to health constraints (Hussain et al., 2020). Lastly, advances in precision livestock farming technologies provide tools to monitor and enhance animal welfare, minimizing stress and promoting optimal milk production through better management and handling practices (Simitzis et al., 2021).

In essence, maintaining robust animal health through disease control, stress management, appropriate nutrition, and precision monitoring is vital for sustaining high milk yield, quality, and dairy farm profitability (Garvey, 2022; Gross, 2022; Kotsampasi et al., 2024; Krömker and Leimbach, 2017; Neculai-Valeanu and Ariton, 2022).

1.4 Importance of data in dairy processing and quality assurance

Dairy production depends heavily on maintaining product quality across the manufacturing process. Quality assurance keeps the nutritional value, taste, and safety of the product as well as the dairy manufacturer's reputation, which may be damaged if the product's food safety or quality were to become compromised. To avoid product recall incidents and audit noncompliance, dairy manufacturers must integrate to ensure optimal quality data management. Quality and technical managers can feel confident with fast, real-time access to end-to-end quality data through the instrument interface and the present system integration. With full visibility, product quality is preserved throughout the manufacturing process.

1.4.1 Role of standardization in quality control

Enterprises can use international or industry standards to confirm that the manufacturing of products and the supply of services stick to the same rules and methods, ensuring consistency in essential aspects of products or services, irrespective of when and where they are manufactured. In the automotive manufacturing industry, strict standards control the dimensions, materials, and processes of manufacturing components, ensuring uniformity and comparability across different models of cars. These standards assist companies in assessing vital qualities and confirm their incorporation into products and services. In global supply chains, many suppliers and manufacturers often work together to create products. By following the same standards, these collaborators may ensure that their products and components meet the same specifications, reducing the risks of inconsistencies in the supply chain, thus contributing to the delivery of high-quality finished products (Cao, 2023).

Standardization also directly impacts product yield and quality. In cheesemaking, monitoring the casein-to-fat ratio through standardization develops fat and protein recoveries, which are the major determinants of yield and profitability. Research into cheese milk standardization determines that adjusting protein and fat concentrations can greatly improve cheese yield and optimize sensory characteristics, demonstrating how front-end control methods affect the result of downstream process (Ali et al., 2021). Process control is an essential element in the dairy industry, playing an important role in maintaining the quality, safety, and efficiency of dairy products. Process control includes monitoring and regulating different operational parameters to follow predefined standards. This improves consistency in the finished product and overall operational efficiency and safety. The primary aims of process control in dairy processing include preserving product quality, assuring consumer safety, and increasing production efficiency. By closely monitoring parameters such as temperature, pressure, and flow rates, dairy producers can achieve precise control over important processes such as pasteurization and homogenization. The future of process control in the dairy industry is expected to be significantly influenced by technological advancements, especially in automation, AI, machine learning (ML), and predictive analytics. These cutting-edge improvements are predicted to develop the precision and efficiency of pasteurization and homogenization processes, so enhancing overall quality control and productivity.

1.4.2 Technological and process interdependence across dairy operations

The dairy industry acts as an integrated system where technological and operational processes across farming, collection, processing, and storage are tightly connected. This interdependence means that changes in one stage directly influence downstream stages, making the entire value chain sensitive to real-time conditions and requiring coordinated process control and technological integration (Fig. 1.1).

Dairy systems are distinguished by sequential operational stages from milk production on farms, collecting and transporting, processing into products, to final storage and distribution. Each stage depends on the results of the previous one, and modern technologies increasingly connect these stages through digital tools and data flows (Lakshmi Priya and Vanathi, 2025a). In the dairy industry,

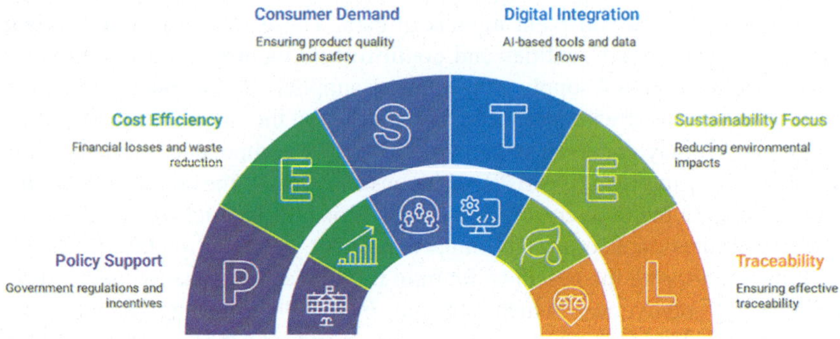

FIGURE 1.1 PESTEL analysis for dairy sector.

better procedures lead to developing the cold chain, reducing spoilage, and ensuring constant quality standards on its products. Furthermore, digital integration facilitates faster coordination and corporate decision-making among all stakeholders (Ersoy et al., 2022). Technological and process interdependence across dairy operations is critical to maintaining product quality, efficiency, and safety throughout the supply chain. Downstream processes in dairy production are highly sensitive to conditions upstream, meaning variations in raw milk quality, handling, and initial processing significantly affect subsequent stages such as fermentation, pasteurization, and product formulation. For example, microbial load and hygiene at milking directly influence the extent of microbial contamination and fermentation dynamics downstream, impacting milk safety and shelf life (Akdeniz and Akalın, 2022).

Microbial and hygiene effects are pivotal in dairy processing. Technologies such as power ultrasound offer dual roles—in microbial inactivation to ensure safety and in activation of beneficial microbes during fermentation, improving the efficiency and quality of fermented dairy products. Maintaining microbial control from raw milk through processing ensures consistent product characteristics and reduced spoilage (Akdeniz and Akalın, 2022; Chandrapala and Leong, 2015). Refrigeration temperature before processing is essential to slow microbial proliferation and enzymatic activity in raw milk. Proper refrigeration preserves milk composition and quality, stabilizes pH, and minimizes spoilage, thereby safeguarding downstream processes including coagulation during cheese making or fermentation for yogurt production. Any deviations in refrigeration can undermine milk quality parameters fundamental to subsequent unit operations (Aydogdu et al., 2023).

Consistent process coordination is vital in dairy operations that are complex and interlinked. Data integration and real-time monitoring technologies play growing roles in synchronizing processing stages, enabling automated control of temperature, pH, flow rates, and microbial activity. Such coordination ensures each process operates under optimal conditions, reducing waste, preventing quality loss, and supporting regulatory compliance. Emerging methods incorporate inline viscosity, compositional analysis, and advanced sensor technologies for holistic process governance (Aydogdu et al., 2023; Cabrera et al., 2025).

Given the sensitivity of downstream stages, the dairy industry depends heavily on coordinated actions throughout all stages of the value chain to preserve product quality, safety, and operational efficiency. Consistent coordination integrates on-farm production, milk collection, transport, and industrial processing through data, timing, and associated operational protocols. Integrated quality and cold-chain management Benchmarking of Indian dairy supply chains revealed that there is no autonomous performance indicator; instead, cold-chain infrastructure, effective traceability, quality management, support for technological developments, and IT systems all demonstrate a strong mutual dependence. Weaknesses in any one area (e.g., insufficient cooling or poor information flow) reduce overall performance (Mor et al., 2018). Corporate-sustainability analysis of Indian dairies finds cold chain as the most essential lever for decreasing waste, financial losses, and environmental impacts; poorly coordinated logistics and distribution (skills, vehicles, temperature control) directly accelerate spoilage and emissions. A review of supply-chain management in the Indian dairy sector also determines that a strong cold chain, digital integration, and efficient logistics are necessary for quality, loss reduction, and competitiveness (Lakshmi Priya and Vanathi, 2025b).

Coordinated, data-driven decision-making AI-based tools such as machine learning, IoT sensors, and blockchain are being used to coordinate production forecasting, inventory, logistics, and quality monitoring in real time, enhancing operational efficiency, food safety, and traceability along the entire dairy supply chain (Serrano-Torres et al., 2025). Data-integration frameworks that automatically collect, clean, and merge multivendor farm data (sensors, management software) show how harmonized data ecosystems can improve welfare and productivity decisions that influence downstream processing schedules, formulation, and quality (Boerman et al., 2025).

1.4.3 Perishability and cold chain dependence as physical constraints

Milk and many fresh dairy products are among the most perishable food commodities because of their high moisture content, rich nutrient composition, and susceptibility to microbial spoilage and physicochemical deterioration. These intrinsic properties create strict time and temperature dependencies, making continuous refrigeration from farm to consumer a critical requirement. Breaks in the cold chain can accelerate spoilage, compromise safety, shorten shelf life, and increase economic losses and waste (Das et al., 2025; Hassoun et al., 2024).

1.4.4 Organizational fragmentation and heterogeneity of dairy value chains

The dairy industry is distinguished by a high degree of organizational fragmentation and structural heterogeneity, which greatly impacts production efficiency, market access, and value chain performance. Dairy value chains include a wide range of operators from subsistence smallholders to industrial processors, each with unique operational scales, resource capacities, and market linkages. This variation impacts how milk is produced, collected, processed, and distributed, often resulting in unequal performance, quality variability, and coordination alterations across the chain (smallholder dairy production systems show diverse cluster behaviors and structural variability) (Nyambo, 2024).

1.4.4.1 Diversity of farm sizes, processing capacities, and market actors

Across regions, dairy industries include many smallholders, fewer medium and large commercial farms, and a mix of cooperatives, private processors, traders, and retailers. Dairy value chains involve operators with widely different farm sizes, resource access, and production objectives. Smallholder farmers often work with limited land, livestock numbers, and links to formal markets, while medium and large operations may utilize more developed technologies and formal supply agreements. This divergence leads to fragmentation in production and milk quality. Smallholders are frequently integrated into informal markets or supply local vendors. However, larger producers provide processors with more standardized raw milk. Research in regions like Africa and Asia reports high proportions of small, fragmented dairy producers with inadequate service linkages, highlighting variability in capabilities and results among dairy actors (Vuvor et al., 2025). In Nairobi's dairy system is divided into seven distinct chain profiles ranging from urban backyard and peri-urban farms to medium and large cooperatives and huge processors. The majority of nodes are separate, small-scale firms (producers, traders, retailers) that operate essentially independently, while supporting even the largest processors through collection centers and cooperatives (Kiambi et al., 2018). In Ethiopia, dairy development around Addis Ababa demonstrates rapid expansion of larger commercial farms and processors, while many small and remote farmers remain isolated from modern buyers and services, with static or reducing productivity (Mintenid et al., 2020).

1.4.4.2 Coexistence of traditional, semiindustrial, and industrial systems

Many dairy value chains have multiple production systems operating at the same time, reflecting important structural and technological differences within the industry. Traditional dairy systems, which are prevalent in many low- and middle-income countries, are primarily based on smallholder production using indigenous breeds, minimal mechanization, and informal market channels. Milk from this system is frequently consumed at home or sold across informal networks, and producers encounter low yields, minimal technological uptake, and poor market connections. This contrasts with modern or specialized systems in which farms and processors are becoming more incorporated into formal markets and utilize improved inputs, breeding, and handling procedures to develop results and quality. Research in Ethiopia shows that traditional dairy producers encounter challenges relative to more specialized systems. Traditional and specialized are the two broad categories of dairy production systems practiced in Ethiopia. Dairy value chains rising from both production systems encounter several challenges (Temesgen et al., 2024). Compared to specialized systems, the traditional system-based dairy value chains face numerous and severe challenges. Traditional dairying is generally smallholder-based and relies on indigenous breeds. Daily milk yield for producers of the traditional system is very low. Whole milk and butter are sold mostly through informal markets. This system produces milk that is either consumed locally or sold as butter. Such structural heterogeneity shows how traditional and more advanced systems coexist within the same national context, causing complexity in value chain performance and technological adoption.

This duality is not limited to African contexts but is also evident in other major dairy-producing countries. In India milk market is highly fragmented, containing of both traditional (consumption at home and traditional milk vendors) and modern (private dairies and cooperatives). Around 60% of the total produced is marketed through various agencies, out of which 36% of milk is handled by the informal sector, and the remaining 24% is disposed of through modern channels (Mohapatra and Pabba, 2022).

1.4.4.3 Geographic dispersion and coordination complexity

Dairy value chains are often geographically dispersed, with production sites remote from processing facilities and most market hubs providing substantial logistical and coordination challenges. In many developing regions, Smallholder farms are dispersed throughout vast rural areas in many developing economies, necessitating the long-distance transportation of milk via milk collecting centers (MCCs) before it reaches processing facilities or metropolitan markets. Such dispersion increases transport costs and transit times, and without sufficient cold chain infrastructure, it accelerates the risk of product spoilage, quality deterioration, and losses. Geographic heterogeneity can further complicate coordination, as producers from remote locations frequently encounter less frequent collection schedules, longer travel times to centralized chilling facilities, and limited access to real-time information about demand and quality requirements. Empirical studies in East Africa demonstrate that distance from collection centers and urban markets significantly impacts milk prices and supply chain efficiency, as farmers farther from nodes in the value chain usually receive lower prices and face higher transaction costs due to their remoteness. These findings highlight how spatial fragmentation of dairy production affects coordination capacity, price transmission, and quality maintenance through the dairy supply chain (Ignowski et al., 2022).

In Nairobi, urban and peri-urban farm systems share similar downstream networks but differ in spatial reach: urban farms primarily serve immediate neighbors or their own consumption, while peri-urban farms supply broader networks and occasionally cooperatives, leading to multiple overlapping, spatially tiered chains (Kiambi et al., 2018).

1.4.5 Data intensity and information asymmetry in dairy systems

1.4.5.1 High volume of biological, operational, and logistical data

Modern dairy systems generate data from multiple sources, containing herd management software, sensor networks, milking equipment, processing plants, and logistics providers. On-the farm, data is collected from herd management software, milking systems, feeding systems, genetic evaluations, environmental monitoring, milk analysis, and diagnostic laboratories, which are often supported by handwritten records (Cabrera et al., 2025). Continuous streams involve milking (volume, flow, conductivity), feeding (diet composition, intake, costs), reproduction, health events, genomics, and processor test outcomes for each shipment (Cabrera et al., 2020). However, this data intensity does not automatically translate into beneficial insight, as many datasets exist in separate formats without standardized structures or interoperability. Studies note that a significant barrier to utilizing this data in dairy farming is the lack of integration and standardization across data sources, which restricts the ability to synthesize information into useful insights for decision-making and predictive analytics.

1.4.5.2 Limited real-time visibility across chain segments

Despite the increasing availability of data from sensors, farm management systems, and herd monitoring technologies, several dairy value chains still lack comprehensive real-time visibility across production, collection, processing, and distribution stages. These restrictions exist because a large portion of data is being generated, such as from milking equipment and animal behavior sensors, which are still used mostly for historical analysis rather than immediate operational decisions. In fact, recent reviews of data analytics in dairy farming demonstrate that while historical data are generally used for model predictions, real-time data usage remains much less frequent (used in around 25% of studies), showing a gap between data generation and time-critical analytics. This gap prevents stakeholders from detecting emerging concerns such as deviations in milk quality, equipment malfunction, or logistic difficulties in real time, leading to less coordinated responses and increasing losses in supply scheduling, quality control, and resource distribution (Palma et al., 2025). Research on data integration challenges further demonstrates that infrastructure restrictions and interoperability issues limit the real-time flow of information. For example, intermittent internet access and fragmented software systems at the farm and collection levels impair sufficient communication of sensor and operational data to central processing systems. Without robust data pipelines and standardized protocols, real-time monitoring platforms cannot work effectively, meaning that decision makers frequently receive updated information only after significant delays, restricting their ability to take quick corrective actions (Cabrera et al., 2025).

1.4.5.3 Structural gaps between data generation and data use

A key challenge in dairy information systems is the gap between data generation and its effective use. While automated systems, sensors, and herd management tools generate large volumes of biological, operational, and performance data, restricted integration, weak infrastructure, and insufficient governance hinder these data from being transformed into useful insights through the value chain. A major contributing factor is the lack of standardized data formats and protocols, which makes it difficult to unify datasets from multiple sources. This causes disconnected repositories, demanding extensive cleaning and conversion, and restricts opportunities for cross-system analytics and coordinated decision-making (Cabrera et al., 2025).

1.4.5.4 Data ownership, privacy, and governance concerns

Another structural limitation originates from uncertainties around data ownership, privacy, and control, which further restrict data use outside its point of origin. Surveys of dairy producers show that many farmers feel powerless over data that is collected from their farms, and a significant majority are not familiar with the terms and implications of data use agreements. For example, up to 69% of farmers surveyed were unfamiliar with data collection standards, and 66% stated they felt they had little control over how their data was used, even when sharing it with external parties. These attitudes reflect a general inability to fully engage in cross-stakeholder data ecosystems, raising the gap between data generation and collaborative usage (Fadul-Pacheco et al., 2022).

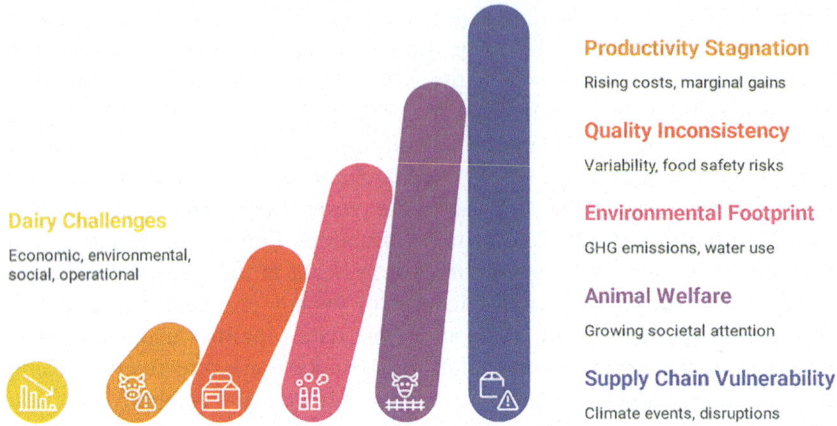

FIGURE 1.2 Dairy sector challenges impact global food system.

1.5 Key challenges confronting the dairy sector

This section will analyze the economic, environmental, social, and operational difficulties that arise from the structural characteristics of dairy systems under current global pressures (Fig. 1.2).

1.5.1 Productivity stagnation and cost volatility

1.5.1.1 Rising input costs

The dairy industry has seen significant increases in key production inputs, particularly feed, which remains the highest cost category for milk production. In the EU, empirical panel analysis (2008–2021) demonstrates that purchased feed and energy costs are major drivers of farmgate milk prices, indicating a high alignment of input and output prices via global commodity markets. Rising marketable input prices must be passed into milk prices to ensure viability, but pass-through is incomplete and uneven across member states (Triatmojo et al., 2025). In the global dairy production industry, feed ingredient prices and availability are extremely volatile; they may influence the composition of the feed ration and affect feed cost and enteric methane (CH_4) emissions (Alqaisi and Schlecht, 2021). Labor is now the second-highest cost after feed in many pasture-based systems; larger Irish herds achieve more hours-per-cow, whereas smaller herds show lower labor efficiency and greater unit labor cost pressure (Deming et al., 2018).

1.5.1.2 Marginal productivity gains under conventional models

Conventional methods of increasing output, which become more concentrated, intensifying, and larger herds, often yield reduced marginal productivity and questionable economic gains: meta-analysis of strategic feeding in Australian grazing systems demonstrates that the theoretical milk response to additional energy rarely occurs in practice due to substitution, body-reserve change, and rumen imbalances; marginal milk responses to supplements are highly context-dependent (Heard et al., 2021). Yield-gap modeling for high-input Dutch herds reported an average gap of only 11% between actual and potential milk yield under controlled conditions, which was explained by limits in intake capacity, protein supply, and heat stress (van der Linden et al., 2021). This implies that in highly efficient systems, further productivity gains within the same paradigm are increasingly gradual. Empirical study on dairy farms in Southern Chile also showed that there is potential for productivity growth, although technical efficiency gains remain small, indicating that several farms are not fully realizing productivity improvements through conventional practices alone. Investments in technology and the implementation of developed production methods were identified as essential drivers for higher productivity if adopted more broadly (Moreira and Bravo-Ureta, 2016).

1.5.1.3 Economic vulnerability of small and medium producers

Small- and medium-sized dairy producers are especially economically dependent because they often operate with limited financial resources and weak market positions, making them less able to withstand shocks compared with larger enterprises. Research on small-sized dairy farms in the United States shows that these operations encounter persistent financial instability, slim profit margins, and restricted access to capital and risk-management tools, which limit their ability to cope (Fig. 1.2) with input cost changes and volatile milk prices. In contrast, larger farms benefit from economies of scale, greater access to financing, and stronger bargaining power, allowing them to stabilize incomes and invest in productivity-developed technologies (Jafri et al., 2024).

Structural analysis of EU farms demonstrates that many smaller units have insufficient production potential, yielding financial indicators that look acceptable on paper but do not produce enough income to improve investment or household consumption, causing exit risk (Poczta et al., 2020). Furthermore, limited utilization of modern technologies due to financial constraints can inhibit the economic performance of medium and small dairy farms. A recent socioeconomic research in the Ecuador–Colombia border area indicated that seasonal input price changes and variable resource availability make long-term financial planning difficult for smaller producers, reducing their capacity to make strategic investments and stabilize operations (Carvajal-Pérez et al., 2024).

1.5.2 Quality inconsistency and food safety risk exposure

The dairy sector faces significant challenges around quality inconsistency and food safety risks due to factors such as microbial contamination, biofilm formation on equipment, and variability in raw milk quality. Psychrotrophic bacteria, capable of surviving refrigeration and producing heat-resistant enzymes, pose a major threat by accelerating milk spoilage and compromising product safety. These contaminants can enter the milk supply through sources including animal health, hygiene in milking environments, water quality, and handling practices. Maintaining high raw milk quality continues to be critical but challenging, as premature deterioration of dairy products remains an ongoing issue in the industry.

Food safety management systems such as Hazard Analysis and Critical Control Points (HACCP) have been implemented to address inherent risks in dairy production, but rigorous controls remain essential in safeguarding consumer health. In addition, biofilm formation on dairy equipment contributes to contamination, product loss, and economic inefficiencies. Advanced cleaning and sanitation techniques including clean-in-place systems, ultrasonic cleaning, enzymatic agents, and nanotechnology-based solutions have been adopted to reduce biofilm-associated risks and support sustainability goals (Chowdhury et al., 2025; Papademas and Bintsis, 2010).

The advent of Dairy 4.0, inspired by Industry 4.0, leverages cutting-edge technologies such as IoT, big data, blockchain, robotics, AI, and 3D printing to transform dairy production from farm to consumer. These innovations facilitate real-time monitoring, enhanced traceability, and process optimization, leading to improved milk quality consistency and reduced food safety risks. IoT sensors enable continuous monitoring of parameters like temperature, humidity, and microbial load, enabling proactive interventions that minimize contamination and spoilage. Big Data analytics and AI support predictive maintenance and quality forecasting, while blockchain improves transparency and trust across the supply chain (Hassoun et al., 2023; Kazi, 2024).

Lean 4.0 practices integrated with digital technologies promote waste reduction and operational efficiency in dairy ecosystems, fostering collaborative value creation by aligning production closer with consumer demands and quality specifications. This digital transformation supports stringent quality management, minimizes process variations, and strengthens food safety assurance throughout dairy operations (Bonamigo et al., 2024). Nonthermal processing technologies like pulsed electric field (PEF) treatments are emerging within Dairy 4.0 to enhance microbial inactivation while maintaining the natural sensory and nutritional attributes of dairy products. Although commercial scale adoption is still evolving, these sustainable interventions promise to decrease food safety risks with minimal impact on product quality (Vashisht et al., 2024).

1.5.3 Environmental footprint and sustainability pressures

1.5.3.1 Greenhouse gas emissions, water use, and nutrient losses

The dairy industry is recognized as an important contributor to GHG emissions and other environmental effects when considered from a full life cycle perspective. Life cycle assessment (LCA) studies show that dairy products generate direct and indirect GHG emissions, with methane (CH_4) from enteric fermentation and nitrous oxide (N_2O) from manure and fertilizer use leading to large shares of the carbon footprint of milk and dairy value chains. Furthermore, dairy production depends on substantial freshwater resources for feed crop irrigation, animal drinking water, and on-farm operations, which cause the sector's high-water footprint. LCA results further demonstrate that dairy activities also promote water eutrophication potential through nutrient losses (e.g., nitrogen and phosphorus) that can leak into water bodies and harm aquatic ecosystems. Such multicategory environmental influences emphasize the complex resource intensity of dairy systems and their challenges in meeting sustainability targets (Guzmán-Luna et al., 2022).

1.5.3.2 Increasing regulatory and societal scrutiny

Dairy production is increasingly evaluated against climate, water, and general sustainability goals, with governments, investors, and consumers demanding measurable improvements and transparency. Regulatory frameworks are increasingly focused on livestock GHG emissions, nutrient losses, and resource use. Global assessments connect ruminant meat and dairy to a disproportionate share of agricultural GHG emissions and land use relative to their contribution to dietary energy, promoting climate- and nitrogen-focused policies, such as emission ceilings, stricter manure and fertilizer regulations, and support for low-impact production systems (Wijerathna-Yapa and Pathirana, 2022). Mapping of dairy-farm sustainability research shows a strong predominance of environmental topics, especially GHG emissions, over social and even economic dimensions, demonstrating how regulation and scientific agendas have converged on environmental performance as a central criterion for sectoral legitimacy (Segerkvist et al., 2020). In parallel, corporate sustainability reporting standards and SDG-aligned policies have increased pressure on dairy supply chains to document and decrease their environmental impact (Ivana and Puvača, 2024; Mwirigi et al., 2025). Societal scrutiny is also developing. Bibliometric analysis since the adoption of the Sustainable Development Goals (SDGs) in 2015 shows a significant increase in publications and public debate around dairy sustainability, driven by concerns about climate change, resource consumption, animal welfare, and rural equity (Mwirigi et al., 2025). Consumer studies shows a positive willingness to pay for dairy products produced under demonstrably sustainable systems (e.g., mountain or extensive systems), implying that environmental and welfare attributes are becoming salient product characteristics and can influence market share and price formation (Nam et al., 2020).

1.5.3.3 Tensions between efficiency, sustainability, and profitability

Efforts to balance productivity development with environmental goals determine structural trade-offs at the farm level and along supply chains. Historical data show that genetic selection and management intensification have more than doubled milk yield per cow while significantly decreasing land, water, and GHG emissions per kilogram of milk, but at the expense of reduced genetic diversity and elevated vulnerability in health, fertility, and robustness traits (Brito et al., 2021). Debates around sustainable intensification emphasize the risk that a narrow focus on environmental efficiency per unit of output can develop high-input, large-scale models that externalize social and ecological costs, marginalize smaller farms, and ignore animal-welfare and rural-livelihood dimensions (Clay et al., 2020; Mwirigi et al., 2025). Systematic mapping of dairy-farm sustainability literature ensures that integrated studies of environmental, economic, and social outcomes remain rare, leaving many trade-offs practically underexplored (Segerkvist et al., 2020). Recent work on pasture-based systems in Germany demonstrates that there is no "one-fits-all" configuration that improves profitability and minimizes environmental effects throughout all categories; instead, management options such as decreased concentrate use, avoiding wetland cultivation, low fertilizer rates, and low replacement with high lifetime yields emerge as partial "win–win" levers, while other dimensions still involve trade-offs (Mennig and Szigeti, 2025). Similarly, dynamic forage systems in intensive Italian dairies improved nitrogen balance, decreased external inputs and carbon footprint, and were at least cost-neutral or cost-saving per hectare. However, it required higher management complexity and some increase in labor per unit of land (Tabacco et al., 2018).

1.5.4 Animal welfare expectations and ethical accountability

1.5.4.1 Growing societal attention to animal health and living conditions

According to recent research, animal welfare has become a major societal issue in livestock production systems, including dairy farming. Public concern about the ethical acceptability of intensive livestock systems and their effect on farm animal welfare has significantly increased over recent decades (Buller et al., 2018; Alonso et al., 2020). Consumers now view welfare-friendly products as healthier, safer, more hygienic, and environmentally friendly, and many express willingness to pay a price premium when they trust that higher welfare is being provided. Surveys in developing countries such as China illustrate that, although detailed knowledge of "animal welfare" remains restricted, the majority of respondents believe farm animals should have enjoyable experiences, be protected by legislation, and be certified for welfare on farms; they also link good welfare to food quality and safety and support pre-slaughter stunning and shorter transport times (Carnovale et al., 2021). At the policy level, global authorities such as the UN Committee on World Food Security and the World Organization for Animal Health (WOAH/OIE) clearly consider animal health and welfare as vital to sustainable livestock systems, calling for implication of the "five freedoms" and related standards.

1.5.4.2 *Market and regulatory demands for welfare assurance*

The move from purely public regulation to a mixed governance system combining legislation and private standards has transformed welfare accountability in the dairy chain (Aumüller et al., year; More et al., 2021). In many regions, effective animal welfare law now integrates with retailer and industry programs that include welfare into quality assurance and supply-chain management. These schemes often demand science-based requirements, on-farm audits, and corrective actions, and are presented as essential prerequisites for market access and corporate social responsibility (Kuo et al., 2025; More et al., 2021). Evaluations of dairy welfare standards in Europe and North America show that strong programs can improve legitimacy when they rely on current science, distinguishing risk assessment from risk management, implementing animal-based indicators and benchmarking, and maintaining system-level review. At the same time, contemporary welfare policy increasingly incorporates ethical principles such as the duty of care, proportionality, and acknowledgment of animals as sentient beings, with legal frameworks progressively changing from forbidding cruelty toward requiring the promotion of positive experiences.

1.5.4.3 *Transparency and verification challenges*

Despite the expansion of assurance schemes, transparency and verifiable credibility remain critical weaknesses. Comparative evaluations of private animal health and welfare standards for dairy production show restricted publicly available evidence that program goals translate into measurable welfare advancements, and considerable diversity in clarity, scope, and evaluation methods. Problems such as vague wording, heterogeneous inspection protocols, and scarce publication of result data make it difficult for people and buyers to assess effectiveness (Kuo et al., 2025; More et al., 2021; Sandøe et al., 2019). Aggregation systems like Welfare Quality® scores have also been challenged for lack of transparency and for allowing major welfare problems to be "hidden" within composite indices, raising ethical questions about how indicators and weights are chosen (Sandøe et al., 2019). Recent research emphasizes that welfare labels must accurately reflect animals' lived experiences, be cost-effective, and align with consumer expectations of "value for money"; lack of transparency, inconsistent criteria, and weak traceability are repeatedly identified as barriers to consumer trust and adoption of higher-welfare products (Tuyttens et al., 2025). Industry-led governance systems in the dairy sector are considered to gain legitimacy when they illustrate accountability, scientific foundation, and broad participation; however, their credibility is undermined where there is heavy industry dominance and limited disclosure of how public input and audit outcomes shape standards and enforcement (More et al., 2021). Ensuring ethically robust, transparent, and independently verifiable welfare assurance is thus vital to achieve modern societal expectations of animal welfare and maintain the social license of the dairy sector.

1.5.5 Supply chain vulnerability, resilience gaps, and market instability

1.5.5.1 Exposure to climate events, geopolitical disruptions, and pandemics

Climate change affects feed availability, animal health, and cold-chain integrity, raising the risk of production shortfalls and food-safety incidents throughout the dairy supply chain. Heat stress, drought-induced feed scarcity, and emerging heat-resistant diseases are predicted to increase processing and refrigeration requirements as well as energy consumption and costs, especially in intensive systems (Malliaroudaki et al., 2022). COVID-19 showed the vulnerability of global and regional dairy chains to sudden movement restrictions, manpower shortages, and changes in consumption patterns; effects occurred through successive waves on both supply and demand, varied according to countries' trade profiles and market structures (Acosta et al., 2021; Kettele and Lechner, 2025). While some organic and economically embedded supply chains in France demonstrated strong buffer and adaptive capacity—autonomy in feed, family labor, and flexible logistics—conventional chains required rapid reconfiguration of product mix and distribution to maintain flows (Kettele and Lechner, 2025; Perrin and Martin, 2021). These examples illustrate that resilience is inconsistent and heavily influenced by value-chain organization and governance.

1.5.5.2 Price volatility and demand uncertainty

Liberalization and the elimination of quota systems have contributed to milk price volatility in Europe, affecting succession, investment planning, and long-term resilience of family-based dairy systems. Empirical study of US dairy futures reveals strong, time-varying volatility and spillover among butter, cheese, and milk contracts, with volatility enhanced during COVID-19 and by information shocks, complicating risk management for producers and processors (Fan et al., 2022, 2025). At the farm level, producers perceive price falls, increasing input costs and negative government actions as major risks alongside climate hazards, indicating a high degree of economic uncertainty (Perrin and Martin, 2021). The consequent instability limits strategic investments in productivity, decarbonization, and animal welfare and exacerbates exit and consolidation trends in many regions.

1.5.5.3 Limited adaptive capacity of fragmented value chains

Dairy farming systems work within complex value chains where resilience cannot be achieved at the farm level alone; poor coordination, asymmetric bargaining power, and fragmented contracting hinder collective responses to shocks (Acosta et al., 2021). Comparative study across five European countries illustrates that national arrangements to maintain volatility, cooperative structures, contracts, and producer organizations differ significantly, but often fail to deliver truly resilient strategies, contributing to limited investment, farm abandonment, and erosion of family-farm identities. Systematic reviews of dairy supply chain risks show the importance of on-farm risks (animal health, feed, climate) alongside off-farm logistics, information, and financial risks, stressing that these are frequently addressed separately rather than through integrated risk-management and resilience frameworks (Wang et al., 2025). Case-based simulation of the Austrian dairy chain demonstrates that resilience measures must be configured effectively through strategic, tactical, and operational levels; stand-alone actions (e.g., backup stocks or additional capacity) can generate unexpected postdisruption inefficiencies if not integrated into a balanced portfolio of redundancy, flexibility, and adaptive decision rules (Kettele and Lechner., 2025).

1.6 Emerging transformation pathways spurred by Food Industry 4.0

1.6.1 Digital enabling technologies

Digital technologies are framed as cognitive enablers that transform dispersed data in the dairy sector into predictive, preventive, and coordinated decision support. Dairy 4.0 relies heavily on AI, which provides superior cognitive skills for managing dairy systems' inherent biological variability, uncertainty, and complexity. AI is a computational technology that aims to mimic, to varying extents, human abilities to recognize their environment, process information, make decisions, and take steps to accomplish certain objectives (Manning et al., 2022; Hassoun et al., 2023). The use of AI in the dairy supply chain is revolutionizing the industry by increasing productivity, sustainability, and quality in processes. Many studies highlight its developed applications, emphasizing its potential to enhance logistics management, enhance traceability, and ensure food safety (Serrano-Torres et al., 2025). AI is transforming the dairy industry, focusing on predictive analytics to improve production, decrease waste, and minimize price volatility (Sinha and Mishra, 2023). Significant roles of AI in the food industry can be generally classified into two main categories: one is food security management, and the other is food quality management (Kumar et al., 2021).

Big data analytics for integrating different, high-volume data streams through value chains: Big data analytics is a vital component of Dairy 4.0 that allows for the integration and analysis of heterogeneous, high-volume data streams from sensors, milking machines, environmental monitors, and supply chain systems. Big data refers to extremely massive and complex data sets that cannot be analyzed or processed using traditional method (Ciccullo et al., 2022; Hassoun et al., 2023). Data integration and analysis are essential for the development of the dairy industry. These processes help dairy farmers and dairy industry to make educated decisions, improve operations, and enhance productivity, ultimately contributing to the overall sustainability of the industry (Cabrera et al., 2025).

Blockchain technology has emerged as a powerful enabler of secure data sharing, transparency, and trust in multiagent dairy supply chain systems, where multiple independent actors—including farmers, processors, distributors, retailers, and regulators—interact and exchange data. By providing a decentralized and immutable ledger, blockchain ensures that information recorded at each stage of the dairy value chain cannot be altered, thereby strengthening data integrity and accountability (Khanna et al., 2022). When integrated with complementary digital technologies such as IoT, blockchain enhances data interoperability, real-time monitoring, auditability, and traceability, while simultaneously reducing transaction costs and information asymmetry across the food supply chain (Casino et al., 2021). This technological convergence enables continuous documentation of critical parameters such as temperature, handling conditions, and transportation status, which are essential for maintaining dairy product quality and safety.

Beyond conventional traceability applications, recent blockchain-based frameworks have been proposed to protect vulnerable consumer groups by supporting the verification of nutritional integrity and the early detection of adulteration and contamination in dairy products (Khanna et al., 2022). In this context, Makarov et al. (2019) developed a blockchain-enabled dairy supply chain management system designed to secure data repositories against unauthorized access, ensure the immutability of transportation and logistics records, and allow retailers, regulators, and consumers to transparently verify supply chain actors and product history from farm to final consumption.

Digital twins as virtual representations for simulation, prediction, and system-level learning: Digital Twins concept model, in which they serve as a bridge connecting physical entities in the real world with their virtual complements in a digital environment, closing the gap between the two. They establish critical linkages between data and information to integrate these products' virtual and real aspects (Peladarinos et al., 2023). Advancements and intelligent interaction: Technological advancements have accelerated the development of Digital twins. Advances in real-time data upload, large-scale data storage, and data fusion technologies support continuous exchange and updates of descriptive data (Botín-Sanabria et al., 2022).

1.6.2 Physical enabling technologies

Physical technologies are positioned as operational enablers that connect digital logic with real-world dairy processes under time- and quality-critical conditions (Fig. 1.3).

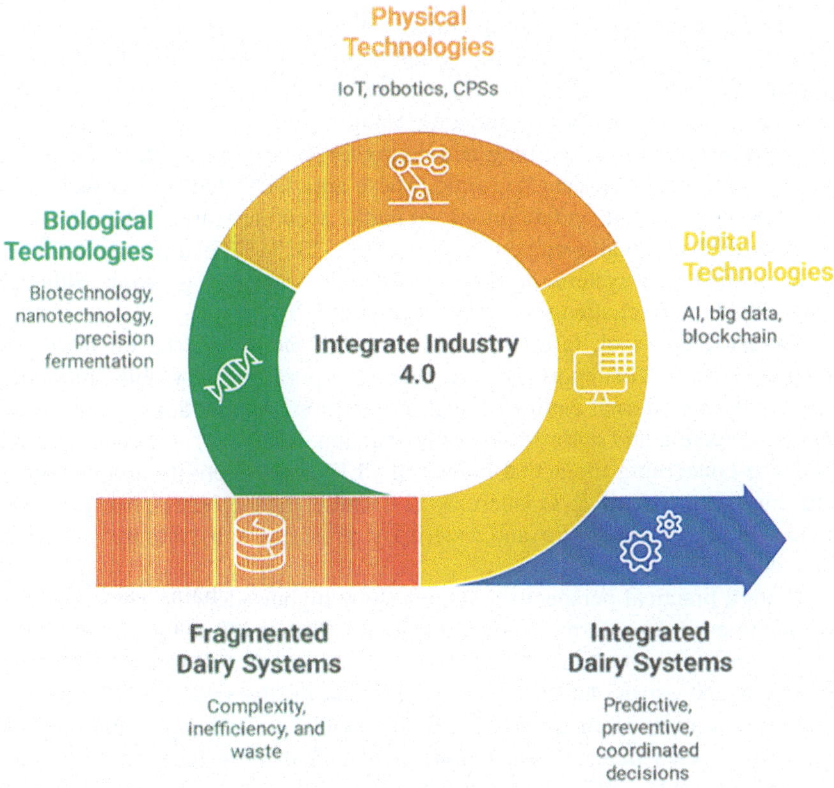

FIGURE 1.3 Transforming the dairy system with Industry 4.0.

1.6.2.1 The Internet of Things and smart sensors for continuous, real-time data acquisition

IoT is a system that connects items via a network to enable data exchange. IoT can develop intelligent environments of powerful tools in various agriculture and livestock areas, allowing the creation, modification, and sharing of information (da Rosa Righi et al., 2020). IoT has great potential to collect and share live data from dairies. In fact, recent research has suggested the use of IoT to automate cow feeding, so that nutrition plans can be designed for each cow and a more accurate milk production forecast can be made for each cow (da Rosa Righi et al., 2020; Hassoun et al., 2023). In the dairy industry, logistics is a vital aspect due to the high perishability of dairy products and the need to distribute these products to all population centers.

Wearable sensor technologies, such as collars equipped with accelerometers, RFID tags, gyroscopes, and GPS, provide noninvasive real-time monitoring of dairy cows' behavior, health, and productivity. These wearable collars gather data on movement, feeding patterns, estrus detection, and disease symptoms, offering farmers detailed insights for timely interventions (Lamanna et al., 2025). Despite challenges like high energy consumption and limited parameter coverage, integrating multiple sensor types in wearable devices is advancing, aiming for comprehensive monitoring with improved battery life and cost efficiency (Lamanna et al., 2025). Beyond wearable sensors, other technologies integrate image, sound, and movement data analyzed with machine learning and computer vision to monitor dairy cattle behavior and health in near real-time. These approaches detect mastitis early, assess body condition, and monitor feed intake accurately, thus enhancing herd management and health outcomes (Liu et al., 2023). The implementation of such sensor-based systems in commercial dairy farms is growing, although some deployment challenges remain (Liu et al., 2023).

Precision livestock farming is evolving beyond simple health alarms to integrated decision-support systems that incorporate sensor and production data alongside external environmental information. Applications include early disease detection and optimization of production efficiency, although scientific evidence concerning the actual impact of PLF on productivity and welfare is still emerging (Kleen & Guatteo, 2023). These technologies also affect the human–animal relationship and veterinary practices substantially (Kleen & Guatteo, 2023).

From a practical perspective, studies show higher adoption rates of digital technologies in dairy farming compared to other livestock sectors, especially in larger herds typically housed in loose housing systems. Automated milk yield recording and estrus detection sensors are among the most widely adopted. However, farmers often weigh factors like benefit-to-cost ratios and the time required for data management before investing in these tools (Abeni et al., 2019; Groher et al., 2020). The broader sustainability impact is significant; IoT and AI applications in dairy farms contribute to environmental goals like reducing GHG emissions by optimizing feed efficiency, manure management, and energy use. Data-driven monitoring underpins better benchmarking for emission reductions and supports policymaking aligned with climate commitments (Neethirajan, 2024).

1.6.2.2 Robotics and automation to reduce manual dependency and improve process consistency

Robotics is an interdisciplinary and multidisciplinary field that integrates developed systems related to mechanical, electrical, and electronics hardware and software (Hassoun et al., 2023). Automation technology is transforming the way milk is produced and processed. The applications of robotics and automation have been widely used in manufacturing industries dealing with specific processes and products (Bd et al., 2018). Robotics and automation technologies have a sector-wide influence, including manufacturing, warehousing, and logistics. Although there are a lot of benefits, the development and implementation of such robotic systems is difficult (Eastwood et al., 2022). Robotic milking can help farmers to decrease their stress and anxiety and improve their resilience (Hassoun et al., 2023). This system also allows for a significant average increase in milking frequency, resulting in greater milk production, while decreasing labor costs. Additional benefits of automatic milking systems include enhanced overall productivity, profitability, and sustainability (Lyons et al., 2022).

1.6.2.3 Cyber-physical systems linking digital intelligence with physical execution

CPSs are important in smart farming, as they integrate computational algorithms with agricultural processes to enable real-time monitoring, control, and automation. These systems connect the physical and digital realms, allowing for a continual flow of data and feedback to promote different farming operations. Unlike technologies like IoT or AI, which primarily deal with data collection, analysis, and prediction, CPSs provide an integrated framework that not only collects information but also supports intelligent decision-making and precise action within a unified system (Montalvo et al., 2025; Zhang et al., 2023).

CPSs in the dairy sector and dairy farms integrate digital intelligence with physical execution through the embedding of sensors, computing, and communication technologies into physical dairy farming processes to optimize productivity and animal health. One prominent example involves CPS-enabled monitoring systems that collect data on animal behavior, health, reproduction, and environmental conditions using sensors attached to dairy cows or installed in the farm environment. These systems transmit heterogeneous data to cloud platforms where machine learning algorithms analyze the information to support real-time decision-making, such as detecting mastitis, monitoring stress or emotions, tracking animal locations via GPS, and optimizing feed and health management (Dineva et al., 2021; Liu et al., 2023). These CPS implementations leverage IoT connectivity with AI and cloud computing in a framework often termed cyber-physical agricultural systems (CPASs), where integration extends into agricultural supply chains for enhanced productivity and sustainability. For instance, cyber-physical agricultural systems combine IoT, cloud storage, and big data analytics to create intelligent networks that control and monitor dairy operations, improving efficiency and reducing human labor requirements (Sharma et al., 2020).

Advanced dairy farm CPS solutions incorporate computer vision and sensor fusion technologies to gather fine-scale data on individual cow behavior and health, enabling precision dairy farming practices. These digital systems facilitate near real-time behavior monitoring and health anomaly detection, improving management decisions and animal welfare. Machine learning models applied to sensor data enable timely detection of diseases and accurate assessment of body conditions, contributing to optimized feeding regimens and overall improved dairy production (Liu et al., 2023). Moreover, cooperative smart farming ecosystems extend CPS applications to networked farms or cooperatives, where data from individual member farms is shared via cloud-supported hubs. This collaborative approach enhances data analysis and AI-supported insights at a cooperative level, benefiting smallholder dairy farmers through shared resources and AI-assisted farming recommendations (Chukkapalli et al., 2020).

In summary, in the dairy sector and dairy farms, CPS examples encompass animal- and environment-sensor-based monitoring systems integrated with AI analytics and cloud computing. These systems enable continuous health, behavior, and environmental monitoring, real-time decision support for disease detection and feeding, animal tracking, and cooperative data sharing for enhanced farm management and productivity (Chukkapalli et al., 2020; Dineva et al., 2021; Liu et al., 2023; Sharma et al., 2020).

1.6.3 Biological enabling technologies: integration of biological innovations with digital analytics and automation

1.6.3.1 Advances in biotechnology for improved biological monitoring and process control

Biotechnology's application in the dairy industry has a significant role that increase milk production, animal health, and food processing. Genetically modified microorganisms like bacteriocins and probiotics have been associated with reducing the risks of type 2 diabetes, metabolic syndrome, and heart disease, as well as better weight management (Amiri et al., 2021). Besides, probiotic products provide microorganisms that are viable, specific, and effective in essential nutritional and physiological systems. *Lactobacillus*, *Bifidobacteria*, *Saccharomyces*, and *Streptococcus* are some of the bacteria found in fermented dairy products. Biopreservation with natural preservatives has a positive impact on consumer health promotion in addition to enhancing food safety and shelf-life as a technological influence (Amiri et al., 2021). Additionally, the application of technology has had a significant effect on managing several challenges encountered by the food industry, including food safety, preservation, nutrition, and allergies, food security, changing consumer needs, environmental concerns, economic viability, and policy reforms. Dairy processing companies worldwide require efficient, low-cost, and automated innovations to boost revenues and meet the demand for a wide range of products. The purpose of this study is to summarize and synthesize various information on current developments in milk production and processing. modern biotechnology has created new and exciting potential in the dairy industry, making milk and milk products more accessible to the poor and meeting the demands of a larger population. Since the dairy industry's primary responsibility is to produce consumers with high-quality, nutritional, and affordable dairy products for consumers, biotechnological intervention at different stages of milk production and processing has become a foregone conclusion. It has given us delicious, nutritious, wholesome, handy, shelf-stable, and safe foods. As long as research and development efforts continue, biotechnology will undoubtedly have a significant impact on the food we eat. It has enormous potential for expanding the variety and quality of food available to humans, particularly healthier and more appealing diets. It also appears likely that it will continue to provide benefits to food processing and safety monitoring as new technologies develop at a faster rate. Furthermore, biotechnological application has an important role in dairy product biopreservation, probiotics manipulation, and production; enzyme production; milk-derived bioactive peptides and other functional ingredients; and starter cultures technology and genetic manipulation (Nurye Gebeyehu, year).

1.6.3.2 Nanotechnology for enhanced sensing, preservation, and functional performance

Nanotechnology is the study, engineering, and application of materials and devices at the nanometer scale. Nanotechnology holds promise as a solution to these problems, providing innovative approaches for extending the shelf life of food, enhancing food safety, and decreasing waste in the food supply chain. nanotechnology can play a crucial role in addressing the global challenges of food security and sustainability. The usage of nanotechnology in food preservation is concentrated on improving the longevity of food products and reducing waste. Here's how nanotechnology can help achieve following goals:

Reducing food waste: By improving packaging and storage, nanotechnology can reduce spoilage and contamination. It can also lead to more efficient food handling by allowing better monitoring of environmental factors like temperature and humidity in food storage, which can help prevent conditions that lead to food spoilage. Nanosensors are an essential component in the application of nanotechnology for food preservation, as they provide real-time monitoring of food conditions. These sensors are integrated into packaging or placed within food products to continuously track factors that affect food quality and safety. Nanotechnology is helping in preserving dairy products and preventing spoilage. As mentioned previously, dairy products are highly susceptible to spoilage due to their high moisture content, which creates a favorable environment for bacterial growth and mold development. Nanotechnology offers promising strategies to mitigate spoilage and extend the shelf life of dairy products such as milk, cheese, and yogurt. In particular, nanoenabled sensing mechanisms, including colorimetric and fluorescent indicators, gas sensors, time–temperature indicators, and antimicrobial nanocomposite films, are increasingly integrated into smart packaging systems to enhance spoilage detection, freshness monitoring, food safety assurance, and microbial control (Muthu et al., 2025).

1.6.3.3 Precision fermentation and bioengineering as emerging control points in biological variability

Precision fermentation represents a novel cell-based production technique that uses synthetic biology and metabolic engineering tools to revolutionize global food production by utilizing "microbial cell factories" to synthesize specified ingredients and produce added-value substances (Gao et al., 2025). Recent advancements in synthetic biology, metabolic engineering, and enzyme engineering serve as the foundation of precision fermentation platforms that produce tailor-made enzymes and functional molecules under extremely controlled conditions. Precision fermentation, combining engineered GRAS microorganisms with controlled bioreactors, is described as an "advanced fermentation system" that enables the targeted biosynthesis of specific proteins, flavors, vitamins, and other compounds with improved yield, stability, and functionality, thereby reducing variability associated with traditional mixed cultures (Boukid et al., 2023; Singh and Kumar, 2025).

1.6.4 Convergence toward Dairy 4.0 as a system-level transformation pathway

The combined adoption of digital, physical, and biological enabling technologies is moving dairy systems from isolated interventions toward integrated, system-level transformation. The concept of Industry 4.0 in agriculture and food systems refers to the integration of digital, physical, and biological sciences into a framework that transforms traditional production toward smart, connected systems. Food Industry 4.0 is defined as a technological paradigm that goes beyond automation of discrete tasks, representing a convergence of digital, physical, and biological domains that allows holistic transformation of food value chains. Such integration is widely acknowledged as critical to transforming agrifood systems by connecting numerous technologies to provide real-time monitoring, decision-making, and coordinated operations across production and supply chains (Hassoun et al., 2024).

This convergence enables continuous data flows, predictive management, and improved alignment of biological processes, physical activities, and decision-making structures. In the agricultural context, it has been demonstrated that digital innovations such as precision farming, IoT, remote sensing, data-driven analytics, AI, and robotics not only improve discrete outcomes (e.g., productivity, sustainability) but also support continuous data flows and predictive capabilities that align biological phenomena with operational processes. By allowing large-scale data capture and real-time analytics across sensors, machines, and biological systems, these technologies facilitate coordinated decision-making that integrates biological variability (e.g., animal health or crop conditions) with physical operations (e.g., milking robots, feeding systems) and strategic management.

Dairy 4.0 thus emerges as an overarching framework that organizes connectivity, automation, and data intelligence across the dairy value chain, setting the foundation for the next chapter. Recent research on Industry 4.0 adoption in the dairy sector underscores how technologies such as AI, IoT, robotics, big data, and blockchain are gradually being integrated throughout the dairy supply chain from farm to table, collectively creating a Dairy 4.0 paradigm. This paradigm is distinguished by connected systems that integrate automated processes, predictive analytics, and digital coordination across production, processing, and distribution. As these technologies converge, they form an overarching framework that improves operational intelligence, synchronizes biological and physical processes, and supports strategic decision-making across the entire dairy value chain (Finger, 2023).

1.7 Conclusion

The dairy industry faces interconnected challenges related to productivity, quality, sustainability, animal welfare, and supply chain resilience, as illustrated by the literature on agricultural digital transformation and dairy systems. Systematic evaluations and empirical studies emphasize persistent inadequacies in production processes, fragmented supply chains, and operational constraints that reduce quality, weaken resilience, and complicate efforts to improve sector-wide performance. Without integrated innovation initiatives, dairy systems will struggle to respond effectively to variable demand, environmental pressures, and animal welfare standards. Structural components of the dairy value chain, such as biological variability, milk's perishability, fragmented data flows, and information asymmetry among operators, limit the effectiveness of incremental, isolated improvements and highlight the need for system-level transformation. Digital adoption remains inconsistent, particularly among small- and medium-sized producers, while sustainability indicators and regulatory frameworks continue to fall behind technological possibilities, reinforcing fragmentation and hindering innovation diffusion. Industry 4.0 enabling technologies develop as systemic responses that promote integration, anticipation, and adaptive management across complex dairy systems, as indicated by recent research on digital transformation in agrifood sectors. Technologies such as AI, big data analytics, IoT, and blockchain have shown potential to strengthen predictive decision-making, enhance traceability and quality assurance, and improve supply chain coordination, thereby tackling multiple interrelated challenges simultaneously. Evidence from this chapter highlights that the integration of digital, physical, and biological technology enablers is a suitable way for enhancing performance while increasing resilience, sustainability, and competitiveness in global food systems where the complexity is continuously increasing. In this perspective, Dairy 4.0 should be understood not merely as a technological upgrade, but as a systemic transformation of biologically variable production systems, reshaping coordination, value creation, and socioeconomic structures across the dairy value chain.

AI disclosure

During the preparation of this work, the author used ChatGPT-5.0 in order to improve readability and language. After using this tool/service, the author reviewed and edited the content as needed and takes full responsibility for the content of the publication.

References

Abeni, F., Petrera, F., Galli, A., 2019. A survey of italian dairy farmers' propensity for precision livestock farming tools. Animals 9 (5), 202. Available at https://doi.org/10.3390/ani9050202.

Acosta, A., McCorriston, S., Nicolli, F., Venturelli, E., Aratchilage, U.G., ArceDiaz, E., et al., 2021. Immediate effects of COVID-19 on the global dairy sector. Agricultural Systems 192. Available at https://doi.org/10.1016/j.agsy.2021.103177.

Adesogan, A.T., Dahl, G.E., 2020. MILK Symposium Introduction: Dairy production in developing countries. Journal of Dairy Science Vol. 103. Elsevier Inc, pp. 9677–9680. Issue 11. Available at https://doi.org/10.3168/jds.2020-18313.

Akdeniz, V., Akalın, A.S., 2022. Recent advances in dual effect of power ultrasound to microorganisms in dairy industry: Activation or inactivation. Critical Reviews in Food Science and Nutrition 62 (4), 889–904. Available at https://doi.org/10.1080/10408398.2020.1830027.

Ali, S.S., Barman, H., Kaur, R., Tomaskova, H., Roy, S.K., 2021. Multi-product multi echelon measurements of perishable supply chain: Fuzzy non-linear programming approach. Mathematics 9 (17). Available at https://doi.org/10.3390/math9172093.

Alonso, M.E., González-Montaña, J.R., Lomillos, J.M., 2020. Consumers' concerns and perceptions of farm animal welfare. Animals Vol. 10 MDPI AG Issue 3. Available at https://doi.org/10.3390/ani10030385.

Alqaisi, O., Schlecht, E., 2021. Feeding models to optimize dairy feed rations in view of feed availability, feed prices and milk production scenarios. Sustainability (Switzerland) 13 (1), 1–15. Available at https://doi.org/10.3390/su13010215.

Amiri, S., Aghamirzaei, M., Mostashari, P., Sarbazi, M., Tizchang, S., Madahi, H., 2021. The impact of biotechnology on dairy industry. Microbial Biotechnology in Food and Health 53–79. Available at https://doi.org/10.1016/B978-0-12-819813-1.00003-7.

Auldist, M.J., Coats, S., Rogers, G.L., McDowell, G.H., 1995. Changes in the composition of milk from healthy and mastitic dairy cows during the lactation cycle. Australian Journal of Experimental Agriculture 35 (4), 427. Available at https://doi.org/10.1071/ea9950427.

Ayalew, W., Wu, X., Tarekegn, G.M., Sisay Tessema, T., Naboulsi, R., Van Damme, R., et al., 2024. Whole genome scan uncovers candidate genes related to milk production traits in Barka cattle. International Journal of Molecular Sciences 25 (11), 6142. Available at https://doi.org/10.3390/ijms25116142.

Aydogdu, T., O'Mahony, J.A., Huppertz, T., Magan, J.D., McCarthy, N.A., 2023. Measuring pH of skim milk and milk permeate at ultra-high temperatures at laboratory and pilot scale. International Dairy Journal 139, 105565.

Bd, M., Ranvir, S., Adil, S., 2018. Robotics: An emerging technology in dairy and food industry: Review. 440 ~ International Journal of Chemical Studies 6 (2), 440–449.

Beckett, E.L., Cassettari, T., Starck, C., Fayet-Moore, F., 2024. Dairy milk: There are alternatives but no equivalents. Food Science & Nutrition 12 (10), 8470–8482. Available at https://doi.org/10.1002/fsn3.4301.

Berry, D.P., Friggens, N.C., Lucy, M., Roche, J.R., 2016. Milk production and fertility in cattle. Annual Review of Animal Biosciences 4 (1), 269–290.

Bednarski, M., Kupczyński, R., 2024. Factors Affecting Milk Productivity, Milk Quality and Dairy Cow Health. Animals : An Open Access Journal from MDPI 14 (24), 3707. Available at https://doi.org/10.3390/ANI14243707.

Boerman, J.P., Brito, L.F., Montes, M.E., Maskal, J.M., Doucette, J., Kalbaugh, K., 2025. Data processing techniques to improve data integration from dairy farms. JDS Communications 6 (3), 339–344. Available at https://doi.org/10.3168/jdsc.2024-0723.

Bojovic, M., McGregor, A., 2023. A review of megatrends in the global dairy sector: what are the socioecological implications? Agriculture and Human Values 40 (1), 373–394. Available at https://doi.org/10.1007/S10460-022-10338-X.

Bokharaeian, M., Toghdory, A., Ghoorchi, T., Ghassemi Nejad, J., Esfahani, I.J., 2023. Quantitative associations between season, month, and temperature-humidity index with milk yield, composition, somatic cell counts, and microbial load: A comprehensive study across ten dairy farms over an annual cycle. Animals 13 (20), 3205. Available at https://doi.org/10.3390/ani13203205.

Bonamigo, A., Nunes, A., Mendes, L.F., Martelotte, M.C., Andrade, H.D.S., 2024. The interrelationship between Lean 4.0 and value co-creation: An empirical study in the dairy sector. International Journal of Lean Six Sigma 15 (7), 1365–1392. Available at https://doi.org/10.1108/IJLSS-06-2022-0141.

Bondan, C., Folchini, J.A., Noro, M., Quadros, D.L., Machado, K.M., González, F.H.D., 2018. Milk composition of holstein cows: A retrospective study. Ciencia Rural 48 (12). Available at https://doi.org/10.1590/0103-8478cr20180123.

Botín-Sanabria, D.M., Mihaita, S., Peimbert-García, R.E., Ramírez-Moreno, M.A., Ramírez-Mendoza, R.A., Lozoya-Santos, Jd.J., 2022. Digital twin technology challenges and applications: A comprehensive review. Remote Sensing 14 (6). Available at https://doi.org/10.3390/rs14061335.

Boukid, F., Ganeshan, S., Wang, Y., Tülbek, M.Ç., Nickerson, M.T., 2023. Bioengineered enzymes and precision fermentation in the food industry. International Journal of Molecular Sciences 24 (12). Available at https://doi.org/10.3390/ijms241210156.

Brito, L.F., Bedere, N., Douhard, F., Oliveira, H.R., Arnal, M., Peñagaricano, F., et al., 2021. Review: Genetic selection of high-yielding dairy cattle toward sustainable farming systems in a rapidly changing world. Animal 15, 100292. Available at https://doi.org/10.1016/j.animal.2021.100292.

Buller, H., Blokhuis, H., Jensen, P., Keeling, L., 2018. Towards farm animal welfare and sustainability. Animals 8 (6), 81. Available at https://doi.org/10.3390/ani8060081.

Cabrera, V.E., Barrientos-Blanco, J.A., Delgado, H., Fadul-Pacheco, L., 2020. Symposium review: Real-time continuous decision making using big data on dairy farms. Journal of Dairy Science 103 (4), 3856–3866. Available at https://doi.org/10.3168/jds.2019-17145.

Cabrera, V.E., Bewley, J., Breunig, M., Breunig, T., Cooley, W., De Vries, A., Fourdraine, R., Giordano, J.O., Gong, Y., Greenfield, R., Hu, H., Lenkaitis, A., Niu, M., Noronha, E.A.F., Sullivan, M., 2025. Data Integration and Analytics in the Dairy Industry: Challenges and Pathways Forward. Animals 15 (3), 329. Available at https://doi.org/10.3390/ANI15030329.

Cai, Z., Dusza, M., Guldbrandtsen, B., Lund, M.S., Sahana, G., 2020. Distinguishing pleiotropy from linked QTL between milk production traits and mastitis resistance in Nordic Holstein cattle. Genetics Selection Evolution 52 (1). Available at https://doi.org/10.1186/s12711-020-00538-6.

Cao, H., 2023. The role of standardization in the operation of enterprise quality management systems. Modern Economics & Management Forum 4 (2), 21. Available at https://doi.org/10.32629/memf.v4i2.1314.

Carnovale, F., Jin, X., Arney, D., Descovich, K., Guo, W., Shi, B., et al., 2021. Chinese public attitudes towards, and knowledge of, animal welfare. Animals 11 (3), 1–27. Available at https://doi.org/10.3390/ani11030855.

Carvajal-Pérez, L.A., Montenegro-Arellano, G.F., Revelo-Ruales, V.W., Terán-Rosero, G.J., Urgilés-Urgilés, G.P., 2024. Socioeconomic determinants of small and medium-sized dairy farms in the Ecuador-Colombia border area. Tropical Animal Health and Production 56 (7). Available at https://doi.org/10.1007/s11250-024-04092-x.

Casino, F., Kanakaris, V., Dasaklis, T.K., Moschuris, S., Stachtiaris, S., Pagoni, M., Rachaniotis, N.P., 2021. Blockchain-based food supply chain traceability: a case study in the dairy sector. International Journal of Production Research 1–13. Available at https://doi.org/10.1080/00207543.2020.1789238.

Chandrapala, J., Leong, T., 2015. Ultrasonic processing for dairy applications: Recent advances. Food Engineering Reviews 7 (2), 143–158. Available at https://doi.org/10.1007/s12393-014-9105-8.

Chowdhury, M.A.H., Reem, C.S.A., Ashrafudoulla, M., Rahman, M.A., Shaila, S., Jie-won Ha, A., et al., 2025. Role of advanced cleaning and sanitation techniques in biofilm prevention on dairy equipment. Comprehensive Reviews in Food Science and Food Safety 24 (3). Available at https://doi.org/10.1111/1541-4337.70176.

Chukkapalli, S.S.L., Mittal, S., Gupta, M., Abdelsalam, M., Joshi, A., Sandhu, R., et al., 2020. Ontologies and artificial intelligence systems for the cooperative smart farming ecosystem. IEEE Access 8, 164045–164064. Available at https://doi.org/10.1109/access.2020.3022763.

Ciccullo, F., Fabbri, M., Abdelkafi, N., Pero, M., 2022. Exploring the potential of business models for sustainability and big data for food waste reduction. Journal of Cleaner Production 340, 130673. Available at https://doi.org/10.1016/j.jclepro.2022.130673.

Cinar, M., Serbester, U., Ceyhan, A., Gorgulu, M., 2015. Effect of somatic cell count on milk yield and composition of first and second lactation dairy cows. Italian Journal of Animal Science 14 (1), 105–108. Available at https://doi.org/10.4081/ijas.2015.3646.

Clay, N., Garnett, T., Lorimer, J., 2020. Dairy intensification: Drivers, impacts and alternatives. Ambio 49 (1), 35–48. Available at https://doi.org/10.1007/S13280-019-01177-Y/TABLES/1.

Connolly, C., Yin, X., Brennan, L., 2023. Impact of Lactation Stage on the Metabolite Composition of Bovine Milk. Molecules 28 (18), 6608. Available at https://doi.org/10.3390/MOLECULES28186608.

Correddu, F., Caratzu, M.F., Lunesu, M.F., Carta, S., Pulina, G., Nudda, A., 2023. Grape, pomegranate, olive, and tomato by-products fed to dairy ruminants improve milk fatty acid profile without depressing milk production. Foods 12 (4). Available at https://doi.org/10.3390/foods12040865.

Das, S., Saha, A., Aggarwal, K., 2025. Natural biopreservatives for dairy products: Microbial and plant-derived solutions and recent advancements. International Dairy Journal 166 (March), 106256. Available at https://doi.org/10.1016/j.idairyj.2025.106256.

Deming, J., Gleeson, D., O'Dwyer, T., Kinsella, J., O'Brien, B., 2018. Measuring labor input on pasture-based dairy farms using a smartphone. Journal of Dairy Science 101 (10), 9527–9543. Available at https://doi.org/10.3168/jds.2017-14288.

Dineva, K., Parvanov, D., Atanasova, T., Mateeva, G., Petrov, P., & Kostadinov, G. (2021). Towards CPS/IoT system for livestock smart farm monitoring international conference automatics and informatics. In *ICAI 2021 Proceedings*. IEEE. https://doi.org/10.1109/ICAI52893.2021.9639460.

Eastwood, C.R., Dela Rue, B., Edwards, J.P., Jago, J., 2022. Responsible robotics design—A systems approach to developing design guides for robotics in pasture-grazed dairy farming. Frontiers in Robotics and AI 9. Available at https://doi.org/10.3389/frobt.2022.914850.

Erem, E., Kilic-Akyilmaz, M., 2024. The role of fermentation with lactic acid bacteria in quality and health effects of plant-based dairy analogues. Comprehensive Reviews in Food Science and Food Safety 23 (4). Available at https://doi.org/10.1111/1541-4337.13402.

Ersoy, P., Börühan, G., Kumar Mangla, S., Hormazabal, J.H., Kazancoglu, Y., Lafcı, Ç., 2022. Impact of information technology and knowledge sharing on circular food supply chains for green business growth. Business Strategy and the Environment 31 (5), 1875–1904. Available at https://doi.org/10.1002/bse.2988.

Estremadoyro, L.J.G., Salome, P.H., Carhuas, J.N., Guzman, S.O., Tacza, A.A., Guillen, M.A.F., et al., 2024. Effects of different seasons on milk quality: A study on two cattle breeds in rainy and drought contexts. World's Veterinary Journal 14 (2), 213–219. Available at https://doi.org/10.54203/scil.2024.wvj26.

Fadul-Pacheco, L., Wangen, S.R., da Silva, T.E., Cabrera, V.E., 2022. Addressing data bottlenecks in the dairy farm industry. Animals 12 (6). Available at https://doi.org/10.3390/ani12060721.

Fan, S.Z., Jump, J., Tse, Y., Yu, L., 2022. Volatility in US dairy futures market. SSRN Electronic Journal. Available at https://doi.org/10.2139/ssrn.4072569.

Fan, Z., Jump, J., Yu, L., 2025. Return comovement and price volatility: a study of the US dairy commodity futures markets. Applied Economics 57 (2), 232–250. Available at https://doi.org/10.1080/00036846.2024.2303614.

Felis, A., Pica-Ciamarra, U., Reyes, E., 2025. Dairy's Development and Socio-Economic Transformation: A Cross-Country Analysis. World 6 (3), 105. Available at https://doi.org/10.3390/WORLD6030105.

Finger, R., 2023. Digital innovations for sustainable and resilient agricultural systems. European Review of Agricultural Economics 50 (4), 1277–1309. Available at https://doi.org/10.1093/erae/jbad021.

Fiorillo, V., Amico, B.M., 2024. Milk Quality and Economic Sustainability in Dairy Farming: A Systematic Review of Performance Indicators. Dairy Vol. 5. Multidisciplinary Digital Publishing Institute (MDPI), pp. 384–402. Issue 3. Available at https://doi.org/10.3390/dairy5030031.

Franceschini, S., Fastré, C., Nickmilder, C., Santschi, D.E., Warner, D., Bahadi, M., et al., 2025. Detection of dairy herd management issues using fatty acid profiles predicted by mid-infrared spectrometry. Animals 15 (11). Available at https://doi.org/10.3390/ani15111575.

Gao, F., Shi, S., Zhao, Y., Yang, D., Liao, X., 2025. Intersecting precision fermentation for global cell-based food production innovation: Challenges and opportunities. Biotechnology Advances 85, 108712. Available at https://doi.org/10.1016/j.biotechadv.2025.108712.

Garvey, M., 2022. Lameness in dairy cow herds: Disease aetiology, prevention and management. Dairy 3 (1), 199–210. Available at https://doi.org/10.3390/dairy3010016.

Gavran, I., Honcharuk, S., Mykhalov, V., Stepanenko, K., Tsimokh, N., 2025. The Impact of Artificial Intelligence on the Production and Editing of Audiovisual Content. Preservation, Digital Technology & Culture 54 (3), 223–235. Available at https://doi.org/10.1515/pdtc-2025-0022.

Golher, D.M., Patel, B.H.M., Bhoite, S.H., et al., 2021. Factors influencing water intake in dairy cows: a review. Int J Biometeorol 65, 617–625. Available at https://doi.org/10.1007/s00484-020-02038-0.

Groher, T., Heitkämper, K., Umstätter, C., 2020. Digital technology adoption in livestock production with a special focus on ruminant farming. Animal 14 (11), 2404–2413. Available at https://doi.org/10.1017/s1751731120001391.

Gross, J.J., 2022. Limiting factors for milk production in dairy cows: Perspectives from physiology and nutrition. Journal of Animal Science 100 (3). Available at https://doi.org/10.1093/jas/skac044.

Guliński, P., 2021. Ketone bodies—Causes and effects of their increased presence in cows' body fluids: A review. Veterinary World 14 (6), 1492–1503. Available at https://doi.org/10.14202/vetworld.2021.1492-1503.

Guzmán-Luna, P., Mauricio-Iglesias, M., Flysjö, A., Hospido, A., 2022. Analysing the interaction between the dairy sector and climate change from a life cycle perspective: A review. Trends in Food Science & Technology 126, 168–179. Available at https://doi.org/10.1016/j.tifs.2021.09.001.

Habimana, V., Nguluma, A.S., Nziku, Z.C., Ekine-Dzivenu, C.C., Morota, G., Mrode, R., et al., 2023. Heat stress effects on milk yield traits and metabolites and mitigation strategies for dairy cattle breeds reared in tropical and sub-tropical countries. Frontiers in Veterinary Science 10. Available at https://doi.org/10.3389/fvets.2023.1121499.

Hamad, A.M.A., Awad, D.A., 2026. Integration of IoT in Dairy Facilities. Next-Generation Technologies in Dairy Processing and Production. IGI Global Scientific Publishing, pp. 139–170.

Hassoun, A., 2024. In: Hassoun, A. (Ed.), Food Industry 4.0: Emerging trends and technologies in sustainable food production and consumption. Academic Press.

Hassoun, A., 2025. In: Hassoun, A. (Ed.), Food Industry 4.0 and food sustainability. Springer.

Hassoun, A., Garcia-Garcia, G., Trollman, H., Jagtap, S., Parra-López, C., Cropotova, J., et al., 2023. Birth of dairy 4.0: Opportunities and challenges in adoption of fourth industrial revolution technologies in the production of milk and its derivatives. Current Research in Food Science 7, 100535. Available at https://doi.org/10.1016/j.crfs.2023.100535.

Hassoun, A., Tarchi, I., Aït-kaddour, A., 2024. Leveraging the potential of fourth industrial revolution technologies to reduce and valorize waste and by-products in the dairy sector. Current Opinion in Green and Sustainable Chemistry 47, 100927. Available at https://doi.org/10.1016/j.cogsc.2024.100927.

Haworth, G.M., Tranter, W.P., Chuck, J.N., Cheng, Z., Wathes, D.C., 2008. Relationships between age at first calving and first lactation milk yield, and lifetime productivity and longevity in dairy cows. Veterinary Record 162 (20), 643–647. Available at https://doi.org/10.1136/vr.162.20.643.

Hayes, E., Wallace, D., O'Donnell, C., Greene, D., Hennessy, D., O'Shea, N., et al., 2023. Trend analysis and prediction of seasonal changes in milk composition from a pasture-based dairy research herd. Journal of Dairy Science 106 (4), 2326–2337. Available at https://doi.org/10.3168/jds.2021-21483.

Heard, J.W., Hannah, M.C., Ho, C.K.M., Wales, W.J., 2021. Predicting immediate marginal milk responses and evaluating the economics of two-variable input tactical feeding decisions in grazing dairy cows. Animals 11 (7). Available at https://doi.org/10.3390/ani11071920.

Hidalgo-Fuentes, B., de Jesús-José, E., Cabrera-Hidalgo, Ad.J., Sandoval-Castilla, O., Espinosa-Solares, T., González-Reza, R.M., et al., 2024. Plant-based fermented beverages: Nutritional composition, sensory properties, and health benefits. Foods 13 (6). Available at https://doi.org/10.3390/foods13060844.

Hill, J., 2024. Science, technology, and innovation in the dairy sector. International Journal of Food Science and Technology 59 (9), 6717–6723. Available at https://doi.org/10.1111/ijfs.17385.

Hussain, S., Hussain, A., Ho, J., Sparagano, O.A.E., Zia, U.-ur-R., 2020. Economic and social impacts of COVID-19 on animal welfare and dairy husbandry in central Punjab, Pakistan. Frontiers in Veterinary Science 7. Available at https://doi.org/10.3389/fvets.2020.589971.

Ignowski, L., Minten, B., Van Campenhout, B., Vandevelde, S., 2022. Trade, value chain technology, and prices: Evidence from dairy in East Africa. Q Open 2 (2). Available at https://doi.org/10.1093/qopen/qoac016.

Ivana, B., Puvača, N., 2024. Economic and Ecological Sustainability of Dairy Production. Journal of Agronomy, Technology and Engineering Management (JATEM 7 (3), 1088–1104. Available at https://doi.org/10.55817/tqxu2115.

Jafri, S.H., Adnan, K.M.M., Baimbill Johnson, S., Talukder, A.A., Yu, M., Osei, E., 2024. Challenges and solutions for small dairy farms in the U.S.: A review. Agriculture (Switzerland) 14 (12). Available at https://doi.org/10.3390/agriculture14122369.

Kazi, K.S.L., 2024. IoT technologies for the intelligent dairy industry. IGI Global, pp. 321–350. Available at 10.4018/979-8-3693-5498-8.ch012.

Keady, M.M., Jimenez, R.R., Bragg, M., Wagner, J.C.P., Bornbusch, S.L., Power, M.L., et al., 2023. Ecoevolutionary processes structure milk microbiomes across the mammalian tree of life. Proceedings of the National Academy of Sciences of the United States of America 120 (28). Available at https://doi.org/10.1073/pnas.2218900120.

Kelly, A.L., Reid, S., Joyce, P., Meaney, W.J., Foley, J., 1998. Effect of decreased milking frequency of cows in late lactation on milk somatic cell count, polymorphonuclear leucocyte numbers, composition and proteolytic activity. Journal of Dairy Research 65 (3), 365–373. Available at https://doi.org/10.1017/S0022029998002921.

Kettele, M., Lechner, G., 2025. Assessing resilience measures in the Austrian dairy supply chain: a case-based simulation integrating strategic, tactical, and operational decisions. International Journal of Production Research 63 (11), 3982–4015. Available at https://doi.org/10.1080/00207543.2024.2432472.

Khanna, A., Jain, S., Burgio, A., Bolshev, V., Panchenko, V., 2022. Blockchain-Enabled Supply Chain platform for Indian Dairy Industry: Safety and Traceability. Foods 11 (17), 2716. Available at https://doi.org/10.3390/FOODS11172716.

Kholif, A.E., Olafadehan, O.A., 2022. Dietary strategies to enrich milk with healthy fatty acids—A review. Annals of Animal Science 22 (2), 523–536. Available at https://doi.org/10.2478/aoas-2021-0058.

Kiambi, S., Alarcon, P., Rushton, J., Murungi, M.K., Muinde, P., Akoko, J., et al., 2018. Mapping Nairobi's dairy food system: An essential analysis for policy, industry and research. Agricultural Systems 167, 47–60. Available at https://doi.org/10.1016/j.agsy.2018.08.007.

Kleen, J.L., Guatteo, R., 2023. Precision livestock farming: What does it contain and what are the perspectives? Animals 13 (5). Available at https://doi.org/10.3390/ani13050779.

Kliem, K.E., Givens, D.I., 2011. Dairy products in the food chain: Their impact on health. Annual review of food science and technology 2, 21–36. Available at https://doi.org/10.1146/annurev-food-022510-133734.

Kotsampasi, B., Karatzia, M.A., Tsiokos, D., Chadio, S., 2024. Nutritional strategies to alleviate stress and improve welfare in dairy ruminants. Animals 14 (17), 2573. Available at https://doi.org/10.3390/ani14172573.

Krömker, V., Leimbach, S., 2017. Mastitis treatment—Reduction in antibiotic usage in dairy cows. Reproduction in Domestic Animals 52 (S3), 21–29. Available at https://doi.org/10.1111/rda.13032.

Kuehn, C., Edel, C., Weikard, R., Thaller, G., 2007. Dominance and parent-of-origin effects of coding and non-coding alleles at the acylCoA-diacylglycerol-acyltransferase (DGAT1) gene on milk production traits in German Holstein cows. BMC Genetics 8 (1). Available at https://doi.org/10.1186/1471-2156-8-62.

Kumar, I., Rawat, J., Mohd, N., Husain, S., Khan, R., 2021. Opportunities of artificial intelligence and machine learning in the food industry. Journal of Food Quality 2021, 1–10. Available at https://doi.org/10.1155/2021/4535567.

Kuo, C., Weary, D.M., Roche, S.M., von Keyserlingk, M.A.G., 2025. Exploring the legitimacy of industry-led farm animal welfare governance using examples of Canadian and United States dairy standards. Animal Welfare 34. Available at https://doi.org/10.1017/awf.2025.17.

Lakshmi Priya, A., Vanathi, R., 2025a. Impact of Supply Chain Management on Productivity of Dairy Industry in India - A Literature Review. Shanlax International Journal of Management 12 (4), 126–134. Available at https://doi.org/10.34293/management.v12i4.8559.

Lakshmi Priya, A., & Vanathi, R. (2025b). A Study on Dairy Supply Chain Management Practices of Dairy Companies with Special Reference to Erode District (Vol. 7, Issue 3). Retrieved from www.ijfmr.com.

Lamanna, M., Bovo, M., Cavallini, D., 2025. Wearable collar technologies for dairy cows: A systematized review of the current applications and future innovations in precision livestock farming. Animals 15 (3), 458. Available at https://doi.org/10.3390/ani15030458.

Leduc, A., Souchet, S., Gelé, M., Le Provost, F., Boutinaud, M., 2021. Effect of feed restriction on dairy cow milk production: A review. Journal of Animal Science 99 (7). Available at https://doi.org/10.1093/jas/skab130.

Lee, J.Y., Kim, I.H., 2006. Advancing parity is associated with high milk production at the cost of body condition and increased periparturient disorders in dairy herds. Journal of Veterinary Science 7 (2), 161–166. Available at https://doi.org/10.4142/jvs.2006.7.2.161.

Lee, S.H., Cho, K.H., Park, M.N., Choi, T.J., Kim, S.D., Do, C.H., 2016. Genetic parameters of milk β-Hydroxybutyric acid and acetone and their genetic association with milk production traits of Holstein cattle. Asian-Australasian Journal of Animal Sciences 29 (11), 1530–1540. Available at https://doi.org/10.5713/ajas.16.0310.

Li, A., Zheng, J., Han, X., Yang, S., Cheng, S., Zhao, J., et al., 2023. Advances in low-lactose/lactose-free dairy products and their production. Foods 12 (13), 2553. Available at https://doi.org/10.3390/foods12132553.

Liu, N., Qi, J., An, X., Wang, Y., 2023. A review on information technologies applicable to precision dairy farming: Focus on behavior, health monitoring, and the precise feeding of dairy cows. Agriculture 13 (10), 1858. Available at https://doi.org/10.3390/agriculture13101858.

Lyons, N.A., Gargiulo, J.I., Clark, C.E.F., Garcia, S.C., 2022. Technology and Robotic Milking in Dairy Production. Encyclopedia of Dairy Sciences: Third Edition, 1 823–829. Available at https://doi.org/10.1016/B978-0-12-818766-1.00289-0.

Ma, Y., Khan, M.Z., Xiao, J., Alugongo, G.M., Chen, X., Chen, T., et al., 2021. Genetic markers associated with milk production traits in dairy cattle. Agriculture (Switzerland) 11 (10). Available at https://doi.org/10.3390/agriculture11101018.

Magro, S., Costa, A., Santinello, M., Penasa, M., De Marchi, M., 2023. Udder health-related traits in cow milk: phenotypic variability and effect on milk yield and composition. Animal 17 (6), 100823. Available at https://doi.org/10.1016/j.animal.2023.100823.

Makarov, E.I., Polyansky, K.K., Makarov, M.E., Nikolaeva, Y.R., Shubina, E.A., 2019. Conceptual approaches to the quality system of dairy products based on the blockchain technology. Studies in Computational Intelligence 826, 1059–1069. Available at https://doi.org/10.1007/978-3-030-13397-9_109/FIGURES/4.

Malliaroudaki, M.I., Watson, N.J., Ferrari, R., Nchari, L.N., Gomes, R.L., 2022. Energy management for a net zero dairy supply chain under climate change. Trends in Food Science and Technology 126, 153–167. Available at https://doi.org/10.1016/j.tifs.2022.01.015.

Manning, L., Brewer, S., Craigon, P.J., Frey, J., Gutierrez, A., Jacobs, N., et al., 2022. Artificial intelligence and ethics within the food sector: Developing a common language for technology adoption across the supply chain. Trends in Food Science and Technology 125, 33–42. Available at https://doi.org/10.1016/j.tifs.2022.04.025.

McNitt, J.I., Lukefahr, S.D., 1990. Effects of breed, parity, day of lactation and number of kits on milk production of rabbits. Journal of Animal Science 68 (6), 1505. Available at https://doi.org/10.2527/1990.6861505x.

Mennig, P., Szigeti, Z., 2025. Synergies and trade-offs between environmental impacts and farm profitability: The case of pasture-based dairy production systems. Journal of Agriculture and Food Research 21, 101798. Available at https://doi.org/10.1016/J.JAFR.2025.101798.

Min, L., Li, D., Tong, X., Nan, X., Ding, D., Xu, B., et al., 2019. Nutritional strategies for alleviating the detrimental effects of heat stress in dairy cows: a review. International Journal of Biometeorology 63 (9), 1283–1302. Available at https://doi.org/10.1007/s00484-019-01744-8.

Mintenid, B., Habte, Y., Tamru, S., & Tesfaye, A. (2020). The transforming dairy sector in Ethiopia. https://doi.org/10.1371/journal.pone.0237456

Mohammadi, H., Farahani, A.H.K., Moradi, M.H., Mastrangelo, S., Di Gerlando, R., Sardina, M.T., et al., 2022. Weighted single-step genome-wide association study uncovers known and novel candidate genomic regions for milk production traits and somatic cell score in Valle del Belice dairy sheep. Animals 12 (9). Available at https://doi.org/10.3390/ani12091155.

Mohapatra, S., R, S., Pabba, A.S., 2022. Analysis of Dairy Value Chains in Organized Sectors of Haryana: A Chain Wide Learning Approach. Indian Journal of Extension Education 96–101. Available at https://doi.org/10.48165/ijee.2022.58420.

Montalvo, A., Camacho, O., Chavez, D., 2025. Cyber-Physical Systems for Smart Farming: A Systematic Review. Sustainability 17 (14), 6393. Available at https://doi.org/10.3390/SU17146393.

Mor, R.S., Bhardwaj, A., Singh, S., 2018. Benchmarking the interactions among performance indicators in dairy supply chainAn ISM approach. Benchmarking: An International Journal 25 (9), 3858–3881. Available at https://doi.org/10.1108/BIJ-09-2017-0254.

More, S.J., Marchewka, J., Hanlon, A., Balzani, A., Boyle, L., 2021. An evaluation of four private animal health and welfare standards and associated quality assurance programmes for dairy cow production. Food Policy 105, 102169. Available at https://doi.org/10.1016/j.foodpol.2021.102169.

Moreira, V.H., Bravo-Ureta, B.E., 2016. Total factor productivity change in dairy farming: Empirical evidence from southern Chile. Journal of Dairy Science 99 (10), 8356–8364. Available at https://doi.org/10.3168/JDS.2016-11055.

Muthu, A., Nguyen, D.H.H., Neji, C., Törős, G., Ferroudj, A., Atieh, R., et al., 2025. Nanomaterials for smart and sustainable food packaging: Nano-sensing mechanisms, and regulatory perspectives. Foods 14 (15). Available at https://doi.org/10.3390/foods14152657.

Mwirigi, D., Fekete-Farkas, M., Borbély, C., 2025. From Cow to Climate—Tracing the Path of Dairy Sustainability: Unveiling the Impact on Sustainable Development Goals Through Bibliometric and Literature Analyses. Animals 15 (7), 931. Available at https://doi.org/10.3390/ANI15070931.

Nam, K., Lim, H., Ahn Il, B., 2020. Analysis of consumer preference for milk produced through sustainable farming: The case of mountainous dairy farming. Sustainability (Switzerland) 12 (7). Available at https://doi.org/10.3390/su12073039.

Neculai-Valeanu, A.S., Ariton, A.M., 2022. Udder health monitoring for prevention of bovine mastitis and improvement of milk quality. Bioengineering 9 (11). Available at https://doi.org/10.3390/bioengineering9110608.

Neethirajan, S., 2024. Net zero dairy farming—Advancing climate goals with big data and artificial intelligence. Climate 12 (2), 15. Available at https://doi.org/10.3390/cli12020015.

Nudda, A., Feligini, M., Battacone, G., Macciotta, N.P.P., Pulina, G., 2003. Effects of lactation stage, parity, β-lactoglobulin genotype and milk SCC on whey protein composition in sarda dairy ewes. Italian Journal of Animal Science 2 (1), 29–39. Available at https://doi.org/10.4081/ijas.2003.29.

Nyambo, D.G., 2024. Uncovering service gaps and patterns in smallholder dairy production systems: A data mining approach. Scientific African 26, e02392. Available at https://doi.org/10.1016/j.sciaf.2024.e02392.

OECD-FAO Agricultural Outlook 2025-2034. (2025). OECD Publishing. https://doi.org/10.1787/601276cd-en

Palma, O., Plà-Aragonés, L.M., Mac Cawley, A., Albornoz, V.M., 2025. AI and data analytics in the dairy farms: A scoping review. Animals 15 (9). Available at https://doi.org/10.3390/ani15091291.

Papademas, P., Bintsis, T., 2010. Food safety management systems (FSMS) in the dairy industry: A review. International Journal of Dairy Technology 63 (4), 489–503. Available at https://doi.org/10.1111/j.1471-0307.2010.00620.x.

Park, Y.W., Haenlein, G.F.W., 2013. Milk and dairy products in human nutrition. Wiley. Available at https://doi.org/10.1002/9781118534168.

Pasandideh, M., Mohammadabadi, M.R., Esmailizadeh, A.K., Tarang, A., 2015. Association of bovine PPARGC1A and OPN genes with milk production and composition in Holstein cattle. Czech Journal of Animal Science 60 (3), 97–104. Available at https://doi.org/10.17221/8074-cjas.

Peladarinos, N., Piromalis, D., Cheimaras, V., Tserepas, E., Munteanu, R.A., Papageorgas, P., 2023. Enhancing smart agriculture by implementing digital twins: A comprehensive review. Sensors 23 (16). Available at https://doi.org/10.3390/s23167128.

Perrin, A., Martin, G., 2021. Resilience of French organic dairy cattle farms and supply chains to the Covid-19 pandemic. Agricultural Systems 190, 103082. Available at https://doi.org/10.1016/J.AGSY.2021.103082.

Poczta, W., Średzińska, J., Chenczke, M., 2020. Economic situation of dairy farms in identified clusters of european union countries. Agriculture (Switzerland) 10 (4). Available at https://doi.org/10.3390/agriculture10040092.

Pollott, G.E., 2004. Deconstructing milk yield and composition during lactation using biologically based lactation models. Journal of Dairy Science 87 (8), 2375–2387. Available at https://doi.org/10.3168/jds.s0022-0302(04)73359-7.

Punia, H., Tokas, J., Malik, A., Sangwan, S., Baloda, S., Singh, N., et al., 2020. Identification and detection of bioactive peptides in milk and dairy products: Remarks about agro-foods. Molecules (Basel, Switzerland) 25 (15), 3328. Available at https://doi.org/10.3390/molecules25153328.

Quesnel, L., MacKay, A., Forsyth, D.M., Nicholas, K.R., Festa-Bianchet, M., 2017. Size, season and offspring sex affect milk composition and juvenile survival in wild kangaroos. Journal of Zoology 302 (4), 252–262. Available at https://doi.org/10.1111/jzo.12453.

da Rosa Righi, R., Goldschmidt, G., Kunst, R., Deon, C., André da Costa, C., 2020. Towards combining data prediction and internet of things to manage milk production on dairy cows. Computers and Electronics in Agriculture 169, 105156. Available at https://doi.org/10.1016/j.compag.2019.105156.

Sabek, A., Li, C., Du, C., Nan, L., Ni, J., Elgazzar, E., et al., 2021. Effects of parity and days in milk on milk composition in correlation with β-hydroxybutyrate in tropic dairy cows. Tropical Animal Health and Production 53 (2). Available at https://doi.org/10.1007/s11250-021-02690-7.

Salomone-Caballero, M., Fresno, M., Álvarez, S., Torres, A., 2024. Effects of parity and somatic cell count threshold on udder morphology, milkability traits, and milk quality in Canarian goats. Animals 14 (9), 1262. Available at https://doi.org/10.3390/ani14091262.

Sandøe, P., Corr, S.A., Lund, T.B., Forkman, B., 2019. Aggregating animal welfare indicators: Can it be done in a transparent and ethically robust way? Animal Welfare 28 (1), 67–76. Available at https://doi.org/10.7120/09627286.28.1.067.

Sanjulián, L., Fernández-Rico, S., González-Rodríguez, N., Cepeda, A., Miranda, J.M., Fente, C., et al., 2025. The role of dairy in human nutrition: Myths and realities. Nutrients 17 (4), 646. Available at https://doi.org/10.3390/nu17040646.

Sarıtaş, S., Mondragon Portocarrero, A.del C., Miranda, J.M., Witkowska, A.M., Karav, S., 2024. Functional yogurt: Types and health benefits. Applied Sciences 14 (24), 11798. Available at https://doi.org/10.3390/app142411798.

Segerkvist, K.A., Hansson, H., Sonesson, U., Gunnarsson, S., 2020. Research on environmental, economic, and social sustainability in dairy farming: A systematic mapping of current literature. Sustainability (Switzerland) 12 (12). Available at https://doi.org/10.3390/SU12145502.

Serrano-Torres, G.J., López-Naranjo, A.L., Larrea-Cuadrado, P.L., Mazón-Fierro, G., 2025. Transformation of the Dairy Supply Chain Through Artificial Intelligence: A Systematic Review. Sustainability (Switzerland) 17 (3), 982. Available at https://doi.org/10.3390/SU17030982/S1.

Shamsuddoha, M., Nasir, T., Hossain, N.U.I., 2023. A Sustainable Supply Chain Framework for Dairy Farming Operations: A System Dynamics Approach. Sustainability 15 (10), 8417. Available at https://doi.org/10.3390/SU15108417.

Sharma, R., Parhi, S., Shishodia, A., 2020. Industry 4.0 applications in agriculture: Cyber-physical agricultural systems (CPASs). Springer Science and Business Media LLC, pp. 807–813. Available at 10.1007/978-981-15-3639-7_97.

Silva Neto, J.B., Brito, L.F., Mota, L.F.M., Magnabosco, C.U., Baldi, F., 2025. Genome-wide association study for feed efficiency indicator traits in Nellore cattle considering genotype-by-environment interactions. Frontiers in genetics 16, 1539056. Available at https://doi.org/10.3389/fgene.2025.1539056.

Simitzis, P., Tzanidakis, C., Tzamaloukas, O., Sossidou, E., 2021. Contribution of precision livestock farming systems to the improvement of welfare status and productivity of dairy animals. Dairy 3 (1), 12–28.

Smith, N.W., Fletcher, A.J., Hill, J.P., McNabb, W.C., 2022. Modeling the contribution of milk to global nutrition. Frontiers in Nutrition 8. Available at https://doi.org/10.3389/fnut.2021.716100.

Singh, A., Kumar, S., 2025. Exploring the Functionality of Microbes in Fermented Foods: Technological Advancements and Future Directions. Fermentation Vol. 11 Multidisciplinary Digital Publishing Institute (MDPI) Issue 6. Available at https://doi.org/10.3390/fermentation11060300.

Sinha, G.K., Mishra, S., 2023. Sustainable Supply Chain Management Practices in the Dairy Industry: A Comparative Study of Leading Dairy Firms and Future Research Directives. Asian Journal of Dairy and Food Research 42 (4), 435–446. Available at https://doi.org/10.18805/AJDFR.DR-2120.

Tabacco, E., Comino, L., Borreani, G., 2018. Production efficiency, costs and environmental impacts of conventional and dynamic forage systems for dairy farms in Italy. European Journal of Agronomy 99, 1–12. Available at https://doi.org/10.1016/J.EJA.2018.06.004.

Temesgen, M., Haji, J., Negassa, A., Galmessa, U., 2024. Evaluation of interventions to improve markets for rural dairy households of Ethiopia: Evidence through system dynamics approach. Heliyon 10 (2), e24129. Available at https://doi.org/10.1016/j.heliyon.2024.e24129.

Triatmojo, A., Calvia, M., Zanon, T., & Fischer, C. (2025). Exploring the impact of input costs on farmgate milk prices in EU countries: An empirical analysis. BIO Web of Conferences, 192, 05004. https://doi.org/10.1051/bioconf/202519205004

Tunick, M.H., Van Hekken, D.L., 2015. Dairy products and health: Recent insights. Journal of Agricultural and Food Chemistry 63 (43), 9381–9388. Available at https://doi.org/10.1021/jf5042454.

Tuyttens, F.A. M., Lawrence, A.B., & Mullan $, S. (2025). A framework for a comprehensive animal welfare label: scientific, logistic, and ethical challenges. 15(2). https://doi.org/10.1093/af/vfaf003.

van der Linden, A., Oosting, S.J., J van de Ven, G.W., Zom, R., van Ittersum, M.K., Gerber, P.J., M de Boer, I.J., 2021. Yield gap analysis in dairy production systems using the mechanistic model LiGAPS-Dairy. Journal of Dairy Science 104, 5689–5704. Available at https://doi.org/10.3168/jds.2020-19078.

Vashisht, P., Singh, L., Mahanta, S., Verma, D., Sharma, S., Saini, G.S., et al., 2024. Pulsed electric field processing in the dairy sector: A review of applications, quality impact and implementation challenges. International Journal of Food Science and Technology 59 (4), 2122–2135. Available at https://doi.org/10.1111/ijfs.16979.

Verardo, V., Gómez-Caravaca, A., Arráez-Román, D., Hettinga, K., 2017. Recent advances in phospholipids from colostrum, milk and dairy by-products. International Journal of Molecular Sciences 18 (1), 173. Available at https://doi.org/10.3390/ijms18010173.

de Vitt, M.G., Signor, M.H., Corrêa, N.G., Breancini, M., Wolschick, G.J., Klein, B., et al., 2024. Combination of phytoactives in the diet of lactating Jersey cows: Effects on productive efficiency, milk composition and quality, ruminal environment, and animal health. Animals 14 (17). Available at https://doi.org/10.3390/ani14172518.

Vuvor, C.M., Epifani, S., Pulina, G., 2025. Exploring opportunities in Africa's dairy production: digital transition as a catalyst for marketing innovation and value chain synergies – a systematic review laying foundations for future research. Frontiers in Animal Science 6, 1632532. Available at https://doi.org/10.3389/FANIM.2025.1632532/BIBTEX.

Wall, E.H., McFadden, T.B., 2008. Use it or lose it: Enhancing milk production efficiency by frequent milking of dairy cows1. Journal of Animal Science 86 (suppl_13), 27–36. Available at https://doi.org/10.2527/jas.2007-0318.

Wang, M., Islam, S., & Yang, W. (2025). Supply chain risks in the dairy industry. Benchmarking. https://doi.org/10.1108/BIJ-12-2023-0874/1258232/SUPPLY-CHAIN-RISKS-IN-THE-DAIRY-INDUSTRY.

Washaya, S., Cheruwa, T., 2025. The effects of parity and lactation stage on milk composition and methane production in Holstein–Friesian cows. International Journal of Dairy Technology 78 (3). Available at https://doi.org/10.1111/1471-0307.70050.

Wilkinson, J.M., Lee, M.R., Rivero, M.J., Chamberlain, A.T., 2020. Some challenges and opportunities for grazing dairy cows on temperate pastures. Grass and Forage Science 75 (1), 1–17.

Wijerathna-Yapa, A., Pathirana, R., 2022. Sustainable Agro-Food Systems for Addressing Climate Change and Food Security. Agriculture (Switzerland) Vol. 12 MDPI Issue 10. Available at https://doi.org/10.3390/agriculture12101554.

Wodajo Tirfie, F., 2023. A review of genetic and non-genetic parameter estimates for milk composition of cattle. Animal and Veterinary Sciences. Available at https://doi.org/10.11648/j.avs.20231103.12.

Yoon, J.T., Lee, J.H., Kim, C.K., Chung, Y.C., Kim, C.-H., 2004. Effects of milk production, season, parity and lactation period on variations of milk urea nitrogen concentration and milk components of Holstein dairy cows. Asian-Australasian Journal of Animal Sciences 17 (4), 479–484. Available at https://doi.org/10.5713/ajas.2004.479.

Zhang, K., Shi, Y., Karnouskos, S., Sauter, T., Fang, H., Colombo, A.W., 2023. Advancements in industrial cyber-physical systems: An overview and perspectives. IEEE Transactions on Industrial Informatics 19 (1), 716–729. Available at https://doi.org/10.1109/TII.2022.3199481.

Chapter 2

Dairy 4.0: concept, enabling technologies, and application areas

Abdo Hassoun[1,2]
[1]*Faculty of Agricultural Engineering, University of Aleppo, Aleppo, Syria,* [2]*Sustainable AgriFoodtech Innovation & Research (SAFIR), Arras, France*

2.1 Introduction

The dairy sector is one of the most complex and sensitive agrifood systems, integrating biological variability, strict safety requirements, cold chain dependency, and high expectations from consumers regarding quality, transparency, and sustainability. Milk and dairy products are highly perishable and biologically active matrices, which makes them particularly vulnerable to safety hazards, quality degradation, fraud, and inefficiencies across the supply chain. At the same time, global dairy systems are under increasing pressure to improve productivity, reduce environmental footprints, ensure animal welfare, and comply with evolving regulatory and societal demands (Falih et al., 2024; Khan et al., 2026; Paçarada et al., 2024; Papadopoulos et al., 2025; Priyashantha, 2025). These challenges have accelerated the adoption of advanced digital and automation technologies, giving rise to the concept of Dairy 4.0.

Dairy 4.0 refers to the systematic digital transformation of the entire dairy value chain, from on farm activities and animal management to milk processing, quality control, cold chain logistics, and market distribution. It builds on the principles of the fourth industrial revolution (Industry 4.0) by integrating cyber-physical systems, interconnected sensors, intelligent data analytics, and autonomous decision support tools. Unlike conventional mechanization or basic automation, Dairy 4.0 enables real-time data exchange, predictive modeling, and adaptive control across physical, digital, and organizational layers of dairy systems, thereby enhancing responsiveness, efficiency, and resilience (Gehlot et al., 2022; Sangode, 2025; Hassoun, Garcia-garcia, et al., 2023; Malik et al., 2023).

Dairy 4.0. DOI: https://doi.org/10.1016/B978-0-443-44730-3.00001-8

At the farm level, precision livestock farming, including precision dairy farming, represents a core pillar of Dairy 4.0. Digital tools such as wearable sensors, automated milking systems, computer vision technologies, and advanced decision support platforms allow continuous monitoring of animal health, behavior, fertility, feeding, and welfare. These technologies support early disease detection, optimized nutrition strategies, and improved reproductive performance, while also reducing labor intensity and subjectivity in farm management. Recent studies emphasize that the transition from traditional livestock practices to data-driven precision systems can significantly improve both productivity and animal welfare outcomes when appropriately implemented (Anastasiou et al., 2025; Matera et al., 2025; Lamanna et al., 2025; Michelena et al., 2025; Neculai-Valeanu et al., 2025).

Beyond primary production, Dairy 4.0 plays a critical role in enhancing quality, safety, and traceability in milk and milk based products. Intelligent sensing systems, real-time monitoring of processing parameters, and data-driven quality control tools enable rapid identification of deviations and potential contamination risks. Smart technologies facilitate a shift from reactive quality assurance toward proactive and preventive food safety management. Evidence from recent research shows that the integration of digital monitoring, predictive analytics, and automated control systems can substantially improve safety assurance and regulatory compliance in dairy processing environments (KS et al., 2025; Patel et al., 2026; Tarapata et al., 2025).

Dairy 4.0 is increasingly conceptualized as an intelligent communication ecosystem, in which information flows seamlessly among animals, farm equipment, processing facilities, logistics operators, and market actors. Technologies such as the Internet of Things (IoT), artificial intelligence (AI), big data, 3D printing, blockchain, and extended reality are central to this ecosystem, enabling secure data exchange, real-time monitoring, transparency, and traceability throughout the dairy supply chain (Gehlot et al., 2022; Sangode, 2025; Anastasiou et al., 2023; Hassoun, Garcia-garcia, et al., 2023).

This chapter aims to provide a comprehensive overview of Dairy 4.0 by clarifying its conceptual foundations, examining key enabling technologies, and discussing major application areas across the dairy value chain. By synthesizing recent high quality research, including systematic reviews and applied studies, the chapter seeks to support researchers, practitioners, and policy-makers in understanding how digital transformation can enhance quality, safety, traceability, efficiency, and sustainability in modern dairy systems.

2.2 The fourth industrial revolution (Industry 4.0)

The fourth industrial revolution, commonly referred to as Industry 4.0, represents a profound transformation in the way production systems are conceived, designed, and managed. Unlike previous industrial revolutions that were driven by singular technological breakthroughs such as steam power, electricity, or automation, Industry 4.0 is defined by the convergence of multiple technological domains into intelligent, interconnected, and adaptive systems. It marks a shift from isolated and linear production models toward highly integrated ecosystems capable of real-time data exchange, autonomous decision-making, and continuous optimization (Bongomin et al., 2025; Folgado et al., 2024; Hassoun, Aït-Kaddour, et al., 2023).

From a historical perspective, Industry 4.0 builds upon the cumulative progress achieved during the earlier industrial revolutions. Industry 1.0 introduced mechanization through water and steam power, enabling the first large-scale industrial production systems. Industry 2.0 was characterized by electrification, mass production, and the emergence of standardized processes, which significantly increased productivity and scalability. Industry 3.0 introduced electronics, information technologies, and programmable automation, laying the foundations for digital control and early forms of smart manufacturing. Industry 4.0 extends this trajectory by embedding intelligence, connectivity, and autonomy directly into machines, processes, and infrastructures, allowing systems to interact, learn, and adapt dynamically (Alarcón and Alarcon, 2025; Karnik et al., 2022; Khan et al., 2025; Malik, Gahlawat, Mor and Hosseinian-Far, 2024; Sangode, 2025; Khan et al., 2023).

A defining characteristic of Industry 4.0 is the tight integration of digital, physical, and biological innovations. Digital technologies constitute the computational backbone of Industry 4.0 and include AI, machine learning, big data analytics, blockchain, cloud and edge computing, and digital twins. Collectively, these technologies enable the collection, storage, processing, and interpretation of large-scale, heterogeneous datasets generated across complex industrial systems, supporting advanced analytics, predictive capabilities, and data-driven decision-making (Aït-Kaddour et al., 2024; Ryalat et al., 2024; Sheik et al., 2025). Physical innovations form the second core pillar of Industry 4.0 and act as the interface between the digital layer and the real environment. These innovations include smart sensors, IoT, robotics and autonomous systems, smart sensors, drones, and cyber-physical systems. Through continuous sensing, connectivity, and actuation, physical technologies allow digital intelligence to be translated into precise and adaptive actions, making the interaction between digital cognition and physical execution central to the operational effectiveness of Industry 4.0 systems (Hassoun et al., 2025; Khan et al., 2025; Qiu et al., 2025; Semercioz-Oduncuoglu and Luning, 2025).

The biological dimension represents a critical yet often underappreciated component of Industry 4.0, particularly in sectors that rely heavily on living systems and biochemical processes. Advances in biotechnology, nanotechnology, genomics, precision fermentation, and bioengineering have significantly expanded the ability to analyze, control, and optimize biological processes with unprecedented precision. These innovations enable deeper understanding of complex biological interactions, improved process consistency, and the development of novel bio-based products and solutions, especially when integrated with digital analytics and physical control systems (Hassoun et al., 2024; Ajaj et al., 2024; Keppler et al., 2026)

Industry 4.0 is also characterized by a strong emphasis on connectivity and interoperability. Interconnected systems enable seamless data flow across organizational boundaries, processes, and hierarchical levels. This connectivity supports horizontal integration across value chains and vertical integration within organizations, linking operational technologies with information technologies and strategic decision-making. Interoperability standards, communication protocols, and data architectures play a crucial role in ensuring that diverse technologies and platforms can function cohesively within Industry 4.0 ecosystems (Javaid et al., 2023; Rehman et al., 2025; Pletsch et al., 2025). Another fundamental aspect of Industry 4.0 is the shift from reactive to predictive and prescriptive management. Traditional industrial systems often rely on corrective actions taken after deviations or failures occur. In contrast, Industry 4.0 systems leverage real-time data, predictive analytics, and simulation tools to anticipate issues before they materialize and to recommend optimal interventions. This proactive approach enhances efficiency, reduces downtime, and improves system resilience. Digital twins and advanced analytics are particularly important in this context, as they allow stakeholders to explore alternative scenarios, assess risks, and optimize system performance without disrupting real operations (Ahmed Murtaza et al., 2024; Hasan and Crawford, 2025; Jia et al., 2022; Singh, 2025; Keshvarinia et al., 2025).

Despite its transformative potential, Food Industry 4.0 faces several technical, economic, and socioorganizational challenges that may hinder widespread and equitable adoption. High investment costs for digital infrastructure, automation, and advanced sensing technologies remain major barriers, particularly for small and medium sized food enterprises, while increased energy demand raises concerns related to operational costs and environmental sustainability. Cybersecurity risks and data governance issues further complicate implementation in increasingly interconnected food systems. Additional challenges stem from the inherent variability of raw materials, biological processes, and heterogeneous supply chains, which limit data standardization, interoperability, and model robustness (Cantarelli da Silva et al., 2025; Hassoun et al., 2025; Neves et al., 2025; Ortega-Gras et al., 2025). The technology centric orientation of Food Industry 4.0 has also raised concerns regarding workforce transformation, skills gaps, potential job displacement, and the erosion of tacit knowledge, as well as unresolved ethical and social issues related to data ownership and unequal access to digital technologies. These limitations have stimulated growing interest in Food Industry 5.0, which seeks to complement Industry 4.0 by emphasizing human centricity, sustainability, and resilience, while Industry 4.0 remains the foundational technological framework for this transition (Hassoun et al., 2024; Romanello and Veglio, 2022).

2.3 Dairy 4.0 and its enabling technologies

The digital transformation of the dairy sector initially gained momentum through the development of precision livestock farming, which focuses on the systematic monitoring and management of animals at the individual level. Precision livestock, especially precision dairy farming, relies on digital technologies to capture real-time data related to cows' health, productivity, fertility, behavior, feeding patterns, and welfare. By integrating sensors, automated milking systems, computer vision, and decision support tools, precision livestock approaches aim to improve farm level performance, reduce uncertainty, and support timely and evidence based management decisions (Besler et al., 2024; Liu et al., 2023b; Palma et al., 2025; Papakonstantinou et al., 2024; Neculai-Valeanu et al., 2025). This animal centered perspective represented a significant shift from traditional herd based management toward data-driven and individualized control strategies.

While precision livestock remains a foundational component of digitalization in dairy farming, recent research and industrial developments indicate that its scope is inherently limited when considered in isolation. Dairy systems extend far beyond animal management and encompass complex interactions among farming practices, milk collection, processing operations, quality and safety control, cold chain logistics, distribution networks, and market interfaces. As a result, the concept of Dairy 4.0 has emerged as a broader and more holistic framework that captures the digital transformation of the entire dairy value chain rather than a single production stage (Espinoza-Sandoval et al., 2024; Gehlot et al., 2022; Kaur et al., 2023; Sangode, 2025).

Dairy 4.0 can be defined as the integrated application of Industry 4.0 principles and technologies across the full dairy chain, from the farm to the final consumer. This includes on-farm data generation, automated milk harvesting, intelligent processing and quality monitoring, digitally controlled cold chains, and data-enabled traceability and market communication. Within this framework, information flows continuously across stages, enabling real-time coordination, transparency, and system level optimization rather than fragmented, stage-specific improvements (Gehlot et al., 2022; Sangode, 2025; Hassoun, Garcia-garcia, et al., 2023). The shift from precision dairy farming to Dairy 4.0 therefore reflects a move from localized digital solutions toward interconnected and interoperable ecosystems that support end-to-end visibility and coordinated decision-making across stakeholders. At present, most reported implementations still rely predominantly on Industry 4.0 technologies, indicating that Dairy 4.0 remains the prevailing operational framework for digital transformation in dairy systems. Looking ahead, a limited number of recent publications have begun to refer to Dairy 5.0, which can be viewed as a future oriented extension where Industry 5.0 principles may progressively complement Dairy 4.0 as the sector matures in human centricity, resilience, and sustainability (Bonamigo et al., 2025; Malik, Mor, Gahlawat, et al., 2025).

The enabling technologies of Dairy 4.0 (Fig. 2.1) largely mirror those of Food Industry 4.0 and can be grouped into digital, physical, and biological domains. Digital technologies such as AI, big data analytics, blockchain, and digital twins support advanced data interpretation, prediction, and decision-making across the dairy chain. Physical technologies, including smart sensors, IoT, robotics, and cyber-physical systems, enable real-time data acquisition and automated control of processes. Biological innovations, such as advances in biotechnology, microbial monitoring, and precision fermentation, further contribute to improved process understanding and consistency, particularly in milk transformation and product development (Espinoza-Sandoval et al., 2024; Malik, Gahlawat, Mor and Hosseinian-Far, 2024).

2.3.1 Artificial intelligence

AI refers to computational systems designed to perform tasks that normally require human intelligence, such as learning, reasoning, pattern recognition, and decision-making. Core types of AI include machine learning, deep learning, and reinforcement learning, each differing in learning mechanisms and levels of autonomy. Machine learning relies on data-driven model training, deep learning uses multilayer neural networks for complex pattern extraction, while reinforcement learning optimizes decisions through reward based interactions (Mandal and Ghosh, 2024; Saha, Padhiary and Chandrakar, 2025; Song et al., 2025). In the dairy sector, AI is increasingly used to analyze large and heterogeneous datasets to support predictive modeling, anomaly detection, and intelligent decision support across production, processing, and supply chain operations (Espinoza-Sandoval et al., 2024; Khanashyam et al., 2025; Malik et al., 2022; Parrenin et al., 2025; Serrano-Torres et al., 2025).

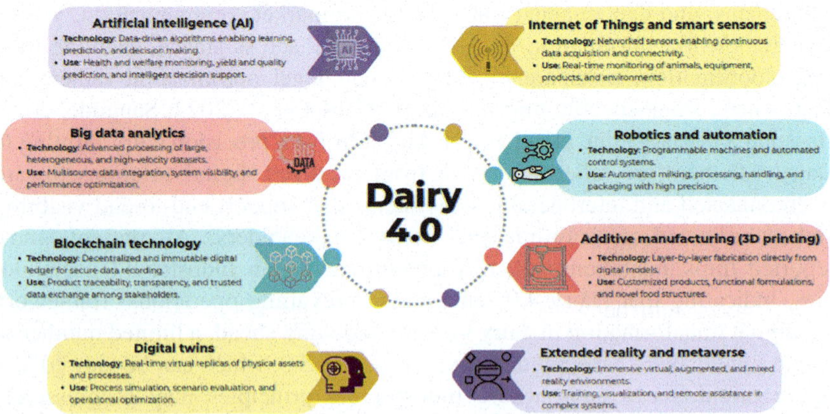

FIGURE 2.1 Core enabling technologies of Dairy 4.0 and their primary technological functions and uses across the dairy value chain.

2.3.2 Big data analytics

Big data analytics refers to the systematic processing and analysis of large-scale, complex, and heterogeneous datasets that exceed the capabilities of traditional data processing tools. It is commonly characterized by the multiple "Vs" of big data, including volume, velocity, variety, veracity, and value, and encompasses descriptive, diagnostic, predictive, and prescriptive analytical approaches. Big data analytics integrates data originating from sensors, machines, information systems, and external sources to generate actionable insights (Asgari, 2025; Ding et al., 2023; Hussein et al., 2025; Siddique et al., 2025). In the dairy sector, big data analytics supports the integration and interpretation of multisource datasets, enabling improved system visibility, performance assessment, and data-driven optimization across interconnected stages of the dairy value chain (Cabrera et al., 2020; Li et al., 2024a; Palma et al., 2025).

2.3.3 Blockchain technology

Blockchain is a decentralized and distributed digital ledger technology that records transactions in immutable, time stamped blocks validated through consensus mechanisms. Blockchain systems are generally classified as public, private, or hybrid, depending on governance structures and access permissions, and can be extended through smart contracts that automate predefined rules and actions. The core features of blockchain include transparency, traceability, data integrity, and resistance to tampering (Astuti and Hidayati, 2023; Manteghi et al., 2023; Pavithra et al., 2025). Blockchain is primarily applied in the dairy sector to enhance trust, transparency, and secure data sharing among multiple stakeholders across complex and geographically dispersed value chains (Tran et al., 2025; Vargas-Bello-Pérez et al., 2025).

2.3.4 Digital twins

Digital twins are dynamic virtual representations of physical assets, processes, or systems that are continuously updated using real-time data. They enable continuous monitoring, simulation, and optimization by integrating sensor data, mathematical models, and data analytics to replicate system behavior under varying conditions (Rashed et al., 2025; Rolofs et al., 2024; Tzachor et al., 2022). Digital twins are increasingly explored in the dairy sector to simulate complex processes, evaluate alternative operational scenarios, and support decision-making and optimization without disrupting real world operations (Feng et al., 2025; Rao and Neethirajan, 2025).

2.3.5 The Internet of Things and smart sensors

IoT refers to networks of interconnected physical objects embedded with sensors, software, and communication technologies that enable real-time data acquisition and exchange. IoT architectures typically include sensing, communication, data processing, and application layers, and may incorporate edge or cloud computing. Smart sensors represent a core component of Industry 4.0, capable of measuring physical, chemical, and biological parameters with high precision (Turgut et al., 2024; Alimadani et al., 2025; Dadhaneeya et al., 2023). In the dairy sector, IoT and smart sensors provide the foundational infrastructure for continuous monitoring, connectivity, and data-driven control across distributed assets and processes (Santos et al., 2025; Tangorra et al., 2024; Papa et al., 2024).

2.3.6 Robotics and automation

Robotics and automation involve the use of programmable machines and automated systems to perform tasks with minimal human intervention. These technologies include industrial robots, autonomous systems, collaborative robots, and automated process control solutions. Robotics systems are often integrated with sensors, AI algorithms, and cyber-physical systems to enable adaptive and intelligent operation (Liberty et al., 2024; Shweta et al., 2025; Woo et al., 2025). In milk, cheese, and other modern dairy production systems, such technologies are increasingly applied to enhance process consistency, operational efficiency, and precision by reducing manual intervention and improving repeatability across production and processing activities (Liu et al., 2023b; Konstantinidis et al., 2023).

2.3.7 Additive manufacturing

Three dimensional printing (3D printing), also known as additive manufacturing, refers to the layer by layer fabrication of objects directly from digital models. In the food domain, common 3D printing techniques include extrusion based printing, inkjet printing, and laser assisted methods, depending on material properties and functional requirements. 3D printing enables precise control over structure, geometry, and composition, supporting customization and design flexibility (Kanwal et al., 2025; Wedamulla et al., 2025; Wang et al., 2025). 3D printing is explored for the development of customized dairy based products, functional formulations, and novel textures, contributing to product innovation and personalization (Dou and Ren, 2024; Tarchi et al., 2024; Lee et al., 2020).

2.3.8 Extended reality and metaverse

Virtual reality, augmented reality, and mixed reality are immersive technologies collectively referred to as extended reality, while the metaverse represents persistent, shared virtual environments integrating these technologies. Virtual reality enables fully immersive digital experiences, augmented reality overlays digital information onto the physical world, and mixed reality allows interaction between real and virtual elements (Bhavadharini et al., 2023; El Jaouhari et al., 2024; Anastasiou et al., 2023; Woo et al., 2025). These technologies are increasingly applied for training, visualization, remote assistance, and system understanding in the dairy sector, enhancing human interaction with complex digital and physical environments (Anastasiou et al., 2023; Woo et al., 2025).

2.4 Applications of Dairy 4.0 across the value chain

This section examines how Dairy 4.0 technologies are applied across the dairy value chain to address key operational, quality, safety, and sustainability challenges. It synthesizes major application domains, including digital quality and safety management, traceability enhancement, and waste reduction and valorization, highlighting the shift from reactive practices toward integrated, data-driven, and predictive system control (Fig. 2.2).

2.4.1 Enhancing produce quality through Dairy Quality 4.0

Dairy quality has traditionally been assessed through a combination of physicochemical, microbiological, and sensory analyses conducted at discrete stages of the value chain. Standard quality determination relies on laboratory testing of parameters such as fat and protein content, acidity, microbial counts,

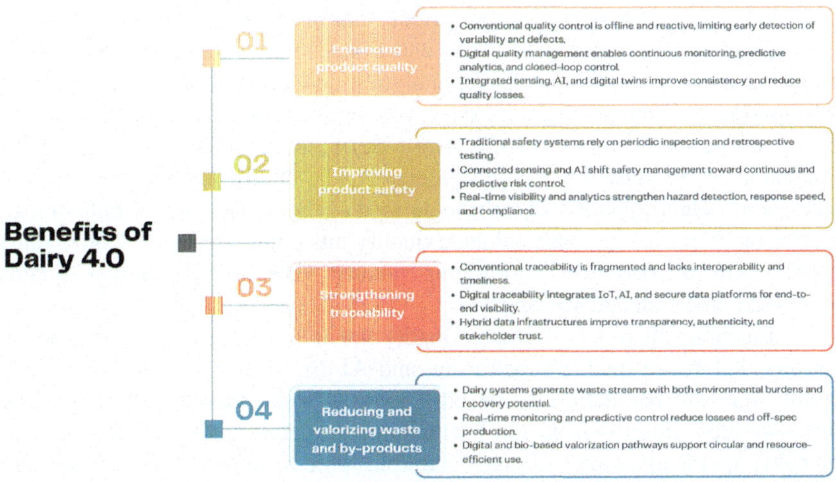

FIGURE 2.2 Key application areas and benefits of Dairy 4.0 across the dairy value chain, including quality enhancement, safety improvement, traceability strengthening, and waste reduction and valorization.

somatic cell count, and selected contaminants, complemented by visual inspection and sensory evaluation. While these methods are scientifically robust and widely standardized, they are largely offline, time consuming, and reactive, meaning that quality deviations are often detected only after they have occurred (Bosakova-Ardenska, 2024; Falih et al., 2024; Priyashantha, 2025). In complex and biologically variable dairy systems, this limits responsiveness, increases variability, and constrains the ability to prevent quality losses proactively.

Against this background, Quality 4.0 has emerged in recent years as a direct derivative of Industry 4.0, aiming to modernize traditional quality management systems by embedding digital connectivity, cyber-physical systems, and data-driven intelligence into quality planning, control, assurance, and improvement. Rather than relying primarily on periodic inspection and corrective actions, Quality 4.0 emphasizes continuous monitoring, predictive analytics, and closed loop decision-making (Attia Hussien Gomaa., 2025; Liu et al., 2023a; Sader et al., 2021). Recent systematic reviews describe Food Quality 4.0 as an Industry 4.0-enabled evolution of quality management, supported by technologies such as IoT, AI, big data analytics, cloud computing, and digital twins, which together enable faster detection of deviations and more effective continuous improvement (Aït-Kaddour et al., 2024; Hassoun, Jagtap, et al., 2023; Semercioz-Oduncuoglu and Luning, 2025).

In the dairy sector, the relevance of Quality 4.0 is particularly pronounced due to the intrinsic biological variability of raw milk, the sensitivity of dairy processing operations, and the strong influence of time and temperature on product quality. Dairy systems also involve multiple interconnected stages, including on farm production, milk collection, processing, storage, distribution, and retail, each of which can affect final product quality. Dairy 4.0 has therefore been conceptualized as an end-to-end digital transformation of the dairy value chain, in which sensors, analytics, and automation enable real-time information exchange and system-level optimization (Espinoza-Sandoval et al., 2024; Gehlot et al., 2022; Kraft et al., 2022; Losacco et al., 2025; Sangode, 2025; Hassoun, Garcia-garcia, et al., 2023). Within this broader framework, Dairy Quality 4.0 can be defined as the application of Industry 4.0 technologies to manage and enhance quality attributes across the entire dairy chain, with a focus on continuous quality visibility, predictive quality assurance, and reduced variability.

A fundamental pillar of Dairy Quality 4.0 is the integration of connected sensing, IoT-based quality monitoring, and AI-driven analytics, which enables quality relevant parameters to be continuously observed rather than assessed intermittently. Smart sensors and IoT architectures generate high frequency data on cold chain temperature, storage conditions, processing variables, and selected physicochemical indicators, while AI based data fusion and pattern recognition techniques transform these heterogeneous data streams into actionable quality information (Ahmed and Shakoor, 2025; Liu et al., 2025; Malik, Mor, Gahlawat, et al., 2025). Recent advances further demonstrate the feasibility of low-cost, AI-enabled IoT sensing for continuous dairy quality monitoring across the farm-to-factory interface. For example, an inexpensive in-tank IoT sensor integrating low-cost spectrometry, embedded machine learning, and narrowband IoT communication has been shown to continuously monitor fat and protein content in raw milk at farms, collection points, and intermediate storage facilities, achieving accuracies of ±0.14% for fat and ±0.07% for protein (Fizza et al., 2025). In addition, AI-supported anomaly detection and predictive models facilitate early identification of process drift and emerging quality deviations before they become critical. In dairy processing, AI has been applied to predict quality trajectories, detect abnormal patterns, and support decision-making in fermentation based products (Choe and Lee, 2025; Palma et al., 2025). For example, AI-driven real-time monitoring and predictive control systems have been proposed for yoghurt fermentation to reduce variability and improve process consistency (Ray et al., 2026; Saetae, 2025). In one recent study, prediction errors for microbial growth and acidification during yoghurt fermentation were reduced by up to 86% compared with traditional kinetic models through a hybrid long short-term memory–support vector regression (LSTM–SVR) framework, demonstrating substantially improved predictive accuracy across industrial temperature ranges. The study further showed that LSTM effectively captured bacterial growth dynamics while SVM accurately modeled pH evolution, confirming the robustness and reliability of AI-driven fermentation monitoring and control (Saetae, 2025).

Another mechanism supporting Dairy Quality 4.0 is the use of digital twins for quality by design and adaptive optimization. In dairy processing, digital twins can strengthen control of heat treatment, fermentation, and other quality sensitive operations, particularly when small parameter deviations can lead to significant quality losses (Kadri and Abdelhadi, 2024; Krupitzer et al., 2022). Recent evidence from a real-time digital twin for fermented dairy (*amasi*) shows that fusing noninvasive computer vision with IoT sensor streams and machine learning can deliver highly accurate acidity tracking, with a fully integrated hybrid feature set, achieving $R^2 > 0.95$ for predicting acidity related variables (Adeleke et al., 2026).

Overall, Dairy Quality 4.0 represents a coherent pathway for enhancing quality consistency, transparency, and robustness across modern dairy systems. Its effectiveness depends not only on the adoption of advanced technologies, but also on their integration into comprehensive quality management frameworks that include clear objectives, validated models, interoperable data infrastructures, and appropriate governance structures.

2.4.2 Improving product safety through Dairy Safety 4.0

Dairy safety has traditionally been managed through preventive programs such as good manufacturing practices (GMP), hazard analysis and critical control points (HACCP), and periodic microbiological and chemical testing conducted at defined control points along the value chain (Dragone et al., 2024; Mureşan et al., 2020; Patel et al., 2026). While these systems have been highly effective in reducing foodborne risks, they remain largely inspection based, retrospective, and dependent on discrete sampling and laboratory analysis. As a result, contamination events, temperature abuse, or hygiene failures are often detected only after they have occurred, limiting the ability to intervene proactively, particularly in complex and high throughput dairy operations.

Within the broader Dairy 4.0 paradigm, safety management is increasingly transitioning toward Dairy Safety 4.0, which applies Industry 4.0 technologies to shift food safety systems from reactive control to predictive and preventive risk management. By integrating IoT-enabled sensing, cyber-physical systems, and AI-driven analytics, Dairy Safety 4.0 enables continuous visibility of safety critical parameters across processing, storage, and distribution stages. Recent reviews emphasize that smart sensing, real-time data acquisition, and advanced analytics can significantly enhance hazard detection, early warning capability, and compliance with food safety regulations in dairy systems (Deshmukh et al., 2025; Hamad and Awad, 2025; Kyaw et al., 2024; Plakantara and Karakitsiou, 2025; Semercioz-Oduncuoglu and Luning, 2025; Trollman et al., 2024). For example, in milk and milk-based products, digital and hyperspectral image analysis, together with IoT-enabled sensing, allows noninvasive and continuous monitoring of safety-relevant attributes, supporting early detection of contamination or abnormal processing conditions. In parallel, the integration of AI, big data analytics, and blockchain technology enhances hazard prediction, data-driven risk assessment, and transparency across the dairy safety chain, strengthening traceability and trust in safety management systems (Patel et al., 2026).

Recent work has demonstrated that inexpensive AI-powered IoT sensors integrating low-cost spectroscopy and machine learning can enable continuous, in-tank monitoring of raw milk composition across farms, transport vehicles, and intermediate depots, providing live visibility of fat and protein content along the farm-to-factory chain. Such systems achieve laboratory-comparable accuracy (±0.14% for fat and ±0.07% for protein) while eliminating delays associated with offline sampling, thereby supporting earlier detection of quality deviations, improved production planning, and proactive safety and quality management in dairy supply chains (Fizza et al., 2025). Machine learning was also used for rapid microbiological safety screening in milk through a low-cost optical sensor array that produces species-specific fluorescence "fingerprints" to discriminate multiple pathogenic and spoilage bacteria in contaminated milk within hours. In that approach, neural-network classification reached 93.8% accuracy after only a 30 minute incubation, and accuracy increased to 98.4% when incubation was extended to 120 minute (multilayer perceptron), demonstrating a practical alternative to slower plate counting and ELISA-type workflows for multi-pathogen identification (Wang et al., 2025).

In contrast to traditional inspection-driven approaches, Dairy Safety 4.0 reframes food safety as a continuously managed risk system in which hazards are anticipated, contextualized, and mitigated in near real time. By combining connected sensing, AI-supported risk modeling, and transparent data infrastructures, this paradigm supports faster incident response, improved regulatory assurance, and greater resilience of dairy safety systems under dynamic production and supply chain conditions.

2.4.3 Strengthening traceability through Dairy Traceability 4.0

Traceability in dairy systems refers to the ability to track and verify the history, location, and movement of milk and dairy products throughout the value chain, from on-farm production to processing, distribution, and retail. Traditional traceability approaches have relied on paper-based records and isolated digital systems within individual supply chain stages, which often lack interoperability, timeliness, and transparency. These limitations can hinder rapid response during safety incidents, reduce consumer confidence, and constrain value chain optimization, particularly in multinational and highly fragmented dairy markets (Bayatzadeh and Talaie, 2025; Rossi et al., 2025; Hassoun, Abdullah, et al., 2024). Under the Dairy 4.0 paradigm, traceability is evolving into a real-time, integrated, and intelligence-driven capability supported by a combination of IoT, smart sensors, cloud connectivity, AI, and distributed ledger technologies, or blockchain. By capturing rich, high-frequency data at every point in the chain, such as milk collection events, processing parameters, transport conditions, and product batching, modern traceability systems enable end-to-end visibility and auditability of dairy flows (Farina et al., 2025; Hassoun, Kamiloglu, et al., 2023; Malik, Gahlawat, Mor and Singh, 2024). These technological innovations have directly driven the emergence of the concept of Dairy Traceability 4.0, which reconceptualizes traceability as a digitally enabled, system-wide function rather than a fragmented record-keeping activity.

Recent work has demonstrated that AI-enabled IoT multisensor systems, integrating physicochemical sensing with machine learning and other AI models, can detect milk adulteration in real time with high accuracy while simultaneously generating trustworthy, time-stamped data streams for digital traceability infrastructures. When coupled with secure data transmission and validation mechanisms, these systems enable validated adulteration events and compositional anomalies to be directly fed into traceability and provenance platforms, thereby strengthening both operational quality assurance and end-to-end product authenticity across the dairy supply chain (Khan et al., 2026; Goyal et al., 2022). For example, an explainable AI-empowered IoT multisensor framework integrates real-time measurements of pH, electrical conductivity, temperature, volatile organic compounds, and key milk constituents with a machine learning-based ensemble technique, achieving approximately

96% accuracy in detecting common adulterants such as urea, starch, sodium bicarbonate, maltodextrin, and formaldehyde. By embedding explainable AI within the decision layer, the system not only enables rapid and automated adulteration detection but also generates transparent and interpretable event-level outputs, which are well suited for direct integration into digital traceability systems and auditable quality records across the dairy value chain (Goyal et al., 2024).

A prominent mechanism for strengthening traceability in Dairy 4.0 is the integration of traceability systems with enabling technologies such as AI, blockchain, and IoT to deliver real-time monitoring, predictive analytics, and secure data management across dairy operations. In this context, blockchain supports transparency, quality, and safety by maintaining a tamper resistant, distributed record of transactions and sensor derived events shared among stakeholders (Vargas-Bello-Pérez et al., 2025; Malik, Mor, Dalal, et al., 2025). For instance, proof-of-concept implementations of hybrid blockchain–AI traceability models have demonstrated enhanced trace data integrity, as well as improved scalability, transaction speed, and system responsiveness, thereby strengthening transparency, accountability, and sustainability in dairy operations (Malik et al., 2026). Other research emphasizes that blockchain-based traceability systems can play a critical role in protected designation of origin (PDO) dairy products, such as feta cheese, where certification, geographic authenticity, and ingredient provenance are central to both regulatory compliance and market differentiation (Tran et al., 2025). In the case of PDO feta cheese production, blockchain-based traceability has been shown to deliver negligible additional environmental impacts while offering clear economic benefits for dairy producers, particularly when consumer willingness to pay a price premium for traceable and certified products is considered. In addition to product-centric traceability, digital tracking at the animal level is emerging as a foundational layer of dairy supply chain visibility. Technologies such as radio-frequency identification (RFID), IoT devices, and biometric data for cattle traceability enable detailed tracking of animal health, movement, and growth performance, which can be synchronized with downstream product records to construct fully integrated traceability chains (Lukacs et al., 2025; Papa et al., 2024).

2.4.4 Reducing and valorizing dairy waste and by-products

Dairy systems generate substantial quantities of waste and by-products across both farming and processing stages, including manure and wastewater on farms, and whey, buttermilk, skim streams, cleaning-in-place effluents, off-spec products, and packaging residues at the factory level. These flows represent a challenge because they intensify environmental pressures through greenhouse gas emissions and resource inefficiencies, yet they also contain recoverable value in the form of proteins, lactose, minerals, bioactive fractions, and energy potential (Selmi et al., 2025; Tarchi et al., 2024; Udourioh et al., 2025).

Recent studies increasingly identify Industry 4.0 technologies as practical enablers for simultaneously limiting avoidable losses and valorizing unavoidable by-products through circular and bioeconomy pathways, thereby improving both environmental and economic performance (Forsan, 2025; Tarchi et al., 2024; Lopes de Sousa Jabbour et al., 2023). At the farm level, life-cycle assessments further show that precision livestock farming tools such as automated milking systems, health monitoring, and heat detection can indirectly reduce waste generation and emissions by improving animal efficiency and lowering disease-related losses (Bianchi et al., 2024; Vlaicu et al., 2024; Lovarelli et al., 2020, 2024).

Beyond valorization, waste reduction increasingly relies on prevention through real-time visibility and predictive control of dairy operations. IoT-enabled sensing systems and connected infrastructures allow continuous monitoring of temperature, hygiene, processing stability, and storage conditions across the dairy value chain, while AI-based analytics enable early identification of process deviations before products become off-spec. This shift from reactive to anticipatory control reduces downgrades, discards, and unnecessary reprocessing, improving overall resource efficiency. This preventive approach emphasizes operational precision, where stable milk quality and process consistency act as control variables that minimize waste formation at its source and support more efficient allocation of energy, water, and inputs across both farming and processing stages (Priyashantha, 2025; Filani et al., 2021; Singh et al., 2024).

In addition to sensing and analytics, additive manufacturing technologies such as 3D printing are emerging as promising Industry 4.0 tools for valorizing food waste and by-products by transforming diverse food by-stream materials into functional, value-added products. Recent evidence demonstrates that food by-products, including nutrient-rich residues and processing by-streams, can be successfully integrated into multiple 3D printing techniques to produce biodegradable packaging, biocomposites, and customized food and nonfood applications, supporting circular economy strategies and sustainable resource recovery (Bareen et al., 2025; Saha, Padhiary, Hoque, et al., 2025). In the dairy sector, milk-derived fractions and dairy by-products have shown strong potential as printable materials due to their favorable gelation behavior, rheological properties, and intrinsic nutritional value. Recent research highlights that dairy-based edible inks, formulated from components such as milk proteins and dairy derivatives, can be structurally tailored through physicochemical or enzymatic modifications to improve printability, shape retention, and nutritional functionality, enabling the production of customized, nutrient-dense 3D-printed dairy foods (Dou and Ren, 2024; Li et al., 2024b; Xu et al., 2025).

Overall, recent work suggests that waste reduction and by-product valorization are increasingly treated as strategic outcomes of digital transformation in dairy operations, with benefits spanning lower disposal costs, improved environmental performance, and stronger circularity across the value chain.

2.5 Conclusion

Dairy 4.0 represents a practical, end-to-end pathway for modernizing the dairy value chain by connecting farm, processing, cold chain, and market operations through interoperable data, cyber-physical systems, and intelligent decision support. By integrating IoT sensing and smart sensors, AI and big data analytics, robotics, digital twins, blockchain, and additive manufacturing applications, Dairy 4.0 enables a shift from periodic and reactive management toward continuous monitoring, predictive control, and transparent coordination across various stakeholders in the dairy sector.

Across application domains, Dairy Quality 4.0 and Dairy Safety 4.0 strengthen proactive quality and hazard management through real-time sensing and advanced analytics, while Dairy Traceability 4.0 enhances provenance assurance, fraud resilience, and stakeholder trust through secure data sharing and end-to-end visibility. In parallel, Dairy 4.0 supports sustainability by reducing avoidable losses through predictive control and enabling circular valorization of unavoidable by-products, including new routes for tailoring dairy derived materials into value added outputs.

However, achieving these benefits at scale depends on more than technology adoption alone. Success requires validated models, robust cybersecurity and data governance, interoperability standards, workforce upskilling, and implementation strategies that remain affordable and accessible for diverse dairy actors. With these conditions in place, Dairy 4.0 can serve as a foundational framework for resilient, quality-driven, and sustainability-oriented dairy systems, while providing a strong bridge toward future innovation that better aligns technology, people, and environmental goals.

AI disclosure

During the preparation of this work, the author used ChatGPT-5.0 in order to improve readability and language. After using this tool/service, the author reviewed and edited the content as needed and takes full responsibility for the content of the publication.

References

Adeleke, I., Nwulu, N., Adebo, O.A., 2026. A digital twin approach for real-time monitoring of amasi acidity using non-invasive computer vision, IoT, and machine learning. Food control 182, 111809. Available at https://doi.org/10.1016/J.FOODCONT.2025.111809.

Ahmed Murtaza, A., Saher, A., Hamza Zafar, M., Kumayl Raza Moosavi, S., Faisal Aftab, M., Sanfilippo, F., 2024. Paradigm shift for predictive maintenance and condition monitoring from Industry 4.0 to Industry 5.0: A systematic review, challenges and case study. Results in Engineering 24, 102935. Available at https://doi.org/10.1016/j.rineng.2024.102935.

Ahmed, N., Shakoor, N., 2025. Advancing agriculture through IoT, Big Data, and AI: A review of smart technologies enabling sustainability. Smart Agricultural Technology 10, 100848. Available at https://doi.org/10.1016/j.atech.2025.100848.

Ajaj, R., Buheji, M., Hassoun, A., 2024. Optimizing the readiness for industry 4.0 in fulfilling the Sustainable Development Goal 1: Focus on poverty elimination in Africa. Frontiers in Sustainable Food Systems 8, 1393935. Available at https://doi.org/10.3389/FSUFS.2024.1393935/BIBTEX.

Alarcón, S., Alarcon, C., 2025. Questioning the concepts of the fourth industrial revolution and Industry 4.0 when describing modernization as a sequential framework. Sustainability 17 (10), 4531. Available at https://doi.org/10.3390/su17104531.

Alimadani, R., Adedeji, A.A., Narimani, M., 2025. A review of IoT applications in food processing and related fields. Journal of Food Processing and Preservation 2025 (1), 3064441. Available at https://doi.org/10.1155/JFPP/3064441.

Anastasiou, E., Balafoutis, A.T., Fountas, S., 2023. Applications of extended reality (XR) in agriculture, livestock farming, and aquaculture: A review. Smart Agricultural Technology 3 (July 2022), 100105. Available at https://doi.org/10.1016/j.atech.2022.100105.

Anastasiou, E., Kasimati, A., Papadopoulos, G., Vatsanidou, A., Gemtou, M., Schwierz, F., et al., 2025. Assessing the agroecological impact of digital tools in livestock production: A systematic review. Smart Agricultural Technology 12, 101480. Available at https://doi.org/10.1016/j.atech.2025.101480.

Asgari, A., 2025. Uncovering the role of big data analytics on the resilience of agri-food supply chains: a systematic literature review. Procedia Computer Science 253, 1631–1639. Available at https://doi.org/10.1016/j.procs.2025.01.225.

Astuti, R., Hidayati, L., 2023. How might blockchain technology be used in the food supply chain? A systematic literature review. Cogent OA, Indonesia Cogent Business and Management 10 (2). Available at https://doi.org/10.1080/23311975.2023.2246739.

Attia Hussien Gomaa, 2025. Quality management excellence in the era of Industry 4.0 (Quality 4.0): A Comprehensive review, gap analysis, and strategic framework. International Journal of Natural-Applied Sciences and Engineering 3 (1), 18–40. Available at https://doi.org/10.22399/ijnasen.22.

Aït-Kaddour, A., Boudalia, S., Özoğul, F., Çiçek, S., Cahyana, Y., Bhat, Z.F., 2024. Food Quality 4.0: Contribution to sustainability Food Industry 4.0: Emerging trends and technologies in sustainable food production and consumption. Elsevier, France, pp. 35–59. Available at https://www.sciencedirect.com/book/9780443155161.

Bareen, M.A., Antonio, D., Corradini, M.G., Rossella, C., Paolo, D., Marta, S., et al., 2025. Enhancing 3D printing performance and product quality through the valorization of food by-products and waste. Comprehensive Reviews in Food Science and Food Safety 24 (5). Available at https://doi.org/10.1111/1541-4337.70267.

Bayatzadeh, S., Talaie, H., 2025. An evaluation of traceability dynamics in dairy supply chains through causal modeling in emerging economies. Supply Chain Analytics 11, 100156. Available at https://doi.org/10.1016/j.sca.2025.100156.

Besler, B.C., Mojabi, P., Lasemiimeni, Z., Murphy, J.E., Wang, Z., Baker, R., et al., 2024. Scoping review of precision technologies for cattle monitoring. Smart Agricultural Technology 9. Available at https://doi.org/10.1016/j.atech.2024.100596.

Bhavadharini, B., Monica, V., Anbarasan, R., Mahendran, R., 2023. Virtual, augmented, and mixed reality as a versatile tool in food consumer behavior evaluation: Recent advances in aroma, taste, and texture incorporation. Comprehensive Reviews in Food Science and Food Safety 22 (6), 4925–4956. Available at https://doi.org/10.1111/1541-4337.13248.

Bianchi, M.C., Gislon, G., Mondini, S., Bava, L., Tamburini, A., Sandrucci, A., et al., 2024. Can technology mitigate the environmental impact of dairy farms? Cleaner Environmental Systems 12, 100178. Available at https://doi.org/10.1016/j.cesys.2024.100178.

Bonamigo, A., da Silva Pereira, E., de Oliveira, J., Antony, J., Garza-Reyes, J., 2025. Lean Service 5.0 in dairy production: An empirical evaluation from the perspective of the sociotechnical and circular economy approach. Business Strategy and the Environment 34 (7), 8134–8151. Available at https://doi.org/10.1002/bse.70015.

Bongomin, O., Mwape, M.C., Mpofu, N.S., Bahunde, B.K., Kidega, R., Mpungu, I.L., et al., 2025. Digital twin technology advancing industry 4.0 and industry 5.0 across sectors. Results in Engineering 26. Available at https://doi.org/10.1016/j.rineng.2025.105583.

Bosakova-Ardenska, A., 2024. Recent trends in computer vision for cheese quality evaluation. Engineering Proceedings. 60 (1). Available at https://doi.org/10.3390/engproc2024060012.

Cabrera, V.E., Barrientos-Blanco, J.A., Delgado, H., Fadul-Pacheco, L., 2020. Symposium review: Real-time continuous decision making using big data on dairy farms. Journal of Dairy Science 103 (4), 3856–3866. Available at https://doi.org/10.3168/jds.2019-17145.

Cantarelli da Silva, R., Bacharini Lima, L., Batistela dos Santos, E., Akutsu, R.de C., 2025. The use of Industry 4.0 and 5.0 technologies in the transformation of food services: An Integrative Review. Foods 14 (24), 4320. Available at https://doi.org/10.3390/foods14244320.

Choe, H.-O., Lee, M.-H., 2025. Artificial intelligence-based anomaly detection technology for equipment condition monitoring in smart farms. Applied Sciences 15 (23), 12843. Available at https://doi.org/10.3390/app152312843.

Dadhaneeya, H., Nema, P.K., Arora, V.K., 2023. Internet of Things in food processing and its potential in Industry 4.0 era: A review. Trends in Food Science & Technology 139, 104109. Available at https://doi.org/10.1016/j.tifs.2023.07.006.

Deshmukh, M.T., Wankhede, P.R., Chakole, N., Kale, P.D., Jadhav, M.R., Kulkarni, M.B., et al., 2025. Towards intelligent food safety: Machine learning approaches for aflatoxin detection and risk prediction. Trends in Food Science and Technology 161. Available at https://doi.org/10.1016/j.tifs.2025.105055.

Ding, H., Tian, J., Yu, W., Wilson, D.I., Young, B.R., Cui, X., et al., 2023. The application of artificial intelligence and big data in the food industry. Foods 12 (24), 4511. Available at https://doi.org/10.3390/foods12244511.

Dou, X., Ren, J., 2024. The application of dairy products and their derivatives as edible inks in 3D printing technology: A review. International Journal of Food Science and Technology 59 (11), 8630–8644. Available at https://doi.org/10.1111/ijfs.17195.

Dragone, R., Grasso, G., Licciardi, G., Di Stefano, D., Frazzoli, C., 2024. Sensors driven system coupled with artificial intelligence for quality monitoring and HACCP in dairy production. Sensing and Bio-Sensing Research 45, 100683. Available at https://doi.org/10.1016/j.sbsr.2024.100683.

Espinoza-Sandoval, O.R., Angeles-Hernandez, J.C., Gonzalez-Ronquillo, M., Ghavipanje, N., Zhang, N., Bayat, A.R., et al., 2024. Dairy farming in the era of artificial intelligence: Trend or a real game changer? Journal of Dairy Research 91 (2), 139–145. Available at https://doi.org/10.1017/S0022029924000426.

Falih, M.A., Altemimi, A.B., Hamed Alkaisy, Q., Awlqadr, F.H., Abedelmaksoud, T.G., Amjadi, S., et al., 2024. Enhancing safety and quality in the global cheese industry: A review of innovative preservation techniques. Heliyon 10 (23). Available at https://doi.org/10.1016/j.heliyon.2024.e40459.

Farina, G., Kocian, A., Brunori, G., Chessa, S., Lai, M.B., Nardi, D., et al., 2025. Interoperable Traceability in agrifood supply chains: enhancing transport systems through IoT sensor data, blockchain, and dataspace. Sensors (Basel, Switzerland) 25 (11), 3419. Available at https://doi.org/10.3390/s25113419.

Feng, C., Reed, K.F., Giordano, J.O., You, F., 2025. Precision decarbonization for clean dairy farming with digital twin and edge intelligence. Nexus 2 (4). Available at https://doi.org/10.1016/j.ynexs.2025.100105.

Filani, O.M., Olajide, J.O., Osho, G.O., 2021. IoT-driven real-time visibility in perishable goods supply chains. APR 2020 | IRE Journals 3 (10), 412–433.

Fizza, K., Banerjee, A., Georgakopoulos, D., Jayaraman, P.P., Yavari, A., Dawod, A., 2025. An inexpensive AI-powered IoT sensor for continuous farm-to-factory milk quality monitoring. Sensors (Basel, Switzerland) 25 (14), 4439. Available at https://doi.org/10.3390/s25144439.

Folgado, F., Calderón, D., González, I., Calderón, A., 2024. Review of Industry 4.0 from the perspective of automation and supervision systems: definitions, architectures and recent trends. Electronics 13 (4), 782. Available at https://doi.org/10.3390/electronics13040782.

Forsan, H.F., 2025. Utilizing the potential of technologies from the fourth industrial revolution to decrease waste and increase the value of byproducts in the dairy industry. In: Aguilar-Zárate, R., Haghi, P., Gómez-García, A.K. (Eds.), Reducing Food Loss and Waste. Springer, pp. 85–101. Available at https://doi.org/10.1007/978-3-031-91693-9_4.

Gehlot, A., Malik, P.K., Singh, R., Akram, S.V., Alsuwian, T., 2022. Dairy 4.0: Intelligent communication ecosystem for the cattle animal welfare with blockchain and IoT enabled technologies. Applied Sciences 12 (14), 7316. Available at https://doi.org/10.3390/app12147316.

Goyal, K., Kumar, P., Verma, K., 2022. Food adulteration detection using artificial intelligence: A systematic review. Archives of Computational Methods in Engineering 29 (1), 397–426. Available at https://doi.org/10.1007/s11831-021-09600-y.

Goyal, K., Kumar, P., Verma, K., 2024. XAI-empowered IoT multi-sensor system for real-time milk adulteration detection. Food control 164 (November 2023), 110495. Available at https://doi.org/10.1016/j.foodcont.2024.110495.

Hamad, A.M.A., Awad, D.A.B., 2025. Integration of IoT in dairy facilities next-generation technologies in dairy processing and production. IGI Global, Egypt, pp. 139–170. Available at https://www.igi-global.com/book/next-generation-technologies-dairy-processing/357881.

Hasan, S.K.K., Crawford, C.W., 2025. A new era for digital twins: Progress and industry adoption. Digital Twin. Available at https://doi.org/10.1080/27525783.2025.2555877.

Hassoun, A., Abdullah, N.A., Aït-kaddour, A., Beşir, A., Zannou, O., Önal, B., et al., 2024. Food traceability 4.0 as part of the fourth industrial revolution: key enabling technologies. Critical Reviews in Food Science and Nutrition 64 (3), 873–889. Available at https://doi.org/10.1080/10408398.2022.2110033.

Hassoun, A., Aït-Kaddour, A., Abu-Mahfouz, A.M., Rathod, N.B., Bader, F., Barba, F.J., et al., 2023. The fourth industrial revolution in the food industry—Part I: Industry 4.0 technologies. Critical Reviews in Food Science and Nutrition 63 (23), 6547–6563. Available at https://doi.org/10.1080/10408398.2022.2034735.

Hassoun, A., Aït-Kaddour, A., Dankar, I., Safarov, J., Ozogul, F., Sultanova, S., 2025. The Significance of Industry 4.0 technologies in enhancing various unit operations applied in the food sector: Focus on food drying. Food and Bioprocess Technology 18 (1), 109–128. Available at https://doi.org/10.1007/s11947-024-03465-2.

Hassoun, A., Garcia-garcia, G., Trollman, H., Jagtap, S., Cropotova, J., Bhat, Z., et al., 2023. Birth of dairy 4.0: Opportunities and challenges in adoption of fourth industrial revolution technologies in the production of milk and its derivatives. Current Research in Food Science 7 (May), 100535. Available at https://doi.org/10.1016/j.crfs.2023.100535.

Hassoun, A., Jagtap, S., Garcia-Garcia, G., Trollman, H., Pateiro, M., Lorenzo, J.M., et al., 2023. Food quality 4.0: From traditional approaches to digitalized automated analysis. Journal of Food Engineering 337. Available at https://doi.org/10.1016/j.jfoodeng.2022.111216.

Hassoun, A., Jagtap, S., Trollman, H., Garcia-Garcia, G., Duong, L.N.K., Saxena, P., et al., 2024. From Food Industry 4.0 to Food Industry 5.0: Identifying technological enablers and potential future applications in the food sector. Comprehensive Reviews in Food Science and Food Safety 23 (6). Available at https://doi.org/10.1111/1541-4337.70040.

Hassoun, A., Kamiloglu, S., Garcia-Garcia, G., Parra-López, C., Trollman, H., Jagtap, S., et al., 2023. Implementation of relevant fourth industrial revolution innovations across the supply chain of fruits and vegetables: A short update on Traceability 4.0. Food chemistry 409. Available at https://doi.org/10.1016/j.foodchem.2022.135303.

Hassoun, A., Mhlanga, D., Rejeb, A., Bhat, Z., Buheji, M., Bigliardi, B., 2025. The role of industry 4.0 in global food security: A promising pathway to ending hunger. Smart Agricultural Technology 11 (March), 100974. Available at https://doi.org/10.1016/j.atech.2025.100974.

Hassoun, A., Tarchi, I., Aït-kaddour, A., 2024. Leveraging the potential of fourth industrial revolution technologies to reduce and valorize waste and by- products in the dairy sector. Current Opinion in Green and Sustainable Chemistry 47, 100927. Available at https://doi.org/10.1016/j.cogsc.2024.100927.

Hussein, J.B., Workneh, T.S., Kassim, A., Ntsowe, K., Melesse, S.F., El-Mesery, H.S., 2025. A review on the impact of big data analytics in transforming agricultural practices, food processing, and preservation strategies. Applied Food Research 5 (2). Available at https://doi.org/10.1016/j.afres.2025.101234.

El Jaouhari, A., Arif, J., Jawab, F., Samadhiya, A., Kumar, A., 2024. Unfolding the role of metaverse in agri-food supply chain security: Current scenario and future perspectives. International Journal of Food Science & Technology 59 (5), 3451–3460. Available at https://doi.org/10.1111/ijfs.16972.

Javaid, M., Haleem, A., Singh, R.P., Suman, R., 2023. An integrated outlook of Cyber–Physical Systems for Industry 4.0: Topical practices, architecture, and applications. Green Technologies and Sustainability 1 (1). Available at https://doi.org/10.1016/j.grets.2022.100001.

Jia, W., van Ruth, S., Scollan, N., Koidis, A., 2022. Hyperspectral Imaging (HSI) for meat quality evaluation across the supply chain: Current and future trends. Current Research in Food Science 5, 1017–1027. Available at https://doi.org/10.1016/j.crfs.2022.05.016.

Kadri, O., Abdelhadi, A., 2024. Optimizing milk pasteurization diagnosis through deep Q-networks and digital twin technology. International Journal of Web Services Research 21 (1), 1–22. Available at https://doi.org/10.4018/ijwsr.366586.

Kanwal, N., Zhang, M., Law, C.L., Khan, S., Huang, J., 2025. Integrating cellular agriculture with 3D food printing: A pathway to customized, sustainable animal-free dairy foods. Food Bioscience 69, 107017. Available at https://doi.org/10.1016/j.fbio.2025.107017.

Karnik, N., Bora, U., Bhadri, K., Kadambi, P., Dhatrak, P., 2022. A comprehensive study on current and future trends towards the characteristics and enablers of industry 4.0. Journal of Industrial Information Integration 27, 100294. Available at https://doi.org/10.1016/j.jii.2021.100294.

Kaur, U., Malacco, V.M.R., Bai, H., Price, T.P., Datta, A., Xin, L., et al., 2023. Invited review: Integration of technologies and systems for precision animal agriculture—A case study on precision dairy farming. Journal of Animal Science 101. Available at https://doi.org/10.1093/jas/skad206.

Keppler, J.K., Werten, M.W.T., Boom, R.M., 2026. Structural deviations between native and precision fermentation-derived food proteins: Techno-functionality and structure formation strategies. Trends in Food Science & Technology 168 (June 2025), 105493. Available at https://doi.org/10.1016/j.tifs.2025.105493.

Keshvarinia, M., MacKenzie, C.A., Zhao, Z., 2025. A simulation-based digital twin model for data-driven decision optimization. Decision Analytics Journal 17, 100646. Available at https://doi.org/10.1016/J.DAJOUR.2025.100646.

Khanashyam, A.C., Jagtap, S., Agrawal, T.K., Thorakkattu, P., Malav, O.P., Trollman, H., et al., 2025. Applications of artificial intelligence in the dairy Industry: From farm to product development. Computers and Electronics in Agriculture 238. Available at https://doi.org/10.1016/j.compag.2025.110879.

Khan, M., Haleem, A., Javaid, M., 2023. Changes and improvements in Industry 5.0: A strategic approach to overcome the challenges of Industry 4.0. Green Technologies and Sustainability 1 (2), 100020. Available at https://doi.org/10.1016/j.grets.2023.100020.

Khan, M.I., Yasmeen, T., Khan, M., Hadi, N.U., Asif, M., Farooq, M., et al., 2025. Integrating industry 4.0 for enhanced sustainability: Pathways and prospects. Sustainable Production and Consumption 54, 149–189. Available at https://doi.org/10.1016/j.spc.2024.12.012.

Khan, M.U., Zhen, L., Tian, J., Farid, A., Chen, Z., Liu, H., 2026. From detection to prevention: A cyber-secure AI ecosystem for mitigating milk adulteration and dairy fraud. Trends in Food Science & Technology 168, 105503. Available at https://doi.org/10.1016/j.tifs.2025.105503.

Konstantinidis, F.K., Balaska, V., Symeonidis, S., Tsilis, D., Mouroutsos, S.G., Bampis, L., et al., 2023. Automating dairy production lines with the yoghurt cups recognition and detection process in the Industry 4.0 era. Procedia Computer Science 217 (2022), 918–927. Available at https://doi.org/10.1016/j.procs.2022.12.289.

Kraft, M., Bernhardt, H., Brunsch, R., Büscher, W., Colangelo, E., Graf, H., et al., 2022. Can Livestock farming benefit from Industry 4.0 technology? Evidence from recent study. Applied Sciences 12 (24), 12844. Available at https://doi.org/10.3390/app122412844.

Krupitzer, C., Noack, T., Borsum, C., 2022. Digital food twins combining data science and food science: System Model, applications, and challenges. Processes 10 (9), 1781. Available at https://doi.org/10.3390/PR10091781.

Ks, P., Basha, S., Chattopadhyay, A., Mahato, K.K., 2025. Recent advancements of light-emitting diodes in dairy industries. Trends in Food Science & Technology 160, 105018. Available at https://doi.org/10.1016/j.tifs.2025.105018.

Kyaw, K.S., Adegoke, S.C., Ajani, C.K., Nwabor, O.F., Onyeaka, H., 2024. Toward in-process technology-aided automation for enhanced microbial food safety and quality assurance in milk and beverages processing. Critical Reviews in Food Science and Nutrition 64 (6), 1715–1735. Available at https://doi.org/10.1080/10408398.2022.2118660.

Lamanna, M., Bovo, M., Cavallini, D., 2025. Wearable collar technologies for dairy cows: A systematized review of the current applications and future innovations in precision livestock farming. Animals: An Open Access Journal from MDPI 15 (3), 458. Available at https://doi.org/10.3390/ANI15030458.

Lee, C.P., Karyappa, R., Hashimoto, M., 2020. 3D printing of milk-based product. RSC Advances 10 (50), 29821–29828. Available at https://doi.org/10.1039/D0RA05035K.

Liberty, J.T., Habanabakize, E., Adamu, P.I., Bata, S.M., 2024. Advancing food manufacturing: Leveraging robotic solutions for enhanced quality assurance and traceability across global supply networks. Trends in Food Science and Technology 153. Available at https://doi.org/10.1016/j.tifs.2024.104705.

Liu, L., Cheng, W., Kuo, H.W., 2025. A Narrative Review on Smart Sensors and IoT solutions for sustainable agriculture and aquaculture practices. Sustainability 17 (12), 5256. Available at https://doi.org/10.3390/SU17125256.

Liu, H.C., Liu, R., Gu, X., Yang, M., 2023a. From total quality management to Quality 4.0. Frontiers of Engineering Management 10 (2), 191–205.

Liu, Na, Qi, Jingwei, An, Xiaoping, Wang, Yuan, 2023b. A Review on Information Technologies Applicable to Precision Dairy Farming: Focus on Behavior, Health Monitoring, and the Precise Feeding of Dairy Cows. Agriculture 13, 1858. Available at 10.3390/agriculture13101858.

Li, C., Zhang, F., Leong Tan, M., Pan Chun, K., Neethirajan, S., 2024a. Net zero dairy farming—Advancing Climate goals with big data and artificial intelligence. Climate 12 (2), 15. Available at https://doi.org/10.3390/CLI12020015.

Li, Guohua, Wang, Bo, Lv, Weiqiao, Mu, Rongyi, Zhong, Yuanliang, 2024b. Effect of induction mode on 3D printing characteristics of whey protein isolate emulsion gel. Food Hydrocolloids 146, 109255. Available at 10.1016/j.foodhyd.2023.109255.

Lopes de Sousa Jabbour, A.B., Frascareli, F.Cd.O., Santibanez Gonzalez, E.D.R., Chiappetta Jabbour, C.J., 2023. Are food supply chains taking advantage of the circular economy? A research agenda on tackling food waste based on Industry 4.0 technologies. Production Planning and Control 34 (10), 967–983. Available at https://doi.org/10.1080/09537287.2021.1980903.

Losacco, C., Pugliese, G., Forte, L., Tufarelli, V., Maggiolino, A., De Palo, P., 2025. Digital Transition as a driver for sustainable tailor-made farm management: An up-to-date overview on precision livestock farming. Agriculture 15 (13), 1383. Available at https://doi.org/10.3390/agriculture15131383.

Lovarelli, D., Bacenetti, J., Guarino, M., 2020. A review on dairy cattle farming: Is precision livestock farming the compromise for an environmental, economic and social sustainable production? Journal of Cleaner Production 262, 121409. Available at https://doi.org/10.1016/j.jclepro.2020.121409.

Lovarelli, D., Bovo, M., Giannone, C., Santolini, E., Tassinari, P., Guarino, M., 2024. Reducing life cycle environmental impacts of milk production through precision livestock farming. Sustainable Production and Consumption 51, 303–314. Available at https://doi.org/10.1016/J.SPC.2024.09.021.

Lukacs, M., Toth, F., Horvath, R., Solymos, G., Alpár, B., Varga, P., et al., 2025. Advanced digital solutions for food traceability: Enhancing origin, quality, and safety through NIRS, RFID, blockchain, and IoT. Journal of Sensor and Actuator Networks 14 (1), 21. Available at https://doi.org/10.3390/jsan14010021.

Malik, M., Gahlawat, V.K., Mor, R.S., 2023. Digital interoperability and transformation using Industry 4.0 technologies in the dairy industry: An SLR and bibliometric analysis. Scientific Journal of Logistics 19 (3), 461–479.

Malik, M., Gahlawat, V.K., Mor, R.S., Dahiya, V., Yadav, M., 2022. Application of optimization techniques in the dairy supply chain: A systematic review. Logistics 6 (4). Available at https://doi.org/10.3390/logistics6040074.

Malik, M., Gahlawat, V.K., Mor, R.S., Hosseinian-Far, A., 2024. Towards white revolution 2.0: Challenges and opportunities for the industry 4.0 technologies in Indian dairy industry. Operations Management Research 17 (3), 811–832. Available at https://doi.org/10.1007/s12063-024-00482-4.

Malik, M., Gahlawat, V.K., Mor, R.S., Singh, M.K., 2024. Unlocking dairy traceability: Current trends, applications, and future opportunities. Future Foods 10. Available at https://doi.org/10.1016/j.fufo.2024.100426.

Malik, M., Mor, A., Dalal, M., Toor, K., Gahlawat, V.K., Yadav, D.K., 2025. Dairy traceability: Pathways to a sustainable and resilient supply chain. In: Nimmanapalli, M., Kumar Singh, R. (Eds.), Trends in Animal-Based Foods. Springer, pp. 137–148. Available at https://doi.org/10.1007/978-3-032-00575-5_7.

Malik, M., Mor, R.S., Gahlawat, V.K., Hassoun, A., Jagtap, S., 2025. Drivers of Industry 5.0 Technologies in dairy industry: An exploratory study. Sustainable Food Technology. Available at https://doi.org/10.1039/d5fb00156k.

Malik, M., Mor, R.S., Gahlawat, V.K., Kumar, V., 2026. Unlocking the potential: Hybrid blockchain and AI-enabled traceability model development and implementation in the dairy industry—Proof-of-concept. Transportation Research Part E: Logistics and Transportation Review 205. Available at https://doi.org/10.1016/j.tre.2025.104552.

Mandal, A., Ghosh, A.R., 2024. Role of artificial intelligence (AI) in fish growth and health status monitoring: A review on sustainable aquaculture. Aquaculture International 32 (3), 2791–2820. Available at https://doi.org/10.1007/s10499-023-01297-z.

Manteghi, Y., Arkat, J., Mahmoodi, A., 2023. The competition between conventional and organic food production in the presence of the blockchain technology. Trends in Food Science and Technology 136 (January), 282–294. Available at https://doi.org/10.1016/j.tifs.2023.03.003.

Matera, R., Pierro, F., Santinello, M., Iraci Fuintino, A., Pacelli, G., Norton, T., et al., 2025. Precision livestock farming in buffalo species: A sustainable approach for the future. Smart Agricultural Technology 11, 101060. Available at https://doi.org/10.1016/j.atech.2025.101060.

Michelena, Á., Fontenla-Romero, Ó., Luis Calvo-Rolle, J., 2025. A review and future trends of precision livestock over dairy and beef cow cattle with artificial intelligence. Logic Journal of the IGPL 33 (4). Available at https://doi.org/10.1093/jigpal/jzae111.

Mureşan, C.C., Marc, R.A.(V.), Jimborean, M., Rusu, I., Mureşan, A., Nistor, A., et al., 2020. Food safety system (HACCP) as quality checkpoints in a spin-off small-scale yogurt processing plant. Sustainability 12 (22), 9472. Available at https://doi.org/10.3390/su12229472.

Neculai-Valeanu, A.S., Sanduleanu, C., Porosnicu, I., 2025. From tradition to precision: Leveraging digital tools to improve cattle health and welfare. Frontiers in Veterinary Science 12, 1549512. Available at https://doi.org/10.3389/FVETS.2025.1549512/BIBTEX.

Neves, A., Alves, A.S.F., Azamfirei Ionita, V., Matias, J.C.O., Teixeira, S., Alcácer, V., et al., 2025. Key drivers and barriers of industry 4.0 for sustainable practices in small and medium-sized enterprises. Sustainable Futures 10. Available at https://doi.org/10.1016/j.sftr.2025.101464.

Ortega-Gras, J.J., Bueno-Delgado, M.V., Puche-Forte, J.F., Garrido-Lova, J., Martínez-Fernández, R., 2025. Exploring Industry 4.0 technologies implementation to enhance circularity in Spanish manufacturing enterprises. Sustainability (Switzerland) 17 (17). Available at https://doi.org/10.3390/su17177648.

Palma, O., Plà-Aragonés, L.M., Mac Cawley, A., Albornoz, V.M., 2025. AI and data analytics in the dairy farms: A scoping review. Animals: An Open Access Journal from MDPI 15 (9). Available at https://doi.org/10.3390/ani15091291.

Papadopoulos, G., Papantonatou, M.-Z., Uyar, H., Kriezi, O., Mavrommatis, A., Psiroukis, V., et al., 2025. Economic and environmental benefits of digital agricultural technological solutions in livestock farming: A review. Smart Agricultural Technology 10, 100783. Available at https://doi.org/10.1016/j.atech.2025.100783.

Papakonstantinou, G.I., Voulgarakis, N., Terzidou, G., Fotos, L., Giamouri, E., Papatsiros, V.G., 2024. Precision livestock farming technology: Applications and challenges of animal welfare and climate change. Agriculture 14 (4), 620. Available at https://doi.org/10.3390/agriculture14040620.

Papa, M., Robson de Medeiros Oliveira, S., Bergier, I., 2024. Technologies in cattle traceability: A bibliometric analysis. Computers and Electronics in Agriculture 227 (P1), 109459. Available at https://doi.org/10.1016/j.compag.2024.109459.

Parrenin, L., Dupuis, A., Danjou, C., & Agard, B. (2025). Machine learning tool for yield maximization in cream cheese production. In H. Dassisti, M., Madani, K., Panetto (Eds.), *Innovative Intelligent Industrial Production and Logistics. IN4PL 2024. Communications in Computer and Information Science* (Vol. 2372, pp. 97–114). Springer. https://doi.org/10.1007/978-3-031-80760-2_6.

Patel, S.S., Bains, A., Kaushik, R., Dhull, S.B., Nagraik, R., Fareed, M., et al., 2026. Leveraging smart technologies to enhance safety in milk and milk-based products. Food control 182. Available at https://doi.org/10.1016/j.foodcont.2025.111871.

Pavithra, T., Rawat, S., Sunil, C.K., 2025. Recent trends in blockchain traceability of food products: A review. Food and Humanity 5 (May), 100666. Available at https://doi.org/10.1016/j.foohum.2025.100666.

Paçarada, R., Hörtenhuber, S., Hemme, T., Wurzinger, M., Zollitsch, W., 2024. Sustainability assessment tools for dairy supply chains: A typology. Sustainability 16 (12), 4999. Available at https://doi.org/10.3390/su16124999.

Plakantara, S.P., Karakitsiou, A., 2025. Transforming agrifood supply chains with digital technologies: A systematic review of safety and quality risk management. Operations Research Forum 6 (3). Available at https://doi.org/10.1007/s43069-025-00511-3.

Pletsch, A.L.B., Stenger, E.A.F., Sehnem, S., 2025. The potential of Industry 4.0 technologies in transforming agricultural and livestock practices: A systematic review. International Journal of Pervasive Computing and Communications 21 (2), 121–155. Available at https://doi.org/10.1108/IJPCC-05-2024-0178.

Priyashantha, H., 2025. World dairy system sustainability: A milk quality perspective. Frontiers in Sustainable Resource Management 4. Available at https://doi.org/10.3389/fsrma.2025.1572962.

Qiu, F., Kumar, A., Hu, J., Sharma, P., Tang, Y.B., Xu Xiang, Y., et al., 2025. A review on integrating IoT, IIoT, and Industry 4.0: A pathway to smart manufacturing and digital transformation. IET Information Security 2025 (1). Available at https://doi.org/10.1049/ise2/9275962.

Rao, S., Neethirajan, S., 2025. Computational architectures for precision dairy nutrition digital twins: A technical review and implementation framework. Sensors (Basel, Switzerland) 25 (16), 4899. Available at https://doi.org/10.3390/s25164899.

Rashed, M.S., Fakhry, S., Satour, R., Abdelkarim, E.A., Sobhy, M., Pathania, S., 2025. IoT and digital twins for ssmart food supply chains. Sustainable Development Goals Series 102, 249–273. Available at https://doi.org/10.1007/978-3-031-76758-6_16.

Ray, A., Sinha, C., Sharma, A.K., Khamrui, K., Hussain, S.A., Dabas, J.K., et al., 2026. AI-driven real-time monitoring and predictive control system for yoghurt fermentation. Food control 182. Available at https://doi.org/10.1016/j.foodcont.2025.111857.

Rehman, Wu, Koondhar, M.A., Afridi, S.K., Albasha, L., Smaili, I.H., Touti, E., et al., 2025. The role of 5G network in revolutionizing agriculture for sustainable development: A comprehensive review. Energy Nexus 17 Available from. Available at https://doi.org/10.1016/j.nexus.2025.100368.

Rolofs, G., Wilking, F., Goetz, S., Wartzack, S., 2024. Integrating digital twins and cyber-physical systems for flexible energy management in manufacturing facilities: A conceptual framework. Electronics 13 (24), 4964. Available at https://doi.org/10.3390/electronics13244964.

Romanello, R., Veglio, V., 2022. Industry 4.0 in food processing: Drivers, challenges and outcomes. British Food Journal 124 (13), 375–390. Available at https://doi.org/10.1108/bfj-09-2021-1056.

Rossi, S., Gemma, S., Borghini, F., Perini, M., Butini, S., Carullo, G., et al., 2025. Agri-food traceability today: Advancing innovation towards efficiency, sustainability, ethical sourcing, and safety in food supply chains. Trends in Food Science & Technology 163, 105154. Available at https://doi.org/10.1016/j.tifs.2025.105154.

Ryalat, M., Franco, E., Elmoaqet, H., Almtireen, N., Al-Refai, G., 2024. The integration of advanced mechatronic systems into Industry 4.0 for smart manufacturing. Sustainability 16 (19), 8504. Available at https://doi.org/10.3390/su16198504.

Sader, S., Husti, I., Daroczi, M., 2021. A review of quality 4.0: Definitions, features, technologies, applications, and challenges. Total Quality Management & Business Excellence. Available at https://doi.org/10.1080/14783363.2021.1944082.

Saetae, D., 2025. Machine learning-based prediction of microbial growth and acidification in yogurt fermentation at industrial temperatures. LWT 231, 118326. Available at https://doi.org/10.1016/j.lwt.2025.118326.

Saha, D., Padhiary, M., Chandrakar, N., 2025. AI vision and machine learning for enhanced automation in food industry: A systematic review. Food and Humanity 4, 100587. Available at https://doi.org/10.1016/j.foohum.2025.100587.

Saha, D., Padhiary, M., Hoque, A., Prasad, G., 2025. 3D printing technology for valorization of food processing wastes and byproducts: A systematic review. Waste Management Bulletin 3 (4), 100192. Available at https://doi.org/10.1016/j.wmb.2025.100192.

Sangode, P.B., 2025. Dairy 4.0: Embracing digital technology for sustainable farming: A review. Agricultural Research Communication Centre. India Asian Journal of Dairy and Food Research 44 (2), 169–178. Available at https://doi.org/10.18805/ajdfr.DR-2242.

Santos, J.M., Garcia, J.M., Dias, J., Martins, J.C., Alvarenga, N., Gonçalves, E.M., et al., 2025. Energy sustainability in the ripening of traditional cheese: Renewable energy sources and Internet of Things based energy monitoring. Dairy 6 (6), 63. Available at https://doi.org/10.3390/dairy6060063.

Selmi, H., Presutto, E., Spano, G., Capozzi, V., Fragasso, M., 2025. Valorising whey: From environmental burden to bio-based production of value-added compounds and food ingredients. Foods 14 (21), 3646. Available at https://doi.org/10.3390/foods14213646.

Semercioz-Oduncuoglu, A.S., Luning, P.A., 2025. Industry 4.0 technologies in quality and safety control systems in food manufacturing: A systematic techno-managerial analysis on benefits and barriers. Trends in Food Science and Technology 163. Available at https://doi.org/10.1016/j.tifs.2025.105144.

Serrano-Torres, G.J., López-Naranjo, A.L., Larrea-Cuadrado, P.L., Mazón-Fierro, G., 2025. Transformation of the dairy supply chain through artificial intelligence: A systematic review. Sustainability (Switzerland) 17 (3). Available at https://doi.org/10.3390/su17030982.

Shweta, A., Mohammed, M., Mohammadi, S., Sharma, U., Sharma, V.S., Rana, M., et al., 2025. Assessment of advancements and applications of robotics, artificial intelligence, and automated technology in the modern food sector. Applied Food Research 5 (2), 101261. Available at https://doi.org/10.1016/j.afres.2025.101261.

Siddique, A., Gupta, A., Sawyer, J.T., Huang, T.S., Morey, A., 2025. Big data analytics in food industry: A state-of-the-art literature review. Npj Science of Food 9 (1). Available at https://doi.org/10.1038/s41538-025-00394-y.

Singh, P., Pandey, S., Manik, S., 2024. A comprehensive review of the dairy pasteurization process using machine learning models. Food control 164 (May), 110574. Available at https://doi.org/10.1016/j.foodcont.2024.110574.

Singh, D., 2025. Harnessing artificial intelligence to safeguard food quality and safety. Journal of Food Protection 88 (11), 100621. Available at https://doi.org/10.1016/j.jfp.2025.100621.

Song, X., Zhang, X., Dong, G., Ding, H., Cui, X., Han, Y., et al., 2025. AI in food industry automation: Applications and challenges. Frontiers in Sustainable Food Systems 9. Available at https://doi.org/10.3389/fsufs.2025.1575430.

Tangorra, F.M., Buoio, E., Calcante, A., Bassi, A., Costa, A., 2024. Internet of Things (IoT): Sensors application in dairy cattle farming. Animals: An Open Access Journal from MDPI 14 (21), 3071. Available at https://doi.org/10.3390/ani14213071.

Tarapata, J., Murphy, T.R., Finnegan, E.W., O'Callaghan, T.F., O'Mahony, J.A., 2025. Approaches for measuring and predicting fouling during thermal processing of dairy solutions. Comprehensive Reviews in Food Science and Food Safety 24 (4). Available at https://doi.org/10.1111/1541-4337.70209.

Tarchi, I., Boudalia, S., Ozogul, F., Câmara, J.S., Bhat, Z.F., Hassoun, A., et al., 2024. Valorization of agri-food waste and by-products in cheese and other dairy foods: An updated review. Food Bioscience 58. Available at https://doi.org/10.1016/j.fbio.2024.103751.

Tran, D., Melissari, F., Le Feon, S., Papadakis, A., Zahariadis, T., Chatzitheodorou, D., et al., 2025. A holistic assessment of blockchain-based traceability systems: The case of protected designation of origin feta cheese production. Computers and Electronics in Agriculture 230, 109863. Available at https://doi.org/10.1016/j.compag.2024.109863.

Trollman, H., Bader, F., Luo, Y., Garcia-Garcia, G., Jagtap, S., Parra-López, C., et al., 2024. Food safety 4.0 food industry 4.0: Emerging trends and technologies in sustainable food production and consumption. Elsevier, United Kingdom, pp. 61–81. Available at https://doi.org/10.1016/B978-0-443-15516-1.00004-9.

Turgut, S.S., Bilge, G., Entrenas, J.A., Pérez-Marín, D., 2024. Significant roles of smart sensors in the modern agriculture and food industry. Food Industry 4.0: Emerging trends and technologies in sustainable food production and consumption. Elsevier, Denmark, pp. 133–170. Available at https://doi.org/10.1016/B978-0-443-15516-1.00008-6.

Tzachor, A., Richards, C.E., Jeen, S., 2022. Transforming agrifood production systems and supply chains with digital twins. Npj Science of Food 6 (1), 47. Available at https://doi.org/10.1038/s41538-022-00162-2.

Udourioh, G.A., Solomon, M.M., Okolie, J.A., 2025. A review of the valorization of dairy industry wastes through thermochemical, biological, and integrated processes for value-added products. Food Science of Animal Resources 45 (2), 375–408. Available at https://doi.org/10.5851/KOSFA.2025.E2.

Vargas-Bello-Pérez, E., Espinoza Sandoval, O.R., Rodríguez-Piñeros, S., Ghavipanje, N., 2025. Blockchain and its future in the supply and value chains of the dairy industry. Journal of Dairy Research 92 (2), 127–134. Available at https://doi.org/10.1017/S0022029925101076.

Vlaicu, P.A., Gras, M.A., Untea, A.E., Lefter, N.A., Rotar, M.C., 2024. Advancing livestock technology: Intelligent systemization for enhanced productivity, welfare, and sustainability. AgriEngineering 6 (2), 1479–1496. Available at https://doi.org/10.3390/agriengineering6020084.

Wang, Y., Feng, Y., Xiao, Z., Luo, Y., 2025. Machine learning supported single-stranded DNA sensor array for multiple foodborne pathogenic and spoilage bacteria identification in milk. Food chemistry 463 (P2), 141115. Available at https://doi.org/10.1016/j.foodchem.2024.141115.

Wang, My, Zhang, My, Luo, Xh, Liang, Lg, Liu, Yn, 2025. 3D printing-driven functional food innovation: Construction of personalized nutritional system and precise regulation mechanism of texture structure. Trends in Food Science and Technology 166 Available from. Available at https://doi.org/10.1016/j.tifs.2025.105389.

Wedamulla, N.E., Choi, Y.J., Zhang, Q., Kim, E.K., 2025. Use of food by-products to design 3D food printing inks: Potential of bioactives extraction and encapsulated 3D food printing ink formulation. Food Research International 213. Available at https://doi.org/10.1016/j.foodres.2025.116376.

Woo, S., Kim, Y., Kim, S., 2025. Converging extended reality and robotics for innovation in the food industry. AgriEngineering 7 (10), 322. Available at https://doi.org/10.3390/AGRIENGINEERING7100322.

Xu, M., Xu, Y., Ji, S., Li, Y., Zhang, C., Zhou, Z., et al., 2025. Structure-printability mechanism of 3D printing ice cream inks affected by hydroxypropyl distarch phosphate with different substitution degrees. Food Hydrocolloids 162, 111013. Available at https://doi.org/10.1016/j.foodhyd.2024.111013.

Chapter 3

Advancing dairy sustainability with Industry 4.0

Zakia Gueboudji[1], Zouheyr Hadri[2], Imane Kouadri[3], Hassen Touati[4,5], Abdo Hassoun[6,7], and Sofiane Boudalia[4,5]

[1]*Department of Microbiology and Biochemistry, Faculty of Natural and Life Sciences, University of Batna 2 Mustapha Ben Boulaid, Batna, Algeria,* [2]*Department of Agricultural sciences, Faculty of Sciences and Technology, University of Relizane, Relizane, Algeria,* [3]*Department of Process Engineering, Laboratory of Industrial Analysis and Materials Engineering, University of May 8 1945 Guelma, Guelma, Algeria,* [4]*Département d'Écologie et Génie de l'Environnement, Université 8 Mai 1945 Guelma, Guelma, Algeria,* [5]*Laboratoire de Biologie, Eau et Environnement, Université 8 Mai 1945 Guelma, Guelma, Algeria,* [6]*Faculty of Agricultural Engineering, University of Aleppo, Aleppo, Syria,* [7]*Sustainable AgriFoodtech Innovation & Research (SAFIR), Arras, France*

3.1 Introduction

Climate change is primarily driven by the continuous rise in atmospheric concentrations of greenhouse gases (GHGs), notably nitrous oxide (N_2O), carbon dioxide (CO_2), and methane (CH_4). These gases intensify the greenhouse effect, leading to profound alterations in global and regional climate systems (Filonchyk et al., 2024; IPCC, 2021). Among the most visible consequences are increasing climate variability, characterized by irregular precipitation regimes, a higher frequency of extreme weather events, and the intensification of floods and droughts, which collectively threaten natural ecosystems and agricultural productivity (Verma et al., 2025; IPCC, 2021). In this context, the agricultural sector plays a dual role, as both a victim of climate change and a significant contributor to GHG emissions. In 2020, agriculture accounted for 5865.47 $MtCO_2e$, representing 12.34% of total global emissions estimated at 47513.15 $MtCO_2e$, when emissions from land-use change and forestry are included (Bagci et al., 2025; Climate Watch, 2021). A substantial proportion of these emissions originates from livestock production systems, which alone contribute more than 83% of total agricultural emissions, with milk and meat production being the dominant sources (Kokemohr et al., 2026;

Dairy 4.0. DOI: https://doi.org/10.1016/B978-0-443-44730-3.00010-9

Climate Watch, 2021). Within livestock-related emissions, enteric fermentation constitutes the most significant share (approximately 5.5 $GtCO_2e$), followed by emissions from manure deposited on pasture (4.5 $GtCO_2e$), highlighting the significant environmental footprint of ruminant-based production systems (Climate Watch, 2021).

Climate projections further indicate that global warming will exacerbate water scarcity, particularly in vulnerable regions. Several studies anticipate a decline in total annual rainfall ranging from 15% to 30%, coupled with the progressive expansion of arid and desert climates (Chen et al., 2023; Christensen et al., 2007). These changes are mainly driven by rising air temperatures and a persistent decrease in precipitation. Alarmingly, the magnitude and spatial extent of these impacts are expected to intensify in the coming decades, becoming severe, widespread, and potentially irreversible if mitigation and adaptation measures are not adequately implemented (IPCC, 2014, 2021; Mariotti et al., 2015; Sahnoune et al., 2013; Zeroual et al., 2016). Such climatic constraints pose major challenges to the dairy sector, which is particularly sensitive to heat stress, water availability, and feed resource fluctuations. Milk represents a strategic food commodity and constitutes an essential component of the diet of nearly 6 billion people worldwide (Gauly and Ammer, 2020; OECD–FAO, 2021). Beyond fluid milk consumption, the dairy industry supports a wide range of value-added products, including cheese, yogurt, ice cream, and butter, which are produced and consumed on a large scale. Consequently, climate-induced disruptions to dairy production systems not only affect farm-level productivity but also threaten the stability of global food supply chains and food and nutrition security (Wreford et al., 2025; FAO, 2013).

Addressing these challenges requires a rapid and profound transformation of the dairy sector toward more sustainable, resilient, and resource-efficient production models. There is an urgent need to adopt production systems that simultaneously reduce environmental pressures, particularly carbon footprints and GHG emissions, while responding to increasing consumer expectations regarding product quality, safety, traceability, and sustainability (Boudalia et al., 2020; Pretty, 2020). In this regard, the combined implementation of emission reduction strategies, the development of healthier and environmentally friendly dairy products, and the integration of advanced digital solutions offer promising opportunities (Neethirajan, 2024; Boudalia et al., 2020; Pretty, 2020). Food Industry 4.0 technologies, including automation, digital monitoring, data analytics, and smart decision-support systems, represent strategic and economically viable pathways to enhance process efficiency, optimize resource use, strengthen sector resilience, and preserve natural resources for future generations (Bousbia et al., 2021; Brini, 2021; IPCC, 2014; Martin et al., 2020; Wainwright et al., 2019).

This chapter aims to assess the role and added value of Food Industry 4.0 technologies in improving the sustainability and performance of the dairy sector. Specifically, it examines how the integration of Industry 4.0 tools and data-driven management frameworks can improve process efficiency, optimize resource utilization, improve product quality and safety, mitigate environmental emissions, and strengthen economic performance across the entire dairy value chain. By adopting a systems-based and sustainability-oriented perspective, this chapter provides scientific and practical insights to support the digital transformation and long-term competitiveness of sustainable dairy production systems.

3.2 Industry 4.0 as a driver of sustainability in the dairy value chain

According to the literature, global milk production reached nearly 968 million tons in 2023, highlighting the large and sustained scale of dairy production worldwide. This volume includes raw milk from all mammalian species, dominated by cow's milk, with significant contributions from buffalo, goats, sheep, and other species (Gueboudji and Gaouar, 2026; FAO, 2024). Milk and dairy products, ranging from highly perishable foods such as fresh milk and cream to more stable products like cheese, butter, milk powder, and functional dairy ingredients including proteins and concentrates, play a fundamental role in human diets (Bousbia et al., 2018). Consequently, increasing attention is being directed toward ensuring that dairy production systems are both sustainable and resilient (Finnegan and Goggins, 2021).

According to the Food and Agriculture Organization of the United Nations, more than four-fifths of the global population regularly consumes dairy products (FAO-GDP, 2018). However, dairy consumption patterns vary considerably across regions. In the Global North, the dairy sector is highly developed, with per capita consumption of fresh dairy products expected to reach 25.2 kg of milk solids by 2030, compared with 12.6 kg per capita in the Global South (OECD–FAO, 2021). This demand is closely linked to the ongoing nutrition transition, characterized by changing lifestyles and dietary preferences, which has driven what is often described as a "livestock revolution" aimed at expanding meat and dairy production (Bojovic and McGregor, 2023). Simultaneously, targeted advertising and promotional campaigns are further accelerating demand for fresh dairy products in rapidly growing markets, particularly in India, Pakistan, and parts of Africa (OECD-FAO, 2020).

As milk production and processing continue to expand worldwide, their environmental footprint is becoming increasingly significant, making the adoption of Industry 4.0 technologies a pressing necessity for the dairy sector (Hassoun et al., 2023c). Technologies such as cyber-physical systems (CPS),

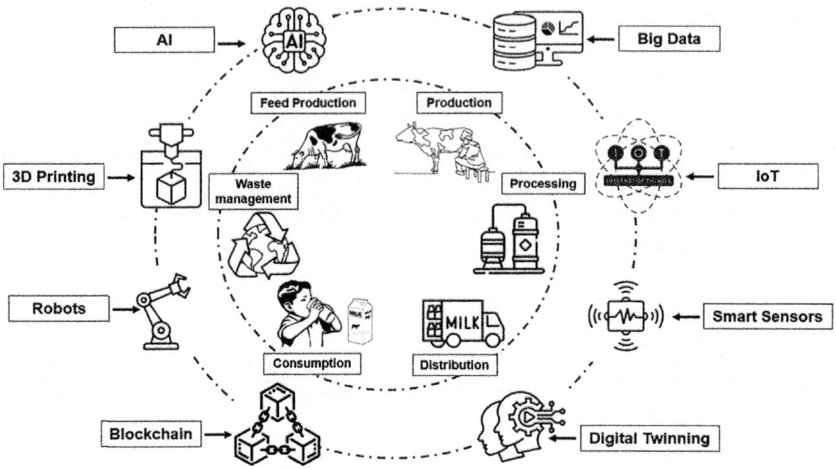

FIGURE 3.1 Application of Industry 4.0 technologies in the dairy sector to enhance sustainability. Schematic representation of digital and automated technologies applied across the dairy value chain, from feed production and milk production to processing, distribution, consumption, and waste management.
Adapted from Hassoun, A., Tarchi, I., Aït-Kaddour, A., 2024. Leveraging the potential of fourth industrial revolution technologies to reduce and valorize waste and by-products in the dairy sector. Current Opinion in Green and Sustainable Chemistry. 47, 100927.

smart sensors, artificial intelligence (AI), 3D printing, big data analytics, the Internet of Things (IoT), digital twins, and blockchain-based traceability tools are now essential to improve operational efficiency and environmental performance across dairy systems (Fig. 3.1; Kazi, 2025) (Malik et al., 2025).

The application of Food Industry 4.0 management frameworks (and its derivative Dairy 4.0) enables dairy processors to better control and optimize critical operations, including energy and water consumption, raw milk quality monitoring, process performance, waste reduction, and GHG mitigation. By enhancing process control, transparency, and system robustness, Industry 4.0 supports the production of safer, healthier, and more sustainable dairy products, while reinforcing the long-term resilience and competitiveness of the sector (Hassoun et al., 2023c; Oztuna Taner and Çolak, 2024). The following sections will examine the role of Industry 4.0 as a driver of sustainability across the entire dairy value chain, as illustrated in Fig. 3.2. These cover the successive stages from feed production and on-farm milk production, through milk transport, processing, and packaging, to distribution, retail, and final consumption, as well as the management and valorization of dairy waste and by-products.

Table 3.1 shows an overview of Industry 4.0 technologies deployed across the dairy value chain.

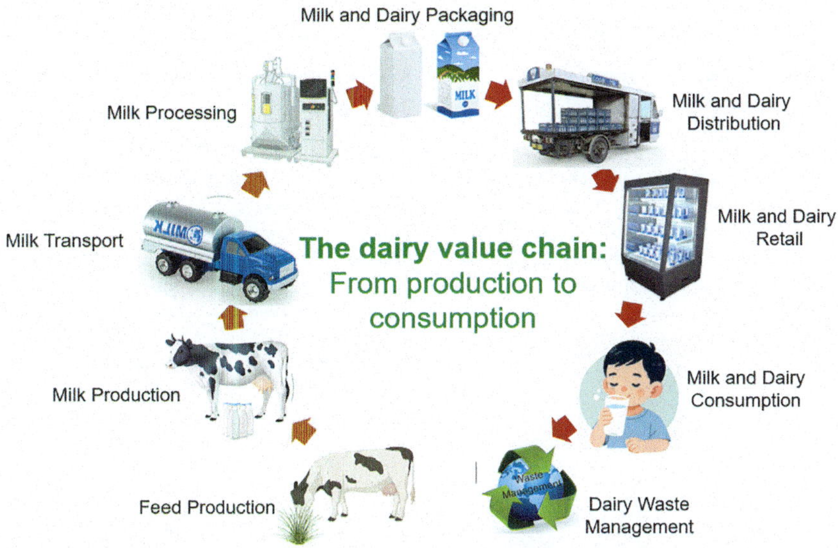

FIGURE 3.2 The dairy value chain from production to consumption (Wang, 2011).
Schematic representation of the dairy value chain, illustrating the main stages from feed and milk production through processing, packaging, distribution, retail, consumption, and waste management.
Adapted from Wang, Y., 2011. U.S. dairy life cycle assessment: From grass to glass, National Academy of Sciences, Rosemont, Illinois, USA.

TABLE 3.1 Industry 4.0 technologies and their impacts on sustainability, in the dairy value chain.

Dairy chain stage	Industry 4.0 technologies	Key sustainability impacts	References
Milk production (farm)	Precision Livestock Farming (PLF), IoT sensors, Automated Milking Systems, AI, big data analytics, Decision Support Systems (DSS)	Improved resource efficiency (feed, water, energy), reduced GHG emissions, optimized herd health and productivity, better animal welfare, lower labor demand, real-time monitoring for early disease detection	Tadele et al. (2025), Kaswan et al. (2024), Palma et al. (2025), Sharma et al. (2007), Grzesiak et al. (2010), Örs et al. (2022), Lee et al. (2024), Lavrijsen-Kromwijk et al. (2024)

(Continued)

TABLE 3.1 (Continued)

Dairy chain stage	Industry 4.0 technologies	Key sustainability impacts	References
Milk processing and transformation	Smart sensors, robotics, AI, digital twins, automation systems, cloud computing, 3D food printing	Enhanced process control, energy and water efficiency, waste use optimization, consistent product quality, improved safety, predictive maintenance, innovation in functional and sustainable products	Dash et al. (2022), Abdurrahman and Ferrari (2025), Akkad and Bányai (2023), Putra et al. (2018), Piana et al. (2025), Kozub et al. (2020)
Packaging	Intelligent packaging, active packaging, smart sensors, RFID, QR codes, nanoparticle-based packaging	Extended shelf life, reduced food spoilage and waste, improved product safety and quality, biodegradable and/or ecofriendly materials, real-time freshness monitoring	Mirza Alizadeh et al. (2021), Shukla et al. (2024), Mkhari et al. (2025), Romero et al. (2021)
Distribution	IoT sensors, GPS, CPS, AI, big data analytics, blockchain	Optimized transportation routes, reduced fuel consumption and emissions, cold-chain integrity, traceability, minimized product losses, real-time monitoring of environmental conditions	Wu et al. (2021), Aamer et al. (2025), Serrano-Torres et al. (2025), Neethirajan (2023), Khanna et al. (2022), Dorothy and Fredrick (2020), Vargas-Bello-Pérez et al. (2025), Makarov et al. (2019)
Consumption and consumer behavior	Smart packaging, mobile apps, blockchain, AI, data-driven nutrition platforms	Informed consumer choices, reduced household food waste, personalized nutrition, transparency and trust in product origin and quality,	Aït-Kaddour et al. (2024), Gao et al. (2020), Tolegenov et al. (2024), El Matbouly et al. (2022), Lukacs et al. (2025),

(*Continued*)

TABLE 3.1 (Continued)

Dairy chain stage	Industry 4.0 technologies	Key sustainability impacts	References
		encouragement of sustainable consumption behaviors	Camel et al. (2024), Hassoun and Galanakis (2025), Kuosmanen et al. (2025), Theodore Armand et al. (2024)
Waste management and by-product valorization	IoT sensors, AI, big data, digital twins, ANN & ML models	Optimized biogas and wastewater treatment, reduced GHG emissions, enhanced energy recovery, improved circular economy practices, efficient reuse of by-products (whey, sludge, manure, etc.)	Udourioh et al. (2025), Hassoun et al. (2023c), Aït-Kaddour et al. (2021), Marycz et al. (2025), Barik and Murugan (2015), Duan et al. (2025)

3.3 Role of Industry 4.0 technologies in sustainable milk production

According to the literature, milk production is a resource-intensive activity that exerts considerable environmental pressures, notably through water and land use and GHG emissions (Cruz-Rivero et al., 2025; Boudebbouz et al., 2023). The environmental performance of dairy systems is determined by a complex interaction of factors, including production intensity, herd size, management strategies, and farm structural characteristics, with feed production and manure management representing major environmental hotspots. However, robust comparisons among dairy production systems remain challenging, as existing studies often report heterogeneous or contradictory outcomes due to differences in methodological frameworks, system boundaries, and regional conditions (Rencricca et al., 2023). In this context, the integration of Industry 4.0 technologies, such as precision sensing, data analytics, automation, and decision-support systems, offers significant potential to enhance sustainability by improving resource-use efficiency, optimizing feed and manure management, and enabling data-driven environmental monitoring and mitigation strategies across dairy value chains (Morrone et al., 2022).

Over the past decade, the deployment of Industry 4.0 technologies in the agri-food sector has expanded markedly, encompassing digital innovations (e.g., AI, big data analytics, and digital twins), physical technologies (robotics and smart sensors), and emerging biological technologies (such as precision fermentation, 3D printing, and recent advances in nanobiotechnology). In response to increasing demands for enhanced supply chain transparency, traceability, and process control, the dairy industry has emerged as one of the fastest adopters of automation and advanced digital solutions (Hassoun et al., 2024).

Digitization constitutes a key enabler of smart dairy farming by facilitating the structured acquisition, integration, and exploitation of data generated by advanced technologies, including electronic animal identification systems, automated milking systems, milk quality analyzers (e.g., Herd Navigator), body condition scoring devices, activity and behavior sensors, precision feeding equipment, microclimate control systems, automated watering units, and digital weighing platforms (Tadele et al., 2025). The progressive implementation of these data-driven and automated control systems enhances operational efficiency, reduces production costs, and improves decision-making processes within dairy production systems (Kozina and Semkiv, 2020). Moreover, the integration of modern digital technologies allows for real-time, centralized visualization of critical production parameters, thereby minimizing the need for continuous human supervision. Evidence from previous studies indicates that the adoption of electronic identification and monitoring systems contributes to the optimization of herd management and production performance at the farm and regional scales (Kaswan et al., 2024).

From a sustainability perspective, precision feeding systems, coupled with real-time animal monitoring, improve feed conversion efficiency and reduce enteric methane emissions and nitrogen excretion, thereby lowering GHG emissions and nutrient losses to soil and water (Brito et al., 2025; Kebreab et al., 2026). Similarly, smart water management systems and automated watering units allow continuous monitoring of water intake and leakage detection, leading to reduced water consumption at the farm level (Et-taibi et al., 2024; Gupta et al., 2020). Energy-efficient automated milking systems and sensor-based climate control technologies further contribute to lower energy demand by optimizing equipment operation and housing conditions (Papakonstantinou et al., 2024; Horváthné Kovács and Zörög, 2025). In addition, digital land-use optimization, achieved through improved herd productivity and health management, can reduce the land footprint per unit of milk produced. Overall, the adoption of electronic identification and integrated monitoring systems significantly enhances herd management efficiency and overall production performance, with measurable cobenefits in terms of reduced GHG emissions, water use, and energy consumption (Qi et al., 2022; Horváthné Kovács and Zörög, 2025).

IoT technologies play a central role in addressing the growing complexity of animal monitoring and management associated with the increasing size of dairy farms and herd expansion. Insufficient control and supervision under these conditions can negatively affect animal welfare, health, productive and reproductive performance, and overall farm profitability. Precision Livestock Farming (PLF) relies on sensor-based systems that enable continuous, real-time monitoring of individual animals, thereby supporting more accurate herd management and performance optimization. When PLF sensors and devices are integrated within IoT architectures, including edge computing, cloud-based platforms, and machine learning (ML) algorithms, they create interconnected systems that facilitate data-driven decision-making and enhance individualized animal management (Tangorra et al., 2024). For example, Posam et al. (2025) proposed an automated dairy farm management system combining IoT and ML for real-time monitoring and control. Interconnected sensors continuously track environmental parameters (temperature, humidity, air quality, rainfall, and fire risk) while cloud-based analytics support automated operations, including feeding, waste management, lighting, fire safety, and climate-responsive housing. An image-based ML model was developed for cattle disease detection, including lumpy skin disease and mastitis, achieving a reported accuracy of 92% based on threefold cross-validation. The system further incorporates geofencing and zonal classification mechanisms to support real-time animal localization and behavior monitoring, illustrating the potential of Industry 4.0 technologies for data-driven and automated dairy farm management.

Additionally, previous studies have demonstrated that crop–animal integration through Livestock Management Systems (LMS) and Agricultural Information Systems (AIS) can provide historical traceability via milking tracking with QR codes and monitoring of animal movements (Erden et al., 2015). Extensive research has explored livestock health monitoring using LoRa IoT, cloud computing, and mobile applications. A Subspace k-Nearest Neighbors (kNN) classifier enables early disease detection and timely alerts, improving animal welfare, reducing economic losses, and enhancing overall farm productivity (Sheik Arafat et al., 2024).

In the dairy sector, the adoption of big data, AI, and ML is playing an increasingly critical role in developing advanced decision support systems (DSS). By processing and analyzing large volumes of heterogeneous data, big data-, AI-, and ML-based tools enable more integrated, predictive, and efficient management of dairy farms (Palma et al., 2025). At the farm level, these technologies support a wide range of operations, including feeding strategies, breeding programs, animal health and behavior monitoring, milking, and resource management, thereby improving decision-making and overall farm performance (Shine and Murphy, 2022). DSS rely on computer-based models and data analysis to optimize resource use, enhance productivity, and improve financial performance on dairy farms. Commercial DSS further demonstrate their importance in modern dairy management by integrating scientific and technological advancements (Palma et al., 2025). Previous studies have highlighted the potential of ML techniques in dairy cattle management. For example, Sharma et al. (2007) reported that artificial neural network (ANN) models can be used to predict first-lactation 305-day milk yield (FLMY305) in Karan Fries crossbred cattle based on partial lactation records and key breeding traits. Similarly, Grzesiak et al. (2010) demonstrated the practical applicability of ML approaches, including multivariate adaptive regression splines (MARS) and ANN, for the identification of cows with insemination problems. Compared with traditional statistical methods, these ML-based approaches provide more conservative and potentially more robust detection outcomes. Furthermore, a noninvasive, video-based animal identification and welfare monitoring approach, inspired by human face recognition pipelines (face detection, cropping, encoding, and matching), was developed using video data from 89 dairy cows in a robotic dairy farm in Australia, achieving an overall identification accuracy of 84% (Dac et al., 2022). The system showed strong potential for integration into automated welfare assessment frameworks, with scalable applications in traceability, welfare monitoring during transport, fraud prevention, and product quality prediction under environmental stress conditions. Overall, big data-, AI-, and ML-enabled systems enhance dairy sustainability by improving resource-use efficiency, reducing input waste, and enabling early health and welfare detection, thereby lowering GHG emissions per unit of milk produced.

Robotic milking systems (RMS) are technologies introduced into dairy farms as part of the digital revolution in agriculture 4.0. According to a recent report, a simulation model comparing robotic and conventional milking systems made the following assumptions: an increase in milk yield of 8.66%, an increase in investment costs of 58.46%, an increase in energy consumption of 36.66%, an increase in feed costs of 1.33%, and a decrease in labor input of 27.84% (Örs et al., 2022). Furthermore, compared to farms not equipped with RMS, Korean farmers using these systems produced 0.10 to 0.11 more calves annually. Additionally, the adopters reduced labor input by 0.15 to 0.30 per head/week while increasing milk output by 2.44 kg to 2.88 kg per head/day (Lee et al., 2024). In Norway, automatic milking systems decreased labor per cow and increased herd sizes (Martinsson et al., 2023). Another study involving 739 Norwegian dairy farmers using automated milking systems reported that adequate training in automatic milking technologies was strongly associated with improved farmer well-being (Hansen et al., 2020). In Germany, researchers found that higher levels of automation were associated with improved emotional states among farmers and a strengthened human–animal bond. Automatic milking systems were also associated with positive effects on cattle behavior and welfare, including a lower prevalence of severe lameness and reduced lower-leg dirtiness, particularly on farms with higher levels of automation (Lavrijsen-Kromwijk et al., 2024). In emerging economies such as India, the integration of robotics with information and communication technologies (ICT) for agricultural production and farm management has been reported to enhance productivity, resource-use efficiency, and sustainability outcomes (Sharma et al., 2022). Recent innovations include robotic on-floor cleaning systems, which have been shown to improve cow welfare while simultaneously reducing ammonia emissions, thereby addressing both animal health and environmental sustainability concerns (Gerrits et al., 2022).

Blockchain technology has been applied to ensure traceability in dairy farms. By enabling a decentralized, immutable, transparent, and reliable system, blockchain allows for the automated recording and monitoring of data throughout the production chain. This real-time traceability enhances decision-making, ensures compliance with safety and quality standards, reduces the risk of errors or fraud, and supports sustainable management practices in the dairy sector (Bosona and Gebresenbet, 2023). Blockchain-enabled systems allow farmers to record detailed animal and farm data, including location, breed, health treatments, and sensor-based monitoring. Data collected during milking, milk handling, and transportation, monitored via sensors and Global Positioning System (GPS) for temperature and real-time tracking, is continuously updated, including handling conditions, additives, and lot identification. Regulatory agencies can verify and certify this information on the network, while consumers can access product details via mobile devices, enabling informed dairy purchases (Shingh et al., 2020). Another blockchain-based system has been developed to track dairy products, enabling farmers to log information on animal births, vaccinations, treatments, and locations, thus allowing consumers and stakeholders to verify the product's origins (Marin et al., 2021). Beyond traceability, blockchain supports scenarios that demand rapid decision-making, coordinated management, or efficient allocation of resources, such as food safety oversight, the control of disease outbreaks, or the distribution of government subsidies. Thanks to its immutable, decentralized, transparent, and securely encrypted structure, blockchain offers robust solutions for managing data, strengthening security, building trust, and safeguarding privacy (Yang et al., 2025).

Fig. 3.3 summarizes the main roles of Industry 4.0 technologies in supporting sustainable milk production.

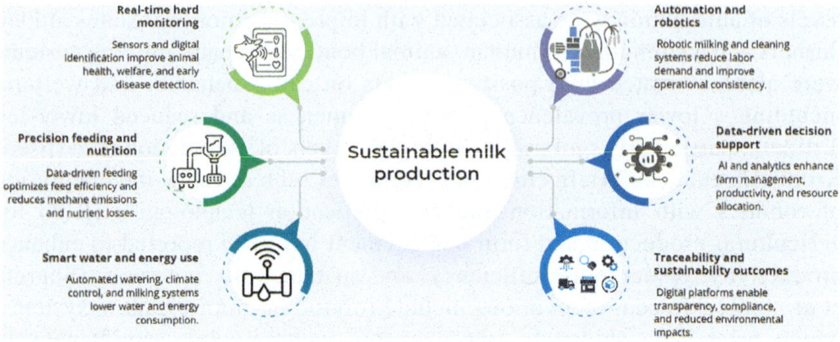

FIGURE 3.3 Key roles of digital and automated technologies in sustainable milk production. Key digital and automated drivers of sustainable milk production, highlighting the roles of real-time herd monitoring, precision feeding, smart water and energy management, automation and robotics, data-driven decision support, and traceability-based sustainability outcomes.

3.4 Role of Industry 4.0 technologies in sustainable milk transformation

Raw milk is transformed into a range of consumable goods, including milk, cheese, butter, yogurt, and ice cream, during the dairy production and processing stages. Milk is first collected at dairy farms from cows, goats, and sheep and then transported to processing plants, where it undergoes essential technological operations to ensure microbiological safety, physicochemical stability, and desired sensory properties. These operations typically include pasteurization, homogenization, fermentation, and coagulation. While homogenization keeps cream from separating and produces smooth, consistent milk, pasteurization gets rid of dangerous bacteria. For fermented dairy products such as yogurt and cheese, fermentation is a critical process driven by specific starter and adjunct bacterial cultures, which play a central role in the development of characteristic aromas, textures, and overall sensory profiles. Dairy production and processing remain a cornerstone of the global food system, playing a critical role in supporting food security, economic development, and human nutrition (Lutsenko et al., 2021; Rudke et al., 2024; Hassoun et al., 2023c).

The dairy sector continues to innovate by integrating sustainable practices and advancing processing technologies to address emerging challenges driven by evolving consumer preferences and growing environmental concerns (Lutsenko et al., 2021; Finnegan and Goggins, 2021). Moreover, various actors in the dairy sector are among the earliest adopters of advanced technologies and automation driven by Industry 4.0. In the dairy sector, Industry 4.0 is primarily enabled by technologies such as AI, robotics, smart sensors, 3D printing, IoT, big data analytics, cloud computing, blockchain, augmented reality, cybersecurity, and digital twins, among others (Hassoun et al., 2023a). These technologies not only drive operational efficiency and automation but also increasingly shape the sustainability performance of dairy value chains. For instance, Bai et al. (2020) assess the sustainability impacts of Industry 4.0 technologies using a UN SDG–based framework and a hybrid decision method. They find that mobile technology has the highest overall sustainability impact, while other technologies (e.g., nanotechnology, simulation, drones) show industry-specific effects.

Digitalization in dairy processing relies on advanced automatic control systems based on microprocessors and microcontrollers, enabling a transition from partial equipment automation to fully automated production lines. Physical processes are continuously mirrored in digital platforms, allowing real-time communication across the value chain and supporting smart, decentralized manufacturing ecosystems (Alcácer and Cruz-Machado, 2019). Automated control systems in milk storage tanks facilitate real-time monitoring of key quality parameters, including protein and fat content and somatic cell count, and support process control during milk normalization (Kozub et al., 2020). Dash et al. (2022) reported that robotics and automation increase process control, save labor costs, and streamline production workflows. In parallel, innovative processing methods such as enzymatic modification, membrane filtration, and micro fluidization create opportunities for waste reduction, improved ingredient functionality, and product innovation. At the organizational level, the integration of digital strategies within collective agri-food systems, as demonstrated in the Parmigiano Reggiano PDO Consortium case, enhances governance, coordination, and system embeddedness through interconnected digital tools (Ciliberti et al., 2024).

IoT technology has been integrated across various areas of manufacturing lines (Akbar et al., 2020). A common application in the dairy supply chain is the use of sensors for real-time monitoring of milk tank levels (Han et al., 2015). By providing real-time data and enhancing traceability, sensor-based technologies improve the efficiency and reliability of the entire dairy sector (Khanna et al., 2022; Tan and Ngan, 2020; Feng, 2017). In cheese production, an IoT- and cloud-based system monitors and controls tanks, tracking temperature, pH, and levels. By enabling remote and autonomous operation, it increases efficiency, reduces waste, and supports sustainable, low-cost production (Piana et al., 2025). Additionally, sensors installed on crates monitor the handling process, ensuring proper hygiene standards and the safe, timely delivery of dairy products from factory storage to distribution centers (Syrmos et al., 2023).

AI can be applied to optimize dairy processes and monitor product quality (Serrano-Torres et al., 2025). Untargeted metabolomics combined with ML has been used to assess the influence of ripening, altimetric zone, and rind inclusion on Parmigiano Reggiano cheese. Ripening was found to strongly influence its chemical profile, with amino acids and lipid derivatives serving as key markers. Random forest models accurately predicted rind levels and authenticated altimetric effects, identifying potential metabolite markers for cheese quality and authenticity (Becchi et al., 2024). Moreover, AI technologies have demonstrated their ability to efficiently identify and remove contaminated products from the market, with applications spanning food processing, smart packaging, and traceability (Nath et al., 2024). An AI-based CNN method was developed to detect melamine and monitor yogurt shelf life, achieving over 91% accuracy for adulteration and 96% for freshness. This rapid, noninvasive approach enhances food safety, reduces chemical testing, minimizes waste, and improves sustainable resource use in dairy production (Pérez-Calabuig et al., 2025).

In another study, a multilayer perceptron ANN model was specifically developed to assess how accurately AI technologies can predict a dairy factory's energy consumption and production efficiency. Production data were collected from the pasteurization process, during which the facility met its energy demands through a combination of lignite coal for boiler operation, electricity for machinery, and steam for heat transfer. Output energy was determined by subtracting the total energy consumed by machines and equipment from the total energy input (Oztuna Taner and Çolak, 2024). Furthermore, a neural network integrated with potentiometric and gas sensor arrays was shown to generate distinct response patterns for fresh, sour, and spoiled milk samples, enabling milk quality classification with an accuracy of 83%, as demonstrated by Putra et al. (2018).

Digital twins and simulation tools play a critical role in dairy milk transformation by enabling continuous synchronization between physical processes and their digital representations. Updated in real time with operational data, digital twins support state monitoring, behavior prediction, and informed decision-making across different time horizons. In dairy processing, they have been applied to optimize Cleaning-In-Place operations by monitoring biofilm formation in heat exchangers and identifying potential microbiological risks, thereby ensuring consistent hygiene, reducing downtime, and improving water and energy efficiency (Abdurrahman and Ferrari, 2025). In addition, the integration of digital twins with big data analytics enables the transformation of conventional in-plant milk-run logistics into CPS, supporting real-time scheduling, routing, and resource planning. This approach has been shown to enhance operational efficiency, flexibility, and energy performance while reducing emissions and downtime compared with conventional methods (Akkad and Bányai, 2023).

One of the most critical stages in dairy product manufacturing is packaging, which plays a key role in protecting products after production and can effectively extend their shelf life (Mirza Alizadeh et al., 2021). Two promising technologies for enhancing food safety and quality are active and intelligent packaging. Active packaging incorporates additives into the packaging material to maintain or improve product quality and shelf life, while intelligent packaging relies on advanced technologies to monitor the condition of the packed food, providing valuable information on its quality during storage and transportation (Shukla et al., 2024).

Intelligent packaging serves as a powerful tool to protect dairy products while also informing consumers about the product's history. For dairy items, intelligent packaging can include indicators (for time, temperature, gas, integrity, and freshness), sensors, and data carriers such as Radio Frequency Identification (RFID) tags or barcodes (Mirza Alizadeh et al., 2021). These packaging materials are often biodegradable and derived from plant-based or food-derived extracts. Nonmigratory active packaging systems function through mechanisms such as oxygen scavenging, moisture absorption, or ethylene removal, whereas active release systems are designed to emit compounds such as carbon dioxide or to deliver antimicrobial and antioxidant agents to the packaged product (Ahmed et al., 2022). Additionally, incorporating nanoparticles such as TiO_2, ZnO, and CuO into sustainable, active packaging made from agri-food waste offers significant potential to improve sustainability, enhance food safety, reduce costs, and provide added benefits, including antibacterial activity, antioxidant properties, and ultraviolet protection (Baraketi and Khwaldia, 2024).

Smart packaging integrates advanced technologies to enhance product quality, safety, and shelf life, offering substantial potential for innovation in the dairy sector. Freshness indicators designed to detect microbial growth and related chemical changes can monitor and communicate food quality during storage and transport. When incorporated into packaging materials, these indicators respond to metabolites and other by-products of microbial activity, producing visible signals that reflect the product's microbiological status (Mkhari et al., 2025). In addition, two intelligent packaging prototypes were developed for cow's milk using high-density polyethylene (HDPE) bottles containing bromothymol blue as a pH indicator, where color changes were triggered by pH variations (Romero et al., 2021). Beyond freshness sensing, smart packaging solutions such as embedded temperature sensors and time–temperature indicators (TTIs) support safer handling of milk and cream during distribution and retail by recording temperature history across the cold-chain. These tools are particularly valuable for identifying temperature excursions that accelerate spoilage or promote microbial growth in highly perishable dairy products (Mkhari et al., 2025).

3.5 Role of Industry 4.0 technologies in sustainable milk distribution

The distribution of milk and dairy products represents one of the most critical stages of the dairy value chain, owing to the high perishability of both raw and processed products such as yogurt, cheese, and ice cream, the strict cold-chain requirements, and the environmental impacts associated with transportation, storage, and packaging (Finnegan and Goggins, 2021; Augustin et al., 2013).

Inefficiencies at this stage can result in significant product losses, increased energy consumption, and higher greenhouse gas emissions (Mangla et al., 2019). Industry 4.0 technologies, including robotics, AI, IoT, big data, and blockchain, provide advanced solutions to these challenges, enabling more efficient and sustainable milk distribution. The full integration of these technologies promises to enhance automation, optimize operational processes, and support environmentally responsible practices across the dairy industry (Hassoun et al., 2023c; Malik et al., 2025).

The use of IoT sensors in refrigerated vehicles, storage facilities, and distribution centers allows for continuous, real-time monitoring of key parameters such as temperature, humidity, vibration, oxygen, gas levels, light, GPS, and transit time. This monitoring enables the early detection of deviations from optimal storage conditions, helping to minimize spoilage and enhance food safety. When integrated with CPS, these technologies can automatically trigger corrective actions, ensuring consistent compliance with quality standards throughout the entire distribution process (Wu et al., 2021; Aamer et al., 2025).

AI and big data analytics play a central role in optimizing dairy logistics in the distribution and delivery of dairy products (Serrano-Torres et al., 2025; Neethirajan, 2023). By analyzing large volumes of data related to demand patterns, traffic conditions, fuel consumption, and delivery schedules, AI-driven tools enable dynamic route planning and load optimization. These approaches significantly reduce transportation distances, fuel use, and carbon emissions, while improving delivery reliability and cost efficiency (Nguyen and Dao, 2015). Huerta-Soto et al. (2023) reported that AI- and ML-based optimization approaches are increasingly applied across the dairy supply chain. Supported by cloud computing and shared data systems, these tools enhance inventory control, logistics, and storage management while reducing waste and product spoilage.

Furthermore, blockchain-based traceability systems significantly enhance transparency, accountability, and trust across dairy distribution networks. In India, the integration of blockchain with smart contracts, IoT technologies, and QR codes has led to the development of an advanced dairy supply chain platform that strengthens traceability, prevents contamination and counterfeiting, preserves nutritional quality, and improves farm profitability (Khanna et al., 2022). Similarly, in Kenya, a blockchain-based model was developed to improve milk delivery systems through the establishment of Milk Delivery Blockchain Manager (MDBM), a decentralized platform that records milk delivery data on the blockchain, ensuring data integrity and providing farmers with secure access to delivery and payment information (Dorothy and Fredrick, 2020).

Blockchain-enabled technologies strengthen milk distribution by securely recording immutable data on product origin, processing, transport conditions, and handling events, thereby supporting regulatory compliance, rapid incident response, and precise product recalls that reduce food waste and economic losses. By improving cold-chain integrity, enhancing food safety, and lowering environmental impacts, these systems increase the resilience and sustainability of dairy supply chains. In addition, blockchain integration further optimizes operations by reducing monitoring and transaction costs, minimizing delays, accelerating deliveries, and lowering transport expenses, while enabling logistics providers to track transportation conditions in real time and retailers to update product availability and sales data efficiently (Makarov et al., 2019; Vargas-Bello-Pérez et al., 2025).

3.6 Role of Industry 4.0 technologies in sustainable milk consumption

Sustainable milk consumption is increasingly shaped by consumer expectations regarding food safety, nutritional quality, environmental responsibility, and ethical production practices (Aït-Kaddour et al., 2024). In China, an emerging economy, consumers generally exhibit limited awareness of sustainable dairy products and tend to associate sustainability primarily with environmental or "ecofriendly" attributes. Nevertheless, empirical evidence indicates a willingness to pay a price premium of approximately 40% for sustainable milk compared with conventional alternatives, particularly among consumers who perceive sustainable milk as higher in quality and among households with children (Gao et al., 2020). In addition, the adoption of Industry 4.0 technologies such as blockchain, mobile applications, and e-commerce platforms has helped bridge the gap between producers and consumers by enhancing transparency, building trust, and enabling more responsible consumption behaviors. This indicates that transparency can effectively promote ethical consumption when information is clear, accessible, and credible (Tolegenov et al., 2024).

Digital traceability platforms, supported by blockchain technology and smart labeling systems such as QR codes or near-field communication (NFC), provide consumers with direct access to detailed information about dairy products (El Matbouly et al., 2022; Wamba and Wicks, 2010). This includes data on farm origin, animal welfare practices, processing methods, nutritional composition, and environmental impacts such as carbon and water footprints (Lukacs et al., 2025; Camel et al., 2024). Access to such information empowers consumers to make informed purchasing decisions and encourages demand for sustainably produced dairy products. Moreover, smart packaging technologies, including digital freshness indicators, play an important role in reducing food waste at the consumer level. Industry 4.0 technologies, such as AI, robotics, big data analytics, IoT, smart sensors, 3D printing, cybersecurity, and blockchain, are increasingly being adopted within the food packaging sector to enhance efficiency, safety, and sustainability (Hassoun et al., 2023b; Aït-Kaddour et al., 2024).

TTIs and sensor-based packaging provide consumers with real-time information on product freshness, enabling better storage and consumption decisions. By doing so, these technologies help reduce the premature disposal of safe dairy products and improve resource efficiency at the household level. Intelligent sensors integrated into active packaging systems can monitor changes in food composition and the packaging micro-environment, such as pH and temperature fluctuations or the accumulation of nitrogenous volatile compounds, to preserve or enhance sensory attributes, safety, and overall product quality (Amin et al., 2022). In contrast, intelligent packaging primarily relies on internal or external sensors to track and communicate information related to a product's condition, history, and safety throughout the supply chain (Trinh et al., 2023).

Industry 4.0 technologies are driving personalized, health-oriented consumption through data-driven nutrition platforms and mobile applications (Hassoun and Galanakis, 2025). By combining consumer preferences, dietary needs, and health data, these tools offer tailored guidance on dairy intake, portion sizes, and product choices, supporting healthier eating habits while promoting sustainability (Kuosmanen et al., 2025). Advances such as AI, 3D food printing, augmented reality, big data analytics, and IoT are transforming the food sector by enabling precision nutrition and personalized diets, supporting the production of functional foods tailored to individual needs, enhancing consumer engagement, and optimizing food monitoring and health-related interventions through data-driven decision making (Hassoun and Galanakis, 2025; Theodore Armand et al., 2024).

Consumer feedback systems enabled by digital platforms allow dairy companies to collect real-time data on consumer satisfaction, preferences, and consumption trends (Serrano-Torres et al., 2025; Malik et al., 2025). This information feeds back into production and distribution planning, enabling better alignment between supply and demand, reducing overproduction, and minimizing environmental impacts. For example, in Ukraine, a consumer behavior model for the milk market was developed via a survey of 450 domestic milk consumers using a 23-item questionnaire. Results show that taste (95%) and naturalness (85%) are top priorities, 49% consider product variety sufficient, while brand familiarity and consistent shelf availability each matter to 25% of respondents (Kulish et al., 2024).

3.7 Role of Industry 4.0 technologies in sustainable dairy waste management

Dairy production and processing generate a wide range of waste streams, including manure, whey, wastewater, sludge, packaging materials, and food losses. If not properly managed, these wastes can lead to significant environmental pressures, such as GHG emissions, nutrient pollution of water bodies, high energy consumption, and inefficient use of valuable resources.

(Udourioh et al., 2025). The adoption of Industry 4.0 technologies provides advanced tools to improve the sustainability, efficiency, and circularity of dairy waste management (Hassoun et al., 2023c).

Smart sensors and IoT technologies enable continuous monitoring of waste generation and treatment processes. Critical parameters, such as organic load, nutrient concentrations, biogas output, energy consumption, and water quality can be tracked in real time. Li et al. (2025) showed that IoT technologies enable low-cost, real-time monitoring of biogas production and greenhouse gas emissions from livestock waste by integrating IoT systems with anaerobic reactors, with data transmitted to mobile devices via the ThingSpeak platform (MathWorks, MATLAB®). During milk processing for the production of cheese, yogurt, butter, and cream, several coproducts are generated, most notably whey and buttermilk. Real-time monitoring of galactooligosaccharide production from cheese whey has been achieved using Fourier-transform infrared (FTIR) spectroscopy sensors combined with chemometric approaches. In addition, various spectroscopy-based sensors can be applied to monitor thermal treatments of dairy products, thereby improving their shelf life, as reported by Aït-Kaddour et al. (2021).

AI and big data analytics enhance the efficiency of waste treatment by optimizing operating conditions in processes such as anaerobic digestion, wastewater treatment, and nutrient recovery systems. Their ability to manage complex, nonlinear, and high-dimensional data improves process control and predictive performance. In addition, soft sensors that integrate microbial, electrochemical, optical, and hybrid technologies enable adaptive, real-time monitoring and control, thereby improving the stability and performance of anaerobic fermentation processes (Marycz et al., 2025). Barik and Murugan (2015) evaluated biogas production from mixtures of Karanja seed cake and cattle dung using pure cattle dung as a control. After 30 days, the mixture containing 25% seed cake and 75% cattle dung produced the highest biogas yield. Moreover, ANN combined with a genetic algorithm successfully predicted and optimized biogas production, highlighting their strong potential for monitoring and optimizing large-scale anaerobic digestion systems. Similarly, data-driven optimization approaches can substantially reduce methane and nitrous oxide emissions, enhance energy recovery from biogas, and promote the reuse of treated water and by-products. In practice, optimized anaerobic digestion systems are capable of reducing methane emissions by up to 50% while generating renewable energy that offsets fossil fuel use. Duan et al. (2025) demonstrated that a Deep Belief Network (DBN) integrated with the Boosted Osprey Optimization Algorithm (BOOA) outperformed other models, delivering higher predictive accuracy and optimization efficiency. The DBN–BOOA framework identified optimal operating conditions that increased biogas production to 31.35 m^3/min, surpassing the performance of alternative approaches.

Digital twins allow dairy processors to simulate different waste management scenarios, such as alternative treatment technologies or valorization pathways for whey and organic residues. In Argentina, computational simulation and digital twins were applied in a small- and medium-sized dairy enterprise to support the introduction of two new cheese products while overcoming constraints related to filling speed, accuracy, and labor costs in a semimanual process. Seven alternative production scenarios combining operators and collaborative robots were evaluated, allowing the selection of solutions that improved productivity, reduced operating time, lowered costs, and reduced waste (Salimbeni et al., 2022). These virtual models support strategic decision-making by identifying solutions that reduce environmental impacts while maximizing resource efficiency and economic returns. From a sustainability perspective, digital twins play a key role in optimizing resource use and improving the efficiency of food processing. They also serve as versatile tools for advancing sustainability initiatives and shaping the future evolution of the food processing industry (Abdurrahman and Ferrari, 2025).

Blockchain-based traceability systems strengthen accountability and trust in waste handling and valorization processes. By securely monitoring waste quantities, treatment methods, and by-product utilization, these systems support certification, sustainability reporting, and collaboration within circular economy initiatives. Hawashin et al. (2025) reported that a private Ethereum-based blockchain ensures accountability, traceability, security, and transparency throughout the supply chain. Leveraging smart contracts and event-based recording, it creates an immutable record of actions and data, fostering more sustainable dairy production. In Vietnam, reliable traceability solutions for the dairy sector were developed using blockchain technology to improve traceability and food safety for domestic dairy products, and consequently reduce waste (Tan and Ngan, 2020). Similarly, other studies have applied blockchain technology to enhance the dairy supply chain by improving traceability, reducing the risk of fraud, increasing the valorization of coproducts, and consequently minimizing waste (Casino et al., 2021; Khanna et al., 2022). Overall, Industry 4.0 technologies are shifting dairy waste management toward resource-efficient, circular practices. By reducing emissions, improving energy and water use, and valorizing by-products, they enhance environmental performance and sustainability (Hassoun et al., 2024).

3.8 Conclusion

The integration of Industry 4.0 technologies across the dairy value chain has demonstrated significant potential to enhance sustainability, efficiency, and resilience in dairy production, processing, distribution, consumption, waste management, and by-products valorization. From PLF and automated milking to AI-driven processing, smart packaging, IoT-enabled logistics, and blockchain-based traceability, digital and CPS are reshaping traditional practices into data-driven, environmentally conscious, and consumer-oriented models. These technologies enable optimized resource use, reduced GHG emissions, improved product quality and safety, and strengthened supply chain transparency, thereby supporting both environmental and economic sustainability.

Looking forward, the dairy sector must continue to embrace digital transformation while addressing challenges related to technology adoption, data interoperability, cybersecurity, cost barriers, and skill development. Future research and innovation should focus on integrating AI, digital twins, IoT, and blockchain with circular economy approaches, renewable energy, and bio-based packaging solutions to create fully sustainable dairy systems. Additionally, consumer engagement through smart labeling, personalized nutrition, and transparent product information will play a key role in promoting responsible consumption and closing the sustainability loop. By fostering collaboration among farmers, processors, technology providers, policymakers, and consumers, the dairy sector can leverage Industry 4.0 not only to enhance productivity and profitability but also to contribute meaningfully to climate resilience, food security, and global sustainability goals.

Acknowledgments

The authors express their sincere appreciation to Professor Moussa Houhamdi for his valuable guidance and constructive feedback on this book chapter.

Funding

This research is part of "Climate-smart livestock production in Arab countries (CSL4MENA)" project, which is funded by the Arab Research & Innovation Co-Funded Alliances (ARICA), the Algerian Ministry of Higher Education and Scientific Research, and the Directorate General for Scientific Research, and Technological Development (DGRSDT).

AI disclosure

During the preparation of this chapter, the authors used ChatGPT solely for language editing purposes, including grammar checking, reformulation, and paraphrasing. The content was fully written, reviewed, and approved by the authors, who take full responsibility for the final version of the chapter

References

Aamer, A.M., Al-Awlaqi, M.A., Rausyan Fikri, M., 2025. Smart food logistics: Design and test of an IoT-based food traceability system. International Journal of Logistics Research and Applications 1–20. Available at https://doi.org/10.1080/13675567.2025.2531796.

Abdurrahman, E.E.M., Ferrari, G., 2025. Digital Twin applications in the food industry: A review. Frontiers in Sustainable Food Systems 9, 2025. Available at https://doi.org/10.3389/fsufs.2025.1538375.

Ahmed, M.W., Haque, M.A., Mohibbullah, M., Khan, M.S.I., Islam, M.A., Mondal, M.H.T., et al., 2022. A review on active packaging for quality and safety of foods: Current trends, applications, prospects and challenges. Food Packaging and Shelf Life 33, 100913. Available at https://doi.org/10.1016/j.fpsl.2022.100913.

Aït-Kaddour, A., Boudalia, S., Özoğul, F., Çiçek, S., Cahyana, Y., Bhat, Z.F., 2024. Chapter 3—Food Quality 4.0: Contribution to sustainability. In: Hassoun, A. (Ed.), Food Industry 4.0. Academic Press, pp. 35–59. Available at https://doi.org/10.1016/B978-0-443-15516-1.00003-7.

Aït-Kaddour, A., Hassoun, A., Bord, C., Schmidt-Filgueras, R., Biancolillo, A., Di Donato, F., et al., 2021. Application of spectroscopic techniques to evaluate heat treatments in milk and dairy products: An overview of the last decade. Food and Bioprocess Technology 14, 781–803. Available at https://doi.org/10.1007/s11947-021-02607-0.

Akbar, M.O., Shahbaz khan, M.S., Ali, M.J., Hussain, A., Qaiser, G., Pasha, M., et al., 2020. IoT for development of smart dairy farming. Journal of Food Quality 2020, 4242805. Available at https://doi.org/10.1155/2020/4242805.

Akkad, M.Z., Bányai, T., 2023. Energy consumption optimization of milk-run-based in-plant supply solutions: An Industry 4.0 approach. Processes 11, 799. Available at https://doi.org/10.3390/pr11030799.

Alcácer, V., Cruz-Machado, V., 2019. Scanning the Industry 4.0: A literature review on technologies for manufacturing systems. Engineering Science and Technology, an International Journal 22, 899–919. Available at https://doi.org/10.1016/j.jestch.2019.01.006.

Amin, U., Khan, M.K.I., Maan, A.A., Nazir, A., Riaz, S., Khan, M.U., et al., 2022. Biodegradable active, intelligent, and smart packaging materials for food applications. Food Packaging and Shelf Life 33, 100903. Available at https://doi.org/10.1016/j.fpsl.2022.100903.

Augustin, M.A., Udabage, P., Juliano, P., Clarke, P.T., 2013. Towards a more sustainable dairy industry: Integration across the farm–factory interface and the dairy factory of the future. International Dairy Journal 31, 2–11. Available at https://doi.org/10.1016/j.idairyj.2012.03.009.

Bagci, A., Sogut, Y., Bozatli, O., Degirmenci, T., 2025. Policy support and agricultural greenhouse gas emissions in BRICS-T countries: The role of financial development, markets, and institutions. Borsa Istanbul Review 25, 886–894. Available at https://doi.org/10.1016/j.bir.2025.05.004.

Bai, C., Dallasega, P., Orzes, G., Sarkis, J., 2020. Industry 4.0 technologies assessment: A sustainability perspective. International Journal of Production Economics 229, 107776. Available at https://doi.org/10.1016/j.ijpe.2020.107776.

Baraketi, S., Khwaldia, K., 2024. Nanoparticles from agri-food by-products: Green technology synthesis and application in food packaging. Current Opinion in Green and Sustainable Chemistry 49, 100953. Available at https://doi.org/10.1016/j.cogsc.2024.100953.

Barik, D., Murugan, S., 2015. An artificial neural network and genetic algorithm optimized model for biogas production from co-digestion of seed cake of karanja and cattle dung. Waste and Biomass Valorization 6, 1015–1027. Available at https://doi.org/10.1007/s12649-015-9392-1.

Becchi, P.P., Rocchetti, G., García-Pérez, P., Michelini, S., Pizzamiglio, V., Lucini, L., 2024. Untargeted metabolomics and machine learning unveil quality and authenticity interactions in grated Parmigiano Reggiano PDO cheese. Food Chemistry 447, 138938. Available at https://doi.org/10.1016/j.foodchem.2024.138938.

Bojovic, M., McGregor, A., 2023. A review of megatrends in the global dairy sector: What are the socioecological implications? Agriculture and Human Values 40, 373–394. Available at https://doi.org/10.1007/s10460-022-10338-x.

Bosona, T., Gebresenbet, G., 2023. The role of blockchain technology in promoting traceability systems in agri-food production and supply chains. Sensors 23, 5342. Available at https://doi.org/10.3390/s23115342.

Boudalia, S., Ben Said, S., Tsiokos, D., Bousbia, A., Gueroui, Y., Mohamed-Brahmi, A., et al., 2020. BOVISOL Project: Breeding and management practices of indigenous bovine breeds: Solutions towards a sustainable future. Sustainability 12, 9891. Available at https://doi.org/10.3390/su12239891.

Boudebbouz, A., Bousbia, A., Khelifa, R., Boussadia, M.I., Ben Chabane, A., Boumendjel, L., et al., 2023. Milk quality and production under climate change uncertainty: Case of the Algerian cattle breed. Journal of Agriculture and Environment for International Development (JAEID) 117, 123–142. Available at https://doi.org/10.36253/jaeid-15000.

Bousbia, A., Boudalia, S., Gueroui, Y., Belaize, B., Meguelati, S., Amrouchi, M., et al., 2018. Nutritional and hygienic quality of raw milk intended for consumption in the region of Guelma, Algeria. Asian Journal of Dairy and Food Research 37, 192–196. Available at https://doi.org/10.18805/ajdfr.DR-123.

Bousbia, A., Boudalia, S., Gueroui, Y., Hadded, K., Bouzaoui, A., Kiboub, D., et al., 2021. Use of multivariate analysis as a tool in the morphological characterization of the main indigenous bovine ecotypes in northeastern Algeria. PLoS One 16, e0255153. Available at https://doi.org/10.1371/journal.pone.0255153.

Brini, R., 2021. Renewable and non-renewable electricity consumption, economic growth and climate change: Evidence from a panel of selected African countries. Energy 223, 120064. Available at https://doi.org/10.1016/j.energy.2021.120064.

Brito, L.F., Heringstad, B., Klaas, I.C., Schodl, K., Cabrera, V.E., Stygar, A., et al., 2025. Invited review: Using data from sensors and other precision farming technologies to enhance the sustainability of dairy cattle breeding programs. Journal of Dairy Science 108 (10), 10447–10474. Available at https://doi.org/10.3168/jds.2025-26554.

Camel, A., Belhadi, A., Kamble, S., Tiwari, S., Touriki, F.E., 2024. Integrating smart Green Product Platforming for carbon footprint reduction: The role of blockchain technology and stakeholders influence within the agri-food supply chain. International Journal of Production Economics 272, 109251. Available at https://doi.org/10.1016/j.ijpe.2024.109251.

Casino, F., Kanakaris, V., Dasaklis, T.K., Moschuris, S., Stachtiaris, S., Pagoni, M., et al., 2021. Blockchain-based food supply chain traceability: A case study in the dairy sector. International Journal of Production Research 59, 5758–5770. Available at https://doi.org/10.1080/00207543.2020.1789238.

Chen, Y., Lu, H., Wu, H., Wang, J., Lyu, N., 2023. Global desert variation under climatic impact during 1982–2020. Science China Earth Sciences 66, 1062–1071. Available at https://doi.org/10.1007/s11430-022-1052-1.

Christensen, J.H., Hewitson, B., Busuioc, A., 2007. Regional climate projections. In: Solomon, S., Qin, D., Manning, M., Chen, Z., Marquis, M., Averyt, K.B., Tignor, M., Miller, H.L. (Eds.), 'Climate change 2007: the physical science basis. Contribution of Working Group I to the Fourth Assessment Report of the Intergovernmental Panel on Climate Change. Cambridge University Press, Cambridge, UK, pp. 847–940.

Ciliberti, S., Frascarelli, A., Polenzani, B., Brunori, G., Martino, G., 2024. Digitalisation strategies in the agri-food system: The case of PDO Parmigiano Reggiano. Agricultural Systems 218, 103996. Available at https://doi.org/10.1016/j.agsy.2024.103996.

Climate Watch, 2021. Historical GHG emissions. World Resources Institute, Washington, DC. Available online. Available at https://www.climatewatchdata.org/ghg-emissions (accessed on 30 May 2022).

Cruz-Rivero, L., Hernández, E.A., Lince-Olguín, E., Mar-Orozco, C.E., López-García, S.A., Cruz-Martínez, P.Y., 2025. Life Cycle assessment applied to milk production and processing: An integrative systematic literature review. Sustainability 17, 1615. Available at https://doi.org/10.3390/su17041615.

Dac, H.H., Gonzalez Viejo, C., Lipovetzky, N., Tongson, E., Dunshea, F.R., Fuentes, S., 2022. Livestock identification using deep learning for traceability. Sensors 22, 8256. Available at https://doi.org/10.3390/s22218256.

Dash, K.K., Fayaz, U., Dar, A.H., Shams, R., Manzoor, S., Sundarsingh, A., et al., 2022. A comprehensive review on heat treatments and related impact on the quality and microbial safety of milk and milk-based products. Food Chemistry Advances 1, 100041. Available at https://doi.org/10.1016/j.focha.2022.100041.

Dorothy, R. & Fredrick, M.A. (2020). Blockchain based milk delivery platform for stallholder dairy farmers in Kenya: Enforcing transparency and fair payment. In *2020 IST-Africa Conference (IST-Africa)* (pp. 1–6). IEEE.

Duan, J., Cao, G., Ma, G., Yazdani, B., 2025. Boosting biogas production through innovative data-driven modeling and optimization methods at NJWTP. Scientific reports 15, 4814. Available at https://doi.org/10.1038/s41598-025-88337-1.

El Matbouly, H., Nikbakhtnasrabadi, F., Dahiya, R., 2022. RFID Near-field Communication (NFC)-Based Sensing technology in food quality control. In: Chandra, P. (Ed.), Biosensing and Micro-Nano Devices: Design Aspects and Implementation in Food Industries. Springer Nature Singapore, Singapore, pp. 219–241. Available at https://doi.org/10.1007/978-981-16-8333-6_9.

Erden, H., Kaya, İ.A., & Çamaşircioğlu, E. (2015). Livestock monitoring system. In 2015 Fourth International Conference on Agro-Geoinformatics (Agro-geoinformatics) (pp. 73–76). https://doi.org/10.1109/Agro-Geoinformatics.2015.7248081.

Et-taibi, B., Abid, M.R., Boufounas, E.M., Morchid, A., Bourhnane, S., Abu Hamed, T., et al., 2024. Enhancing water management in smart agriculture: A cloud and IoT-Based smart irrigation system. Results in Engineering 22, 102283. Available at https://doi.org/10.1016/J.RINENG.2024.102283.

FAO-GDP (2018). Climate change and the global dairy cattle sector—The role of the dairy sector in a low-carbon future. Rome. 36 pp. Licence: CC BY-NC-SA- 3.0 IGO.

FAO (2013). Climate-smart agriculture: Sourcebook. Retrieved from: http://www.fao.org/3/i3325e/i3325e00.htm, Rep. No. 9251077207.

FAO (2024). Food and Agriculture Organization of the United Nations (FAO): FAOSTAT statistical database. Retrieved from: https://www.fao.org/faostat/ .

Feng, T. (2017). A supply chain traceability system for food safety based on HACCP, blockchain & Internet of things. In *2017 International Conference on Service Systems and Service Management* (pp. 1–6). https://doi.org/10.1109/icsssm.2017.7996119.

Filonchyk, M., Peterson, M.P., Zhang, L., Hurynovich, V., He, Y., 2024. Greenhouse gases emissions and global climate change: Examining the influence of CO2, CH4, and N2O. Science of the total environment 935, 173359. Available at https://doi.org/10.1016/j.scitotenv.2024.173359.

Finnegan, W., Goggins, J., 2021. Chapter 6—Environmental impact of the dairy industry. In: Galanakis, C.M. (Ed.), Environmental impact of agro-food industry and food consumption. Academic Press, pp. 129–146. Available at https://doi.org/10.1016/B978-0-12-821363-6.00004-7.

Gao, Z., Li, C., Bai, J., Fu, J., 2020. Chinese consumer quality perception and preference of sustainable milk. China Economic Review 59, 100939. Available at https://doi.org/10.1016/j.chieco.2016.05.004.

Gauly, M., Ammer, S., 2020. Review: Challenges for dairy cow production systems arising from climate changes. Animal 14, s196–s203. Available at https://doi.org/10.1017/S1751731119003239.

Gerrits, B., Mes, M., Schuur, P., & Andringa, R. (2022). A Simulation Model for Cooperative Robotics in Dairy Farms. In *2022 Winter Simulation Conference (WSC)* (pp. 831–842). https://doi.org/10.1109/wsc57314.2022.10015409.

Grzesiak, W., Zaborski, D., Sablik, P., Żukiewicz, A., Dybus, A., Szatkowska, I., 2010. Detection of cows with insemination problems using selected classification models. Computers and Electronics in Agriculture 74, 265–273. Available at https://doi.org/10.1016/j.compag.2010.09.001.

Gueboudji, Z., & Gaouar, S.B. S. (2026). Epigenetics and microbiome dynamics in dairy animal productivity. In Semir Bechir Suheil Gaouar et al. (Eds), *Management and productivity of dairy animals under African breeding conditions*. Springer Nature. https://doi.org/10.1007/978-3-032-15624-2.

Gupta, A.D., Pandey, P., Feijóo, A., Yaseen, Z.M., Bokde, N.D., 2020. Smart water technology for efficient water resource management: A review. Energies 13 (23), 6268. Available at https://doi.org/10.3390/EN13236268.

Han, Y., Feng, Y., Liu, L., Ju, J., Wang, Z., 2015. IOT-enabled quality management process innovation and analytics in China's. Dairy Industry: A Data Flow Modeling Perspective.

Hansen, B.G., Bugge, C.T., Skibrek, P.K., 2020. Automatic milking systems and farmer wellbeing–exploring the effects of automation and digitalization in dairy farming. Journal of Rural Studies 80, 469–480. Available at https://doi.org/10.1016/j.jrurstud.2020.10.028.

Hassoun, A., Galanakis, C.M., 2025. Industry 4.0 technologies promote healthy and nutritious food. Discover Food 5, 333. Available at https://doi.org/10.1007/s44187-025-00639-5.

Hassoun, A., Aït-Kaddour, A., Abu-Mahfouz, A.M., Rathod, N.B., Bader, F., Barba, F.J., et al., 2023a. The fourth industrial revolution in the food industry—Part I: Industry 4.0 technologies. Critical Reviews in Food Science and Nutrition 63, 6547–6563. Available at https://doi.org/10.1080/10408398.2022.2034735.

Hassoun, A., Boukid, F., Ozogul, F., Aït-Kaddour, A., Soriano, J.M., Lorenzo, J.M., et al., 2023b. Creating new opportunities for sustainable food packaging through dimensions of industry 4.0: New insights into the food waste perspective. Trends in Food Science & Technology 142, 104238. Available at https://doi.org/10.1016/j.tifs.2023.104238.

Hassoun, A., Garcia-Garcia, G., Trollman, H., Jagtap, S., Parra-López, C., Cropotova, J., et al., 2023c. Birth of dairy 4.0: Opportunities and challenges in adoption of fourth industrial revolution technologies in the production of milk and its derivatives. Current Research in Food Science 7, 100535. Available at https://doi.org/10.1016/j.crfs.2023.100535.

Hassoun, A., Tarchi, I., Aït-Kaddour, A., 2024. Leveraging the potential of fourth industrial revolution technologies to reduce and valorize waste and by-products in the dairy sector. Current Opinion in Green and Sustainable Chemistry 47, 100927. Available at https://doi.org/10.1016/j.cogsc.2024.100927.

Hawashin, D., Salah, K., Jayaraman, R., Ahmad, R.W., Kanhere, S., Damiani, E., 2025. Enhancing sustainability in dairy industry: Blockchain-based waste reduction. Sustainable Futures 9, 100529. Available at https://doi.org/10.1016/j.sftr.2025.100529.

Horváthné Kovács, B., Zörög, Z., 2025. Digital livestock farming in climate-smart agriculture: An overview to advance the SDGs. Discover Sustainability 6, 940. Available at https://doi.org/10.1007/s43621-025-01866-7.

Huerta-Soto, R., Ramirez-Asis, E., Tarazona-Jiménez, J., Nivin-Vargas, L., Norabuena-Figueroa, R., Guzman-Avalos, M., et al., 2023. RETRACTED: Predictable inventory management within dairy supply chain operations. International Journal of Retail & Distribution Management 53, 1–17. Available at https://doi.org/10.1108/ijrdm-01-2023-0051.

IPCC, 2014. Climate Change 2014: Synthesis Report. Contribution of Working Groups I, II and III to the Fifth. In: Core Writing Team, Pachauri, R.K., Meyer, L.A. (Eds.), Assessment Report of the Intergovernmental Panel on Climate Change. IPCC, Geneva, Switzerland, pp. 151.

IPCC, 2021. In: Masson-Delmotte, V., Zhai, P., Pirani, A., Connors, S.L., Péan, C., Berger, S. (Eds.), Climate Change 2021: The physical science basis. Contribution of Working Group I to the Sixth Assessment Report of the Intergovernmental Panel on Climate Change. IPCC.

Kaswan, S., Chandratre, G.A., Upadhyay, D., Sharma, A., Sreekala, S.M., Badgujar, P.C., et al., 2024. CHAPTER 4—Applications of sensors in livestock management☆ In: Tarafdar, A., Pandey, A., Gaur, G.K., Singh, M., Pandey, H.O. (Eds.), Engineering applications in livestock production. Academic Press, pp. 63–92. Available at https://doi.org/10.1016/B978-0-323-98385-3.00004-9.

Kazi, K.S.L., 2025. IoT technologies for the intelligent dairy industry: A new challenge. In: Thandekkattu, S.G., Vajjhala, N.R. (Eds.), Designing sustainable Internet of Things solutions for smart industries. IGI Global Scientific Publishing, Hershey, PA, USA, pp. 321–350. Available at https://doi.org/10.4018/979-8-3693-5498-8.ch012.

Kebreab, E., Pressman, E.M., Lucas, K.R.G., 2026. Review: Sustainable Ruminant Production: Circular systems, nutrition, and methane mitigation. Animal 101772. Available at https://doi.org/10.1016/j.animal.2026.101772.

Khanna, A., Jain, S., Burgio, A., Bolshev, V., Panchenko, V., 2022. Blockchain-enabled supply chain platform for Indian dairy industry: Safety and traceability. Foods 11, 2716. Available at https://doi.org/10.3390/foods11172716.

Kokemohr, L., Mittenzwei, K., Kuhn, T., 2026. Greenhouse gas abatement costs of Norwegian dairy farms. Agricultural Systems 233, 104592. Available at https://doi.org/10.1016/j.agsy.2025.104592.

Kozina, A.M., Semkiv, L.P., 2020. Sustainable development of dairy farming through the use of digital technologies. IOP Conference Series: Earth and Environmental Science 613, 012061. Available at https://doi.org/10.1088/1755-1315/613/1/012061.

Kozub, Y.A., Komlatsky, V.I., Khoroshailo, T.A., 2020. About some automated processes in the production of dairy products. IOP Conference Series: Materials Science and Engineering 862, 032021. Available at https://doi.org/10.1088/1757-899x/862/3/032021.

Kulish, T., Sokil, Y., Legeza, D., Sokil, O., Budnikevich, I., Diyora, B., 2024. Digitalization of consumers' behavior model in the dairy market. In: In: Semenov, A., Yepifanova, I., Kajanová, J. (Eds.), Data-centric business and applications: Modern trends in financial and innovation data processes 2023 Volume 1. Springer Nature Switzerland, Cham, pp. 187–205. Available at https://doi.org/10.1007/978-3-031-54012-7_8.

Kuosmanen, S., Koponen, S., Konttinen, H., Niva, M., 2025. Consumers' capabilities, opportunities, and motivations to consume dairy and plant-based alternatives in a dairy-rich food culture. Future Foods 12, 100686. Available at https://doi.org/10.1016/j.fufo.2025.100686.

Lavrijsen-Kromwijk, L., Demba, S., Müller, U., Rose, S., 2024. Impact of automation level of dairy farms in Northern and Central Germany on dairy cattle welfare. Animals 14, 3699. Available at https://doi.org/10.3390/ani14243699.

Lee, Y.-G., Han, K., Chung, C., Ji, I., 2024. Effects of smart farming on the productivity of Korean dairy farms: A case study of robotic milking systems. Sustainability 16, 9991. Available at https://doi.org/10.3390/su16229991.

Li, A., Pandey, A., Pandey, P., 2025. Application of IoT in monitoring greenhouse gas emissions in anaerobic reactors. Energies 18, 6191. Available at https://doi.org/10.3390/en18236191.

Lukacs, M., Toth, F., Horvath, R., Solymos, G., Alpár, B., Varga, P., et al., 2025. Advanced digital solutions for food traceability: Enhancing origin, quality, and safety through NIRS, RFID, blockchain, and IoT. Journal of Sensor and Actuator Networks 14, 21. Available at https://doi.org/10.3390/jsan14010021.

Lutsenko, M., Halai, O., Legkoduh, V., Lastovska, I., Borshch, O., Nadtochii, V., 2021. Milk production process, quality and technological properties of milk for the use of various types of milking machines. Acta Scientiarum. Animal Sciences 43. Available at https://doi.org/10.4025/actascianimsci.v43i1.51336.

Makarov, E.I., Polyansky, K.K., Makarov, M.E., Nikolaeva, Y.R., Shubina, E.A., 2019. Conceptual approaches to the quality system of dairy products based on the blockchain technology. In: Popkova, E.G. (Ed.), Ubiquitous computing and the Internet of Things: Prerequisites for the development of ICT. Springer International Publishing, Cham, pp. 1059–1069. Available at https://doi.org/10.1007/978-3-030-13397-9_109.

Malik, M., Mor, R.S., Gahlawat, V.K., Hassoun, A., Jagtap, S., 2025. Drivers of Industry 5.0 technologies in dairy industry: An exploratory study. Sustainable Food Technology 3, 1556–1568. Available at https://doi.org/10.1039/d5fb00156k.

Mangla, S.K., Sharma, Y.K., Patil, P.P., Yadav, G., Xu, J., 2019. Logistics and distribution challenges to managing operations for corporate sustainability: Study on leading Indian diary organizations. Journal of Cleaner Production 238, 117620. Available at https://doi.org/10.1016/j.jclepro.2019.117620.

Marin, M.-P., Marin, I., & Vidu, L. (2021). Learning about the reduction of food waste using blockchain technology. arXiv preprint arXiv:2101.02026.

Mariotti, A., Pan, Y., Zeng, N., Alessandri, A., 2015. Long-term climate change in the Mediterranean region in the midst of decadal variability. Climate Dynamics 44, 1437–1456. Available at https://doi.org/10.1007/s00382-015-2487-3.

Martin, G., Barth, K., Benoit, M., Brock, C., Destruel, M., Dumont, B., et al., 2020. Potential of multi-species livestock farming to improve the sustainability of livestock farms: A review. Agricultural Systems 181, 102821. Available at https://doi.org/10.1016/j.agsy.2020.102821.

Martinsson, E., Hansson, H., Mittenzwei, K., Storm, H., 2023. Evaluating environmental effects of adopting automatic milking systems on Norwegian dairy farms. European Review of Agricultural Economics 51, 128–156. Available at https://doi.org/10.1093/erae/jbad041.

Marycz, M., Turowska, I., Glazik, S., Jasiński, P., 2025. Artificial intelligence in anaerobic digestion: A review of sensors, modeling approaches, and optimization strategies. Sensors 25, 6961. Available at https://doi.org/10.3390/s25226961.

Mirza Alizadeh, A., Masoomian, M., Shakooie, M., Zabihzadeh Khajavi, M., Farhoodi, M., 2021. Trends and applications of intelligent packaging in dairy products: A review. Critical Reviews in Food Science and Nutrition 62, 383–397. Available at https://doi.org/10.1080/10408398.2020.1817847.

Mkhari, T., Adeyemi, J.O., Fawole, O.A., 2025. Recent advances in the fabrication of intelligent packaging for food preservation: A review. Processes 13, 539. Available at https://doi.org/10.3390/pr13020539.

Morrone, S., Dimauro, C., Gambella, F., Cappai, M.G., 2022. Industry 4.0 and Precision Livestock Farming (PLF): An up to date overview across animal productions. Sensors 22, 4319. Available at https://doi.org/10.3390/s22124319.

Nath, P.C., Mishra, A.K., Sharma, R., Bhunia, B., Mishra, B., Tiwari, A., et al., 2024. Recent advances in artificial intelligence towards the sustainable future of agri-food industry. Food Chemistry 447, 138945. Available at https://doi.org/10.1016/j.foodchem.2024.138945.

Neethirajan, S., 2023. Artificial Intelligence and sensor technologies in dairy livestock export: Charting a digital transformation. Sensors 23, 7045.

Neethirajan, S., 2024. Net zero dairy farming—Advancing climate goals with big data and artificial intelligence. Climate 12, 15. Available at https://doi.org/10.3390/cli12020015.

Nguyen, T.H. D. & Dao, T.M. (2015). Novel approach to optimize milk-run delivery: A case study. In *2015 IEEE International Conference on Industrial Engineering and Engineering Management (IEEM)* (pp. 351–355). IEEE. https://doi.org/10.1109/ieem.2015.7385667.

OECD-FAO, 2020. OECD-FAO Agricultural Outlook 2020-2029. OECD Publishing, Paris. Available at https://doi.org/10.1787/aa3fa6a0-en.

OECD–FAO (2021). OECD-FAO Agricultural Outlook 2021–2030. Table C.5—World dairy projections: Milk, butter and cheese. OECD–FAO. Retrieved from https://www.oecd.org/content/dam/oecd/en/publications/reports/2021/07/oecd-fao-agricultural-outlook-2021-2030_31d65f37/19428846-en.pdf .

Örs, A., Oğuz, C., Semin, A., Skvortsov, E., 2022. The effect of robotic milking systems on economic performance of dairy farms with a simulation Model. New Medit 21. Available at https://doi.org/10.30682/nm2202g.

Oztuna Taner, O., Çolak, A.B., 2024. Dairy factory milk product processing and sustainable of the shelf-life extension with artificial intelligence: A model study. Frontiers in Sustainable Food Systems 8, 2024. Available at https://doi.org/10.3389/fsufs.2024.1344370.

Palma, O., Plà-Aragonés, L.M., Mac Cawley, A., Albornoz, V.M., 2025. AI and data analytics in the dairy farms: A scoping review. Animals 15, 1291. Available at https://doi.org/10.3390/ani15091291.

Papakonstantinou, G.I., Voulgarakis, N., Terzidou, G., Fotos, L., Giamouri, E., Papatsiros, V.G., 2024. Precision livestock farming technology: Applications and challenges of animal welfare and climate change. Agriculture 14, 620. Available at https://doi.org/10.3390/agriculture14040620.

Pérez-Calabuig, A.M., Pradana-López, S., Cancilla, J.C., Mena, M.L., Torrecilla, J.S., 2025. AI-powered optical quality control of yogurt: Detecting melamine adulteration and monitoring shelf-life. Applied Food Research 5, 101510. Available at https://doi.org/10.1016/j.afres.2025.101510.

Piana, V.J., Pelisser, C., Silveira, M.S., Endres, C.M., Finco, D.A., Gonçalves, A.L., 2025. Use of Internet of Things for monitoring, automation, and performance in industrial production of cheese. Research, Society and Development 14 e6714649054-e6714649054.

Posam, S., Reddy G, L.K., Cm, V., 2025. Automated dairy farm management system powered by IoT and ML. Results in Engineering 28, 107508. Available at https://doi.org/10.1016/j.rineng.2025.107508.

Pretty, J., 2020. New opportunities for the redesign of agricultural and food systems. Agriculture and Human Values 37, 629–630. Available at https://doi.org/10.1007/s10460-020-10056-2.

Putra, M.A., Rivai, M., & Arifin, A. (2018). Milk assessment using potentiometric and gas sensors in conjunction with neural network. In *2018 International Seminar on Intelligent Technology and Its Applications (ISITIA)* (pp. 409–412). https://doi.org/10.1109/isitia.2018.8710944.

Qi, Y., Han, J., Shadbolt, N.M., Zhang, Q., 2022. Can the use of digital technology improve the cow milk productivity in large dairy herds? Evidence from China's Shandong Province. Frontiers in Sustainable Food Systems 6, 2022. Available at https://doi.org/10.3389/fsufs.2022.1083906.

Rencricca, G., Froldi, F., Moschini, M., Trevisan, M., Ghnimi, S., Lamastra, L., 2023. The environmental impact of permanent meadows-based farms: A comparison among different dairy farm management systems of an Italian cheese. Sustainable Production and Consumption 37, 53–64. Available at https://doi.org/10.1016/j.spc.2023.02.012.

Romero, A., Sharp, J.L., Dawson, P.L., Darby, D., Cooksey, K., 2021. Evaluation of two intelligent packaging prototypes with a pH indicator to determine spoilage of cow milk. Food Packaging and Shelf Life 30, 100720. Available at https://doi.org/10.1016/j.fpsl.2021.100720.

Rudke, C.R.M., Camelo-Silva, C., Rudke, A.R., Prudencio, E.S., de Andrade, C.J., 2024. Trends in dairy products: New ingredients and ultrasound-based processing. Food and Bioprocess Technology 17, 811–827. Available at https://doi.org/10.1007/s11947-023-03153-7.

Sahnoune, F., Belhamel, M., Zelmat, M., Kerbachi, R., 2013. Climate change in Algeria: Vulnerability and strategy of mitigation and adaptation. Energy Procedia 36, 1286–1294. Available at https://doi.org/10.1016/j.egypro.2013.07.145.

Salimbeni, S., Romera, N., Redchuk, A., 2022. Digital Twin in a Dairy Factory. In: Matt, D.T., Vidoni, R., Rauch, E., Dallasega, P. (Eds.), Managing and implementing the digital transformation. Springer International Publishing, Cham, pp. 103–113.

Serrano-Torres, G.J., López-Naranjo, A.L., Larrea-Cuadrado, P.L., Mazón-Fierro, G., 2025. Transformation of the dairy supply chain through artificial intelligence: A systematic review. Sustainability 17, 982. Available at https://doi.org/10.3390/su17030982.

Sharma, A.K., Sharma, R.K., Kasana, H.S., 2007. Prediction of first lactation 305-day milk yield in Karan Fries dairy cattle using ANN modeling. Applied Soft Computing 7, 1112–1120. Available at https://doi.org/10.1016/j.asoc.2006.07.002.

Sharma, A., Devadas, V.S., Sharma, H., Bhagya, K.D., Pandey, H., Soni Gaytri, S., et al., 2022. Advancement of agricultural technology in farming of India. Bhartiya Krishi Anusandhan Patrika 36, 313–319. Available at https://doi.org/10.18805/BKAP390.

Sheik Arafat, I., Priya, C., Premkumar, R., & S.M, H.N. (2024). Real Time Cattle Health Monitoring and Early Disease Detection Using IoT and Machine Learning. In *2024 5th International Conference on Electronics and Sustainable Communication Systems (ICESC)* (pp. 450–455). https://doi.org/10.1109/icesc60852.2024.10690107.

Shine, P., Murphy, M.D., 2022. Over 20 years of machine learning applications on dairy farms: A comprehensive mapping study. Sensors 22, 52. Available at https://doi.org/10.3390/s22010052.

Shingh, S., Kamalvanshi, V., Ghimire, S., Basyal, S., 2020. Dairy supply chain system based on blockchain technology. Asian Journal of Economics, Business and Accounting 14, 13–19. Available at https://doi.org/10.9734/ajeba/2020/v14i230189.

Shukla, N., Tripathi, A., Tiwari, A., Pophaly, S., Sharma, K., Sahu, S., 2024. From farm to table: Advancements in smart packaging for dairy and food products. Biological Forum–An International Journal 16, 33–45.

Syrmos, E., Bechtsis, D., Vlachos, D., Papapanagiotakis, G., Todi, T., Papaspyropoulou, M., et al., 2023. An IoT monitoring system for dairy products. In: Rabie, K., Kharel, R., Mohjazi, L., Elganimi, T., Khan, W.U. (Eds.), IoT as a service. Springer Nature Switzerland, Cham, pp. 12–22.

Tadele, E., Worku, D., Yigzaw, D., Muluneh, T., Melese, A., 2025. Precision of dairy farming: navigating challenges and seizing opportunities for sustainable dairy production in Africa. Frontiers in Animal Science 6, 2025. Available at https://doi.org/10.3389/fanim.2025.1541838.

Tan, A., Ngan, P.T., 2020. A proposed framework model for dairy supply chain traceability. Sustainable Futures 2, 100034. Available at https://doi.org/10.1016/j.sftr.2020.100034.

Tangorra, F.M., Buoio, E., Calcante, A., Bassi, A., Costa, A., 2024. Internet of Things (IoT): Sensors application in dairy cattle farming. Animals 14, 3071. Available at https://doi.org/10.3390/ani14213071.

Theodore Armand, T.P., Nfor, K.A., Kim, J.-I., Kim, H.-C., 2024. Applications of Artificial Intelligence, Machine Learning, and Deep Learning in Nutrition: A Systematic Review. Nutrients 16, 1073. Available at https://doi.org/10.3390/nu16071073.

Tolegenov, B., Rafif, K., Rafi, M., Herawati, R., Abdullah, A., Sari, L.K., 2024. Digital transparency and consumer awareness: Digital platforms for educating consumers about product origins and sustainability. AL-MUZARA'AH 12, 277–287. Available at https://doi.org/10.29244/jam.12.2.277-287.

Trinh, B.M., Chang, B.P., Mekonnen, T.H., 2023. The barrier properties of sustainable multiphase and multicomponent packaging materials: A review. Progress in Materials Science 133, 101071. Available at https://doi.org/10.1016/j.pmatsci.2023.101071.

Udourioh, G.A., Solomon, M.M., Okolie, J.A., 2025. A review of the valorization of dairy industry wastes through thermochemical, biological, and integrated processes for value-added products. Food Science of Animal Resources 45, 375–408. Available at https://doi.org/10.5851/kosfa.2025.e2.

Vargas-Bello-Pérez, E., Espinoza Sandoval, O.R., Rodríguez-Piñeros, S., Ghavipanje, N., 2025. Blockchain and its future in the supply and value chains of the dairy industry. Journal of Dairy Research 92, 127–134. Available at https://doi.org/10.1017/s0022029925101076.

Verma, K.K., Song, X.-P., Kumari, A., Jagadesh, M., Singh, S.K., Bhatt, R., et al., 2025. Climate change adaptation: Challenges for agricultural sustainability. Plant, Cell & Environment 48, 2522–2533. Available at https://doi.org/10.1111/pce.15078.

Wainwright, W., Glenk, K., Akaichi, F., Moran, D., 2019. Conservation contracts for supplying Farm Animal Genetic Resources (FAnGR) conservation services in Romania. Livestock Science 224, 1–9. Available at https://doi.org/10.1016/j.livsci.2019.03.016.

Wamba, S.F. & Wicks, A. (2010). RFID deployment and use in the dairy value chain: Applications, current issues and future research directions. In *2010 IEEE International Symposium on Technology and Society* (pp. 172–179). IEEE. https://doi.org/10.1109/istas.2010.5514642.

Wang, Y., 2011. U.S. dairy life cycle assessment: From grass to glass. National Academy of Sciences, Rosemont, Illinois, USA.

Wreford, A., Guenther, M., Saunders, J.T., 2025. Evaluating the economic impacts of climate change on Aotearoa–New Zealand's Dairy Sector. Australian Journal of Agricultural and Resource Economics. Available at https://doi.org/10.1111/1467-8489.70062.

Wu, P.-J., Lin, L.-T., Huang, C.-C., 2021. Diagnosing the service quality of perishable-food logistics: Temperature-sensitive milk delivery. Asia Pacific Journal of Marketing and Logistics 34, 1943–1955. Available at https://doi.org/10.1108/apjml-04-2021-0255.

Yang, Y., Lin, M., Lin, Y., Zhang, c., wu, c., 2025. a survey of blockchain applications for management in agriculture and livestock Internet of Things. Future Internet 17, 40. Available at https://doi.org/10.3390/fi17010040.

Zeroual, A., Assani, A.A., Meddi, M., 2016. Combined analysis of temperature and rainfall variability as they relate to climate indices in Northern Algeria over the 1972–2013 period. Hydrology Research 48, 584–595. Available at https://doi.org/10.2166/nh.2016.244.

Part II

Enhancing dairy operations using Industry 4.0 technologies

Chapter 4

Enhancing dairy product quality through Industry 4.0 technologies

Mouna Boulares[1], Atef Lakoud[1], Ines Tarchi[2], Olfa Ben Moussa[1], Imen Mahmoudi[1], Olfa Rebai[3], Sami Fattouch[3], and Abderrahmane Aït-Kaddour[4]

[1]*Research Laboratory: Technological Innovation and Food Security LR22-AGR01, Higher Institute of Food Industries (ESIAT), University of Carthage, Tunis, Tunisia,* [2]*National School of Engineers of Sfax (ENIS), Laboratory of Electrochemistry and Environment, University of Sfax, Sfax, Tunisia,* [3]*Department of Biological Engineering and Chemistry, National Institute of Applied Sciences and Technology (INSAT), Laboratory Ecochimie (LR21ES02), University of Carthage, Tunis, Tunisia,* [4]*VetAgro Sup, Departement of Sustainable Foods, Management and Innovation/ Clermont-Auvergne, INRAE, VetAgro Sup, Cheese Research Unit, Lempdes, France*

4.1 Introduction

Quality management in the dairy sector has traditionally relied on laboratory analyses and food safety standards to ensure product quality and safety. Despite their effectiveness, these practices were largely reactive and time-consuming (Farid et al., 2025). Nowadays, the dairy industry is entering a new era of transformation, driven by the convergence of digital, physical, and biological technologies within the framework of Industry 4.0. In fact, agrifood systems, under digitalization, are being restructured into a dynamic ecosystem powered by data-driven processes (Vahdanjoo et al., 2025). Emerging technologies such as the Internet of Things (IoT), smart sensors, artificial intelligence (AI), and advanced dashboards now enable real-time monitoring of milk quality and processing conditions into every stage of dairy production, reducing the dependence on offline laboratory testing and enabling proactive interventions, and as consequence transforming reactive control into continuous, predictive quality management (Akano et al., 2024; Mhapsekar et al., 2025).

In this context, the technological revolution known as Dairy 4.0 redefines the concept of "quality," which has traditionally been associated with safety, sensory attributes, and traceability across the supply chain (Gottschald, 2024).

Dairy 4.0. DOI: https://doi.org/10.1016/B978-0-443-44730-3.00008-0

It extends this notion to cover a wider range of dimensions such as health and nutrition personalization of food (Agrawal et al., 2025), consumer trust through safeguards implemented during production (Balakrishnan et al., 2025), and sustainability achieved via energy efficiency and circular production systems (Kazancoglu et al., 2022). Therefore, Dairy 4.0 should not be limited to the incorporation of new quality dimensions. Instead, it represents a systemic digital transformation of the dairy sector, built upon interconnected cyber-physical systems that enable real-time data acquisition, continuous monitoring, and data-driven decision-making. This transformation extends across the entire value chain, ensuring that quality, sustainability, and efficiency are integrated at every stage of production. Dairy 4.0 adopts digital technologies to foster sustainable farming practices (Sangode, 2025). Its role in advancing cattle welfare through innovative communication systems has also been highlighted (Gehlot et al., 2022). Furthermore, the opportunities and challenges associated with adopting fourth industrial revolution technologies across milk production and dairy processing position Dairy 4.0 as a model for more sustainable, transparent, and resilient dairy systems (Hassoun et al., 2023).

This concept of Dairy 4.0 represents the integration of Industry 4.0 technologies such as IoT, AI, blockchain, robotics, and digital twins into dairy farming and processing (Hassoun et al., 2023) to enhance monitoring and efficiency across the value chain. For example, it was reported that IoT-enabled smart farming systems to improve milk quality monitoring and supply chain visibility (Akbar et al., 2020; Mhapsekar et al., 2025). Also, blockchain platforms was proved to reinforce food safety and traceability in dairy operations (Khanna et al., 2022; Bhatt and Emdad, 2025). Moreover, AI was reported as a tool that contributes to predictive food safety models and personalized nutrition strategies (Balakrishnan et al., 2025; Agrawal et al., 2025). Digital ecosystems, combined with the shift from reactive quality control to a proactive quality approach enabled by real-time data integration and predictive analytics (Akano et al., 2024), integrate quality into every stage of production. This transition from quality control to quality by design ensures that safety, sustainability, and consumer trust are integral to the production process rather than treated as a final checkpoint.

In an era where digital infrastructures and intelligent systems are reshaping the agrifood landscape, technologies of the industry 4.0 are emerging as a catalyst for elevating the quality of dairy products. The purpose of this chapter is to elucidate how advanced digital tools, ranging from sensor-based monitoring to interconnected data platforms, are redefining quality assurance practices across the dairy sector. By presenting illustrative case studies, the chapter demonstrates how these technologies embed quality control and continuous improvement into every stage of the dairy value chain, fostering more precise, transparent, and predictive forms of oversight. It also examines the challenges, opportunities, and future directions associated with the integration of Industry 4.0 technologies into dairy quality management.

4.2 Redefining quality: beyond safety and shelf life

Traditionally, food quality is centered on three key aspects: safety, sensory appeal, and composition. Safety is of paramount importance, ensuring food was free from harmful microbiological, chemical, or physical hazards, thus protecting public health (Hassoun et al., 2023). Sensory attributes generally referring to flavor, aroma, texture, and appearance strongly influence consumer acceptance and market success. While composition referred to nutritional content and product consistency, generally guaranteeing compliance with labeling and expected benefits. These three pillars shaped the classic model of food quality guiding industry and regulatory standards throughout the 20th century.

In recent decades, however, this narrow framework has proven insufficient to address the complexity of modern food systems and the evolving expectations of consumers and regulators. One of the most notable additions to the concept of quality has been traceability. Originally conceived as a tool to improve recall efficiency in cases of contamination, traceability has become a broader mechanism for ensuring authenticity and transparency. Consumers increasingly perceive traceability as a guarantee of food integrity, associating it with safety, trustworthiness, and even ethical practices (Aung and Chang, 2014). Beyond consumer perception, traceability systems provide regulators and producers with powerful tools to manage risks, verify claims, and support international trade by ensuring harmonized standards of accountability. This transition highlights how modern quality definitions extend beyond immediate product attributes toward systemic qualities of the entire supply chain. The concept of food quality has evolved significantly within regulatory frameworks such as the Codex Alimentarius. Created jointly by the FAO and WHO in 1963, the Codex initially focused on basic hygiene, compositional standards, and safety to protect consumers and enable international trade. While these standards were effective for simpler food systems of the past, they have become insufficient for today's complex, global, and technology-driven supply chains. Modern food systems demand not only compliance with standards but also agility, transparency, and predictive management to address emerging challenges effectively (FAO, 2023).

This has spurred the adoption of new regulatory paradigms. Standards such as ISO 22000:2018 illustrate this evolution by integrating risk-based thinking into Food Safety Management Systems (FSMS). Unlike earlier frameworks that emphasized end-product testing and compliance, ISO 22000:2018 promotes a continuous improvement logic grounded in the Plan-Do-Check-Act (PDCA) cycle (Rihawi, 2024). The aim is not only to safeguard public health but also to ensure resilience, adaptability, and international credibility of food production systems. Such frameworks reflect a broader regulatory philosophy: food quality and safety are no longer static checklists but dynamic processes requiring continuous monitoring and optimization.

Digital technologies are also emerging as key contributors in redefining the paradigm of quality assurance, particularly through the digital transformation of Hazard Analysis and Critical Control Points (HACCP). Traditional HACCP systems have relied largely on manual monitoring and data recording, which are prone to human error and inefficiencies.

The modern digital HACCP integrates IoT sensors, automated data logging, and cloud-based analytics, allowing real-time monitoring of critical parameters such as temperature, humidity, and pH. This reduces the risk of human error and produces tamper-proof digital records that facilitate auditing (Zarid, 2025). Meanwhile, blockchain technology supports Smart Quality Assurance by creating immutable supply chain ledgers that ensure full traceability and validation of product claims, while enabling automation through smart contracts and IoT integration. These innovations mark a shift from reactive compliance to proactive, predictive, and adaptive quality management. Applications like smart milk quality analysis that use sensors and machine learning exemplify these advancements by enabling instant quality grading and alerts, boosting producers' efficiency and regulatory compliance (Shubham et al., 2023).

Beyond regulation, new dimensions of quality have emerged in response to broader societal demands. Personalization has become a defining feature of the contemporary food system. Advances in nutrigenomics and consumer analytics make it possible to tailor food products to individual health profiles, dietary restrictions, or lifestyle preferences. Technologies such as 3D food printing exemplify this trend, enabling customization of nutrient composition, portion sizes, and even sensory characteristics (Pramudito et al., 2025). This shift reflects a move away from the "one-size-fits-all" model that characterized industrial food production, toward a consumer-centered approach ("one-size-per-need") in which quality is defined relative to individual needs and expectations.

Equally critical is the incorporation of sustainability as a fundamental aspect of quality. Growing concerns about climate change, the depletion of natural resources, and environmental harm have shifted sustainability from a secondary issue to a key quality benchmark. Within the dairy industry, the concept of Dairy 4.0 illustrates how advanced technologies are being leveraged to tackle these challenges. Innovations such as intelligent farming systems, precision feeding, automated waste management, and energy-efficient production processes work together to lower environmental impacts and optimize the use of resources (Hassoun et al., 2023). Such practices not only improve ecological performance but also enhance consumer trust by aligning food production with broader societal values of responsibility and stewardship.

TABLE 4.1 Comparison of traditional versus modern food quality dimensions.

Dimension	Traditional focus	Modern extension	References
Safety	Compliance with microbiological and chemical hazards	Digital HACCP, predictive monitoring	FAO (2017), Zarid (2025)
Sensory	Taste, texture, appearance	Customization via consumer analytics, 3D printing	Hassoun et al. (2023)
Composition	Nutritional content, labeling compliance	Functional foods, nutrigenomics personalization	Pramudito et al. (2025)
Traceability	Limited recall management	Blockchain, full supply chain transparency	Hassoun et al. (2023)
Sustainability	Marginal	Central to Dairy 4.0, circular economy, ecoefficiency	Hassoun et al. (2023)

These developments demonstrate that food quality has evolved into a multidimensional and dynamic construct. Traditional parameters, safety, sensory appeal, and compositional integrity, remain essential, but they are now complemented by traceability, personalization, and sustainability as shown in Table 4.1.

Regulatory shifts, technological innovations, and consumer expectations are converging to redefine quality as a holistic promise that extends beyond the product itself to encompass the entire food system. Quality is no longer understood merely as the absence of defects or risks; it is increasingly seen as the embodiment of trust, functionality, transparency, and responsibility from production to consumption and beyond.

4.3 Smart technologies and emerging approaches for enhancing "Dairy Quality 4.0"

Smart technologies, spearheaded by AI, have revolutionized the Dairy Quality 4.0 landscape in recent years (Becchi et al., 2025; Džermeikaitė et al., 2025). This section examines the integration of Industry 4.0 tools, where AI serves as the primary analytical engine coupled with IoT, robotics, and blockchain, to enhance the quality of industrial dairy products. Beyond traditional processing, this framework encompasses emerging biotechnologies such as precision fermentation and explores how digital twins and advanced sensors bridge the gap between precision farming and industrial excellence.

The dairy industry is thus entering a new era in which quality is no longer verified only at the end but actively managed throughout the entire production process. This emerging transformation is increasingly shaped by both consumers and regulatory pressures.

With the widespread use of social media and instant access to information, consumers now expect clear proof of nutritional value, ethical sourcing, and sustainability, while being able to hold brands accountable (Campbell and Feldpausch, 2022). In addition, evolving policies and stricter regulations are reinforcing this shift by imposing new standards for traceability, emissions reporting, and data integrity, making proactive quality design not just an innovative option but a mandatory part of dairy operations (Campbell and Feldpausch, 2022).

Consequently, Quality 4.0 is framed here as a multidimensional concept, integrating consumer-driven personalization and sustainability metrics as core pillars of modern dairy value chains. Beyond traditional standards, modern approaches in dairy quality integrate AI-assisted tools with real-time life cycle assessment and sustainability scoring to produce products that are not only nutritious and sensorially optimized but also health-promoting (Schiano et al., 2020). Indeed, with a continuous focus on environmental impact, advances in AI and digital monitoring tools are helping producers to control the sensory attributes of dairy products and meticulously optimize formulations through mathematical modeling to, meet individual preferences (Ben Amara et al., 2024).

4.3.1 The Internet of Things

IoT refers to a connected system that enables sensing, communicating and analyzing data allowing optimizing workflows (Kazancoglu et al., 2022). In the modern dairy industry, this technology transforms traditional production into a "smart" environment where every process is monitored. IoT architecture can be represented as shown in Fig. 4.1 based on Mhapsekar et al. (2025), Morgan et al. (2021), and Akbar et al. (2020) studies. It includes a physical layer that englobes sensors providing data such as temperature probes and milk flow meters, but also the actuating equipment that enables to control industrial or farming activities. Measurements data collected from sensors and smart devices is transmitted using communication protocols such as Wi-Fi, Ethernet, 5 G, and ZigBee.

Data is then stored in clouds, visualized and processed using AI tools such as Machine Learning algorithms to have a better exploitation of results for taking decisions or emitting command signals that travel backward through the communication layer to the actuating and regulation systems of the physical

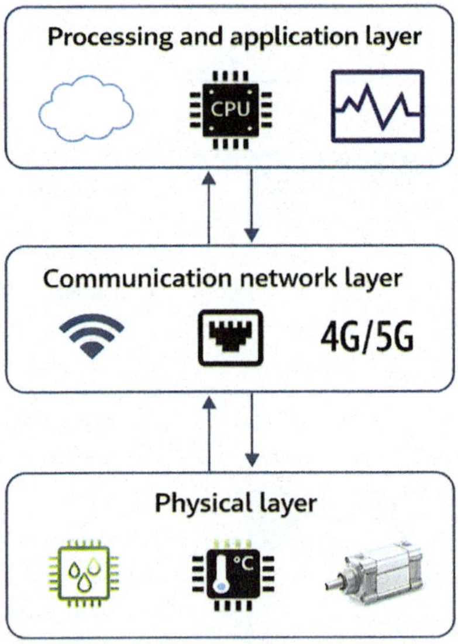

FIGURE 4.1 Internet of things framework. This figure presents a layered Internet of Things (IoT) architecture, showing how physical devices such as sensors and actuators collect data, transmit it through communication networks layer, and deliver it to processing and application layers for analysis, visualization, and control.

layer (Akbar et al., 2020). This real-time feedback loop allows for autonomous adjustments to processing parameters, ensuring consistent product safety and operational efficiency (Mhapsekar et al., 2025). As a complement to IoT, edge computing, which enables to process data locally without the need of a centralized system, can fasten response times and enhance the quality of dairy products by reducing latency in critical operations (Durgun, 2024).

4.3.2 Machine learning frameworks within artificial intelligence

It is essential to clarify that AI and Machine Learning (ML) are not equivalent, although they are occasionally conflated. ML is a specialized subset of AI focused on developing algorithms to achieve AI's broader objective: emulating human intelligence (Pardo et al., 2025). While most current food-related applications rely on established ML techniques, other paradigms such as reinforcement learning are still emerging in this field. Within the dairy sector, various ML approaches, including deep learning architectures, have been deployed and can be categorized as illustrated in Fig. 4.2 (Balakrishnan et al., 2025; Durgun, 2024; Mhapsekar et al., 2025).

In the context of food safety, Decision Trees (DT) and Random Forests (RF) are frequently employed. Support Vector Machines (SVM) have proven effective for detecting adulteration, whereas Convolutional Neural Networks

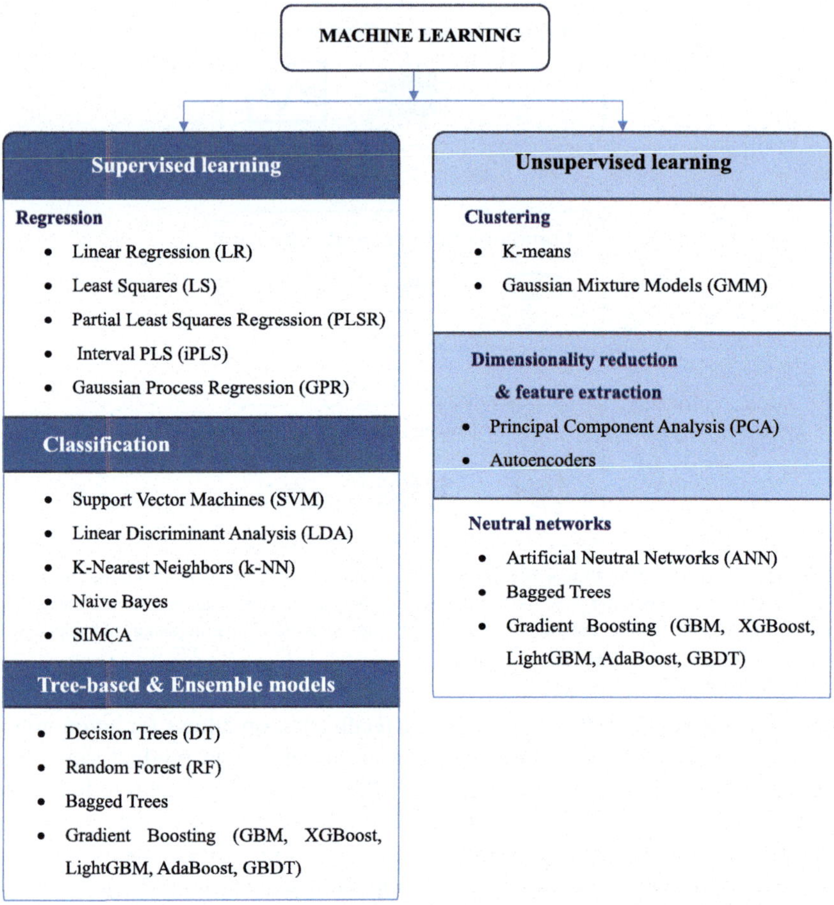

FIGURE 4.2 **Supervised and unsupervised machine learning techniques.** Conceptual framework of machine learning paradigms illustrating supervised and unsupervised approaches, associated learning tasks, and representative algorithms relevant to food science and food process engineering applications.

(CNNs) serve as powerful tools for image processing and quality assessment. Overall, supervised learning where algorithms are trained on labeled data with known outcomes remains the most commonly used approach in dairy applications, in contrast to unsupervised learning (Balakrishnan et al., 2025).

This context is illustrated by several case studies. For instance, Bisutti et al. (2024) combined Fourier Transform Infrared Spectroscopy (FTIR) with an Auto-ML framework using models such as RF, Gradient Boosting Machines (GBM), and multilayer feedforward Artificial Neural Networks (ANN) to assess farm milk quality based on mineral composition. In their study, Habsari et al. (2022) applied ANN and K-means clustering, integrated with an IoT system, to classify raw milk into three grades based on pH, temperature, odor, turbidity, color, fat, and taste indicators.

Farah et al. (2021) employed differential scanning calorimetry alongside RF, GBM, and Multilayer Perceptron (MLP) models to detect adulterants such as whey, starch, urea, and formaldehyde in bovine milk. Gutiérrez et al., (2020) used biosensors coupled with SVM models to identify antibiotic residues in milk. Hosseini et al. (2021) demonstrated that the interval Partial Least Squares (iPLS) model performed well in assessing anionic surfactant adulteration using Near-Infrared spectroscopy. Durgun (2024) leveraged edge computing and the Extreme Gradient Boosting (XGBoost) algorithm for real-time assessment of protein and fat contents in fresh milk, enabling rapid decision-making for process optimization. Finally, Loddo et al. (2025) integrated computer vision with ML models such as Convolutional Neural Networks (CNNs), Random Forest (RF), k-Nearest Neighbors (k-NN), Support Vector Machines (SVM), and Gradient Boosting (GB), to predict the optimal curd-cutting time in cheese processing.

4.3.3 Robotics and automation

The food industry is changing more and more due to automation and robotics, which enhances output, cleanliness, and product quality. One noteworthy example is Fanuc's use of robotic "pick-and-place" systems in cheese packaging plants. These robots handle delicate goods like mozzarella or soft cheeses using sophisticated 3D vision, guaranteeing high throughput while upholding a sterile environment that is unmatched by manual handling.

Beyond packaging, recent advancements have integrated robotics directly into the core of dairy processing and animal welfare monitoring. According to Singh and Sharma (2026), the latest generation of robotic milking systems (RMS) utilizes embedded sensors to provide granular data on milk flow and mammary health, acting as a primary gatekeeper for raw milk quality. These robotic vision systems can identify flaws, color changes, or foreign particles in dairy products; automated measurement systems monitor pH, fat content, and moisture in real time to guarantee that product specifications are being followed. Furthermore, the use of collaborative robots (cobots) in the dairy sector has shown that human–robot interaction can optimize secondary packaging without compromising safety standards (Hassoun et al., 2023).

The synergy between hardware and data is now central to the industry, as highlighted by Hassoun et al. (2022). Autonomous mobile robots (AMRs) equipped with environmental sensors are increasingly used to maintain barn hygiene and monitor feed consistency, directly influencing the chemical stability of the final dairy product.

Robots connected to sensors offer continuous feedback, facilitating quick corrective actions and total traceability. Automated systems handle milk cartons, cheese wrapping, or yogurt cups in sterile environments, significantly minimizing contamination risks. Moreover, automated inoculation, stirring, and temperature control systems maintain consistent microbial growth conditions, which are vital for safety. These parameters are supported by advanced Clean-In-Place (CIP) and Sterilize-In-Place (SIP) protocols, where robotic systems clean tanks and pipelines without dismantling, thereby maintaining the highest hygienic standards of Dairy 4.0 (Hassoun et al., 2022).

4.3.4 Digital twins and simulations

Digital twins are virtual copies of real systems, processes, or products. They are becoming more common in dairy production to model production lines, fermentation processes, and supply chain dynamics. One concrete example of the application of digital twin technology can be found at Arla Foods, where digital twins have been implemented to optimize production processes. In particular, virtual replicas of spray dryers used for milk powder production allow the company to simulate the impact of variations in parameters such as humidity and temperature on the final product. This approach enables process optimization without interrupting ongoing production, thereby reducing operational risks and improving efficiency (Abdurrahman and Ferrari, 2025). For instance, in yogurt process, a digital twin can model temperature profiles, starter culture activity, and pH evolution. This lets producers test different recipes, fermentation times, or environmental conditions without Risking batch failure, which improves flavor, texture, and shelf life. Recently highlighted this proactive approach, showing that digital twins can help reduce waste in fermentation units by predicting microbial changes before they happen. These digital replicas can anticipate microbial risks and ensure quality proactively rather than reactively.

4.3.5 Advanced sensors and spectroscopy

Advanced sensors and spectroscopic techniques are increasingly recognized as transformative tools in modern food systems, particularly within the dairy industry where quality, authenticity, and safety are critical (Singh, 2025). Conventional laboratory techniques are often time-consuming and rely on destructive sampling. In contrast, spectroscopic approaches such as near-infrared (NIR), mid-infrared (MIR), Raman spectroscopy, and hyperspectral imaging enable rapid, noninvasive, and high-precision analytical measurements.

These technologies allow producers to monitor chemical composition, detect falsification, and assess freshness in real time, and as consequence reduce reliance on lengthy laboratory procedures and support more agile decision-making across the supply chain (Ziani et al., 2025).

Among notable applications, it is interesting to mention the use of NIR spectroscopy in milk powder production, where sensors can predict protein and fat content during processing. This capability allows manufacturers to adjust parameters in real time to meet nutritional and regulatory standards, improving efficiency and reducing waste (Singh, 2025; Ziani et al., 2025). Raman spectroscopy has been employed to detect adulterants such as melamine in milk, providing a molecular fingerprint that enables rapid identification of contaminants at very low concentrations. Moreover, hyperspectral imaging has shown considerable effectiveness in monitoring cheese production, providing combined spatial and spectral information that enables the detection of defects, including irregular ripening or microbial contamination, while preserving product authenticity and enhancing high-value branding approaches (Ziani et al., 2025).

The integration of sensors and spectroscopy with artificial intelligence and machine learning is opening new frontiers. Large datasets generated by these technologies can be processed to predict quality attributes with remarkable accuracy. For example, machine learning models trained on spectral data have been used to classify milk samples according to geographic origin, supporting traceability initiatives. Also, AI-enhanced spectroscopy has been applied to monitor spoilage in yogurt, enabling more accurate shelf-life prediction compared to traditional approaches (Ziani et al., 2025). These innovations demonstrate how advanced sensors, when combined with digital technologies, has the potential to transform supply chains by enabling predictive quality management and reducing economic losses associated with spoilage or fraud.

Despite these advances, the sector continues to face persistent challenges. The cost of implementing sophisticated sensors and spectroscopic systems can be prohibitive for small-scale producers, and calibration across diverse food matrices requires careful standardization. Moreover, integrating sensor data into blockchain or digital platforms raises questions about interoperability and data governance. Nevertheless, current trends in research and industrial adoption indicate that these obstacles will be progressively overcome, particularly as demand for transparency and sustainability in food systems continues to grow.

These advancements highlight how advanced sensors and spectroscopy could transform quality assurance in the dairy industry and other fields (Ziani et al., 2025).

4.3.6 Blockchain

Blockchain is defined as a distributed and immutable ledger system that guarantees data integrity, security, and transparency (ISO/TC 307, 2024). While ISO/TC 307 emphasizes the technology's architecture and standards, the National Institute of Standards and Technology (NISTIR 8202, 2018) promotes its decentralized and transactional nature.

In the dairy sector, blockchain is a game-changing technology to improve supply chain transparency, product quality, and authenticity while increasing customer confidence (Malik et al., 2024). In complex supply chains, it enhances visibility from milk collection to delivery, making it especially useful for high-value products where provenance is paramount. Finally, blockchain is enhanced by internal and external traceability systems that guarantee complete supply chain visibility, promote accountability, and uphold rigorous standards for food safety and quality.

Blockchain enables real-time recording and verification of all supply chain stages, including farm production, processing, packaging, and distribution (Varavallo et al., 2022). By securely documenting every transaction in chronological order, a tamper-proof and fraud-resistant product history is created. This approach guarantees that stakeholders and customers receive transparent information about a product's origin, handling, and quality, successfully lowering the possibility of adulteration, fraud, or mislabeling.

For high-value or Protected Designation of Origin (PDO) cheeses, blockchain-based traceability not only boosts consumer confidence but also facilitates regulatory compliance and more effective recall management (Varavallo et al., 2022). Ultimately, by offering a safe, verifiable record of every transaction, blockchain transforms traditional record-keeping into a powerful tool for building trust and ensuring authenticity across the entire dairy value chain.

Throughout the dairy supply chain, transparency and traceability are guaranteed by blockchain, a decentralized, unchangeable ledger that serves as a potent deterrent against food fraud. Food fraud, the deliberate act of misleading consumers by unlawfully altering or misrepresenting ingredients for profit, is a serious problem in the dairy industry. Common fraudulent practices include adulterating milk with inexpensive alternatives like water or melamine to fake protein content, or falsely labeling inferior cheese with a PDO status (Dursun and Gunalay, 2026).

These vulnerabilities make a robust system like blockchain essential; every stage, from raw milk collection to the distribution of the finished product, is securely and impenetrably documented. Due to the distributed and immutable nature of these records, any data modification is extremely challenging as each transaction is chronologically connected to the previous one. Consequently, blockchain radically rethinks trust by providing regulators and customers with access to verifiable product histories.

This automated traceability strengthens the integrity and authenticity of dairy products through unprecedented transparency (Dursun and Gunalay, 2026).

The application of blockchain in the dairy sector is no longer confined to theoretical concepts; real applications demonstrate how it can improve traceability through an innovative "phygital" (physical + digital) approach. A leading example is Parmigiano-Reggiano Protected Designation of Origin (PDO), which utilizes edible microchip p-chips embedded in the casein label of each cheese wheel. This p-chip provides an unalterable link between the physical product and its digital record on the blockchain, allowing stakeholders to trace the milk's origin and curing time, thereby eliminating reliance on unverified authenticity claims (Patel, 2023).

Similarly, the Fontina PDO supply chain operates on Algorand's "green" blockchain to ensure transparency with a low carbon footprint, documenting every event from mountain pasture collection to final packaging with immutable timestamps (Varavallo et al., 2022). This system is further enhanced by Automated Governance, where Internet of Things (IoT) sensors connected to Smart Contracts track temperature, humidity, and transportation conditions. If a deviation, such as a temperature spike, is detected, the Smart Contract executes automatically to trigger actions like flagging the shipment or disallowing its qualification for PDO status, ensuring quality and safety are maintained without human intervention (Nalini et al., 2024).

4.4 Biological innovation

During recent decades, significant new frontiers in food microbe engineering have been reached. Powerful new toolkits, provided by precision fermentation and synthetic biology, are enabling the manipulation of microorganisms at molecular, cellular, and process-wide levels. The microorganisms traditionally responsible for cheese and yogurt fermentation are now being programmed into highly efficient processing units, boosted with new engineered functions.

Through these techniques, these microscopic living factories are currently guided to design dairy products with unprecedented consistency, create novel flavors and textures, extend shelf life, and produce sustainable animal-free dairy alternatives (Schiano et al., 2020).

4.4.1 Engineered microbial platforms

Recent advances in synthetic biology advanced the quick transition from conventional *Lactococcus lactis*, *Streptococcus thermophilus*, and *Lactobacillus* species to genetically optimized microbial strains capable of delivering precisely targeted quality products with unprecedented efficiency and predictability (Xie et al., 2024).

CRISPR-Cas9 gene editing technique has emerged as the principal technology for rational strain design in lactic acid bacteria and heterologous expression hosts. Xie et al. (2024) provide comprehensive documentation of CRISPR applications in dairy fermentation, including: (i) phage resistance enhancement through CRISPR immunity system engineering, which is critical for large industrial-scale fermentation stability; (ii) metabolic pathway redesign to optimize acid production, flavor compound biosynthesis, and stress tolerance; and (iii) elimination of undesirable gene clusters responsible for off-flavor formation or antibiotic resistance markers. Song et al. (2017) reported that CRISPR-Cas9 D10A nickase-assisted genome editing in *Lactobacillus casei* was a critical advancement for regulatory compliance and consumer acceptance.

More recently, the endogenous Cas9-based genome editing systems in *Lacticaseibacillus rhamnosus* GG, the most widely studied probiotic strain, has enabled precision modification for enhanced therapeutic functionality and manufacturing robustness (Xie et al., 2025). In parallel, AI-guided strain evolution is gaining traction, where machine learning models help predict beneficial genetic variants and support data-driven optimization through screening and cultivation pipelines (Džermeikaitė et al., 2025). These systems are expected to reduce strain development cycles from 3–5 years to just 6–12 months, significantly lowering R&D capital expenditure and accelerating time-to-market for novel functionalities.

Besides, multiomics integration including genomics, transcriptomics, proteomics, and metabolomics enabled holistic understanding of strain performance under industrial conditions, thus, facilitating rational design of robust production organisms (Becchi et al., 2025).

4.4.2 From molecular design to industrial applications

The progression from molecular design to industrial implementation is well illustrated in the evolution of lactic acid bacteria engineering. Both naturally occurring and genetically modified strains are selected or developed for specific technological and economic objectives (Soumya and Nampoothiri, 2021). Building on insights from natural genetics, precision fermentation applies direct metabolic engineering to the same microorganism species to enhance industrial performance. One major application involves redesigning exopolysaccharide biosynthesis pathways in *Lactococcus lactis* and *Streptococcus thermophilus* to fine-tune yogurt texture while improving viscosity and preventing syneresis bypassing the need for external stabilizers (Soumya and Nampoothiri, 2021).

Similarly, flavor compound production can be accelerated through targeted metabolic optimization of amino acid and lipid metabolism as well as key volatiles synthesis. Engineered cultures were able to produce key flavor molecules such as acetaldehyde for yogurt, diacetyl for buttery notes, methanethiol for aged cheese character, and fruity branched-chain esters. These targeted interventions enable flavor maturation timelines to shorten from traditional 6- to 24-month aging cycles to just a few weeks (McAuliffe et al, 2019).

Together, these approaches, which integrate natural strain diversity with molecular design and precise genomic modifications, offer a scalable framework for producing high-value fermented dairy products. This framework delivers enhanced functionality and strong consumer appeal while optimizing manufacturing efficiency and economic returns.

In this context, it is important to highlight the role of molecular bioengineering in extending the shelf life and improving the safety of innovative dairy products. CRISPR-modified strains of *Lactococcus lactis* have been engineered to constitutively or inducibly express antimicrobial systems, including enhanced bacteriocin and organic acid cocktails production

(Haryani et al., 2024). These authors reported that CRISPR/Cas9-mediated disruption of the lactate dehydrogenase (ldh) gene in various lactic acid bacteria strains redirected carbon flux from lactate toward antimicrobial compounds, significantly intensifying inhibitory activity against foodborne pathogens, with *Enterococcus faecalis*-ldhΔ showing up to 78% higher bactericidal potentials compared to wild-type strains.

This metabolic bioengineering approach enables the development of economically viable biopreservatives and functional food ingredients while improving food safety (Haryani et al., 2024). Beyond antimicrobial activity, lactic acid bacteria play a crucial role in the release of bioactive peptides encrypted in bovine β-casein, which exhibit antihypertensive, antioxidative, antimicrobial, and immunomodulatory activities. While these peptides are naturally liberated during gastrointestinal digestion or microbial fermentation, yields vary considerably among strains due to differences in proteolytic systems.

To overcome these limitations, synthetic biology and molecular design have been employed to enhance targeted discovery and production of bioactive peptides (Haryani et al., 2024). Complementary cell-free systems, including lactic acid bacteria lysates and culture supernatants (CFCS), offer rapid prototyping and high-throughput screening platforms. CFCS have been shown to exhibit antibacterial and antifungal properties via organic acids, fatty acids, and proteinaceous compounds (Mani-López et al., 2022). These cell-free systems may contain bacteriocins like leucocin, pediocin, and sakacin that effectively inhibit *Listeria monocytogenes* in milk.

Besides, multiomics strategies integrating genomics, transcriptomics, proteomics, and metabolomics provide a systems-level understanding of strain performance and regulatory bottlenecks, enabling rational optimization of lactic acid bacteria for both antimicrobial and bioactive peptide production (Kahraman Ilıkkan, 2025). These advances in molecular design, genome bioengineering, and cell-free approaches pave the way for next-generation functional fermented dairy products that combine at once the extended shelf life, the intrinsic safety, bioactivities production, and enhanced sensory and economic value.

Numerous reports showed that molecular design is useful to produce animal-free dairy proteins via recombinant *Saccharomyces cerevisiae*, *Komagataella phaffi* (*Pichia pastoris*), and filamentous fungi (*Aspergillus oryzae*, *Trichoderma reesei*). These engineered cells can produce dairy proteins with molecular fidelity indistinguishable from bovine sources (Ajayeoba and Ijabadeniyi, 2025). Despite rapid progress, several challenges remain unresolved. Biosafety concerns persist over horizontal gene transfer (HGT), particularly the dissemination of antibiotic resistance markers to gut microbiota (Stirling and Silver, 2020). Therefore, postfermentation inactivation via heat, UV, or purification is therefore recommended to eliminate viable engineered cells (EFSA et al., 2025).

Consumer acceptance studies across Brazil, Germany, India, the United Kingdom, and the United States report >50% purchase intent for precision-fermentation cheese, especially when framed around sustainability and taste; European markets exhibit higher Genetically Modified Organism (GMO) skepticism, necessitating transparency tools such as QR-based strain documentation and third-party safety and environmental verification (Szejda et al., 2021). Even though the progression from molecular design to full-scale industrial deployment is no longer theoretical, with engineered microbial strains moving from laboratory proof-of-concept (POC) to commercial manufacturing, their transition into market-ready products is still constrained by nonscientific barriers. Patent restrictions, especially the ongoing disputes over CRISPR rights, make it hard for small- and medium-sized companies to get access to key technologies. As a result, more people are now pushing for shared licensing systems and patent pools (Tara et al., 2024).

4.5 Dairy personalization

Recent reports are highlighting the speedy growth of the AI in personalized nutrition market, which reached USD 4.13 billion in 2024 and is projected to achieve USD 20.98 billion by 2034 (17.8% CAGR), with alternative estimates indicating USD 10.37 billion by 2029 (23.2% CAGR), driven by wearable biosensors, Electronic Health Records (EHR) integration, direct-to-consumer genomics, and consumer willingness to pay 15–40% premiums for health-optimized dairy products (Insight Ace Analytic, 2024; The Business Research Company, 2024). These technologies are revolutionizing nutrition, replacing rigid, outdated food pyramids with intelligent systems. Driven by deep learning and computer vision, they can now adapt dietary guidance to make it more personalized.

To illustrate this shift, the trending markets showed that 69% of Americans are seeking personalized regimens when selecting products, while direct-to-consumer channels are accounting for 65% of market share, thus, reflecting strong subscription model preference and generating premium customer lifetime values in comparison with conventional retail approaches (CRN, 2024).

4.5.1 Consumer data for flavor/nutrition tailoring

AI is fundamentally redefining food quality, shifting it from a static discreet set of compliance metrics to a dynamic and personalized variables that adapt to individuals' biometric feedback in real time (Zheng et al., 2024). This shift was driven by the sophisticated platforms that integrate multiomics data. Using genomics to look at specific gene variants, known as Single Nucleotide Polymorphism (SNP) profiling, it could be possible to understand traits like lactose intolerance, which affects about 35% and 90% of people of European and East Asian descent, respectively. Such bioinformatic technologies also look for genetic predispositions to casein allergies, like *IL4/IL13* markers identification, *FTO/MTHFR* variations in nutrient metabolism and *TAS1R2/TAS1R3/TAS2R38* taste receptors diversity.

These digital platforms incorporate metabolomics by utilizing data from Continuous Glucose Monitors (CGMs), which capture 288 measurements daily. This rich dataset was used by Zeevi et al. (2015) to build machine learning (ML) models, then refined by (Berry et al., 2020), to predict blood sugar response to a meal with a remarkable correlation of 0.68 to 0.76, allowing for the precise optimization of macronutrient. The unique ecosystem of the gut could be analyzed via microbiomics, utilizing genetic sequencing methods such as 16 S rRNA and shotgun metagenomics.

This analysis allows for the identification of specific probiotic strains, such as *Lactobacillus rhamnosus* GG or *Bifidobacterium longum*, that are well-suited to an individual's personal microbiome, as they are selected for their compatibility with the native microbial community. This central biological data is then enriched with behavioral insights from purchase history, which can predict flavor preferences with 72%–78% accuracy, activity tracker data, and even regional taste databases.

This integrated data empowers digital formulation engines that create truly personalized foods. As noted by Sutehall and Pitsiladis (2025), the future of personalized recommendations in sports nutrition lies in next-generation bioinformatics and AI. For an endurance athlete, such biocomputed approach might dynamically design a drink with a precise 4:1 carbohydrate to protein ratio, containing 25 to 35 g of whey protein and 300 to 500 gm of sodium, while strengthening trainers with 1:1 ratio, 35–50 g protein, 3–4 g leucine, and 5 g creatine. Moreover, for someone on a ketogenic diet, optimized formulations could be recommended where 70%–80% of calories come from fat, using Medium-Chain Triglyceride (MCT) oil to help achieve optimal ketone levels, specifically 0.5 to 3.0 mM β-hydroxybutyrate. Micronutrient fortification is equally precise, with vitamin D3 dosages from 1000 to 4000 International Units personalized to an individual's measured blood levels, or iron supplements of 18 to 25 mg ferrous bisglycinate recommended based on specific needs. The probiotics included are chosen for proven benefits. For instance, *Lactobacillus rhamnosus* GG has been shown in a meta-analysis by Liu et al. (2013) to reduce the frequency of respiratory infections in children by 20%–38%. A fermented milk, designed to stabilize blood sugar, might contain 20 g of resistant starch and 8 g of inulin. In addition, sports recovery yogurt was formulated containing 35 g whey isolate, 5 g branched-chain amino acids (BCAA) in a 2:1:1 ratio (Leucine: Isoleucine: Valine), 200 mg anthocyanins, and 400 mg sodium. An optimized product for seniors could feature easy to digest protein hydrolysate (25 g), bone supporting vitamin K_2 (180 μg), vitamin D_3 (1000 IU), and minimal lactose (<0.5 g lactose).

4.5.2 Adaptive manufacturing and cocreation tools

A synergy between smart digital tools and flexible factories creates a powerful framework for personalized food design. Knoerzer and Muthukumarappan (2021) emphasized the key role of the Computational Fluid Dynamics (CFD) modeling to simulate a product's texture, predict shelf-life, and optimize thermal processes. By modeling how heat moves and liquids mix, CFD helps companies perfect a recipe for safety and quality long before they make a physical batch, saving significant time and resources.

At the same time, consumers are stepping into the role of cocreator through mobile nutrition apps. These apps provide digital interfaces that let people customize their food preferences, a trend explored by Mortazavi and Gutierrez-Osuna (2023). Nowadays, this shift is powering a prosperous market. The global digital personalized nutrition sector was valued at USD 669.3 million in 2024 and is projected to grow to over USD 3094 million by 2034, representing a 16.6% annual growth rate. This expansion is fueled by AI.

Agrawal et al. (2025) projects that by 2025 more than half of all diet and nutrition apps will use AI to offer personalized meal advice and analyze eating habits. Through these apps, users can fine-tune settings for sweetness, fat content, fruit levels, and flavors.

Companies like Nutrigenomix and Viome then turn these preferences into reality, delivering custom products each month and continuously refining the formulas based on the user's latest health data. On the factory level, this personalization is made possible by flexible production systems. These Flexible Manufacturing Systems (FMS) are built with an automated modular design using interconnected workstations, robots, and conveyors that can be quickly rearranged to meet new production demands through real-time monitoring, thus, ensuring efficiency and precision (Singh, 2025). Multilane filling machines from Tetra Pak and GEA can run several unique recipes at the same time. According to Derossi et al. (2020), these smart robotic systems can make instant adjustments, allowing manufacturers to respond quickly to consumer tastes while reducing waste and cost. Nevertheless, such innovative operational landscape necessitates careful attention to regulation and ethics. Protecting people's sensitive health data is fundamental, requiring strict adherence to laws like Europe General Data Protection Regulation (GDPR) and the Health Insurance Portability and Accountability Act (HIPAA). To build trust, researchers explored how blockchain technology could create secure and transparent systems for managing data integrity and user consent. These digital safeguards complement legal protections against genetic discrimination, such as the US Genetic Information Nondiscrimination Act (GINA), which remain critically important in the era of personalized nutrition.

Furthermore, any health claims made by these AI-powered nutrition systems must meet the strict evidence standards of bodies like the European Food Safety Authority (EFSA). In some cases, the algorithms themselves may be regulated as medical device software by the US Food and Drug Administration (FDA).

4.6 Sustainability as a quality metric

Sustainability is increasingly being treated not just as a component of social responsibility, but as a concrete, quantifiable, and verifiable dimension of product quality. This trend is reflected in the purchasing decisions of consumers, retailers, and institutional procurement agencies, who are integrating these metrics into their evaluation processes. Empirical market data substantiates the value of this integration showing a 15%–20% higher willingness-to-pay among consumers when sustainability metrics are prominently displayed as primary quality indicators.

Furthermore, brands that achieve sustainability differentiation have been found to capture volume share gains of 20%–30% in competitive categories, just within 24 months (Canavari and Coderoni, 2020). The key condition for this premium is the transparent communication of environmental impacts.

4.6.1 Life cycle assessment, carbon/water footprint, blockchain-powered ecolabels

The technological infrastructure for the recent sustainability shifts is moving beyond retrospective Life Cycle Assessment (LCA) calculations and being more and more built on real-time environmental foot printing. For instance, industry 4.0 sensor networks are based on IoT-enabled data acquisition in dairy cattle farming while monitoring energy consumption in kilowatt-hours for processes like pasteurization and homogenization with high-pressure pumps operating at 150–250 bar (Tangorra et al., 2024). These innovative devices are also able to track water use via volumetric flow meters, quantifying 2–5 L of water per liter of milk processed, refrigerant leakage of hydrofluorocarbons (HFCs) like R404A with a global warming potential of 3922, and waste streams through mass balance accounting. Furthermore, cloud-based analytics platforms like Microsoft Azure IoT, Amazon Web Services (AWS) IoT Greengrass, Siemens MindSphere, and Schneider Electric EcoStruxure consolidate and harmonize these data streams. This enables real-time LCA dashboards that calculate batch-specific carbon footprints within 2–4 hours postproduction. Subsequently, machine learning algorithms identify hotspots, such as excessive pasteurization temperatures that increase energy use by 8%–12% and enable predictive optimization.

A comprehensive set of principal metrics is determined to measure the environmental impact of dairy farming. Three scores could point out the carbon footprint: (i) the direct emissions from the farm is categorized as Scope 1; (ii) the indirect emissions from purchased electricity are classified as Scope 2, with grid emission factors varying between 50 and 800 g CO_2-eq/kWh; and (iii) most significantly, Scope 3 value chain emissions are estimated to represent 70%–85% of the total impact. Enteric methane from dairy cattle is included as responsible for 120–150 kg CO_2-eq per cow annually (Mazzetto et al, 2022). Industry benchmarks indicate footprints of 0.9–1.5 kg CO_2-eq per liter for fluid milk and 8–12 kg CO_2-eq per kilogram for cheese. Recent assessment of smallholder dairy farms demonstrates that carbon footprint varies significantly based on farm management practices, with major determinants including feed quality and herd productivity (Mech et al., 2023).

Besides, the water footprint comprises multidimensional accounting of (i) *blue water* at 2–5 L per liter of milk processed, (ii) *gray water* for pollution dilution, and (iii) *green water* representing 95%–98% of the total in grazing-based systems, summing to 1–2 m^3 per liter of milk. Additional metrics include land use impact measured in m^2·year per kilogram product, eutrophication potential characterized as kg PO_4-equivalent, and acidification potential as kg SO_2-equivalent.

Furthermore, blockchain-based verification ensures immutability and transparency through cryptographic hashing of feed origin certificates and audits for renewable energy adoption in farms, as well as transport manifests with Global Positioning System (GPS) tracking (Varavallo et al., 2022). The use of methane-reducing feed additives can be tracked to automatically generate carbon credits, which can reduce methane by 25%–30%. They also manage premium payments of 8%–12% for low-footprint milk and enforce supply chain exclusion for noncompliance with thresholds like a carbon intensity above 1.2 kg CO_2-eq per liter. Khanna et al. (2022) reported that the blockchain-enabled platforms designed for the Indian dairy industry have demonstrated enhanced safety and traceability across complex supply chains which, useful in quality assurance and fraud prevention, as well as sustainability verification. The IBM Food Trust with Nestlé tracked 700+ farms in New Zealand and demonstrated a 23% increase in consumer trust and a 15% willingness-to-pay premium. The Arla Foods Climate Check Program also tracks over 9000 European farms, providing benchmarking and sustainability premiums of €0.02–0.04 per liter of milk, representing €3,000–6,000 in annual income increases per farm with a carbon intensity below 1.0 kg CO_2-eq/kg.

4.6.2 Case studies: QR code-enabled environmental scoring

Innovated dairy packaging integrates QR codes and similar machine-readable links to provide access to comprehensive and digitally verified sustainability profiles. It can display the cradle-to-gate carbon footprints calculated using the ISO 14067:2018 standard, water intensity metrics as contextualized with a water stress index, and digital verification of ecolabels such as EU Organic, Rainforest Alliance, and Carbon Neutral Certified.

Recent experimental studies supported the impact of such labels demonstrating that ecoscores significantly influence consumers' perceptions of environmental sustainability and increase purchase intentions for foods with better environmental ratings (Taillie et al., 2024). This transparency is more and more operationalized through standardized multiattribute scoring systems.

For instance, the French Eco-Score is structured to provide an A-E grade based on a weighted algorithm where 40% of the score is allocated to climate change impact, 25% to water pollution, 15% to air pollution, 10% to resource depletion, and 10% to biodiversity impact. In 2024, this label has been voluntarily adopted by 35% of the French retailers. Likewise, the UK's Foundation Earth system is based on climate impact rating where green color attributed to values under 1.0 kg CO_2-eq per kilogram, amber for 1.0–2.5, and red for higher values. According to its 2024 Impact Assessment, 68% of shoppers already using these labels, where a switch to lower-impact alternatives is already made by 42% of them. However, recent comprehensive reviews noted that these ecolabeling schemes lack standardization and are challenging regarding the communication of complex environmental trade-offs (Tiboni-Oschilewski et al., 2024).

These transparency tools make sustainable production more profitable and create enforceable market standards. Regarding carbon-neutral dairy products, price premiums of 12%–25% are commanded, as illustrated by specific brand examples. Furthermore, in recent research, it was revealed that willingness-to-pay for traceable products is significantly influenced by geographical and price-informed consumer preferences, with perceived quality and trust being enhanced by transparency (Nasekos et al., 2025). Major retailers, Stores and Supermarkets are now requiring companies to be more transparent about their environmental impact. Furthermore, the strategic advantage of verifiable, low-carbon production will be solidified by the EU's Carbon Border Adjustment Mechanism (CBAM), which is scheduled for full implementation in 2026 and through which tariffs (about €15–35 per ton CO_2-eq) will be imposed on high-emission imports, therefore, solidifying the strategic advantage of verifiable, low-carbon production.

4.7 Smart dairy quality matrix: an integrated framework

The Smart Dairy Quality Matrix (SDQM) represents a novel, integrated framework designed to enhance quality management in the dairy sector by linking four interdependent pillars: Safety, Sensory, Sustainability, and Personalization, within the digital and analytical capabilities of Industry 4.0. Traditional quality systems often consider these dimensions separately; however, the SDQM treats them as interconnected, requiring simultaneous monitoring, quantification, and optimization (Table 4.2). This systemic approach is increasingly relevant as dairy industries face stringent regulatory requirements, elevated consumer expectations, environmental sustainability challenges, and a rising demand for personalized nutrition.

The SDQM places safety at the forefront, a prerequisite for market access and consumer trust. Dairy products are exposed to microbiological hazards, including pathogens and spoilage organisms, as well as chemical contaminants such as mycotoxins, pesticide residues, heavy metals, and veterinary drugs.

TABLE 4.2 Integrating Smart Dairy Quality Matrix (SDQM) pillars with digital tools and performance indicators.

SDQM pillar	Main objective	Tools/related technologies (Industry 4.0)	Proposed KPIs
Safety	Prevent microbiological and chemical risks	IoT, biosensors, NIR, predictive AI, international databases (Revelou et al., 2025)	% Critical Control Points (CCPs) monitored in real time; microbial loads (CFU/mL); recall rates (Akano et al., 2024; Marinagi et al., 2023)
Sensory	Maintain perceived quality (taste, texture, aroma) (Ruiz-Capillas and Herrero, 2021)	E-noses, e-tongues, spectral imaging, rheological analysis, preference algorithms (Martinez-Velasco et al., 2023; Yixuan, 2024)	Sensory panel scores; instrumental indices (viscosity, flavor profile); consumer satisfaction
Sustainability	Reducing environmental footprint and optimizing resources (Ivana and Puvača, 2024)	Digital twins, blockchain, environmental sensors, PLF (Precision Livestock Farming) (Bhatt and Emdad, 2025)	Carbon footprint (kg CO_2/kg milk); water consumption; coproduct recovery (%)
Personalization	Adapt products to individual needs (Singh, 2025)	Omics (genomics, metabolomics), AI, blockchain traceability, 3D food printing (Singh, 2025)	Share of personalized sales (%); consumer adoption rate; validated health indicators

Traditional safety approaches, relying on HACCP and periodic somatic cell count monitoring, are often reactive and may miss emerging risks.

Industry 4.0 technologies enable a shift toward proactive, predictive safety management. Biosensors, NIR, and rapid molecular diagnostics allow real-time contaminant detection across the dairy chain (Revelou et al., 2025), while big data analytics and artificial intelligence support early-warning systems by detecting anomalies before product quality is compromised (Hassoun et al., 2023).

Franceschini et al. (2022) illustrated the use of milk records-including yield, somatic cell counts, and Fourier Transform Mid-Infrared (FT-MIR) spectral traits, to monitor overall dairy cow health, highlighting the value of indirect quantitative traits for large-scale surveillance.

The sensory pillar acknowledges that consumer acceptance depends not only on safety but also on flavor, aroma, texture, and appearance. Traditional sensory evaluation relies on human panels and basic instrumental methods, which are limited in scale and can be subjective (Ruiz-Capillas and Herrero, 2021). Advances in understanding the dairy matrix emphasize the interactions among proteins, fats, minerals, and microstructures, which shape both nutritional bioavailability and sensory perception. Industry 4.0 tools, including electronic noses (e-noses), electronic tongues (e-tongues), texture analyzers, rheological sensors, spectral imaging, and artificial vision systems, enable real-time, multidimensional assessment of sensory attributes (Martinez-Velasco et al., 2023). ML algorithms integrate these data with consumer preference analytics from digital platforms, allowing predictive control over sensory deviations and preemptive corrective actions (Yixuan, 2024).

Sustainability addresses the environmental impact of dairy production, encompassing greenhouse gas emissions, water and energy use, and waste generation (Ivana and Puvača, 2024). Within the SDQM, sustainability is embedded as a measurable, core component of quality. Key performance indicators include carbon footprint per kilogram of milk, water-use intensity, energy efficiency ratios, and by-product valorization rates. Precision livestock farming technologies, environmental sensors, automated feeding systems, and digital twins of processing facilities enable continuous monitoring and optimization of resources and processes. Blockchain ensures traceability and credibility of sustainability metrics (Bhatt and Emdad, 2025).

Notably, sustainability is synergistic with other pillars: efficient resource use can reduce contamination risks, and valorizing by-products can enhance functional dairy formulations. The personalization pillar reflects consumer demand for tailored nutrition, accounting for individual variability in metabolism, genetics, and lifestyle. Within SDQM, personalization is operationalized through omics data (genomics, metabolomics, microbiomics), wearable health devices, and consumer analytics (Singh, 2025), which are processed using AI to design customized dairy formulations, including modifications in protein and fat structures, incorporation of bioactives, and development of lactose-free or allergen-free variants.

Digital manufacturing technologies, such as modular mixing and 3D food printing, facilitate personalized production at scale, with blockchain ensuring transparency of nutrient content and provenance (Hassoun et al., 2023). A central strength of the SDQM is its translation of the four pillars into operational Key Performance Indicators (KPIs), quantifiable metrics such as efficiency, quality, compliance, and sustainability, that serve as benchmarks to evaluate performance and guide strategic decision-making. Safety KPIs include real-time monitoring of critical control points, microbial load measurements, and recall rates (Akano et al., 2024).

Sensory KPIs integrate trained-panel scores, instrumental indices, and consumer satisfaction metrics. Sustainability KPIs encompass carbon and water footprints, energy use per unit, and by-product valorization. Personalization KPIs track sales of tailored products, adherence to recommended consumption, and validated health outcomes. Each KPI incorporates metadata on measurement frequency, data provenance, and uncertainty, ensuring comparability across contexts. Additionally, studies on supply chain resilience (SCRes) demonstrate links between SCRes elements and nonfinancial KPIs, such as supplier rejection rates, consumer satisfaction, on-time delivery ratios, and productivity metrics (Marinagi et al., 2023).

For visualization and interpretation, SDQM recommends radar charts or matrix dashboards. Radar charts display normalized indices of safety, sensory, sustainability, and personalization, highlighting strengths and weaknesses for a product. Matrix dashboards map KPIs across the supply chain, revealing trade-offs, such as a product with high sensory acceptance and personalization that may have higher environmental impacts. Visual tools also facilitate governance: safety thresholds remain nonnegotiable, multiobjective optimization uses transparent weighting, and iterative validation cycles compare predictions with laboratory and sensory data.

Taking the example of plain yogurt, a SDQM (Safety, Sensory, Sustainability, and Personalization) profile can be constructed by combining scientific data and reasoned assumptions. Sustainability is estimated at 60% based on a life cycle assessment reporting 2.92 kg CO_2-equivalent per kilogram of yogurt (Ghinea and Leahu, 2020), which represents a moderate environmental impact compared to other products. Sensory performance is assigned 80% according to consumer studies showing high acceptability of taste and texture, while not perfect enough to justify a full score. Safety is rated 85% because microbial risks are low under standard manufacturing practices, but advanced monitoring is not always applied. Personalization receives 40% since plain yogurt is rarely tailored to individual nutritional needs, though minor variations like fat content or lactose-free options exist.

In the radar chart (Fig. 4.3), each axis corresponds to a pillar, and the points reflect the estimated scores. Connecting the points forms a closed polygon that visually summarizes the yogurt's qualitative footprint, highlighting high safety and sensory quality, moderate sustainability, and limited personalization.

This approach provides an intuitive multidimensional view of food quality and demonstrates how SDQM can integrate both measured data and informed assumptions for product evaluation.

In conclusion, the SDQM consolidates concepts from dairy matrix science, food safety, sustainability assessment, and personalized nutrition into a unified,

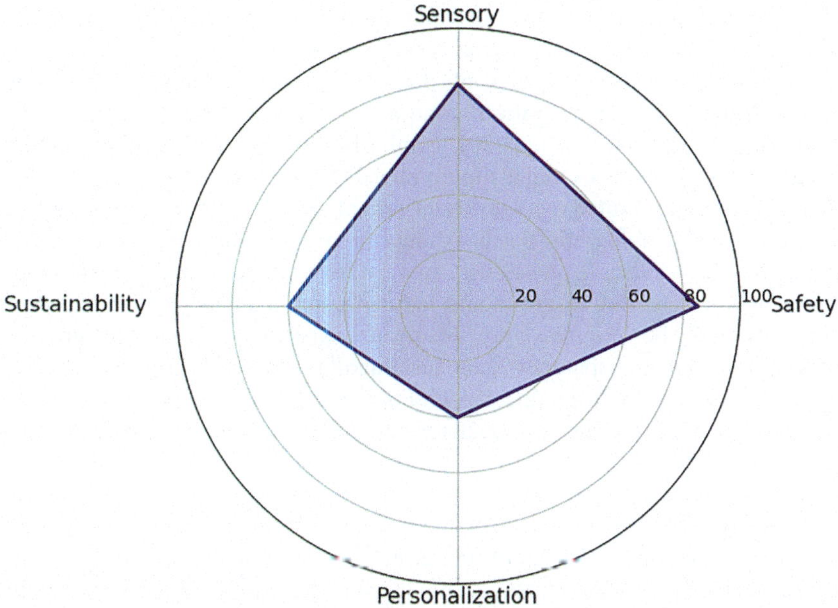

FIGURE 4.3 **SDQM radar of plain yogurt based on estimated scores.** This radar chart compares the relative emphasis of four key dimensions (safety, sensory quality, sustainability, and personalization) using a normalized score scale between 0 and 100. It visually and rapidly highlights strengths and gaps across these criteria within a single system or evaluation framework.

technology-enabled framework. Its novelty lies in integrating Industry 4.0 tools, IoT sensors, biosensors, digital twins, AI, and blockchain, into a system that renders quality multidimensional, measurable, and actionable. For researchers, SDQM provides a roadmap for validating KPIs, establishing benchmarks, and exploring consumer perceptions. For industry practitioners, it offers a pragmatic architecture to ensure food safety, optimize sensory performance, reduce environmental impacts, and deliver personalized dairy solutions, enhancing competitiveness and consumer trust in a data-driven food system.

4.8 Challenges, ethics, and future directions

The integration of digital technologies into the dairy sector presents significant opportunities to enhance efficiency, safety, and sustainability. However, it raises critical ethical, technical, and governance challenges that must be addressed to ensure inclusive and equitable innovation. Artificial intelligence and precision tools such as IoT-based sensors and big data analytics enable evidence-driven decision-making that improves productivity, optimizes resource use, and mitigates environmental impacts.

However, unresolved issues of data governance, including ownership, access rights, and monetization continue to compromise trust in digital platforms (Jouanjean et al., 2020). Similarly, advancing digital traceability in agrifood systems faces major limitations, notably the absence of harmonized standards and the limited availability of reliable datasets (Patidar et al., 2023).

Additionally, integrating sensor data into blockchain or other digital infrastructures raises questions of interoperability and governance. Nevertheless, scientific progress and industrial applications suggest that these obstacles will be progressively resolved, particularly with the growing expectations for transparency and sustainability in food systems. Overall, these progressions highlight the capacity of advanced sensors and spectroscopy to transform quality assurance practices in the dairy sector and related domains (Ziani et al., 2025).

Equity and access also represent a major concern. Small- and medium-sized enterprises (SMEs) often lack the financial resources, technical expertise, or institutional support to adopt advanced digital solutions, exacerbating the digital divide (Omowole et al., 2024). In contrast, multinational corporations can invest in proprietary platforms, consolidating competitive advantages and reinforcing global inequities.

The FAO (2023) emphasizes the need to establish inclusive digital ecosystems and capacity-building initiatives to ensure that smallholder farmers and developing economies are not excluded from the benefits of Industry 4.0.

Looking to the future, several emerging research priorities are redefining innovation in the dairy sector. Nanobiosensors, integrated into IoT platforms or smart packaging, enable real-time monitoring of pathogens, contaminants, and chemical residues, thereby enhancing product safety and consumer confidence (Kulkarni et al, 2022). Autonomous dairies supported by robotic milking and real-time animal health monitoring promise to optimize both productivity and animal welfare. Quantum computing is also expected to accelerate modeling capacities for food safety, supply chain optimization, and sustainability forecasting. Yet, the successful deployment of these technologies will depend on robust academia–industry partnerships that align technological progress with ethical values, sustainability goals, and societal needs (Ahmed et al., 2022).

Ultimately, the future of smart dairy systems will be determined not only by technological advances but also by the ability to embed ethics, transparency, and equity into their development. Approaches such as privacy-preserving data frameworks and federated learning can reconcile innovation with confidentiality, ensuring that sensitive data are protected while enabling knowledge sharing (Vahdanjoo et al., 2025).

4.9 Conclusion

The evolution of the dairy industry through Industry 4.0 concept highlights the importance of advanced technologies in the enhancement of dairy products. These technologies lead to ensure the transition from conventional safety and shelf-life concerns toward a multidimensional version of quality by integrating new essential elements to capture the complexity of modern production systems. The integration of smart technologies including IoT, machine learning, blockchain, robotics, digital twins, advanced sensors, and spectroscopy offers new opportunities to enhance quality management through precision, automation, and real-time monitoring. The proposed Smart Dairy Quality Matrix provides a comprehensive framework that combines these innovations, enabling multidimensional assessment and evidence-based decision-making. Moreover, emerging approaches in quality design highlight the need for adaptive and resilient strategies that balance technological advancement with sustainability, transparency, and equity. All these perspectives highlight a fundamental change where quality is redefined as a dynamic concept driven by innovation, responsibility, and global challenges.

Finally, the vision for Dairy 4.0 is to integrate transparency and ethical innovation into the dairy sector to reframe the relationship between producers and consumers and to ensure sustainable, data-driven, and trustworthy food systems for the future.

AI disclosure

During the preparation of this chapter, the authors used ChatGPT (developed by OpenAI) to assist in developing the chapter. The authors subsequently reviewed and revised the content generated by the tool, and assume full responsibility for the content of the publication.

References

Abdurrahman, E.E.M., Ferrari, G., 2025. Digital Twin applications in the food industry: a review. Frontiers in Sustainable Food Systems 9. Available at https://doi.org/10.3389/fsufs.2025.1538375.

Agrawal, K., Goktas, P., Kumar, N., Leung, M.F., 2025. Artificial intelligence in personalized nutrition and food manufacturing: A comprehensive review of methods, applications, and future directions. Frontiers in Nutrition 12. Available at https://doi.org/10.3389/fnut.2025.1636980.

Ahmed, F., Fattani, M.T., Ali, S.R., Enam, R.N., 2022. Strengthening the bridge between academic and the industry through the academia-industry collaboration plan design model. Frontiers in Psychology 13. Available at https://doi.org/10.3389/fpsyg.2022.875940.

Ajayeoba, T.A., Ijabadeniyi, O.A., 2025. Transforming food for the future: Precision fermentation as a key to sustainability, nutrition, and health. SSRN Electronic Journal. Available at https://doi.org/10.2139/ssrn.5119686.

Akano, O.A., Hanson, E., Nwakile, C., Esiri, A.E., 2024. Designing real-time safety monitoring dashboards for industrial operations: A data-driven approach. Global Journal of Research in Science and Technology 2 (2), 001–009. Available at https://doi.org/10.58175/gjrst.2024.2.2.0070.

Akbar, M.O., Shahbaz Khan, M.S., Ali, M.J., Hussain, A., Qaiser, G., Pasha, M., et al., 2020. IoT for development of smart dairy farming. Journal of Food Quality 2020. Available at https://doi.org/10.1155/2020/4242805.

Aung, M.M., Chang, Y.S., 2014. Traceability in a food supply chain: Safety and quality perspectives. Food Control 39, 172–184. Available at https://doi.org/10.1016/j.foodcont.2013.11.007.

Balakrishnan, P., Anny Leema, A., Jothiaruna, N., Assudani, P.J., Sankar, K., Kulkarni, M.B., et al., 2025. Artificial intelligence for food safety: From predictive models to real-world safeguards. Trends in Food Science & Technology 163, 105153. Available at https://doi.org/10.1016/j.tifs.2025.105153.

Becchi, P.P., Rocchetti, G., Lucini, L., 2025. Advancing dairy science through integrated analytical approaches based on multi-omics and machine learning. Current Opinion in Food Science 63, 101289. Available at https://doi.org/10.1016/j.cofs.2025.101289.

Ben Amara, S., Lakoud, A., Mahmoudi, I., Ben Tekaya, I., Amri, A., Snoussi, A., et al., 2024. Optimization of the industrial production process of Tunisian date paste for sustainable food systems. Processes 12 (10), 2083. Available at https://doi.org/10.3390/pr12102083.

Berry, S.E., Valdes, A.M., Drew, D.A., Asnicar, F., Mazidi, M., Wolf, J., et al., 2020. Human postprandial responses to food and potential for precision nutrition. Nature Medicine 26 (6), 964–973. Available at https://doi.org/10.1038/s41591-020-0934-0.

Bhatt, G.D., Emdad, A., 2025. Blockchain's role in environmental sustainability: Bibliometric insights from scopus. Green Technologies and Sustainability 3 (4), 100236. Available at https://doi.org/10.1016/j.grets.2025.100236.

Bisutti, V., Flavio, L., Mota, M., Giannuzzi, D., Toscano, A., Amalfitano, N., et al., 2024. Infrared spectroscopy coupled with machine learning algorithms for predicting the detailed milk mineral profile in dairy cattle. Food Chemistry 461, 140800. Available at https://doi.org/10.1016/j.foodchem.2024.140800.

Campbell, C.G., Feldpausch, G.L., 2022. Invited review: The consumer and dairy food waste: An individual plus policy, systems, and environmental perspective. Journal of Dairy Science 105 (5), 3736–3745. Available at https://doi.org/10.3168/jds.2021-20994.

Canavari, M., Coderoni, S., 2020. Consumer stated preferences for dairy products with carbon footprint labels in Italy. Agricultural and Food Economics 8 (1). Available at https://doi.org/10.1186/s40100-019-0149-1.

Council for Responsible Nutrition, 2024. CRN survey shows consistent supplement usage with increase of specialty product use over time. Retrieved from https://crnusa.org/newsroom/crn-survey-shows-consistent-supplement-usage-increase-specialty-product-use-over-time (accessed Nov 04, 2025).

Derossi, A., Husain, A., Caporizzi, R., Severini, C., 2020. Manufacturing personalized food for people uniqueness. An overview from traditional to emerging technologies. Critical Reviews in Food Science and Nutrition 60 (7), 1141–1159. Available at https://doi.org/10.1080/10408398.2018.1559796.

Durgun, M., 2024. Real-time milk quality control using multi-spectral sensing and edge computing: Advancing on-site detection of milk components with XGBoost. Applied Sciences 14 (23), 10916. Available at https://doi.org/10.3390/app142310916.

Dursun, E., Gunalay, Y., 2026. The Role of Blockchain in Food Supply Chains: Enhancing Traceability. In: Khojasteh, Y., Mousavi Jahan Abadi, S.M. (Eds.), International Series in Operations Research & Management Science. Springer, pp. 169–201. Emerging Technologies in Supply Chains, chapter 0. Available at https://doi.org/10.1007/978-3-032-01218-0_7.

Džermeikaitė, K., Šidlauskaitė, M., Antanaitis, R., Anskienė, L., 2025. Enhancing genomic selection in dairy cattle through artificial intelligence: Integrating advanced phenotyping and predictive models to advance health, climate resilience, and sustainability. Dairy 6 (5), 50. Available at https://doi.org/10.3390/dairy6050050.

Elumalai, A., Natarajan, V., 2025. Advancements in analytical technologies for ensuring food quality and authentication: A comprehensive review. Journal of Food Composition and Analysis 139, 107075. Available at https://doi.org/10.1016/j.jfca.2024.107075 Available at.

EFSA Panel on Genetically Modified Organisms (GMO), Casacuberta, J., Barro, F., Braeuning, A., de Maagd, R., Epstein, M.M., et al., 2025. New developments in biotechnology applied to animals: An assessment of the adequacy and sufficiency of current EFSA guidance for animal risk assessment. EFSA Journal 23.

FAO, 2017. Trade and food standards. Food and Agriculture Organization of the United Nations; World Trade Organization.

FAO, 2023). The State of Food and Agriculture. FAO. Available at https://doi.org/10.4060/cc7724en

Farah, J.S., Cavalcanti, R.N., Guimarães, J.T., Balthazar, C.F., Coimbra, P.T., Pimentel, T.C., et al., 2021. Differential scanning calorimetry coupled with machine learning technique: An effective approach to determine the milk authenticity. Food Control 121. Available at https://doi.org/10.1016/j.foodcont.2020.107585.

Franceschini, S., Grelet, C., Leblois, J., Gengler, N., Soyeurt, H., Crowe, M., et al., 2022. Can unsupervised learning methods applied to milk recording big data provide new insights into dairy cow health? Journal of Dairy Science 105 (8), 6760–6772. Available at https://doi.org/10.3168/jds.2022-21975.

Gehlot, A., Malik, P.K., Singh, R., Akram, S.V., Alsuwian, T., 2022. Dairy 4.0: Intelligent communication ecosystem for the cattle animal welfare with blockchain and IoT enabled technologies. Applied Sciences 12 (14), 7316. Available at https://doi.org/10.3390/app12147316.

Ghinea, C., Leahu, A., 2020. Life cycle assessment of fermented milk: Yogurt production. Ovidius University Annals of Chemistry 31 (1), 49–54. Available at https://doi.org/10.2478/auoc-2020-0010.

Gottschald, M., 2024. Advancing food safety through digital traceability, interoperability, harmonized data and collaborative partnerships. Journal fur Verbraucherschutz und Lebensmittelsicherheit 19 (3), 257–258. Available at https://doi.org/10.1007/s00003-024-01522-8.

Gutiérrez, P., Godoy, S.E., Torres, S., Oyarzún, P., Sanhueza, I., Díaz-García, V., et al., 2020. Improved antibiotic detection in raw milk using machine learning tools over the absorption spectra of a problem-specific nanobiosensor. Sensors 20 (4552). Available at https://doi.org/10.3390/s20164552.

Habsari, W., Udin, F., & Arkeman, Y., 2022. An analysis and design of fresh milk smart grading system based on internet of things. *IOP Conference Series: Earth and Environmental Science, 1063*(1), 012059. https://doi.org/10.1088/1755-1315/1063/1/012059.

Hassoun, A., Jagtap, S., Garcia-Garcia, G., Trollman, H., Tokatlı, F., Gomez, B., et al., 2022. Food quality 4.0: From traditional approaches to digitalized automated analysis. Journal of Food Engineering 335, 111216. Available at https://doi.org/10.1016/j.jfoodeng.2022.111216.

Haryani, Y., Abdul Halid, N., Goh, S.G., Nor-Khaizura, M.A.R., Md Hatta, M.A., Sabri, S., et al., 2024. Efficient metabolic pathway modification in various strains of lactic acid bacteria using CRISPR/Cas9 system for elevated synthesis of antimicrobial compounds. Journal of Biotechnology 395, 53–63. Available at https://doi.org/10.1016/j.jbiotec.2024.09.002.

Hassoun, A., Garcia-Garcia, G., Trollman, H., Jagtap, S., Parra-López, C., Cropotova, J., et al., 2023. Birth of dairy 4.0: Opportunities and challenges in adoption of fourth industrial revolution technologies in the production of milk and its derivatives. Current Research in Food Science 7, 100535. Available at https://doi.org/10.1016/j.crfs.2023.100535.

He, Y., Zhang, L., Wang, X., 2023. Digital twin-driven smart manufacturing in the dairy industry: A review of fermentation process optimization. Journal of Dairy Science and Technology 106 (2), 145–162.

Hosseini, E., Ghasemi, J.B., Daraei, B., Asadi, G., Adib, N., 2021. Near-infrared spectroscopy and machine learning-based classification and calibration methods in detection and measurement of anionic surfactant in milk. Journal of Food Composition and Analysis 104, 104170. Available at https://doi.org/10.1016/j.jfca.2021.104170.

Ivana, B., Puvača, N., 2024. Economic and ecological sustainability of dairy production. Journal of Agronomy, Technology and Engineering Management (JATEM) 7 (3), 1088–1104. Available at https://doi.org/10.55817/tqxu2115.

Insight Ace Analytic (2024). Global AI in personalized nutrition market—By type, application, end user, provider, and region, trends, industry competition analysis, revenue and forecast to 2034. InsightAce Analytic Pvt. Ltd.

ISO/TC 307. (2024). ISO 22739:2024—Blockchain and distributed ledger technologies—Vocabulary. Organisation Internationale de Normalisation. Retrieved from https://www.iso.org/standard/73771.

Jouanjean, M.-A., Casalini, F., Wiseman, L., Gray, E., 2020. Issues around data governance in the digital transformation of agriculture. OECD Food, Agriculture and Fisheries Papers 146. Available at https://doi.org/10.1787/53ecf2ab-en.

Kahraman Ilıkkan, Ö., 2025. Lactobacillomics as a new notion in lactic acid bacteria research through omics integration. World Journal of Microbiology and Biotechnology 41 (2), 68. Available at https://doi.org/10.1007/s11274-025-04285-y.

Kazancoglu, Y., Ozbiltekin-Pala, M., Sezer, M.D., Kumar, A., Luthra, S., 2022. Circular dairy supply chain management through Internet of Things-enabled technologies. Environmental Science and Pollution Research. Available at https://doi.org/10.1007/s11356-021-17697-8.

Khanna, A., Jain, S., Burgio, A., Bolshev, V., Panchenko, V., 2022. Blockchain-enabled supply chain platform for indian dairy industry: Safety and traceability. Foods 11 (17), 2716. Available at https://doi.org/10.3390/foods11172716.

Knoerzer, K., Muthukumarappan, K. (Eds.), 2021. Innovative Food Processing Technologies: A Comprehensive Review. Elsevier ISBN: 978-0-12-815781-7.

Kulkarni, M.B., Ayachit, N.H., Aminabhavi, T.M., 2022. Recent advancements in nanobiosensors: Current trends, challenges, applications, and future scope. Biosensors 12 (10). Available at https://doi.org/10.3390/bios12100892.

Liu, S., Hu, P., Du, X., Zhou, T., Pei, X., 2013. *Lactobacillus rhamnosus* GG supplementation for preventing respiratory infections in children: A meta-analysis of randomized, placebo-controlled trials. Indian Pediatrics 50 (4), 377–381. Available at https://doi.org/10.1007/s13312-013-0123-z.

Loddo, A., Di Ruberto, C., Armano, G., Manconi, A., 2025. Detecting coagulation time in cheese making by means of computer vision and machine learning techniques. Computers in Industry 164, 104173. Available at https://doi.org/10.1016/j.compind.2024.104173.

Malik, M., Gahlawat, V.K., Mor, R.S., Singh, M.K., 2024. Unlocking dairy traceability: Current trends, applications, and future opportunities. Future Foods 10, 100426. Available at https://doi.org/10.1016/j.fufo.2024.100426.

Mani-López, E., Arrioja-Bretón, D., López-Malo, A., 2022. The impacts of antimicrobial and antifungal activity of cell-free supernatants from lactic acid bacteria in vitro and foods. Comprehensive Reviews in Food Science and Food Safety 21 (1), 604–641. Available at https://doi.org/10.1111/1541-4337.12872.

Marinagi, C., Reklitis, P., Trivellas, P., Sakas, D., 2023. The impact of Industry 4.0 technologies on key performance indicators for a resilient Supply Chain 4.0. Sustainability 15 (6), 5185. Available at https://doi.org/10.3390/su15065185.

Martinez-Velasco, J.D., Filomena-Ambrosio, A., Garzón-Castro, C.L., 2023. Technological tools for the measurement of sensory characteristics in food: A review. Food Research 12, 340. Available at https://doi.org/10.12688/f1000research.131914.1.

Mazzetto, A.M., Falconer, S., Ledgard, S., 2022. Mapping the carbon footprint of milk production from cattle: A systematic review. Journal of Dairy Science 105 (12), 9713–9725. Available at https://doi.org/10.3168/jds.2022-22117.

McAuliffe, O., Kilcawley, K., Stefanovic, E., 2019. Symposium review: Genomic investigations of flavor formation by dairy microbiota. Journal of Dairy Science 102 (1), 909–922. Available at https://doi.org/10.3168/jds.2018-15385.

Mech, A., Devi, G.L., Sivaram, M., Sirohi, S., Dhali, A., Kolte, A.P., et al., 2023. Assessment of carbon footprint of milk production and identification of its major determinants in smallholder dairy farms in Karnataka, India. Journal of Dairy Science 106 (12), 8847–8860. Available at https://doi.org/10.3168/jds.2022-22153.

Mhapsekar, R., Kilbane, D., Davy, S., Abraham, L., Fenelon, M., O'Shea, N., 2025. A systematic review of the internet of things and artificial intelligence applications in milk quality monitoring and analysis. International Journal of Dairy Technology 78 (3). Available at https://doi.org/10.1111/1471-0307.70049.

Morgan, J., Halton, M., Qiao, Y., Breslin, J.G., 2021. Industry 4.0 smart reconfigurable manufacturing machines. Journal of Manufacturing Systems 59, 481–506. Available at https://doi.org/10.1016/j.jmsy.2021.03.001.

Mortazavi, B.J., Gutierrez-Osuna, R., 2023. A review of digital innovations for diet monitoring and precision nutrition. Journal of diabetes science and technology 17 (1), 217–223. Available at https://doi.org/10.1177/19322968211041356.

National Institute of Standards and Technology, 2018. *Blockchain technology overview* (NISTIR 8202). Gaithersburg, MD: U.S. Department of Commerce. https://doi.org/10.6028/NIST.IR.8202.

Nalini, P., Balusamy, R., Sivakumar, C., 2024. Smart contracts and IoT based blockchain framework for decentralized monitoring of dairy supply chain operations. Journal of Computational Analysis and Applications 33 (1A), 561–579.

Nasekos, J., Collart, A.J., Canales, E., Thayer, A., 2025. Got traceability? A geographical-price-informed choice experiment assessing consumer preferences for blockchain-verified traceability in cow's milk. Journal of Food Distribution Research 56 (2), 1–32.

Omowole, B.M., Olufemi-Phillips, A.Q., Ofodile, O.C., Eyo-Udo, N.L., Ewim, S.E., 2024. Barriers and drivers of digital transformation in SMEs: A conceptual analysis. International Journal of Scholarly Research in Science and Technology 5 (2), 019–036. Available at https://doi.org/10.56781/ijsrst.2024.5.2.0037.

Pardo, D., Castillo, M., Mulayim, M.O., Cerquides, J., 2025. Machine learning in cheese-making: Methods, applications, and the future. Food Engineering Reviews. Available at https://doi.org/10.1007/s12393-025-09420-x.

Patel, P., 2023. Sand-size microchips take on fake foods and drugs. IEEE Spectrum. Available at https://spectrum.ieee.org/food-fraud-microchip.

Patidar, A., Sharma, M., Agrawal, R., Sangwan, K.S., 2023. Supply chain resilience and its key performance indicators: An evaluation under Industry 4.0 and sustainability perspective. Management of Environmental Quality: An International Journal 34 (4), 962–980. Available at https://doi.org/10.1108/MEQ-03-2022-0091.

Pramudito, P., Rachim, F., Abduh, M.S., Maulana, A., Salim, C., 2025. Food product quality: Literature study on the dimensions of food product quality. International Journal of Social Science and Human Research 8 (6). Available at https://doi.org/10.47191/ijsshr/v8-i6-52.

Revelou, P.K., Tsakali, E., Batrinou, A., Strati, I.F., 2025. Applications of machine learning in food safety and HACCP monitoring of animal-source foods. Foods 14 (6). Available at https://doi.org/10.3390/foods14060922 Multidisciplinary Digital Publishing Institute (MDPI).

Rihawi, B., 2024. The impact of ISO 22000: 2018 on food facilities performance with multiple production lines. CYTA - Journal of Food 22 (1). Available at https://doi.org/10.1080/19476337.2024.2431281.

Ruiz-Capillas, C., Herrero, A.M., 2021. Sensory analysis and consumer research in new product development. Foods 10 (3), 582. Available at https://doi.org/10.3390/FOODS10030582.

Sangode, P.B., 2025. Dairy 4.0: Embracing digital technology for sustainable farming: A review. Asian Journal of Dairy and Food Research 44 (2), 169–178. Available at https://doi.org/10.18805/ajdfr.DR-2242.

Schiano, A.N., Harwood, W.S., Gerard, P.D., Drake, M.A., 2020. Consumer perception of the sustainability of dairy products and plant-based dairy alternatives. Journal of Dairy Science 103 (12), 11228–11243. Available at https://doi.org/10.3168/jds.2020-18406.

Shubham, M., Vibhuti, T., Pooja, K., Bibve, M.P.S., 2023. Smart milk quality analysis & grading using IOT. International Journal of Research Publication and Reviews 4 (11), 2022–2025. Available at www.ijrpr.com.

Singh, D., 2025. Harnessing Artificial Intelligence to Safeguard Food Quality and Safety. Journal of Food Protection 88, 100621. Available at https://doi.org/10.1016/j.jfp.2025.100621.

Song, X., Huang, H., Xiong, Z., Ai, L., Yang, S., 2017. CRISPR-Cas9 D10A nickase-assisted genome editing in *Lactobacillus casei*. Applied and Environmental Microbiology 83 (22), e01259-17. Available at https://doi.org/10.1128/AEM.01259-17.

Soumya, M.P., Nampoothiri, K.M., 2021. An overview of functional genomics and relevance of glycosyltransferases in exopolysaccharide production by lactic acid bacteria. International Journal of Biological Macromolecules 184, 1014–1025. Available at https://doi.org/10.1016/j.ijbiomac.2021.06.131.

Stirling, F., Silver, P.A., 2020. Controlling the implementation of transgenic microbes: Are we ready for what synthetic biology has to offer? Molecular Cell 78 (4), 614–623. Available at https://doi.org/10.1016/j.molcel.2020.03.034.

Sutehall, S., Pitsiladis, Y., 2025. Personalized nutrition for the enhancement of elite athletic performance. Scandinavian Journal of Medicine & Science in Sports 35 (4), e70044. Available at https://doi.org/10.1111/sms.70044.

Szejda, K., Bryant, C.J., Urbanovich, T., 2021. US and UK consumer adoption of cultivated meat: A segmentation study. Foods 10 (5), 1050. Available at https://doi.org/10.3390/foods10051050.

Taillie, L.S., Wolfson, J.A., Prestemon, C.E., Bercholz, M., Ewoldt, L., Ruggles, P.R., et al., 2024. The impact of an eco-score label on US consumers' perceptions of environmental sustainability and intentions to purchase food: A randomized experiment. PLoS One 19 (6), e0306123. Available at https://doi.org/10.1371/journal.pone.0306123.

Tangorra, F.M., Buoio, E., Calcante, A., Bassi, A., Costa, A., 2024. Internet of Things (IoT): Sensors application in dairy cattle farming. Animals 14 (21), 3071. Available at https://doi.org/10.3390/ani14213071.

Tara, A., Singh, P., Gautam, D., Tripathi, G., Uppal, C., Malhotra, S., et al., 2024. CRISPR-mediated editing of β-lactoglobulin (BLG) gene in buffalo. Scientific Reports 14 (1), 14822. Available at https://doi.org/10.1038/s41598-024-65359-9.

The Business Research Company, 2024. Artificial Intelligence (AI). In *Personalized Nutrition Market Report 2024-2034*. The Business Research Company.

Tiboni-Oschilewski, O., Abarca, M., Santa Rosa Pierre, F., Rosi, A., Biasini, B., Menozzi, D., et al., 2024. Strengths and weaknesses of food eco-labeling: A review. Frontiers in Nutrition 11, 1381135. Available at https://doi.org/10.3389/fnut.2024.1381135.

Vahdanjoo, M., Sørensen, C.G., Nørremark, M., 2025. Digital transformation of the agri-food system. Current Opinion in Food Science 63, 101287. Available at https://doi.org/10.1016/j.cofs.2025.101287.

Varavallo, G., Caragnano, G., Bertone, F., Vernetti-Prot, L., Terzo, O., 2022. Traceability platform based on green blockchain: An application case study in dairy supply chain. Sustainability 14 (6), 3321. Available at https://doi.org/10.3390/su14063321.

Xie, Z., McAuliffe, O., Jin, Y.S., Miller, M.J., 2024. Invited review: Genomic modifications of lactic acid bacteria and their applications in dairy fermentation. Journal of Dairy Science 107 (11), 8749–8764. Available at https://doi.org/10.3168/jds.2024-24989.

Xie, Z., Jin, Y.S., Miller, M.J., 2025. Exploiting endogenous CRISPR-Cas9 system for functional engineering of probiotic lacticaseibacillus rhamnosus GG (LGG). Available at https://doi.org/10.1101/2025.01.21.634116.

Yixuan, Z., 2024. Utilizing machine learning algorithms for consumer behaviour analysis. Applied and Computational Engineering 49 (1), 213–219. Available at https://doi.org/10.54254/2755-2721/49/20241186.

Zarid, M., 2025. The green HACCP approach: Advancing food safety and sustainability. Sustainability 17 (17), 7834. Available at https://doi.org/10.3390/su17177834.

Zeevi, D., Korem, T., Zmora, N., Israeli, D., Rothschild, D., Weinberger, A., et al., 2015. Personalized nutrition by prediction of glycemic responses. Cell 163 (5), 1079–1094. Available at https://doi.org/10.1016/j.cell.2015.11.001.

Zheng, J., Wang, J., Shen, J., An, R., 2024. Artificial intelligence applications to measure food and nutrient intakes: Scoping review. Journal of Medical Internet Research 26, e54557. Available at https://doi.org/10.2196/54557.

Ziani, I., Bouakline, H., El Guerraf, A., El Bachiri, A., Fauconnier, M.L., Sher, F., 2025. Integrating AI and advanced spectroscopic techniques for precision food safety and quality control. Trends in Food Science & Technology 156, 104850. Available at https://doi.org/10.1016/j.tifs.2024.104850.

Chapter 5

Ensuring dairy safety with Industry 4.0

Samuel Imathiu[1], Isaac Osuga[2], Eddy Owaga[3], and Peter Kahenya[1]
[1]*Department of Food Science and Technology, Jomo Kenyatta University of Agriculture and Technology, Nairobi, Kenya,* [2]*Department of Animal Sciences, Jomo Kenyatta University of Agriculture and Technology, Nairobi, Kenya,* [3]*Institute of Food Bioresources Technology, Dedan Kimathi University of Technology, Nyeri, Kenya*

5.1 Introduction

For centuries, milk and milk products (collectively referred to as dairy) have been recognized as vital for human nutrition and well-being. Worldwide, dairy forms part of the diet for majority of the people, with around six billion people relying on it as a food source (FAO, 2013). Dairy products are nutrient-dense foods that supply energy, high-quality protein, and a wide range of essential micronutrients such as calcium, magnesium, potassium, zinc, and phosphorus in readily bioavailable forms (Rizzoli, 2014). For this reason, milk has generally been considered a complete food (Górska-Warsewicz et al., 2019). Despite the unpredictable climate variability, global milk production has been increasing in the recent past (Bai et al., 2023), though at a slow pace. In 2024, for example, global liquid milk production rose by 1.4%, reaching 982 million tons, while the global trade in dairy products was 86 million tons (milk equivalent), representing a 0.2% increase in production compared to 2023.

Despite increased global production and demand for dairy products, the dairy industry faces challenges related to product safety. The unique nutrient composition and perishability of the dairy products make them highly susceptible to microbial and chemical adulteration, posing significant food safety challenges. Microbiological hazards remain the most critical, with pathogens such as *Listeria monocytogenes*, *Salmonella enterica*, *Escherichia coli* O157:H7, *Staphylococcus aureus*, and *Campylobacter jejuni* frequently being implicated in dairy-related foodborne outbreaks (Owusu-Kwarteng et al., 2020). These risks are particularly pronounced in raw milk and artisanal dairy products, where inadequate pasteurization and unhygienic handling amplify the contamination routes (De Klerk and Robinson, 2022). Beyond pathogens,

Dairy 4.0. DOI: https://doi.org/10.1016/B978-0-443-44730-3.00012-2

residues of veterinary drugs, heavy metals, and mycotoxins add complexity to dairy safety (Akinyemi et al., 2021). Of increasing concern is the role of dairy systems in disseminating antimicrobial resistance through the misuse of antibiotics in livestock, thereby posing a major threat to both animal and human health. Climate variability further exacerbates these risks by influencing pathogen ecology, feed contamination and cold-chain reliability. Addressing dairy safety, therefore, requires integrated strategies that combine good agricultural and hygienic practices, innovative detection and risk prediction technologies, and robust regulatory frameworks to safeguard public health (Azizsafaei et al., 2022).

Science-based mitigation strategies are required in the dairy value chain to ensure product quality and safety. Embracing innovation and emerging technologies is becoming increasingly important to address the current global challenges facing many food industry sectors, including the dairy industry. Growing literature shows that the adoption of technologies of the fourth industrial revolution (Industry 4.0) has promising potential to bring about breakthroughs, new insights and unlock advancement opportunities across many areas of food quality and safety in the manufacturing sector (Hassoun et al., 2023). Industry 4.0 refers to the intelligent integration of machines and processes through cyber-physical systems that utilize embedded and networked technologies to enable autonomous control (Xu et al., 2021). In dairy safety, Dairy 4.0 refers to the use of cyber-physical systems such as IoT devices, blockchain, robotics, and advanced big data analytics to create intelligent, networked processes that enable real-time monitoring, automated control and traceability across the dairy supply chain, thereby reducing microbial, chemical, and other risks and ensuring product quality and safety. The adoption of Industry 4.0 technologies within the food industry, though gaining pace, remains at an early stage. Evidence from published literature indicates that Dairy 4.0 adoption in developed countries is relatively advanced, with approximately 15-50% of dairy farms integrating at least one form of digital or automated technology, depending on the region (Gabriel and Gandorfer, 2023). In contrast, adoption levels in developing countries where production systems are predominantly smallholder-based remain substantially lower, generally lower than 10% for most advanced Dairy 4.0 technologies. Fig. 5.1 shows the large disparity that exists in selected Dairy 4.0 technologies adoption between the developed and the developing countries. High adoption of automation and digital technologies in Europe and North America, moderate adoption in Latin America and parts of Asia, and very low adoption in Africa and South Asia mirror global patterns.

Industry 4.0 is characterized by the adoption of smart systems and advanced technologies throughout the entire value chain, fostering intelligent manufacturing, and production processes that optimize efficiency, enhance food quality and significantly reduce food losses. Emerging tools such as the Internet of Things (IoT) sensors, blockchain, and artificial intelligence (AI) are

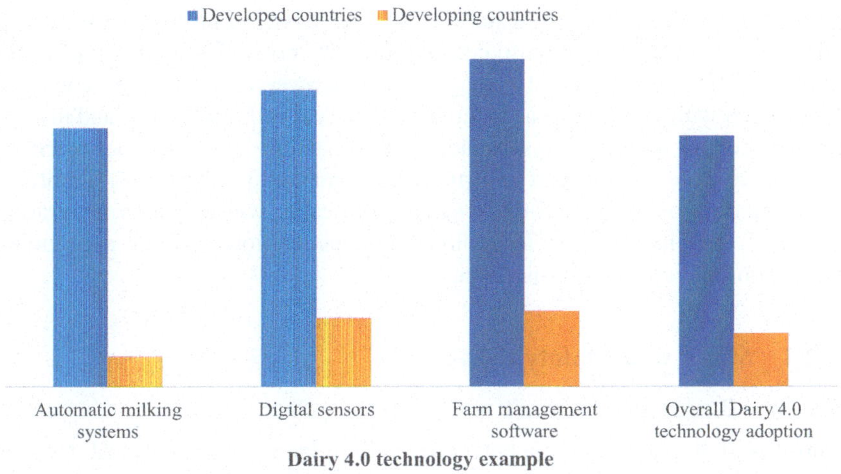

FIGURE 5.1 Schematic illustration of the adoption gap for Dairy 4.0 technologies between developed and developing countries based on patterns documented in the literature (Gargiulo et al., 2018; Groher et al., 2020; Silvi et al., 2021).

driving this transformation by strengthening transparency, predictive risk management and quality assurance in dairy production. IoT-enabled sensors monitor critical parameters like temperature and humidity throughout the supply chain, ensuring optimal storage conditions and reduced spoilage risks (Khanna et al., 2022). Blockchain technology provides a transparent record of each dairy product's journey, from farm to consumer, increasing accountability and consumer trust. Additionally, AI-driven analytics facilitate real-time decision-making, enabling proactive identification and mitigation of potential safety hazards in dairy production, handling, processing, and supply (Ferris et al., 2020).

In an era where a single food contamination event can spark global alarm, Industry 4.0 offers the precise, connected tools needed to transform dairy food safety from a reactive to a predictive approach. The purpose of this chapter is to highlight key dairy safety risks and explore how Industry 4.0 technologies have been integrated into the dairy sector to enhance dairy safety across the production and supply chain. Further, the chapter highlights the challenges, opportunities, and future trends in Industry 4.0 adoption in dairy safety management.

5.2 Global dairy safety landscape

Globally, the dairy sector is a vital source of nutrition and economic livelihood, yet it faces complex safety challenges due to milk's highly perishable nature, which necessitates robust management throughout the entire value chain, from production to consumption. In terms of the burden of milk-borne diseases, a substantial disparity exists between high-income countries (associated with issues such as listeriosis outbreaks and food fraud) and low- to middle-income countries (associated with systemic vulnerabilities such as inadequate cold chains and informal dairy markets). Therefore, a holistic understanding of the global dairy safety landscape, which allows integrating scientific, regulatory, and technological considerations within appropriate socioeconomic contexts, is necessary.

5.2.1 Major dairy safety risks

5.2.1.1 Biological hazards

Considered the most significant dairy value chain risk, biological hazards (pathogens) include harmful microorganisms such as bacteria, viruses, and parasites that cause foodborne illnesses to consumers (Wang et al., 2025). Key pathogens of importance in dairy include *Listeria monocytogenes*, *Salmonella* spp., *Escherichia coli* O157:H7, *Campylobacter* spp., and *Staphylococcus aureus* (Oliver et al., 2005). Dairy product contamination may occur through infected animals, poor milking hygiene, contaminated water or inadequate storage and transport conditions. Consumption of raw and inadequately processed dairy products enhances the risk of biological hazards. In high-income countries where pasteurization and stringent product quality and safety monitoring are implemented, the incidence of disease outbreaks is generally low. Even so, outbreaks of *Listeria monocytogenes* in ready-to-eat dairy products remain a persistent concern. Conversely, in low- and middle-income countries, limited cold chain infrastructure, reliance on informal markets, and insufficient regulatory enforcement contribute to a high prevalence of biological hazards in dairy products (El-Zamkan and Hameed, 2016). Effective control of biological hazards requires a combination of good agricultural and hygienic practices at the farm level, pasteurization, application of hazard analysis and critical control point (HACCP) systems at processing level, and continuous surveillance to protect consumer health and maintain dairy sector resilience. The key biological hazards in the dairy value chains are presented in Table 5.1

5.2.1.2 Chemical hazards

Chemical hazards in the dairy value chain arise from contaminants that compromise food safety and public health. Major concerns include agricultural contaminants such as veterinary drug residues, pesticides and naturally occurring toxins such as aflatoxin M1, which is a metabolite of aflatoxin B1 commonly found in contaminated animal feed in tropical countries (van Asselt et al., 2017). Aflatoxin M1 is particularly problematic in low- and middle-

TABLE 5.1 Key biological hazards in the dairy value chain.

Pathogen	Primary source	Associated dairy products	Common incident/outbreak scenario	References
Campylobacter jejuni	Cattle intestines, manure	Raw milk, raw cheeses	Localized outbreaks caused by the consumption of unpasteurized milk, often affect children or individuals on at-risk farms.	Jaakkonen et al. (2019)
Salmonella spp.	Cattle intestines, farm environment	Raw milk, cheese (especially soft cheeses), butter	Large-scale outbreaks caused by postpasteurization contamination or pasteurization failure, can affect a wide geographic area.	Flink and Nyberg (2020)
Listeria monocytogenes	Ubiquitous in the environment (soil, water), persists in cool, damp dairy factories	Soft cheeses, ice cream, and cream cheese	Postpasteurization contamination in the processing plant. Causes severe, widespread outbreaks due to its long survival period and ability to grow at refrigeration temperatures.	Gonzales-Barron et al. (2023)
Shiga-toxin producing *E. coli* (STEC), e.g., *E. coli* O157:H7	Cattle manure	Raw milk, raw cheeses, and sometimes yogurt	Outbreaks linked to raw milk consumption. It can cause severe complications like Hemolytic Uremic Syndrome (HUS) in children.	Flink and Nyberg (2020)
Brucella spp.	Infected animals	Raw milk, raw cheeses	Occupational hazard for livestock handlers. Outbreaks occur from eating raw cheeses or traditional unpasteurized products.	Bardenstein et al. (2023)
Coxiella burnetii	Animal birth products, manure from infected animals	Raw milk	Primarily an airborne occupational hazard, but it can be tansmitted via raw milk.	Valkovska et al. (2021)

income countries, where poor feed storage practices and limited regulatory oversight increase exposure risks (Kafle et al., 2024). Consumption of aflatoxin-contaminated food over a long period of time is associated with carcinogenic, immunotoxic, and hepatotoxic effects (van Asselt et al., 2017). Antimicrobial drug misuse can result in antimicrobial residues in dairy products, which can contribute to antimicrobial resistance in consumers. Environmental pollutants such as heavy metals, including lead and cadmium, cleaning and disinfectant residues, and dioxins can also enter the dairy value chain through contaminated water, soil, or packaging materials (Souto et al., 2025).

In high-income countries with strict monitoring systems, compliance with maximum residue limits and testing protocols has generally mitigated the occurrence of chemical safety risks. However, there are emerging concerns about intentional, economically motivated incidents of adulteration, such as the melamine scandal in China in 2008 (Pei et al., 2011). Melamine continues to be detected in infant formulas and other dairy products in developed countries such as the United States (Zhu and Kannan, 2018). Effective risk management requires integrated approaches encompassing feed quality control, residue monitoring, adherence to Codex Alimentarius standards, and strengthening regulatory frameworks to ensure consumer protection and international trade integrity. The main chemical hazards in the dairy value chains are presented in Table 5.2.

5.2.1.3 *Physical hazards*

Physical hazards in the dairy value chain refer to foreign objects that can pose safety risks during production, processing or packaging. Common contaminants include glass fragments from broken light fixtures or jars, metal shavings or screws from worn equipment, stones or soil from the farm environment, and plastic pieces from packaging materials or tools (van Asselt et al., 2017). These hazards often come from equipment failure, poor handling practices, or environmental contamination at the farm or processing facility.

Although less frequent than biological or chemical hazards, physical contaminants can result in consumer injury, product recalls and erosion of market confidence. In high-income countries, detection technologies such as metal detectors, x-ray scanners and improved packaging systems have minimized their occurrence. Conversely, in low- and middle-income countries, limited investment in processing technologies and quality assurance systems increases the likelihood of foreign material contamination. Preventive measures include strict adherence to good manufacturing practices, application of HACCP systems and regular equipment inspection and maintenance. Strengthening worker training and adopting modern detection technologies are essential for reducing physical hazard risks and safeguarding consumer health and the dairy industry's reputation (Joomun et al., 2024).

TABLE 5.2 Key chemical hazards in the dairy value chain.

Chemical hazard	Primary source in value chain	Reason for contamination	Health risk and incident type	References
Antibiotic drug residues	On-farm treatment of cows for bacterial diseases	Milk from treated cows is collected before the mandatory withdrawal period has ended.	Antimicrobial resistance, allergic reactions. Incidents include rejection of entire tanker loads in dairies.	Ghimpeţeanu et al. (2022)
Mycotoxins, e.g., aflatoxin M1	Animal feed	Animals consume feed, e.g., cereals and legumes contaminated with molds such as aflatoxin-producing *Aspergillus* spp. and metabolize aflatoxin B1 to aflatoxin M1, which is secreted in milk.	Aflatoxin is a carcinogen, which can cause liver damage. It can lead to widespread product recalls due to exceeding regulatory limits.	Ushkalov et al. (2020)
Cleaning and sanitizing agents	Processing facility: clean-in-place systems	Inadequate rinsing of equipment after cleaning with acids, caustics, or iodine-based sanitizers.	Acute poisoning, gastrointestinal illness. It can lead to localized consumer complaints and product recalls.	McCarthy et al. (2018)
Melamine	Processing/collection of milk. It is usually intentional adulteration	An adulterant added to diluted milk to artificially inflate the protein content reading in quality tests.	Kidney failure, kidney stones and death in infants. It can lead to massive, intentional fraud with fatal consequences.	Xiu and Klein (2010)

(*Continued*)

TABLE 5.2 (Continued)

Chemical hazard	Primary source in value chain	Reason for contamination	Health risk and incident type	References
Environmental contaminants (dioxins, PCBs, heavy metals)	On-farm: environment and feed	Contaminated soil, water or feed leads to bioaccumulation of toxins in the milk fat.	Carcinogens, endocrine disruptors, and organ damage. It can lead to long-term exposure risk, leading to localized food chain shutdowns.	Souto et al. (2025)
Pesticides and herbicides	On-farm: feed and environment	Residues from crops used as cattle feed, or from pasture treatment, are transferred to milk.	Neurotoxins, carcinogens. Detection during routine monitoring leading to product issue holds.	Schopf et al. (2023)
Allergens	Processing plant: cross-contamination	Inadequate cleaning between production runs of different products (e.g., plain milk followed by a nut-flavored milk product).	Life-threatening anaphylactic shock in allergic individuals. Leads to major recalls due to mislabeling or cross-contamination.	Hooda et al. (2025)

5.2.1.4 Allergenic hazards

Dairy products are among the nine major food allergens (Adedeji et al., 2024), posing a high risk to sensitized individuals who may experience reactions ranging from hives to life-threatening anaphylaxis upon exposure. The primary allergenic components are proteins, specifically casein and whey, which retain their allergenicity even after undergoing processing methods such as heating or fermentation (Hooda et al., 2025). Risks arise primarily from cross-contamination in facilities processing multiple products and from mislabeling or undeclared milk ingredients in products marketed as dairy-free or vegan.

In high-income countries, strict allergen labeling laws and monitoring systems aim to protect consumers yet mislabeling and undeclared allergens remain leading causes of food recalls (Hooda et al., 2025). In low- and middle-income countries, weak regulatory enforcement and informal markets exacerbate exposure risks, as products may lack clear allergen information. Effective management requires compliance with Codex Alimentarius guidelines on allergen declaration, robust traceability systems and good manufacturing practices to prevent cross-contamination. Advances in rapid allergen detection methods and consumer education also play vital roles. Ultimately, addressing allergenic hazards is essential not only for protecting vulnerable populations but also for fostering consumer trust and enabling equitable participation in international dairy trade.

5.2.2 Dairy supply chain vulnerabilities

Dairy products' perishability and sensitivity to quality changes are directly associated with the complex supply chains and risk exposure from farm-to-fork. Since milk spoils quickly and passes through many hands, even minor lapses can trigger safety, economic or public health concerns.

5.2.2.1 On-farm vulnerabilities

At the farm level, dairy safety risks often come from poor hygiene and animal health practices. Inadequate milking hygiene, contaminated udder surfaces, and mastitis infections can introduce *Staphylococcus aureus*, *E. coli* O157:H7, and *Listeria monocytogenes* into raw milk. The indiscriminate use of veterinary drugs and failure to follow withdrawal periods can lead to antibiotic residues that threaten product safety and may promote antimicrobial resistance (van Asselt et al., 2017). Feed contamination, especially from moldy or poorly stored feeds allows aflatoxin B1 to enter the food chain through milk as aflatoxin M1 (Ushkalov et al., 2020).

5.2.2.2 Milk collection and bulking centers

Milk collection and bulking points constitute critical handling stages where risks occur. In many developing countries, milk from multiple smallholders is pooled without prior microbial or compositional testing (Nyokabi et al., 2024).

The absence of cooling tanks or unreliable electricity accelerates bacterial multiplication during transport. Economic pressures may often lead to adulteration practices such as dilution with water or addition of preservatives such as hydrogen peroxide to mask spoilage and maintain volume-based income . These actions not only compromise safety but also erode consumer trust in the local dairy systems.

5.2.2.3 Milk processing and packaging

Processing facilities present both opportunities and risks for milk safety. When correctly managed, pasteurization for example, eliminates pathogens and extends product shelf life. However, poor equipment sanitation, ineffective cleaning-in-place systems or suboptimal pasteurization parameters can allow pathogen survival or recontamination. For example, *Listeria monocytogenes* persists in dairy environments through biofilm formation, leading to post-pasteurization contamination (Flink and Nyberg, 2020), which can result to listeriosis outbreaks. In small-scale plants, manual handling and inadequate segregation between raw and processed products zones further elevates the risk.

5.2.2.4 Distribution and retail

The integrity of the cold chain is indispensable for safeguarding dairy safety during distribution and retail operations. Any disruption in temperature control, such as inconsistent refrigeration or inadequate monitoring, creates conditions that favor rapid microbial proliferation and subsequent spoilage. Such fluctuations may also lead to enzymatic and physicochemical changes that compromise product texture, flavor and nutritional quality. Packaging defects such as leakage, weak seals or exposure to direct sunlight intensify these risks by accelerating oxidation, increasing moisture loss, and permitting the entry of contaminants. In many developing countries markets, milk is commonly transported and sold through informal channels where regulatory oversight is limited (Wang et al., 2025). Under these conditions, the absence of rigorous quality checks, reliable documentation and effective recall mechanisms hampers the rapid isolation and removal of compromised products from circulation. Studies have shown that even minor deviations from optimal storage conditions can significantly reduce the shelf life of milk and other perishable dairy products leading to both economic loss and public health risk (Nyokabi et al., 2024). These vulnerabilities highlight the urgent need for systematic monitoring, accountability, and improved coordination across the distribution and retail chain.

5.2.2.5 Governance and market structure

The resilience of a dairy supply chain is ultimately shaped by its governance architecture. Fragmented production systems with numerous smallholders and weak institutional coordination make it challenging to enforce hygiene and safety standards. In many developing countries, inspection agencies operate with limited laboratory capacity and insufficient personnel, reducing the frequency and effectiveness of compliance monitoring (Grace, 2015). Conversely, developed countries systems maintain comprehensive regulatory frameworks aligned with Codex Alimentarius and/or ISO standards, supported by automated testing, digital traceability and efficient recall mechanismsa.

5.2.3 Safety disparities in developed and developing economies

In developed economies, vulnerabilities in the dairy supply chain are predominantly process-based, arising from technological malfunctions, human error, or economic fraud (Yang et al., 2023). These risks are effectively mitigated through a combination of automation, rigorous regulatory frameworks, routine quality audits and consumer-driven accountability mechanisms, often supported by internationally recognized standards such as Codex, HACCP, and ISO 22000. Such systems enable rapid detection and correction of deviations in processing, packaging, and distribution, thereby maintaining high levels of product safety and consistency. In contrast, developing economies face more systemic and structural challenges. Poor infrastructure, limited access to modern processing and storage facilities, informal dairy trade, and weak governance collectively heighten the risk of contamination and compromise product quality (Feyisa et al., 2024). These limitations make monitoring, traceability, and enforcement of safety standards difficult, increasing both public health risks and economic losses throughout the dairy value chain.

Addressing these disparities demands multidimensional approaches. Strengthening veterinary and extension services, expanding cold-chain capacity and formalizing smallholder supply chains can substantially enhance safety outcomes (Feyisa et al., 2024). Moreover, integrating accessible rapid testing, establishing traceability across actors and building consumer awareness can foster accountability throughout the dairy value chain. Ultimately, the resilience of the dairy value chain depends not only on technological interventions but also on institutional coherence and cross-sector collaboration (Joomun et al., 2024). A risk-based food-safety culture embedded within both industry and public institutions remains the cornerstone of sustainable dairy development (Spagnoli et al., 2024).

5.3 Core Industry 4.0 technologies and their relevance to dairy safety

The coming decades will see a sharp increase in demand for animal-derived foods driven by population growth, higher income gains, and urbanization, placing greater strain on dairy systems and accelerating the need for efficient innovative-driven solutions (Hassoun et al., 2023). Industry 4.0 approaches, when applied thoughtfully across production, processing, and distribution, would improve supply-chain resilience, reduce losses, and strengthen safety governance while supporting sustainable livelihoods. Reviews of recent literature indicate growing evidence that digitalization and data-driven practices can help reconcile rising demand with food-safety and sustainability goals, particularly when accompanied by institutional support and inclusive deployment strategies (de Oliveira et al., 2024).

The concept of Industry 4.0 originated in Germany and basically describes the wave of high-tech innovations that emerged between 2011 and 2015 (Kagermann and Wahlster, 2022). It is now widely used to describe fully automated and interconnected industrial production processes. Recently, Dairy 4.0 concept has been described (Hassoun et al., 2023). Dairy 4.0 is essentially the implementation of the advanced digital technologies and concepts of Industry 4.0 specifically within the dairy farming, processing and supply chain sectors (Fig. 5.2).

The advances in computing power and digital technologies have reached the dairy value chain, where they have evolved into a sector-specific paradigm referred to as Dairy 4.0 (Rose and Chilvers, 2018). Its pillars include AI, big data analytics, robotics, IoT and blockchain, all of which are leveraged to enhance efficiency, sustainability and productivity in dairy value chains (Gehlot et al., 2022). Examples of Industry 4.0 technology applications in the dairy industry are shown in Table 5.3.

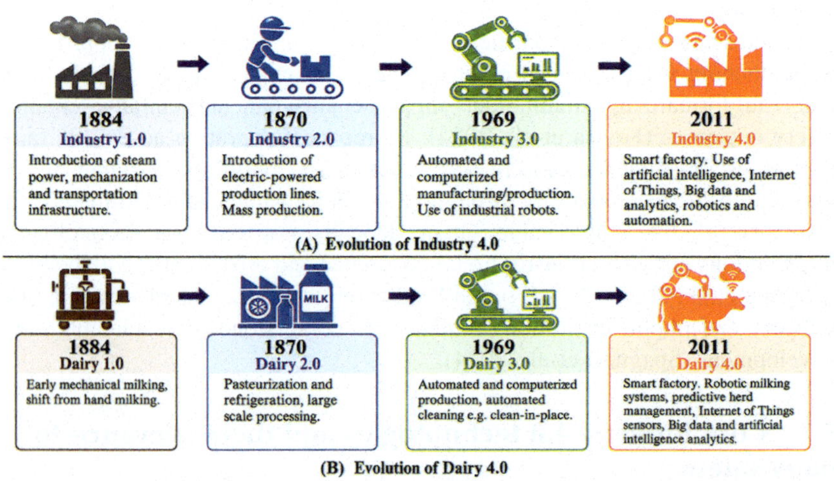

FIGURE 5.2 Evolution of industry and dairy. Comparison of the four industrial revolutions (Industry 1.0 through 4.0) and the corresponding technological advancements in the dairy sector (Dairy 1.0 through 4.0).

TABLE 5.3 Examples of Industry 4.0 technology applications in the dairy industry.

Industry 4.0 technology	Study objective	Relevance to dairy industry	References
Internet of Things (IoT)	Real-time monitoring of temperature, pH, and spoilage. Progress and challenges of sensor technologies for dairy food safety monitoring.	IoT devices like the Narrow-band Internet of Things (NB-IoT) are integrated with smart collar sensors to detect estrus periods with greater precision. Through it, dairy farmers in northwest of China increased their annual revenue by 50% after adopting this technology. IoT applications extend to milk quality testing, particularly in detecting adulteration and yeast spoilage. Monitoring spoilage involves the use of spectral, electrochemical, pH, temperature, and humidity sensors.	Habsari et al. (2022)
	Application of quaphotomics and NIR water spectral patterns in dairy production. Assessment mid-infrared spectroscopy for detecting milk adulterants and contaminants. Assessment of the ability of an online NIR spectrometer to predict key milk quality parameters continuously	Business Process Modeling Notation (BPMN), use case diagrams, and DHT22 sensors (a low-cost humidity and temperature sensor) were used to track milk temperature in real time. Identified that spectral sensors (NIR, FTIR, hyperspectral) consistently reduce detection times and improve accuracy compared to conventional laboratory methods, supporting safer milk supply chains. Showed that water spectral fingerprints can detect subtle adulteration and spoilage, offering a novel monitoring approach for milk freshness. Confirmed MIR spectroscopy can detect common adulterants such as urea, starch, and sucrose at low concentrations, providing robust quality assurance tools for processors. Demonstrated strong correlations ($R^2 > 0.9$) between predicted and laboratory-measured milk composition, proving the system's effectiveness	Habsari et al. (2022), Fernández González et al. (2024), Giovinazzo et al. (2024), Ceniti et al. (2023), Iweka et al. (2023)

(*Continued*)

TABLE 5.3 (Continued)

Industry 4.0 technology	Study objective	Relevance to dairy industry	References
	in an automatic milking system.	for automated, continuous quality monitoring and integration into Industry 4.0 frameworks.	
Big data	Forecasting spoilage, shelf-life, and contamination risks	Research utilizing the Temperature Humidity Index big data (THI-BD) has shown that heat stress negatively impacts dairy cattle, hence leading to reduced milk production Research assessed the global health status of dairy cows using big data from milk records; milk yield, somatic cell count and Fourier transform mid-infrared (FT-MIR)-based predictors related to milk composition and animal health status. The results indicated that the quantitative traits obtained reflect some of the major health disorders in dairy farming and big data should continuously be used to monitor a large herd. A study explored the challenges faced by dairy farms in detecting hyperketonemia, a metabolic disorder affecting cows during the transition to lactation. The results suggested that big data can analyze the massive datasets available to farms such as milk production, composition, records of cow management, and genomics to monitor the onset of postpartum hyperketonemia.	Ekine-Dzivenu et al. (2020), Franceschini et al. (2022)
Artificial intelligence (AI)	Smart quality control and anomaly detection	AI can mine data from multiple sources and simplify repetitive and difficult decision-making tasks in dairy farming AI has been used in precision livestock farming for milk capture technology. If the milk production reduces by 20%, a farmer will be informed promptly in order to decide on the next course of action, which can include troubleshooting the feed bunk or calling a veterinarian. Cornell researchers used rumination monitoring systems daily to decide which cow needs an immediate medical attention.	Nogoy et al. (2021), de Oliveira et al. (2024), de Oliveira et al. (2024), Gromski et al. (2015), Loddo et al. (2022), Ji et al. (2022), Mhapsekar

TABLE 5.3 (Continued)

Industry 4.0 technology	Study objective	Relevance to dairy industry	References
		ML-driven milk quality assessment has been extended to species identification, and as such, the authenticity of dairy products from different breeds is assured. Techniques such as Partial Least Squares Discriminant Analysis (PLS-DA) and k-Nearest Neighbors (k-NN) have been used to differentiate milk from various animal sources. Machine learning techniques have been used to determine cheese quality during the ripening process Data acquired from automatic milking machines has been used to predict the value of important parameters such as daily milk yield, milk composition, and milking frequency. For instance, this study developed a machine learning framework using data from a robotic dairy farm to predict such parameters. Deep learning models were primarily employed in multiclass classification tasks like identifying different types of adulteration of milk. The data from the IoT sensors is uploaded and subsequent analysis of body temperature is done through deep learning method to detect diseases, estrus, parturition, and mastitis.	et al. (2025), Mhapsekar et al. (2025)
Blockchain	Tracking milk from farm to consumer	A study proposed a dairy logistics supply ecosystem with blockchain technology to provide fresh and safe dairy products at fair prices by employing smart contracts with crowdfunding functionality.	Varavallo et al. (2022)

(Continued)

TABLE 5.3 (Continued)

Industry 4.0 technology	Study objective	Relevance to dairy industry	References
		A blockchain-based traceability solution for the Fontina PDO cheese supply chain was deployed, which demonstrated the potential to lower environmental impact and reduce transaction costs.	
Automation and robotics	Hygienic milking and processing systems	The widely known application of robotics is the automatic milking system, which employs laser-guided detection of teat placement. Sensors on the milking systems will collect live data on daily milk yield, milk composition, and milking frequency Beyond milking, robotics has also been applied in other areas of dairy farming, such as self-driving vehicles to automate feed delivery and floor cleaning, thereby improving operational efficiency.	Espinoza-Sandoval et al. (2024), Espinoza-Sandoval et al. (2024), Espinoza-Sandoval et al. (2024)

5.3.1 Internet of Things-enabled smart monitoring systems for dairy health and milk quality

The Internet of Things (IoT) refers to a network of interconnected physical objects embedded with sensors, software, and communication technologies that enable the collection, exchange and analysis of data. In agriculture and particularly in dairy farming, IoT has emerged as a key enabler of smart farming, allowing farmers to monitor livestock and machinery in real time, thereby improving efficiency and sustainability. Specifically in the dairy sector, IoT technologies have been applied to animal health monitoring, reproductive management and milk quality assurance, offering farmers actionable insights that optimize productivity and animal welfare (Rutten et al., 2013). In such cases, data is uploaded to cloud servers, where machine learning algorithms analyze it and provide real-time insights into an animal's health and overall condition. For instance, to optimally produce milk, a cow's temperature must be continuously monitored to accurately predict its estrus period. Farmers in Northwest China, for example, have reported a 50% increase in revenue after adopting IoT devices such as Narrowband IoT (NB-IoT) smart collar sensors that enable estrus detection with greater precision.

IoT applications in milk quality assurance and fraud detection are evolving rapidly alongside animal monitoring systems. Advanced sensor arrays and "electronic tongue" platforms, incorporating potentiometric and spectroscopic technologies have demonstrated high sensitivity in detecting common milk adulterants and differentiating among milk grades. These systems provide low-cost, near-real-time screening capabilities that complement conventional laboratory analyses and enhance traceability across the dairy value chain (Perez-Gonzalez et al., 2024). When the continuous data streams generated by such sensors are coupled with machine learning algorithms, particularly artificial neural networks and chemometric models, automated grading, authentication, and anomaly detection become possible at milk collection centers and throughout the cold chain (Behkami et al., 2019).

Smart sensors based on spectral data are central to dairy safety as they provide rapid, nondestructive and highly automated assessment of chemical contaminants, adulterants, and microbial contaminants. Techniques such as Raman spectroscopy, infrared spectroscopy, and hyperspectral imaging are being embedded into smart sensing platforms capable of continuous monitoring and real-time decision support in dairy processing (Joolaei Ahranjani et al., 2025). These sensors detect food safety hazards and/or quality changes with high sensitivity, thereby reducing reliance on traditional laboratory-based methods that are time-consuming (Fernández González et al., 2024). Raman spectroscopy has demonstrated strong capability for detecting nitrogen-rich adulterants and other chemical contaminants in milk, offering high sensitivity suitable for in-line screening (Meenakshi et al., 2025). Infrared spectroscopy enables robust quantification of milk constituents and reliable detection of adulteration when combined with chemometric modeling (Ceniti et al., 2023). Similarly, near-infrared sensors provide rapid multiparameter monitoring and are increasingly integrated into Industry 4.0 systems for automated quality assessment (Iweka et al., 2023). Hyperspectral imaging extends these capabilities by coupling spatial and spectral information, allowing sensitive detection of food safety hazards such as melamine in milk powder and supporting machine-vision-based safety control (Patel et al., 2024). Collectively, these spectral smart sensors form the foundation of data-rich, connected and predictive dairy safety systems, which aligns with Industry 4.0 goals of traceability, automation, and early risk detection.

5.3.2 Big data and predictive analytics for forecasting spoilage, shelf-life, and contamination risk

Big data and predictive analytics are transforming dairy production and quality management by enabling continuous assessment of spoilage dynamics, shelf-life, and contamination risks across the dairy value chain. In modern dairy farms, sensors attached to animals and processing systems generate large, heterogeneous datasets that include physiological, environmental, and product-quality indicators. AI and statistical modeling techniques convert these massive data streams into actionable insights, making them invaluable for managing complex, data-intensive systems such as large herds or multisite cold chains (Rutten et al., 2013). A practical example is the Temperature-Humidity Index Big Data (THI-BD) model, which integrates weather, physiological, and milk-yield datasets to predict heat stress impacts on dairy performance. Such predictive models not only safeguard productivity and animal welfare but also indirectly influence milk stability and compositional quality under thermal stress conditions (Kulaz and Ser, 2022).

Beyond on-farm monitoring, big data approaches have found increasing application in predicting spoilage and estimating dairy products shelf-life during storage and distribution. By analyzing time-temperature profiles, microbial counts, and physicochemical parameters collected through IoT-based sensors, predictive analytics platforms can identify deviations that signal early spoilage or safety risks. For instance, machine learning models trained on historical temperature and microbial growth data can dynamically estimate remaining shelf-life and alert distributors before products exceed safety thresholds (Ji et al., 2022). These predictive systems enhance dairy cold-chain integrity and minimize dairy products losses through proactive decision-making.

At the industrial level, dairy processors are leveraging big data architectures to harmonize information from production plants, logistics and retail outlets (Hassoun, 2025). Companies such as Nestlé have invested in advanced predictive analytics tools to optimize quality consistency and shelf-life forecasting across global supply chains. These systems integrate sensor data, digital twins, and cloud-based dashboards to continuously monitor temperature abuse, product age, and contamination risk in transit. Ultimately, big data and predictive analytics form the analytical foundation for data-driven food safety management. By correlating multisource datasets including microbial counts, storage temperatures, and chemical markers, dairy stakeholders can anticipate spoilage, evaluate contamination probabilities, and implement preventive interventions rather than relying on reactive controls (Serrano-Torres et al., 2025). As the sector advances toward fully digitized quality assurance, these tools will be critical in ensuring traceability, reducing waste and maintaining consumer confidence in dairy products.

5.3.3 Artificial intelligence and machine learning applications in dairy health monitoring and quality control

Artificial intelligence (AI) refers to the ability of machines to learn patterns and make predictions by combining computer science with robust datasets to enable problem-solving. When an AI model is exposed to new data, it performs machine learning (ML), which involves identifying patterns and improving performance over time without being explicitly programmed (Cabrera et al., 2025). To improve precision, the model must be continuously trained on relevant large, structured collections of naturally occurring text or spoken language data to refine its predictive capabilities (Habsari et al., 2022). As additional computational layers are introduced, the system advances to deep learning, a more complex level of ML that uses artificial neural networks (ANN) inspired by the structure of the human brain, thereby yielding more accurate and context-aware predictions.

In livestock management, by analyzing the combined impacts of physiological and environmental factors, AI systems can recommend optimal treatment strategies for cows, improving both milk quality and quantity. AI can also employ predictive models to detect early signs of anomalies within a herd and provide farmers with necessary information to help make decisions on preventive measures that can reduce disease spread, minimize treatment costs, and reduce incidences of low productivity in affected livestock. For example, the Dairy Brain Project has leveraged real-time data from milking equipment, feed intake systems, and health-tracking devices attached to cows (Wang et al., 2023). Additionally, unsupervised ML techniques such as clustering and autoencoders enhance disease detection by identifying anomalies without predefined labels, making AI-driven systems more adaptable to emerging threats . Recently, a study demonstrated that ML algorithms applied to automated milking system data achieved prediction accuracies of 62%–74% for clinical mastitis, thus offering farmers with early intervention options and reduced treatment costs (Faris et al., 2024).

In milk handling and processing, AI models equipped with cognitive and decision-making layers have been developed that are capable of performing multiple tasks simultaneously, for example, identifying adulteration, quantifying milk composition and classifying milk sources. This versatility makes them more effective than conventional laboratory approaches. ML-driven milk quality assessment has also been extended to species identification, ensuring the authenticity of dairy products. Techniques such as Partial Least Squares Discriminant Analysis (PLS-DA) and k-Nearest Neighbors (k-NN) have been applied to differentiate milk from various animal sources, addressing concerns around mislabeling, authenticity and transparency (Gromski et al., 2015). Beyond herd health, ML combined with computer vision have been employed in dairy product quality control. An example is the automatic monitoring system that has successfully classified ripening stages in cheeses, reducing human subjectivity and ensuring consistent product quality (Loddo et al., 2022).

5.3.4 Blockchain technology for dairy supply chain traceability and transparency

Blockchain is a permanent digital ledger used to record transactions and track assets within a business network, providing a single and reliable source of traceability in which transactions are grouped into blocks, which are then cryptographically linked, so that if an error occurs, a new record is added while previous entries remain unchanged and visible to all relevant parties. In the dairy industry, blockchain technology is being applied to improve traceability and transparency by enabling milk and dairy products to be tracked from farm to consumer. Its key strengths include data interoperability, cost reduction, transparency, auditability, integrity, and authenticity. By building intelligent ecosystems for milk supply chains, blockchain can help address the rising global demand for food in a more sustainable and accountable manner. For example, a green blockchain-based traceability platform applied in the Fontina PDO cheese supply chain in Italy achieved low energy-consumption, reduced transaction cost and real-time data-sharing between producers and consumers (Varavallo et al., 2022). Similarly, a blockchain-enabled food-traceability platform integrating smart contracts, IoT sensors, and QR-code-driven tracking in the Indian dairy industry showed improvements in information-sharing, inventory management and distribution efficiency, and enhanced food-safety monitoring and adulteration detection (Khanna et al., 2022). These findings underscore blockchain as a transformative tool for building resilient, sustainable and consumer-centric dairy ecosystems.

At the operational level, blockchain architectures in dairy supply chains are often layered with IoT for data capture, cloud/off-chain databases for storage, and smart contracts for workflow automation. Real-time entries such as milk collection time-stamp, truck temperature log, and processing-plant batch number get hash-linked to the ledger, while downstream consumers or regulators scan QR codes to verify the chain-of-custody information and assurance certificates (Khanna et al., 2022). This improves transparency and fosters consumer trust by making provenance and authenticity verifiable. Additionally, immutable blockchain logs offer auditability and reduce the risk of fraud and product substitution, especially relevant for products vulnerable to adulteration or origin-mislabeling.

5.3.5 Automation and robotics in hygienic milking and processing systems

Robotics, as an inherently interdisciplinary field combining mechanical, electrical, electronic, and software systems is increasingly employed in dairy farming to design, deploy and operate advanced autonomous machines. These systems address critical challenges such as labor shortages, human error, precision in farm operations, sustainability, and profitability (John et al., 2016). On modern dairy farms, robotic milking systems (often called automatic milking systems) facilitate cluster-based milking, allow cows to self-select milking times and reduce dependency on manual labor. For example, a teat-cup attachment robot developed in China achieved an attachment success rate of 98% while dynamically adapting to teat movement significantly reducing labor input and improving milking throughput (Wang et al., 2023). Beyond mere attachment automation, the integration of sensors on robotic milking systems enables real-time monitoring of milking frequency, milk yield, milk composition and animal health indicators. Farms adopting robotics report reductions in labor input, enhanced energy-use efficiency and higher operational performance (Lee et al., 2024). Robotic systems have also been retrofitted with barn-cleaning robots, autonomous feed-delivery vehicles and manure-handling robots, contributing to hygiene improvement, lower pathogen risk, and reduced environmental contamination.

Milking robots often include automated teat-cleaning, pre- and postmilking disinfection, and sterile cup attachment, which reduce cross-contamination risks and support milk-quality assurance (Wang et al., 2023). In processing systems, robotics perform tasks such as packaging, sorting, palletizing, and cleaning all with high consistency and minimal human contact further elevating hygiene standards. The coupling of robotics with sensor networks and cloud platforms enables continuous data flows, prompt anomaly detection, and proactive maintenance (Palanisamy et al., 2025).

5.4 Challenges and future perspectives of Industry 4.0 in dairy safety

Despite its transformative potential to enhance dairy safety through dynamic risk assessment, automated quality control, real-time monitoring, and end-to-end product traceability, the adoption of Industry 4.0 technologies in the dairy sector remains constrained by significant challenges. Addressing these obstacles is crucial to fully realize the benefits of smart dairy operations, including improved food safety, reduced spoilage, enhanced productivity, and strengthened consumer trust.

The emergence of Dairy Safety 4.0 concept is a direct consequence of the technological and conceptual foundations laid by Industry 4.0, which denotes the adoption of Industry 4.0 tools (discussed in section 5.3) across the dairy value chain aimed at improving efficiency, sustainability and product quality (Hassoun et al., 2023). Industry 4.0's influence on dairy safety began with real-time data acquisition and analytics using technologies discussed in Section 5.3. This real-time monitoring detects anomalies such as temperature deviations indicative of safety and spoilage risk or somatic cell count fluctuations signaling mastitis much faster than traditional manual inspections. Such capabilities mitigate contamination risks promptly and enable interventions before safety breaches occur (Malik et al., 2025). The integration of Industry 4.0 technologies has enabled continuous monitoring of milk composition, automated detection of anomalies and predictive maintenance of processing equipment capabilities that form the backbone of Dairy Safety 4.0 (Hassoun, 2025).

Smart sensors and IoT devices, for example, have revolutionized on-farm and in-plant data collection. Real-time monitoring, for example, IoT sensors and wearable collars now provide continuous data on physiological parameters such as rumination and temperature, and milk quality, for example, somatic cell count and pH at every stage (Neculai-Valeanu et al., 2025). These tools allow for real-time tracking of animal health, milking hygiene, storage conditions and processing parameters. Such data streams support early detection of safety risks and enable rapid corrective actions. The dairy sector's adoption of these technologies mirrors the broader Industry 4.0 trend of embedding intelligence into physical systems to enhance reliability and efficiency (Sangode, 2025). Artificial intelligence and machine learning have further accelerated the transition toward Dairy Safety 4.0. AI-driven analytics can identify small patterns in milk quality data, predict spoilage risks, and optimize processing workflows. AI-driven predictive analytics (AI and ML models) are increasingly used to forecast disease outbreaks like mastitis and detect anomalies in the processing line before contaminated products enter the supply chain (Semercioz-Oduncuoglu and Luning, 2025). These capabilities reduce human error and enhance the precision of safety interventions. The dairy industry is increasingly relying on AI to manage variability in milk quality and to ensure compliance with stringent safety standards across global markets. Collectively, these technological advances have catalyzed a paradigm shift from reactive to predictive and preventive safety management in the dairy sector. The concept of Dairy Safety 4.0 formalizes this shift by framing dairy safety as an integrated, data-driven ecosystem rather than a set of isolated quality checks.

5.4.1 Challenges of adopting Industry 4.0 in the dairy industry

5.4.1.1 Limited digital infrastructure

Limited digital infrastructure and poor connectivity in many rural areas, especially in developing countries where dairy farms are situated, hinder the widespread adoption of advanced technologies. As a result, farmers often face difficulties in accessing and utilizing tools such as precision farming and automated milking systems (Dadi et al., 2021). Insufficient broadband coverage, unstable internet connectivity, and inadequate access to advanced information and communication technologies restrict farmers from fully exploiting digital innovations (Zambon et al., 2019). Precision livestock farming, automated milking systems, real-time monitoring sensors, and data-driven decision-support tools all depend heavily on reliable connectivity and robust digital networks. Without this foundation, the integration of technologies such as IoT, AI, and cloud-based platforms is highly constrained. Furthermore, the digital divide not only slows technology adoption but also exacerbates socioeconomic inequalities, as smallholder and remote farmers are disproportionately excluded from opportunities for productivity gains, animal welfare improvements, and enhanced food safety assurance. Addressing these infrastructural gaps is therefore essential for enabling equitable access to smart farming solutions and unlocking the full potential of Industry 4.0 in transforming dairy production systems.

5.4.1.2 High cost of implementing new technologies

The high cost of implementing advanced technologies represents a major barrier to adoption, particularly for small and medium-sized dairy enterprises (SMEs), which are typically constrained by limited financial and capital resources (Bahn et al., 2021). Investments in precision farming equipment, automated milking systems, robotics and digital platforms require substantial upfront expenditure as well as ongoing maintenance and training costs. For SMEs, these expenses may exceed available budgets or fail to yield an immediate return on investment, making adoption less feasible compared to larger farms that benefit from economies of scale. Larger enterprises can spread technology costs across higher production volumes, thereby reducing per-unit costs and accelerating productivity gains, while smaller farms often struggle to justify the financial outlay. This imbalance contributes to a widening digital divide within the dairy sector, where technologically advanced large farms gain competitive advantages in efficiency, product quality and market access, while SMEs risk being left behind. Over time, such disparities may reinforce structural inequalities in the dairy industry, potentially threatening the viability of smaller farms unless targeted financial support, cooperative models, or policy incentives are introduced to level the playing field.

5.4.1.3 *Need for skilled workforce*

Companies seeking to adopt Industry 4.0 solutions must ensure they have adequately trained human resources and provide the necessary tools and infrastructure for staff development (Goller et al., 2021). This requirement can be particularly difficult for smaller dairy farms and firms, which often struggle to attract qualified personnel or finance comprehensive training programs. In addition, the increasing reliance on connected devices, sensors, and data analytics introduces significant concerns regarding data privacy and cybersecurity.

5.4.1.4 *Limited or lack of collaboration and coordination among supply chain stakeholders*

In the dairy industry, the successful implementation of Industry 4.0 technologies relies heavily on strong collaboration and coordination among supply chain stakeholders, including farmers, processors, distributors and technology providers. Effective integration of advanced tools such as automated milking systems, real-time monitoring sensors, and data analytics platforms requires alignment across multiple nodes of the supply chain to ensure interoperability, data sharing and process standardization. Without coordinated efforts, technological investments may fail to achieve their intended benefits due to mismatched processes, inconsistent data management or gaps in operational capacity (Hassoun et al., 2023). Collaborative frameworks, such as integrated farm-to-fork platforms and stakeholder partnerships, can facilitate knowledge transfer, optimize resource utilization, and enhance decision-making across the chain. Furthermore, fostering trust and transparency among stakeholders is essential, particularly when sharing sensitive production data or implementing joint technology initiatives. Therefore, the degree of supply chain collaboration is a key determinant of the efficiency, scalability, and sustainability of Industry 4.0 adoption in the dairy sector.

5.4.1.5 *Cybersecurity and data privacy concerns*

Modern dairy plants rely on interconnected devices to monitor temperature, pH, microbial levels, and processing flow in real time. Any cyberattack, malware, or system breach can disrupt production, compromise food safety, or result in contamination going undetected. The complexity of cyber-physical systems in processing lines increases vulnerability. Many sensors, automated control units and data networks lack standardized security protocols, making them prone to unauthorized access or operational interference (Cabrera et al., 2025). Additionally, the sharing of sensitive production data with suppliers, distributors, and regulatory authorities raises questions about data ownership, privacy, and compliance, particularly under fragmented regulatory frameworks. Automated dairy operations increasingly rely on networked refrigeration systems, pasteurizers, sorting robotics, etc. Cyberattacks against operational technology (OT) systems threaten both supply chain continuity and food safety (Neethirajan, 2024). Risks range from tampering with storage temperatures to disrupting distribution flows. Resilience requires defense-in-depth security, cyber-aware design and OT risk assessments integrated into food-safety audits.

Addressing these challenges requires robust cybersecurity measures, including encryption, multifactor authentication and secure network architectures. By prioritizing cybersecurity and data privacy, dairy processors can ensure the integrity of operations, maintain compliance with food safety regulations, and protect consumer trust in dairy product quality.

5.4.1.6 Sensor reliability, standardization, and validation

One of the critical challenges in implementing Industry 4.0 for dairy safety lies in the reliability, standardization and validation of sensor technologies. Dairy environments present harsh operating conditions, with constant exposure to variable temperatures, high humidity, and rigorous cleaning processes such as clean-in-place. These factors can compromise sensor accuracy and durability, leading to measurement drift or false readings. Unreliable sensors undermine the integrity of real-time monitoring systems, making it difficult to detect microbial hazards, chemical contaminants, or indicators of spoilage in a timely manner (Fernández González et al., 2024).

Equally important is the issue of standardization. Industry 4.0 landscape is characterized by diverse technologies, ranging from IoT devices and biosensors to blockchain-enabled tracking and AI-based analytics. However, the absence of universal standards for sensor calibration, data formats, and communication protocols makes it challenging to integrate these tools into a coherent, interoperable system (Cabrera and Fadul-Pacheco, 2021). Without standardization, data comparability across supply chain actors becomes limited, reducing the potential of Industry 4.0 to enhance dairy traceability and safety assurance.

Validation represents an additional barrier to widespread adoption. New-generation sensors must be rigorously validated against gold-standard laboratory techniques such as microbial culture, polymerase chain reaction, or chromatographic assays. Many promising technologies, including biosensors, remain in the experimental or pilot stage, lacking sufficient validation data to gain regulatory acceptance (Fernández González et al., 2024). Until sensor systems are proven to be both accurate and robust, regulatory agencies and industry stakeholders will remain cautious in relying on them for critical food safety decisions.

5.4.1.7 Ethical and regulatory considerations

The adoption of Industry 4.0 technologies in the dairy sector introduces regulatory and ethical complexities. Traditional food quality and safety regulations were not designed for digital technologies, making it necessary to update validation protocols and establish harmonized standards (Serrano-Torres et al., 2025). Policymakers must establish robust frameworks to ensure the ethical application of tools such as artificial intelligence, particularly in areas such as food quality, safety and traceability. This calls for the development of clear guidelines for responsible technology use and the continuous updating of regulations to keep pace with rapid innovations (Hassoun, 2025). Effective regulations must evolve in tandem with technological change, embedding both technical validation and ethical safeguards to ensure safe, fair and trustworthy dairy systems in the current digital era.

5.4.2 Future perspectives and strategic opportunities

The integration of Industry 4.0 technologies such as IoT, AI, blockchain, and robotics into dairy production offers transformative opportunities to enhance dairy safety. Real-time monitoring systems can detect contaminants and pathogens promptly, enabling immediate corrective actions and reducing contamination risks. Blockchain ensures traceability and transparency, bolstering consumer trust (Feng et al., 2020). Industry 4.0 empowers the food industry to swiftly and efficiently share information with consumers, suppliers, and regulatory authorities. By leveraging digital technologies, these businesses can strengthen traceability across the supply chain, thereby promoting greater transparency and accountability. Moreover, predictive analytics can forecast potential safety breaches, allowing for proactive measures. Strategically adopting these technologies can lead to more resilient, efficient, and safer dairy supply chains, aligning with global trends towards digitalization and sustainability in food systems.

Building on these advances, recent studies suggest that the future of digitally enabled dairy safety will move beyond Industry 4.0 toward a more human-centric, resilient, and sustainability-oriented approach, often described as Industry 5.0. This transition emphasizes that technological efficiency alone is insufficient; rather, digital systems must be designed to enhance human decision-making, societal trust and long-term system robustness (Singh et al., 2025). In this context, the concept of Dairy Safety 5.0 is emerging as an evolution of Dairy Safety 4.0, integrating advanced digital tools with ethical, social and environmental considerations.

A defining feature of Dairy Safety 5.0 is human-machine collaboration. Although AI-driven analytics and automation can significantly improve hazard detection and process optimization, critical food safety decisions such as initiating product recalls or responding to contamination events still require human oversight and contextual judgment. As a result, growing attention is being given to explainable and transparent AI systems that enable quality managers, regulators, and inspectors to clearly interpret, verify, and when necessary, override algorithmic decisions (Agostinho et al., 2023). This approach strengthens decision accountability and helps address ethical concerns related to algorithmic bias, responsibility attribution, and excessive reliance on opaque or noninterpretable computational models.

Within the evolution from Dairy Safety 4.0 to Dairy Safety 5.0, resilience has emerged as a strategic pillar shaping future dairy safety frameworks. Dairy Safety 4.0 has largely focused on centralized, highly automated digital systems to enhance monitoring and control. However, climate variability, cyber risks and global supply chain disruptions have revealed the vulnerabilities of such configurations. Dairy Safety 5.0 responds to these challenges by prioritizing resilience through the adoption of advanced digital architectures such as edge computing, decentralized sensor networks, and digital twins. These technologies enable adaptive risk management by supporting real-time scenario analysis, early warning capabilities and rapid system recovery following disruptions (Gehlot et al., 2022). Such resilience-oriented capabilities are essential for maintaining robust dairy safety assurance under increasingly uncertain and dynamic operational conditions.

Sustainability-oriented digitalization further broadens the scope of Dairy Safety 5.0. In addition to reducing food losses and improving resource efficiency, future systems increasingly address social sustainability, which includes equitable access to digital safety tools for smallholder farmers and small and medium-sized enterprises. Ethical data governance, fair benefit sharing, and inclusive regulatory frameworks are therefore essential to prevent digital exclusion and to ensure that safety-enhancing technologies strengthen, rather than fragment dairy value chains (Manning et al., 2023).

Collectively, these emerging trends suggest that the future of dairy safety lies in sociotechnical integration, where ethical AI, resilient digital infrastructures, and inclusive governance mechanisms reinforce consumer confidence and regulatory legitimacy. Dairy Safety 5.0 thus represents a strategic opportunity to reframe digital innovation not merely as a technological upgrade, but as a holistic transformation aligned with public health, sustainability, and societal trust.

5.5 Conclusion

The transformation of the dairy sector through Industry 4.0 technologies represents a fundamental advancement toward achieving safer, more sustainable, and efficient food systems. As global demand for milk and dairy products rises, maintaining food safety across the entire value chain becomes more challenging due to persistent biological, chemical, physical, and allergenic hazards, as well as differing regulatory capacities among countries. Industry 4.0 includes a range of digital technologies such as IoT, AI, big data analytics, blockchain, robotics, and automation that work together to improve monitoring, analysis, and control throughout the dairy supply chain. IoT sensors enable continuous measurement of important environmental and process parameters such as temperature, humidity, and microbial contamination, allowing early detection and action. When combined with AI and big data analytics, these systems support predictive modeling and better decision-making related to product quality, herd health, and process efficiency. Blockchain technology provides transparency and traceability by maintaining permanent records of product movement from farm to consumer, which enhances accountability and builds consumer trust. Automation and robotics further increase hygiene, accuracy, and consistency, while also decreasing human error in processing and handling operations.

Despite these opportunities, the transition to Dairy 4.0 faces several structural challenges. Inadequate digital infrastructure, especially in rural and developing areas, restricts access to high-speed connectivity necessary for real-time data exchange. High investment costs, limited technical expertise, and cybersecurity issues also hinder widespread adoption. Additionally, the lack of standardized protocols and changing regulatory frameworks complicates system integration and interoperability. To ensure an inclusive and effective digital transformation, three strategic priorities are essential. First, promote digital inclusivity through investment in broadband infrastructure, affordable technologies, and capacity-building initiatives. Second, guide technological adoption with policy harmonization and ethical governance, ensuring data integrity, privacy, and compliance with international food safety standards such as those of the Codex Alimentarius. Third, develop collaborative innovation ecosystems involving academia, industry, and government to cocreate adaptable digital solutions suitable for diverse production contexts.

Finally, the convergence of Industry 4.0 with sustainability goals improves both food safety and environmental responsibility. Innovative technologies boost resource efficiency, reduce losses, and cut greenhouse gas emissions by optimizing energy and feed use. The future of the dairy sector, therefore, relies on combining technological innovation with social equity, ethical safeguards, and strong governance. Through this inclusive approach, Dairy 4.0 will ensure safer products, build consumer trust, and help create a resilient and sustainable global food system.

AI disclosure

During the preparation of this chapter, the authors used ChatGPT (developed by OpenAI) to assist in developing the chapter outline. The authors subsequently reviewed and revised the content generated by the tool, and assume full responsibility for the content of the publication.

References

Adedeji, A.A., Priyesh, P.V., Odugbemi, A.A., 2024. The magnitude and impact of food allergens and the potential of AI-based non-destructive testing methods in their detection and quantification. Foods 13 (7). Available at https://doi.org/10.3390/foods13070994.

Agostinho, C., Dikopoulou, Z., Lavasa, E., Perakis, K., Pitsios, S., Branco, R., et al., 2023. Explainability as the key ingredient for AI adoption in Industry 5.0 settings. Frontiers in Artificial Intelligence 6. Available at https://doi.org/10.3389/frai.2023.1264372.

Akinyemi, M.O., Ayeni, K.I., Ogunremi, O.R., Adeleke, R.A., Oguntoyinbo, F.A., Warth, B., et al., 2021. A review of microbes and chemical contaminants in dairy products in sub-Saharan Africa. Comprehensive Reviews in Food Science and Food Safety 20 (2), 1188–1220. Available at https://doi.org/10.1111/1541-4337.12712.

Azizsafaei, M., Hosseinian-Far, A., Khandan, R., Sarwar, D., Daneshkhah, A., 2022. Assessing risks in dairy supply chain systems: A system dynamics approach. Systems 10 (4), 114. Available at https://doi.org/10.3390/systems10040114.

Bahn, R.A., Yehya, A.A.K., Zurayk, R., 2021. Digitalization for sustainable agri-food systems: Potential, status, and risks for the Mena region. Sustainability (Switzerland) 13 (6). Available at https://doi.org/10.3390/su13063223.

Bai, Z., Liu, C., Wang, H., Li, C., 2023. Evolution characteristics and influencing factors of global dairy trade. Sustainability 15 (2), 931. Available at https://doi.org/10.3390/su15020931.

Bardenstein, S., Grupel, D., Blum, S.E., Motro, Y., Moran-Gilad, J., 2023. Public and animal health risks associated with spillover of Brucella melitensis into dairy farms. Microbial Genomics 9 (4). Available at https://doi.org/10.1099/mgen.0.001014.

Behkami, S., Zain, S.M., Gholami, M., Khir, M.F.A., 2019. Classification of cow milk using artificial neural network developed from the spectral data of single- and three-detector spectrophotometers. Food Chemistry 294, 309–315. Available at https://doi.org/10.1016/j.foodchem.2019.05.060.

Cabrera, V.E., Fadul-Pacheco, L., 2021. Future of dairy farming from the Dairy Brain perspective: Data integration, analytics, and applications. International Dairy Journal 121. Available at https://doi.org/10.1016/j.idairyj.2021.105069.

Cabrera, V.E., Bewley, J., Breunig, M., Breunig, T., Cooley, W., De Vries, A., et al., 2025. Data integration and analytics in the dairy industry: Challenges and pathways forward. Animals 15 (3), 329. Available at https://doi.org/10.3390/ani15030329.

Ceniti, C., Spina, A.A., Piras, C., Oppedisano, F., Tilocca, B., Roncada, P., et al., 2023. Recent advances in the determination of milk adulterants and contaminants by mid-infrared spectroscopy. Foods 12 (15). Available at https://doi.org/10.3390/foods12152917.

Dadi, V., Nikhil, S.R., Mor, R.S., Agarwal, T., Arora, S., 2021. Agri-Food 4.0 and innovations: Revamping the supply chain operations. Production Engineering Archives 27 (2), 75–89. Available at https://doi.org/10.30657/pea.2021.27.10.

De Klerk, J.N., Robinson, P.A., 2022. Drivers and hazards of consumption of unpasteurised bovine milk and milk products in high-income countries. PeerJ. 10. Available at https://doi.org/10.7717/peerj.13426, https://peerj.com/articles/13426/.

de Oliveira, F.M., Ferraz, G.Ae.S., André, A.L.G., Santana, L.S., Norton, T., Ferraz, P.F.P., 2024. Digital and precision technologies in dairy cattle farming: A bibliometric analysis. Animals 14 (12). Available at https://doi.org/10.3390/ani14121832.

Ekine-Dzivenu, C.C., Mrode, R., Oyieng, E., Komwihangilo, D., Lyatuu, E., Msuta, G., et al., 2020. Evaluating the impact of heat stress as measured by temperature-humidity index (THI) on test-day milk yield of small holder dairy cattle in a sub-Sahara African climate. Livestock Science 242, 104314. Available at https://doi.org/10.1016/j.livsci.2020.104314.

El-Zamkan, M.A., Hameed, K.G.A., 2016. Prevalence of *Campylobacter jejuni* and *Campylobacter coli* in raw milk and some dairy products. Veterinary World 9 (10), 1147–1151. Available at https://doi.org/10.14202/vetworld.2016.1147-1151.

Espinoza-Sandoval, O.R., Angeles-Hernandez, J.C., Gonzalez-Ronquillo, M., Ghavipanje, N., Zhang, N., Bayat, A.R., et al., 2024. Dairy farming in the era of artificial intelligence: Trend or a real game changer? Journal of Dairy Research 91 (2), 139–145. Available at https://doi.org/10.1017/S0022029924000426.

Faris, D.N., Gad, A.M., El-Tarabany, M.S., Ramadan, S.I., Afifi, G.G., Manaa, E.A., 2024. Machine learning algorithms for clinical mastitis prediction in a dairy herd use automated milking system data. Journal of Advanced Veterinary Research 14 (6), 975–981. Available at https://www.advetresearch.com/index.php/AVR/article/view/1901.

Feng, H., Wang, X., Duan, Y., Zhang, J., Zhang, X., 2020. Applying blockchain technology to improve agri-food traceability: A review of development methods, benefits and challenges. Journal of Cleaner Production 260, 121031. Available at https://doi.org/10.1016/j.jclepro.2020.121031.

Fernández González, A., Badía Laíño, R., Costa-Fernández, J.M., Soldado, A., 2024. Progress and challenge of sensors for dairy food safety monitoring. Sensors 24 (5), 1383. Available at https://doi.org/10.3390/s24051383.

Ferris, M.C., Christensen, A., Wangen, S.R., 2020. Symposium review: Dairy Brain—Informing decisions on dairy farms using data analytics. Journal of Dairy Science 103 (4), 3874–3881. Available at https://doi.org/10.3168/jds.2019-17199.

Feyisa, B.W., Haji, J., Mirzabaev, A., 2024. Adoption of milk safety practices: Evidence from dairy farmers in Ethiopia. Agriculture & Food Security 13 (1). Available at https://doi.org/10.1186/s40066-024-00479-z.

Flink, C., Nyberg, K., 2020. Occurrence of *Campylobacter* spp., *Salmonella* spp. and shiga toxin-producing *Escherichia coli* in inline milk filters from Swedish dairy farms. Journal of Food Safety 40 (1). Available at https://doi.org/10.1111/jfs.12726.

Food and Agriculture Organization (FAO), 2013. Milk and Dairy Products in Human Nutrition. FAO, Rome, Italy. Available at https://www.fao.org/4/i3396e/i3396e.pdf.

Franceschini, S., Grelet, C., Leblois, J., Gengler, N., Soyeurt, H., 2022. Can unsupervised learning methods applied to milk recording big data provide new insights into dairy cow health? Journal of Dairy Science 105 (8), 6760–6772. Available at https://doi.org/10.3168/jds.2022-21975.

Gabriel, A., Gandorfer, M., 2023. Adoption of digital technologies in agriculture—An inventory in a european small-scale farming region. Precision Agriculture 24 (1), 68–91. Available at https://doi.org/10.1007/s11119-022-09931-1.

Gargiulo, J.I., Eastwood, C.R., Garcia, S.C., Lyons, N.A., 2018. Dairy farmers with larger herd sizes adopt more precision dairy technologies. Journal of Dairy Science 101, 5466–5473. Available at https://doi.org/10.3168/jds.2017-13324.

Gehlot, A., Malik, P.K., Singh, R., Akram, S.V., Alsuwian, T., 2022. Dairy 4.0: Intelligent communication ecosystem for the cattle animal welfare with blockchain and IoT enabled technologies. Applied Sciences 12 (14), 7316. Available at https://doi.org/10.3390/app12147316.

Ghimpețeanu, O.M., Pogurschi, E.N., Popa, D.C., Dragomir, N., Drăgotoiu, T., Mihai, O.D., et al., 2022. Antibiotic use in livestock and residues in food—a public health threat: A review. Foods 11 (10). Available at https://doi.org/10.3390/foods11101430.

Giovinazzo, S., Brambilla, M., Cattaneo, T.M.P., Lazzari, A., Bisaglia, C., Muncan, J., et al., 2024. Aquaphotomics and NIR water spectral patterns in dairy production: A review exploring potentials and challenges. Journal of Dairy Research 91 (3), 249–261. Available at https://doi.org/10.1017/S0022029924000608.

Goller, M., Caruso, C., Harteis, C., 2021. Digitalisation in agriculture: Knowledge and learning requirements of German dairy farmers. International Journal for Research in Vocational Education and Training 8 (2), 208–223. Available at https://doi.org/10.13152/IJRVET.8.2.4.

Gonzales-Barron, U., Cadavez, V., Guillier, L., Sanaa, M., 2023. A critical review of risk assessment models for *Listeria monocytogenes* in dairy products. Foods 12 (24), 4436. Available at https://doi.org/10.3390/foods12244436.

Górska-Warsewicz, H., Rejman, K., Laskowski, W., Czeczotko, M., 2019. Milk and dairy products and their nutritional contribution to the average Polish diet. Nutrients 11 (8), 1771. Available at https://doi.org/10.3390/nu11081771.

Grace, D., 2015. Food safety in low and middle income countries. International Journal of Environmental Research and Public Health 12 (9), 10490–10507. Available at https://doi.org/10.3390/ijerph120910490.

Groher, T., Heitkämper, K., Umstätter, C., 2020. Digital technology adoption in livestock production with a special focus on ruminant farming. Animal 14, 2404–2413. Available at https://doi.org/10.1017/S1751731120001391.

Gromski, P.S., Muhamadali, H., Ellis, D.I., Xu, Y., Correa, E., Turner, M.L., et al., 2015. A tutorial review: Metabolomics and partial least squares-discriminant analysis—A marriage of convenience or a shotgun wedding. Analytica Chimica Acta 879, 10–23. Available at https://doi.org/10.1016/j.aca.2015.02.012.

Habsari, W., Udin, F., Arkeman, Y., 2022. An analysis and design of fresh milk smart grading system based on internet of things. IOP Conference Series: Earth and Environmental Science 1063 (1), 012059. Available at https://doi.org/10.1088/1755-1315/1063/1/012059.

Hassoun, A., Garcia-Garcia, G., Trollman, H., Jagtap, S., Parra-López, C., Cropotova, J., Bhat, Z., Centobelli, P., Aït-Kaddour, A., 2023. Birth of dairy 4.0: Opportunities and challenges in adoption of fourth industrial revolution technologies in the production of milk and its derivatives. Current Research in Food Science 7, 100535. Available at https://doi.org/10.1016/j.crfs.2023.100535.

Hassoun, A., 2025. Food sustainability 4.0: Harnessing fourth industrial revolution technologies for sustainable food systems. Discover Food 5 (1). Available at https://doi.org/10.1007/s44187-025-00461-z.

Hooda, A., Vikranta, U., Duary, R.K., 2025. Principles of food dairy safety: Challenges and opportunities. Food Engineering Series 16, 35–65. Available at https://doi.org/10.1007/978-3-031-75834-8_2.

Iweka, P., Kawamura, S., Mitani, T., Kawaguchi, T., 2023. Online near-infrared spectroscopy for the measurement of cow milk quality in an automatic milking system. Engineering Proceedings 56 (1). https://doi.org/10.3390/ASEC2023-16302.

Jaakkonen, A., Castro, H., Hallanvuo, S., Ranta, J., Rossi, M., Isidro, J., Lindström, M., Hakkinen, M., Elkins, C.A., 2019. Longitudinal study of shiga toxin-producing *Escherichia coli* and *Campylobacter jejuni* on Finnish Dairy farms and in raw milk. Applied and Environmental Microbiology 85 (7). Available at https://doi.org/10.1128/aem.02910-18.

Ji, B., Banhazi, T., Phillips, C.J.C., Wang, C., Li, B., 2022. A machine learning framework to predict the next month's daily milk yield, milk composition and milking frequency for cows in a robotic dairy farm. Biosystems Engineering 216, 186–197. Available at https://doi.org/10.1016/j.biosystemseng.2022.02.013.

John, A.J., Clark, C.E.F., Freeman, M.J., Kerrisk, K.L., Garcia, S.C., Halachmi, I., 2016. Review: Milking robot utilization, a successful precision livestock farming evolution. animal 10 (9), 1484–1492. Available at https://doi.org/10.1017/s1751731116000495.

Joolaei Ahranjani, P., Dehghan, K., Esfandiari, Z., Joolaei Ahranjani, P., 2025. A systematic review of spectroscopic techniques for detecting milk adulteration. Critical Reviews in Analytical Chemistry 1–32. Available at https://doi.org/10.1080/10408347.2025.2477535.

Joomun, A.B.Z., Wallace, C.A., Aumjaud, B., Ramful-Baboolall, D., 2024. Food safety culture and climate prevailing in micro and small food manufacturing enterprises in Mauritius and Rodrigues. International Journal of Food Science and Technology 59 (2), 685–702. Available at https://doi.org/10.1111/ijfs.16851.

Kafle, S., Paudel, M., Shrestha, C., Kathayat, K.B., Sapkota, R.C., Tiwari, A., Subedi, D., 2024. Aflatoxin M1 contamination in Dairy Milk in Kathmandu, Nepal. Toxins 16 (11). Available at https://doi.org/10.3390/toxins16110468.

Kagermann, H., Wahlster, W., 2022. Ten years of Industrie 4.0. Science 4 (3), 26. Available at https://doi.org/10.3390/sci4030026.

Khanna, A., Jain, S., Burgio, A., Bolshev, V., Panchenko, V., 2022. Blockchain-enabled supply chain platform for Indian dairy industry: Safety and traceability. Foods 11 (17), 2716. Available at https://doi.org/10.3390/foods11172716.

Kulaz, E., Ser, G., 2022. A meta-analysis of heat stress in dairy cattle: The increase in temperature humidity index affects both milk yield and some physiological parameters. Czech Journal of Animal Science 67 (6), 209–217. Available at https://doi.org/10.17221/1/2022-cjas.

Lee, Y.-G., Han, K., Chung, C., Ji, I., 2024. Effects of smart farming on the productivity of Korean dairy farms: A case study of robotic milking systems. Sustainability 16 (22), 9991. Available at https://doi.org/10.3390/su16229991.

Loddo, Andrea, Ruberto, Cecilia Di, Armano, Giuliano, Manconi, Andrea, 2022. Automatic monitoring cheese ripeness using computer vision and artificial intelligence. IEEE Access 10, 122612–122626. Available at https://doi.org/10.1109/access.2022.3223710.

Malik, M., Mor, R.S., Gahlawat, V.K., Hassoun, A., Jagtap, S., 2025. Drivers of Industry 5.0 technologies in dairy industry: An exploratory study. Sustainable Food Technology 3 (5), 1556–1568. Available at https://doi.org/10.1039/d5fb00156k.

Manning, L., Brewer, S., Craigon, P.J., Frey, J., Gutierrez, A., Jacobs, N., Kanza, S., Munday, S., Sacks, J., Pearson, S., 2023. Reflexive governance architectures: Considering the ethical implications of autonomous technology adoption in food supply chains. Trends in Food Science and Technology 133, 114–126. Available at https://doi.org/10.1016/j.tifs.2023.01.015.

McCarthy, W.P., O'Callaghan, T.F., Danahar, M., Gleeson, D., O'Connor, C., Fenelon, M.A., Tobin, J.T., 2018. Chlorate and other oxychlorine contaminants within the dairy supply chain. Comprehensive Reviews in Food Science and Food Safety 17 (6), 1561–1575. Available at https://doi.org/10.1111/1541-4337.12393.

Meenakshi, Das, S., Nanda, O., Kumari, A., Chhibber, K.D., Singh, D., et al., 2025. Surface-enhanced Raman spectroscopy and machine learning-based analysis of adulteration in milk using microwave-synthesized silver nanoparticles. Food Analytical Methods 18 (12), 2819–2830. Available at https://doi.org/10.1007/s12161-025-02900-0.

Mhapsekar, R., Kilbane, D., Davy, S., Abraham, L., Fenelon, M., O'Shea, N., 2025. A systematic review of the internet of things and artificial intelligence applications in milk quality monitoring and analysis. International Journal of Dairy Technology 78 (3). Available at https://doi.org/10.1111/1471-0307.70049.

Neculai-Valeanu, A.S., Sanduleanu, C., Porosnicu, I., 2025. From tradition to precision: Leveraging digital tools to improve cattle health and welfare. Frontiers in Veterinary Science 12. Available at https://doi.org/10.3389/fvets.2025.1549512.

Neethirajan, S., 2024. Artificial intelligence and sensor innovations: Enhancing livestock welfare with a human-centric approach. Human-Centric Intelligent Systems 4 (1), 77–92. Available at https://doi.org/10.1007/s44230-023-00050-2.

Nogoy, K.M.C., Park, J., Chon, S.I., Sivamani, S., Park, M.J., Cho, J.P., Hong, H.K., Lee, D.H., Choi, S.H., 2021. Precision detection of real-time conditions of dairy cows using an advanced artificial intelligence hub. Applied Sciences (Switzerland) 11 (24). Available at https://doi.org/10.3390/app112412043.

Nyokabi, N.S., Korir, L., Lindahl, J.F., Phelan, L., Gemechu, G., Berg, S., Mihret, A., Wood, J.L.N., Moore, H.L., 2024. Exploring the adoption of food safety measures in smallholder dairy systems in Ethiopia: Implications for food safety and public health. Food Security 16 (2), 423–435. Available at https://doi.org/10.1007/s12571-024-01439-y.

Oliver, S.P., Jayarao, B.M., Almeida, R.A., 2005. Foodborne pathogens in milk and the dairy farm environment: Food safety and public health implications. Foodborne Pathogens and Disease 2 (2), 115–129. Available at https://doi.org/10.1089/fpd.2005.2.115.

Owusu-Kwarteng, J., Akabanda, F., Agyei, D., Jespersen, L., 2020. Microbial safety of milk production and fermented dairy products in Africa. Microorganisms 8 (5), 752. Available at https://doi.org/10.3390/microorganisms8050752.

Palanisamy, Y., Kadirvel, V., Ganesan, N.D., 2025. Recent technological advances in food packaging: Sensors, automation, and application. Sustainable Food Technology 3 (1), 161–180. Available at https://doi.org/10.1039/d4fb00296b.

Patel, D., Bhise, S., Kapdi, S.S., Bhatt, T., 2024. Non-destructive hyperspectral imaging technology to assess the quality and safety of food: A review. Food Production, Processing and Nutrition 6 (1). Available at https://doi.org/10.1186/s43014-024-00246-4.

Pei, X., Tandon, A., Alldrick, A., Giorgi, L., Huang, W., Yang, R., 2011. The China melamine milk scandal and its implications for food safety regulation. Food Policy 36 (3), 412–420. Available at https://doi.org/10.1016/j.foodpol.2011.03.008.

Perez-Gonzalez, C., Garcia-Hernandez, C., Garcia-Cabezon, C., Rodriguez-Mendez, M.L., Dias, L., Martin-Pedrosa, F., 2024. Analysis of milk adulteration by means of a potentiometric electronic tongue. Journal of Dairy Science 107 (11), 9135–9144. Available at https://doi.org/10.3168/jds.2024-25140.

Rizzoli, R., 2014. Dairy products, yogurts, and bone health. The American Journal of Clinical Nutrition 99 (5), 1256S. Available at https://doi.org/10.3945/ajcn.113.073056.

Rose, D.C., Chilvers, J., 2018. Agriculture 4.0: Broadening responsible innovation in an era of smart farming. Frontiers in Sustainable Food Systems 2. Available at https://doi.org/10.3389/fsufs.2018.00087.

Rutten, C.J., Velthuis, A.G.J., Steeneveld, W., Hogeveen, H., 2013. Invited review: Sensors to support health management on dairy farms. Journal of Dairy Science 96 (4), 1928–1952. Available at https://doi.org/10.3168/jds.2012-6107.

Sangode, P.B., 2025. Dairy 4.0: Embracing digital technology for sustainable farming: A review. Asian Journal of Dairy and Food Research 44 (2), 169–178. Available at https://doi.org/10.18805/ajdfr.DR-2242.

Schopf, M.F., Pierezan, M.D., Rocha, R., Pimentel, T.C., Esmerino, E.A., Marsico, E.T., Verruck, S., 2023. Pesticide residues in milk and dairy products: An overview of processing degradation and trends in mitigating approaches. Critical Reviews in Food Science and Nutrition 63 (33), 12610–12624. Available at https://doi.org/10.1080/10408398.2022.2103642.

Semercioz-Oduncuoglu, A.S., Luning, P.A., 2025. Industry 4.0 technologies in quality and safety control systems in food manufacturing: A systematic techno-managerial analysis on benefits and barriers. Trends in Food Science & Technology 163, 105144. Available at https://doi.org/10.1016/j.tifs.2025.105144.

Serrano-Torres, G.J., López-Naranjo, A.L., Larrea-Cuadrado, P.L., Mazón-Fierro, G., 2025. Transformation of the dairy supply chain through artificial intelligence: A systematic review. Sustainability 17 (3), 982. Available at https://doi.org/10.3390/su17030982.

Silvi, R., Pereira, L.G.R., Paiva, C.A.V., Tomich, T.R., Teixeira, V.A., Sacramento, J.P., et al., 2021. Adoption of precision technologies by Brazilian dairy farms: The farmer's perception. Animals 11 (12), 3488. Available at https://doi.org/10.3390/ani11123488.

Singh, K.A., Patra, F., Ghosh, T., Mahnot, N.K., Dutta, H., Duary, R.K., 2025. Advancing food systems with industry 5.0: A systematic review of smart technologies, sustainability, and resource optimization. Sustainable Futures 9, 100694. Available at https://doi.org/10.1016/j.sftr.2025.100694.

Souto, M.R.S., Pimenta, A.M., Catarino, R.I.L., Leal, M.F.C., Simões, E.T.R., 2025. Heavy metals in milk and dairy products: Safety and analysis. Pollutants 5 (3). Available at https://doi.org/10.3390/pollutants5030029.

Spagnoli, P., Defalchidu, L., Vlerick, P., Jacxsens, L., 2024. The relationship between food safety culture maturity and cost of quality: An empirical pilot study in the food industry. Foods 13 (4), 571. Available at https://doi.org/10.3390/foods13040571.

Ushkalov, V., Danchuk, V., Midyk, S., Voloshchuk, N., Danchuk, O., 2020. Mycotoxins in milk and dairy products. Food Science and Technology 14 (3). Available at https://doi.org/10.15673/fst.v14i3.1786.

Valkovska, L., Mališevs, A., Kovaļenko, K., Bērziņš, A., Grantiņa-Ieviņa, L., 2021. Coxiella burnetii DNA in milk, milk products, and fermented dairy products. Journal of Veterinary Research 65 (4), 441–447. Available at https://doi.org/10.2478/jvetres-2021-0055.

van Asselt, E.D., van der Fels-Klerx, H.J., Marvin, H.J.P., van Bokhorst-van de Veen, H., Groot, M.N., 2017. Overview of food safety hazards in the European dairy supply chain. Comprehensive Reviews in Food Science and Food Safety 16 (1), 59–75. Available at https://doi.org/10.1111/1541-4337.12245.

Varavallo, G., Caragnano, G., Bertone, F., Vernetti-Prot, L., Terzo, O., 2022. Traceability platform based on green blockchain: An application case study in dairy supply chain. Sustainability 14 (6), 3321. Available at https://doi.org/10.3390/su14063321.

Wang, C., Ding, F., Ling, L., Li, S., 2023. Design of a teat cup attachment robot for automatic milking systems. Agriculture 13 (6), 1273. Available at https://doi.org/10.3390/agriculture13061273.

Wang, M., Islam, S., Yang, W., 2025. Supply chain risks in the dairy industry. Benchmarking: An International Journal. Available at https://doi.org/10.1108/bij-12-2023-0874.

Xiu, C., Klein, K.K., 2010. Melamine in milk products in China: Examining the factors that led to deliberate use of the contaminant. Food Policy 35 (5), 463–470. Available at https://doi.org/10.1016/j.foodpol.2010.05.001.

Xu, X., Lu, Y., Vogel-Heuser, B., Wang, L., 2021. Industry 4.0 and Industry 5.0—Inception, conception and perception. Journal of Manufacturing Systems 61, 530–535. Available at https://doi.org/10.1016/j.jmsy.2021.10.006.

Yang, M., Işık, C., Yan, J., 2023. Analysis of collaborative control of dairy product supply chain quality based on evolutionary games: Perspectives from government intervention and market failure. Heliyon 9 (12), e23024. Available at https://doi.org/10.1016/j.heliyon.2023.e23024.

Zambon, I., Cecchini, M., Egidi, G., Saporito, M.G., Colantoni, A., 2019. Revolution 4.0: Industry vs. agriculture in a future development for SMEs. Processes 7 (1), 36. Available at https://doi.org/10.3390/pr7010036.

Zhu, H., Kannan, K., 2018. Continuing occurrence of melamine and its derivatives in infant formula and dairy products from the United States: Implications for environmental sources. Environmental Science & Technology Letters 5 (11), 641–648. Available at https://doi.org/10.1021/acs.estlett.8b00515.

Chapter 6

Enhancing traceability in the dairy supply chain using Industry 4.0 technologies

Abderrahmane Aït-Kaddour[1], Jasur Safarov[2], Oumayma Boukria[3], Mohd Nizam Lani[4], Sofiane Boudalia[5,6], and Shakhnoza Sultanova[2,7]
[1]*VetAgro Sup, Departement of Sustainable Foods, Management and Innovation/Clermont-Auvergne, INRAE, VetAgro Sup, Cheese Research Unit, Lempdes, France,* [2]*Tashkent State Technical University named after Islam Karimov, Tashkent, Uzbekistan,* [3]*University Mohammed IV Polytechnic (UM6P), African Sustainable Agriculture Research Institute (ASARI), College of Agriculture and Environmental Sciences (CAES), Ben Guerir, Morocco,* [4]*Faculty of Food Science and Agrotechnology, Universiti Malaysia Terengganu, Kuala Nerus, Terengganu, Malaysia,* [5]*Département d'Écologie et Génie de l'Environnement, Université 8 Mai 1945 Guelma, Guelma, Algeria,* [6]*Laboratoire de Biologie, Eau et Environnement, Université 8 Mai 1945 Guelma, Guelma, Algeria,* [7]*Tashkent City Khakimiyat, Tashkent, Uzbekistan*

6.1 Introduction

The Food and Agriculture Organization of the United Nations (FAO) identified the concept of food traceability as the capacity to trace a food product throughout different stages such as production, processing, and distribution (FAO, 2025). The primary goal of traceability is to detect promptly and precisely the products that could lead to a risk and prevent health hazards, minimize wastage of resources, and prevent loss of money along the supply chain. In addition to food safety, traceability is important in mitigating the challenges of food fraud, food adulteration, and food authenticity, which continue to gain prominence in food systems across the globe (Fanelli et al., 2021; Galvez et al., 2018; Islam and Cullen, 2021; Qian et al., 2020). Therefore, food traceability may be regarded as a strategic requirement as it has a close relationship with consumer trust, food safety, operational integrity, and brand reputation.

With the increased globalization, interconnectedness, and data-driven nature of food supply chains, traditional traceability frameworks of food, both based on handwritten documentation, batch-level traceability or batch

Dairy 4.0. DOI: https://doi.org/10.1016/B978-0-443-44730-3.00014-6

identifier, and disjointed information systems, are no longer adequate to satisfy the technical challenges of the current food systems. These restrictions are especially high in the dairy industry, which is perishable, has a high temperature control requirement, and a high microbial risk, where end-to-end and real-time supply chain transparency is needed.

There have been many reports of adulteration and contamination in the dairy industry, which highlight the weaknesses of the current traceability system. One of the best-known cases of the systematic failure of traceability is the melamine adulteration crisis in China (Pigini and Conti, 2017). In addition to food safety failures, dairy production has a strong environmental impact in terms of land use, water consumption, and the release of greenhouse gases. Traceability systems, in this respect, provide a good tool for monitoring and reporting, as well as mitigation of environmental impacts and their result of supporting the implementation of more sustainable practices. Together, these problems (i.e., adulteration incidents and environmental impact) are indicative of the necessity of an efficient traceability system (Msalik et al., 2024) to provide food safety, establish consumer confidence, encourage regulatory compliance and promote sustainability (Khanna et al., 2022) within the dairy industry.

The advent of Industry 4.0 presents a new paradigm in technological capabilities, with the help of which the dairy traceability systems may be fundamentally restructured. The intersection of technologies such as smart sensors, artificial intelligence (AI), blockchain, and real-time analytics can be applied to offer a transformational prism through which traceability in the dairy supply chain can be reconsidered. Traceability within Industry 4.0 can be predictive and adaptive and built into the entire lifecycle of dairy products, on the pasture-to-packaging-to-fridge of consumers, instead of being a reactive tool to trace the origin of a food safety incident once it occurs.

This chapter explains how the 4.0 Industrial Revolution technologies can transform the structure and purpose of the traceability systems in the dairy supply chain. It presents the Traceability 4.0 concept, which is a relatively new conceptual framework, and it combines both the technological, regulatory, and operational aspects of digital transformation in the dairy sector. Besides, the chapter also gives original implementation situations and critically assesses the impediments and facilitators to its extensive use.

6.2 The traditional dairy supply chain: traceability gaps and challenges

As a system to guarantee food safety, environmental accountability, regulatory adherence, and consumer confidence, food traceability has become a new pillar of contemporary agrifood systems (Guldiken et al., 2021). To address such goals, regulatory systems were developed across the globe to formalize the traceability requirements. In the European Union, the General Food Law (Regulation EC 178/2002) has created the legal framework requiring all food business operators to record the essential traceability information, such as the product and lot number, supplier and customer details, country of origin, packaging and transportation information, as well as logistical records (Bosona and Gebresenbet, 2013) Fig. 6.1. Similar regulation programs have been established in other countries, including the United States (Food Safety Modernization Act, FSMA), China (Food Safety Law of China, FSL), and Japan (Food Sanitation Act, FSA), where standardized data collection and management are regarded as a way of enhancing food safety, traceability, and transparency across the supply chain (Zhang and Bhatt, 2014).

The operational level Hazard Analysis and Critical Control Points (HACCP) system is an internationally established standard of food safety management. HACCP has seven principles that depend on the core concepts aimed at identifying the possible hazards and total control of the entire process to be achieved by having structured documentation. The utilization of this framework has become more effective due to the growing use of digital information systems and software that is specific to HACCP (McMeekin et al., 2006). However, even with the current developments, several small and medium-sized enterprises (SMEs) still operate on manual or semidigital systems, resulting in inefficiencies and reduced reliability of data (Chryssochoidis et al., 2009; Nishantha et al., 2010).

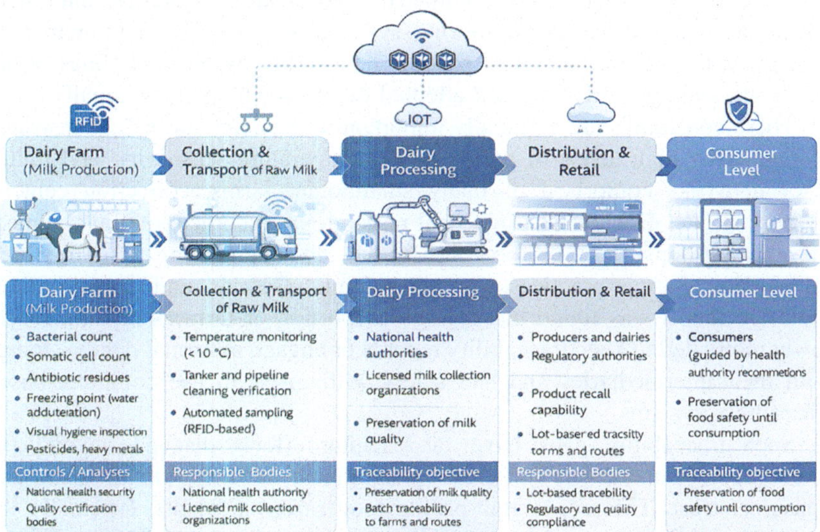

FIGURE 6.1 **Example of Industry 4.0/Dairy 4.0 integrated control and traceability framework along the dairy supply chain.** This figure illustrates key analytical controls, responsible bodies, and traceability objectives throughout the entire supply chain, from milk production at the farm level to the consumer level. It also integrates, in its upper part, Industry 4.0 technologies that can be implemented in each step of the supply chain to improve the control and traceability of milk and dairy products.

The success of any traceability system is significantly influenced by the quality, structure, and availability of data that gets registered. Four essential characteristics of traceability information, as defined in Bosona and Gebresenbet (2013) and McEntire et al. (2010), are *breadth* (the amount and extent of recorded information), *depth* (the degree of traceability upstream and downstream), *precision* (data accuracy and information integrity), and *access* (ease and speed of data retrieval or sharing). Data on traceability, after it is created, should be validated, standardized, and transformed into formats that are useful in operational decisions and compliance with the regulations.

Traditional food traceability systems normally have two main models that handle traceability data. The former, which is commonly called the one-step-up, one-step-down model, captures the necessary data of the product's direct supplier and customer, and provides a more simplified and pragmatic way. Aggregated information flow model is the second; it involves the detailed data across the entire supply chain and introduces transparency, yet it needs sophisticated data integration instruments (Bosona and Gebresenbet, 2013; Zhang and Bhatt, 2014). Although both models continue to be very popular, their efficiency is usually limited due to human involvement, a scanty information system, and small availability of real-time data.

The trade-off in the traditional traceability systems is dependent on human input; hence, they are slow, error-prone, and poorly integrated. Despite the new technologies, including automated identification systems, intelligent databases, and web-based monitoring systems that have been implemented to address the performance and product quality management (Folinas et al., 2006), the dairy industry has not adopted the new technologies equally.

These constraints are especially important in the dairy supply chain, which is marked by high levels of perishability, temperature sensitivity, and multi-actor networks comprised of many smallholder farms, collection centers, processors, and distributors. Empirical research has recorded low involvement of some groups of producers in formal documentation and structured information-sharing practices. As an example, Berman and Swani (2010) indicated that it was found that a large proportion of farmers in China were lowly motivated to keep traceability records or engage in formal data exchange with their intermediaries and had a low willingness to use computer-based traceability systems.

Such traceability gaps have far-reaching effects that go beyond the inefficiency of operations and create a significant threat to human health and consumer trust. The recent high-profile food safety incidents, especially the melamine adulteration crisis, have shown the extent to which insufficient traceability may delay the product recall, impede root-cause investigations, and increase the public health effects. Gossner et al. (2009) reported that an estimated 300,000 infants and young children in China were impacted by the melamine contamination, resulting in renal and urinary tract diseases, kidney stones included, and led to several deaths (Berman and Swani, 2010).

Together, these issues highlight the limitations of the conventional traceability systems in the dairy supply chain in terms of structural constraints. The partial nature of the data collection process reduced real-time visibility, and the lack of adequate integration among the stakeholders decreases the possibility of responsibility for safety incidents, regulatory requirements and sustainability goals. Such deficiencies provide an additional indication of the critical need for a paradigm shift toward more connected, digital and resilient traceability models that can handle the increasing complexity of contemporary dairy supply chains.

6.3 Reconceptualizing traceability with Industry 4.0

With the emergence of Industry 4.0 technologies, it has become possible to radically transform the conceptualization and operationalization of traceability into agrifood supply chains, and the dairy industry (Table 6.1). As noted above, historical traceability systems in the dairy industry have been based on manual traceability systems using paper-based logs, simple barcode scanning, which offer limited visibility and are subject to errors, delays, and failure to capture all the information. Industry 4.0 technologies enable overcoming these restrictions with the help of cyber-physical systems, such as Internet of Things (IoT), cloud computing, big data analytics, and blockchain integration into the traceability architecture, forming a dynamic and real-time monitoring ecosystem.

The increased agreement between international institutions emphasizes digital technologies as the foundation of the modernization of food quality and safety systems. Food and Agriculture Organization (FAO), in its Strategic Priorities of Food Safety (2022–2031) (FAO, 2023), recommends a digital revolution of food control mechanisms in the world, which the author emphasizes are automation, integration of data and their real-time control as the key to effective world governance. Similarly, in a parallel position, the World Health Organization (World Health Organization, 2022) recognizes digital transformation as a catalyst force that is changing the way food safety is handled, monitored and implemented globally. In line with these international efforts the European Commission (2021) predicts that by 2030, more than 90% of SMEs in Europe will be at least at a foundational level of digital capability, a change that will be signaled by a continent-wide transformation of the agrifood ecosystem to a more intelligent, connected, and transparent state.

Though Industry 4.0 technologies in the dairy industry are only at an initial implementation phase, a number of successful implementations have proven that they can be used to transform traceability and transparency. Carrefour, Nestlé, Arla Foods, and Maersk are the largest corporations that have initiated blockchain-, IoT-, and QR-based solutions to improve the visibility of their supply chains and consumer confidence. As an example, Carrefour has launched a traceability system based on blockchains to provide consumers with the history of product origin via QR code (Ryan, 2022), and Nestlé has

TABLE 6.1 Comparison between classical traceability system and Traceability 4.0 system.

Feature/ attribute	Classical traceability system	Traceability 4.0 system
Data capture method	Manual logs, paper forms, spreadsheets, basic barcodes	Automated sensors, IoT devices, RFID (Radio Frequency Identification), QR codes, cloud-based data collection
Data timeliness	Delayed, batch-based reporting	Real-time monitoring across all supply chain stages
Visibility	Limited to immediate suppliers and buyers	End-to-end visibility from farm to consumer
Data accuracy & integrity	Prone to errors, missing information, tampering	High accuracy, tamper-proof via blockchain and automated validation
Data analysis	Minimal, mostly descriptive	Predictive and prescriptive analytics using AI and Big Data
System interoperability	Low; often incompatible across stakeholders	High; integrated platforms allow seamless data exchange across all actors
Decision-making support	Reactive (post-incident)	Proactive (anticipate risks, optimize operations)
Traceability scope	"Where" only (location of product)	Multidimensional: "Where, What, How, When, Why"
Regulatory compliance	Manual verification; time-consuming audits	Automatic reporting; instant compliance tracking
Cost & maintenance	Low-tech, but labor-intensive and error-prone	Investment-heavy initially, but reduces long-term inefficiencies and recalls
Example technologies	Paper records, spreadsheets, barcode scanners	IoT sensors, RFID, blockchain, AI, cloud platforms, mobile dashboards

used AI- and IoT-based systems to track milk collection and logistics (Usman, 2022). Equally, Arla Foods has built a digital tracking platform based on blockchain to ensure transparency in sourcing dairy products (Best, 2019), and IoT and blockchain integration have been used by Maersk to enhance logistics visibility of dairy products that are perishable goods (Aritua et al., 2021). These projects demonstrate how digital technologies could turn old-fashioned traceability with its paperwork into dynamic systems, which are driven by data (Varavallo et al., 2022).

In practice, IoT devices will be able to constantly record vital parameters in the form of temperature, time/date of collection, conditions of transportation, bacterial load, and processing phases, creating a stream of data along the supply chain (Hassoun et al., 2024). These large volumes of data can be analyzed using Big Data Analytics (BDA) to identify trends, predict quality deviations, and aid in making predictive choices, and therefore, traceability is not the responsive reporting tool but the active control mechanism (Bag and Pretorius, 2022; Rajput and Singh, 2020). Blockchain technology will improve the visibility and irreversibility of data, whereby the entire supply chain participants, farmers, processors, distributors, and regulators can refer to a safe and traceable product history. Meanwhile, AI-based algorithms can streamline logistics, identify anomalies, and enhance operational efficiency and smart sensors and mobile dashboard can offer actionable information on farm and processing levels.

The adoption of traceability, together with Industry 4.0 technologies in the dairy industry, is still unequal and disparate, regardless of these developments. Although big companies have come a long way, they have not been used broadly among SMEs and smallholder farms. Lack of digital literacy, financial challenges and infrastructure make many farmers stick to manual documentation or simple barcode systems (Berman and Swani, 2010). This makes the overall digital maturity of the dairy supply chain uneven and prevents the achievement of end-to-end traceability.

Recent research highlights the opportunities and the existing constraints of digital integration in agrifood systems. Chatterjee et al. (2024) examined the Indian food and beverage industry, and the study revealed that the adoption of Industry 4.0 tools has demonstrated efficiency. Romanello and Veglio (2022) investigated organizational factors and obstacles to Industry 4.0 adoption in an Italian food processor, and Konfo et al. (2023) talked about overall issues in the agrifood industry and its adoption opportunities. Senturk et al. (2023) explored the benefits and shortcomings of IoT in agriculture with respect to the issue of data management and infrastructure failures, whereas Kaur et al. (2022) studied the way to use the blockchain and IoT to foster transparency and traceability in the food supply chain. Nonetheless, many of these studies are theoretical, whereas little empirical evidence has been provided in the context of complex dairy supply chains, involving many collection locations and variable sizes of production (Agrawal et al., 2022; Lei et al., 2023; Sahu et al., 2022).

To conclude, Industry 4.0 technologies, including IoT, AI, blockchain, and Big Data analysis, have shown great potential to improve traceability in the dairy sector; however, such technologies are still used only in pilot projects or tech-savvy companies on a large scale. The shift to Traceability 4.0 will need not only technological integration but also institutional harmonization of capacity building and investment in interoperability schemes that will ensure the connection between smallholder producers and industrial-scale processes.

Integrating traceability with Industry 4.0 technologies allows dairy supply chains to move beyond merely recording the "where" of a product.

Traceability 4.0 enables answering such important questions as what, when, how, and why, which offers a multidimensional digital track that follows each batch of milk or dairy product on its way to the customer. Inculcating the operational ecosystem with traceability via real-time monitoring, automated capture of data, and interoperable platform, Industry 4.0 turns the supply chain into resilient, transparent, and data-oriented networks that can guarantee food safety, quality assurance, and even regulatory compliance.

6.4 Technological enablers of Traceability 4.0 in dairy

The concept of Traceability 4.0 in the dairy industry is based on the digital technologies integration (i.e., IoT, blockchain, AI) that altogether improve the supply chain's transparency, safety, and efficacy. Conventional traceability systems, often fragmented, fail to adequately respond to the vulnerability of raw milk to chemical, microbiological, and thermal hazards, or to satisfy growing consumer requirements for authenticity and sustainability (Durgun, 2024; Sowmya and Ponnusamy, 2021).

To meet these requirements, the dairy sector is adopting state-of-the-art digital solutions that provide secure, transparent, real-time tracking of dairy products from the farmer to the consumer (Table 6.2). Based on generating this transformation, Industry 4.0 tools offer specific solutions, IoT and smart sensors facilitate real-time quality control (Durgun, 2024; Sowmya and Ponnusamy, 2021), and blockchain keeps data safe with records that cannot be changed (Group and Kasten, 2019; Khanna et al., 2022; Vasanthraj et al., 2024). AI and ML (Machine Learning) generate predictive insights for risk prevention and reduction (Agiomavriti et al., 2024; Cockburn, 2020; Devi et al., 2025; Lasser et al., 2021), while digital twins allow process simulation and optimization (Hosseini-Motlagh et al., 2025; Zhang et al., 2023), and cloud–edge integration ensures scalability, enabling immediate responses at farm or transport level while supporting large-scale analytics (Abbas et al., 2025; Alonso et al., 2020; Mhapsekar et al., 2022).

Collectively, these technologies enable industry to identify the origin and magnitude of safety concerns, while reducing reliance on noncompliant products and minimizing risks of negative outcomes (Rossi et al., 2025). Traceability 4.0 also advertises product quality, authenticity, and efficient recalls, and incentive mechanisms and constitutes the technological underpinnings of sustainable dairy production, compliance to regulation, and consumer confidence (Cabrera et al., 2020; Lai et al., 2025; Malik et al., 2024).

6.4.1 Internet of Things and smart sensors

The fundamental technology of real-time traceability is IoT and smart sensors. They form connected systems to capture, collect, and relay chemical, biological, as well as environmental information along the dairy supply chain

TABLE 6.2 Traceability technologies and their potential applications in the dairy industry.

Technology	Objective	Function	Models mentioned	References
IoT & smart sensors	Continuous monitoring of animal behavior and health	Detect estrus, lameness, mastitis; track rumination and movement	Wearable IoT sensors (collars, RFID, accelerometers)	Alsaaod et al. (2019), Arago et al. (2022), Cabrera et al. (2025)
	Automated, data-driven decision-making	Integration of physiological and environmental data streams	ML algorithms in "Dairy Brain"	Cabrera and Fadul-Pacheco (2021), da Rosa Righi et al. (2020), Neethirajan (2020)
	Early microbial spoilage detection	Identify VOCs (acetic acid, ethanol, acetaldehyde)	Gas sensor networks; pattern recognition	Hussain et al. (2016), Kadam and Shinde (2018), Phukkaphan et al. (2021), Putra et al. (2018)
	Detect pathogens, antibiotics, spoilage	High-specificity detection of contaminants, nondestructive, rapid biosensing	Enzyme sensors; immunosensing, bulk and surface acoustic wave (BAW, SAW)	Hu et al. (2015), Länge (2022), Poghossian et al. (2019), Renny et al. (2005)
	Maintain safe processing conditions	Continuous monitoring of pasteurization, fermentation, storage parameters	Threshold-based triggers; digital logging	Escobet et al. (2019), Madhukara et al. (2024), Mhapsekar et al. (2022)
	Real-time compositional assessment	Measure fat, protein, density, temperature	Cloud-based analytics	Drekalovic (2021), Lovarelli et al. (2022), Suresh Babu et al. (2018)

(*Continued*)

TABLE 6.2 (Continued)

Technology	Objective	Function	Models mentioned	References
	Classify milk and *detect adulteration*	Measure pH, turbidity, color, *temperature, fat content*	Sensor fusion; Fuzzy *classifiers*	Habsari et al. (2022), Rajakumar et al. (2018), Tangorra et al. (2024)
	Cold-chain monitoring and logistics optimization	Tracking of temperature, pH, tilt, shock, route	GPS loggers, cloud-linked sensors, RFID + Near-Field Communication (NFC) tags	Dasaklis et al. (2019), Malik et al. (2025), Mehannaoui et al. (2025), Vara Martínez et al. (2018)
	Maintain cold-chain integrity and prevent microbial growth	Continuous monitoring of thermal conditions in milk tanks, collection trucks, and storage units	Real-time IoT monitoring; cloud dashboards	Malik et al. (2025), Mania et al. (2018), Mehannaoui et al. (2025), Tan and Ngan (2020)
	Ensure correct handling during transport and detect physical stress	Monitoring of shock, vibration, tilt; reconstruction of transport conditions along transit routes	Accelerometer-based motion analysis; GPS mapping	de las Morenas et al. (2014)
	End-to-end freshness and traceability	Temperature & humidity tracking in packaging	Smart packaging with RFID + chemical sensors	Facchini et al. (2024), Høyer et al. (2019), Tan and Ngan (2020)
Blockchain	Protect analytical integrity	Stores lab tests, sampling history & milk quality metrics on-chain to avoid falsification	Blockchain + IoT + QR code	Casino et al. (2021), Niya et al. (2021), Rambim and Awuor (2020)
	Real-time authentication	Records pH, temperature & handling data automatically; provides "virtual quality stamp" validation	IoT-Blockchain integrated sensing systems (Long-Range "LoRa" + IoT nodes + smart contracts)	Caldarelli et al. (2020), Fonseca and Garrocho (2024), Kumar and Kumar (2025), Mendonça et al. (2023)

	Detect and trace adulteration	Logs anomalies detected by sensors; enables tamper-proof fraud alerts	Blockchain audit trail + IoT sensors	Padma et al. (2023), Sezer et al. (2018)
	Enhance transparency, prevent fraud, ensure data immutability across dairy chain	Records all production, quality, logistics and transformation events on-chain; smart contracts automate validation; consumer access	Blockchain-based traceability platform	Fang and Stone (2021), Khanna et al. (2022), Kumar and Kumar (2025), Lopes et al. (2020), Nukapeyi et al. (2024), Vasanthraj et al. (2024), Zhou and Nanseki (2023)
AI/ML	Adulteration detection	The models are used to detect the presence of adulterants, classify adulteration types or levels, and simultaneously identify multiple adulterants in milk	CNN, SVM, LDA, RF, Decision Tree, Naive Bayes, ANN, MLP, GBM	Amsaraj et al. (2021), Anand et al. (2019), Aqeel et al. (2024, 2025), Moharkar and Patnaik (2019), Sowmya and Ponnusamy (2021), Tohidi et al. (2018)
	Multiadulterant identification and quantification	Models classifying several adulterants simultaneously and quantify their concentration in milk samples	CNN, PLS-DA, SVM, LS-SVM, ANN, SVR	Amsaraj et al. (2021), Anand et al. (2019), Campuzano et al. (2020), Lanjewar et al. (2024), Malik and Shahzadi (2025), Pérez-Calabuig et al. (2023)
	Milk composition prediction (fat, protein, lactose, SNF)	Models accurately predict major compositional parameters of milk using regression on spectral data	PLSR, SVR, GPR, PCR	Agiomavriti et al. (2024), Alwadi et al. (2024), Bittal et al. (2025), Ji et al. (2022), Mu et al. (2020), Vishnu and Kumar (2024)
	Detection of antibiotics and chemical contaminants	Models detecting the presence of antibiotic residues or surfactants using biosensor or NIR-based ML systems	SVM, ML	Gutiérrez et al. (2020)

(*Continued*)

TABLE 6.2 (Continued)

Technology	Objective	Function	Models mentioned	References
Cloud–edge integration	Monitoring animal behavior and health	Real-time data processing, reducing the need for extensive cloud resources, and continuous monitoring of cows in farms	Edge computing-DRN-YOLO, IoT-based system, Fog computing-AI	Alonso et al. (2020), Kang et al. (2024), Mhapsekar et al. (2025), Shen et al. (2021), Yu et al. (2022)
	Real-time monitoring	Monitoring of milk quality and animal health, allowing for immediate action to be taken when issues are detected	Edge-centric Industrial Internet of Things, AI, split edge intelligence model, Tensorflow	Mhapsekar et al. (2025), Shen et al. (2021), Yu et al. (2022)
	Improving Traceability	Ensuring dairy products traceability from farm to consumer, enhancing transparency and trust	Edge-IoT platform, Distributed Ledger Technologies, blockchain, AI	Abbas et al. (2025), Alonso et al. (2020)
	Adulteration detection	Real-time classification of milk adulterants, providing dairy processors with immediate insights into milk quality	Edge-AI-FTIR, CNNs, Edge Sensor node	Campuzano et al. (2020), Durgun (2024), Malik and Brem (2021), Mhapsekar et al. (2022)

(Bittal et al., 2025; Malik et al., 2024; Priyadharshini et al., 2024). These also enable the detection, monitoring, and control of items remotely through network infrastructures, so they bring the physical world closer to information systems and enhance their accuracy, efficiency, and responsiveness (Geethanjali, 2017; Priyadharshini et al., 2024; Vidhya et al., 2023).

The biosensors and wearables constantly monitor temperature, rumination, estrus, and heart rate at the farm level, whereas smart collars and RFID track the identity and movements of animals. This information drives digital herd management systems, which further improve animal feeding, milking rate and welfare performance measures.

The IoT sensors are also important in the production stage to streamline feed management and resource use, as well as create the first layer of traceability, which is building information continuously on the animals and raw milk (Iwasaki et al., 2019; Lavanya et al., 2023; Umapathi et al., 2025). The sensors allow tracking animal productivity, food consumption, and environmental factors in real time, whereas data analysis can be applied to streamline feed management and use of available resources (Cabrera and Fadul-Pacheco, 2021; Chung et al., 2020; Jaeger et al., 2019). Multisensor systems will measure the pH, conductivity, turbidity, and redox potential to regulate the activity of enzymes, determine spoilage, or assess the quality of milk (Madhukara et al., 2024; Marathe et al., 2025; Yang et al., 2020). Moreover, the presence of volatile compounds associated with microbial spoilage is detected by gas and color sensors, whereas the detection of antibiotic residues or adulterants such as urea and starch is made possible via biosensors (Hu et al., 2015; Mottram et al., 2007; Poghossian et al., 2019; Renny et al., 2005).

To ensure the cold chain, thermal sensors with GPS and cloud-linked recorders are deployed during collection and transportation to determine the temperature, pH, and tank vibrations of the milk (Chakurkar et al., 2018; Sun et al., 2016). Any deviation immediately sends an alert, and necessary preventive steps can be implemented prior to contamination occurring (Baluprithviraj et al., 2024; Mehannaoui et al., 2025; Priyadharshini et al., 2024). Smart contracts built on IoT can also be used to check the conditions of transportation before it is delivered automatically (Gehlot et al., 2022). The accuracy of the quality standards is ensured by these real-time data streams, and the creation of a digital presence of each batch of milk increases the level of transparency and consumer confidence.

6.4.2 Blockchain technology

Blockchain also supports IoT by protecting the integrity, immutability and decentralization of data. A transaction is tracked in a distributed ledger at every critical control point, like milking, pasteurization, packaging, or distribution (Group and Kasten, 2019; Khanna et al., 2022; Lin et al., 2018; Vasanthraj et al., 2024). After release, these data entries cannot be changed further, ensuring complete traceability and traceable audit trails.

On the production level, blockchain technology offers a safe registry that may store the specifics of herd health, antibiotics consumption, and carbon footprint, thus facilitating ethical and sustainable milk production (Malik et al., 2024). Paper records have been substituted with blockchain records through systems like Milk Delivery Blockchain Manager that have been implemented in Kenya to remove falsification of records and to ensure farmers receive fair remuneration (Rambim and Awuor, 2020).

Dairy cooperatives in Europe have combined blockchain with QR codes, which provide consumers with the opportunity to scan a cheese package and get immediate access to its origin, processing history, and certification information (Marin et al., 2019; Tan and Ngan, 2020).

Together with the IoT, blockchain records sensor measurements over time and authenticates them via smart contracts, which eventually automates the process of meeting temperature, hygiene, and delivery requirements (Fang and Stone, 2021; Mendonça et al., 2023). This kind of synergy provides the visibility of the entire supply chain and provides a quick and focused recall, reducing wastage and financial losses.

Moreover, blockchain-based networks are decentralized, which reduces centralized databases and single points of failure that cause cybersecurity issues (Niya et al., 2021), providing a strong infrastructure to facilitate trustworthy data exchange between producers, processors, and regulators.

6.4.3 Artificial intelligence, machine learning, and neural networks

The AI and ML in traceability 4.0 offer predictive intelligence whereby massive and diverse data streams are transformed into valuable insights (Mu et al., 2020; Nukapeyi et al., 2024). In farms, AI algorithms based on the signal measured by IoT sensors are used to predict milk yield and identify the early signs of diseases such as mastitis or lameness. Gait abnormalities can be detected with more than 95% accuracy using deep learning models that examine either accelerator or video data (Taneja et al., 2020; Wu et al., 2020). Similarly, it involves the daily yield and composition prediction based on environmental and physiological data by models like the Random Forest or XGBoost (Alwadi et al., 2024; Ji et al., 2022).

Regarding processing and quality control, AI systems can analyse spectroscopic, biosensor, or e-nose data to quantify fat, protein, and lactose or detect adulterants such as starch, sucrose, and peroxide (Gomes, 2016; Ma et al., 2013; Sivalingam and Rayappan, 2012). NIR or MIR spectroscopy, when combined with ML classifiers, can be integrated into production line for rapid and nondestructive quality assessment (Bahri et al., 2019; Modi et al., 2025; Tangorra et al., 2024). In addition, unsupervised learning model such as autoencoders or Gaussian mixture models can be used to identify abnormal spectral patterns and automatically identify potential contamination events (Vasafi et al., 2021; Xue and Zhao, 2024).

In general, AI enhances the speed and objectivity of decision-making and turns the traceability of a mere passive recording into intelligent and self-correcting quality control.

6.4.4 Digital twins and cloud–edge integration

A digital twin is a dynamic virtual device that is in perpetual motion and is a virtual model of a real-life process or system and is updated in real-time by real data. In dairy production, digital twins can achieve the simulation of pasteurization, fermentation, or packaging processes, which gives the operators the opportunity to test the hypothetical conditions and optimize the processes (Escobet et al., 2019; Niaki et al., 2017). As an example, the brining time or temperature variations can be predicted in a twin model to determine the impact on the texture of cheese and its shelf-life.

Together with the AI analysis and advanced computerization, these systems allow self-diagnosing the cases of inefficiency and automatically modifying the common operating parameters. Data processing in real-time is done on state-of-the-art devices at the farm or truck level, and a cloud server is utilized to coordinate data across multiple sources to be able to do a massive analysis and regulatory reporting (Abbas et al., 2025; Alonso et al., 2020).

This kind of two-sided design balances local responsiveness with global scalability and, therefore, forms the computational backbone of traceability 4.0 and helps maintain federated data governance in which each actor maintains control over the information they generate as they add to the overall transparency.

All these technological catalysts are reinventing dairy traceability as a system that is interconnected, transparent, and smart. Visibility is available through IoT and sensors, integrity is presented by blockchain, forecasting and automation are offered by AI, simulations are conducted by digital twins, and scalability is ensured by cloud-edge integration.

The merging of these technologies in Traceability 4.0 signals a paradigm shift in the monitoring of the food safety and sustainability process by the attention to data rather than compliance, which will form the basis of a stable and more transparent future of the dairy industry.

6.5 Designing a Dairy Traceability 4.0 architecture

The shift toward a more digital and intelligent Traceability 4.0 solution necessitates the move beyond the single application of technology in isolated deployments to a functional, coherent, and scalable architectural platform (Fig. 6.2). Based on the technological enablers in Section 6.4, the proposed architecture will group the process of data acquisition, transmission, secure

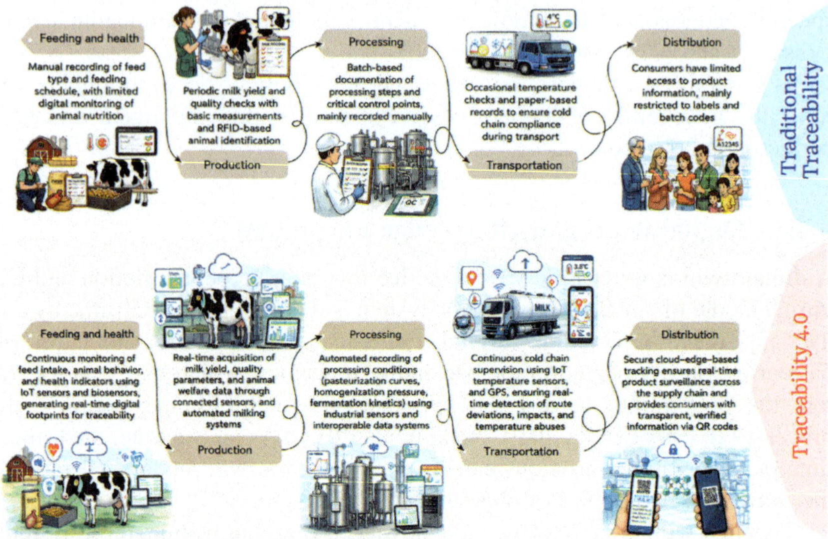

FIGURE 6.2 From traditional dairy traceability to Traceability 4.0 architecture. This figure highlights the difference between the conventional dairy traceability and the Dairy Traceability 4.0 architectures. The presentation in the same figure of those two architectures gives a clear overview of the highest differences in the monitoring/feeding, milk and dairy processing, transportation, and distribution steps.

data storage, analytics, and user interaction into interrelated layers that facilitate real-time monitoring, data integrity, and informed decision making in all phases of the dairy value chain, including feeding and production as well as processing, transportation, and distribution.

6.5.1 Sensor and data acquisition layer

This layer in the system is the interface point between the physical dairy value chain and the digital traceability system. It includes distributed sensing facilities installed on farms and retail stores, where direct monitoring of critical physical, chemical, biological, and operational parameters of importance to milk quality and safety, as well as animal welfare, is monitored.

At this level, production activities like milking, emptying into storage tanks, transportation environments, and processing operations are captured in a systematic manner and digitized into structured digital information. These data streams are comprehensive, frequent, and accurate, and are the determinants of the strength of downstream traceability, analytics, and decision-support functions.

The shift toward smart dairy monitoring, as pointed out in the literature, is determined by the possibility to connect heterogeneous sensing devices into interoperable networks that can generate synchronized and high-resolution datasets in geographically dispersed settings (Cannas et al., 2020; Mania et al., 2018).

6.5.2 Connectivity, data transmission, and edge/cloud layer

After acquisition, data will have to be transferred, preprocessed, and stored in a wide variety of operational settings, including highly interconnected processing plants, remote rural farms, and mobile transport units. This layer will provide a reliable transfer of information between data acquisition locations and centralized analytical and storage facilities (Alonso et al., 2020; Yu et al., 2022).

Local data filtering, anomaly detection, and alert generation at the edge level allow for reducing latency and bandwidth needs and achieving a timely operational response (Kang et al., 2024; Mhapsekar et al., 2024). The aggregate information is then moved to the cloud platform to be stored over time, cross-site connected, as well as interoperable with enterprise-level systems, such as Enterprise Resource Planning (ERP), Laboratory Information Management Systems (LIMS), and herd management systems (Tan & Ngan, 2020; Cabrera et al., 2021).

This hybrid edge-cloud architecture promotes scalability, data harmonization, encryption, and system resilience, which is needed to ensure continuity and reliability in real-time traceability architectures.

6.5.3 Blockchain and database layer

This layer provides the structural backbone for data integrity, transparency, and trust across the traceability system. The key events and process milestones are documented in the form of time-stamped and tamper-resistant records, which are verifiably traceable across any organizational boundary.

Although the ledger facilitates tracing and auditing of key control points, databases with high frequency sensor data, laboratory data, spectral fingerprints, and documentation needed to conduct advanced analysis and reporting are associated (Fonseca and Garrocho, 2024; Patelli and Mandrioli, 2020). This is a balanced strategy that meets the transparency needs and the performance, scalability and depth of analysis.

6.5.4 AI analytics and decision-support layer

The analytics layer converts both recorded and real-time information into meaningful, actionable information that can be used to facilitate operational control, risk management, and continuous improvement. This layer allows identifying deviations in advance, predicting the loss of quality, and determining a failure to meet safety limits, based on historical and additional real-time data at the production, processing, and logistics levels.

Digital twins built into this layer give dynamic virtual representations of physical processes and logistics flows, allowing the analysis of scenarios, optimization of processes, and the assessment of corrective strategies under different conditions (Cabrera and Fadul-Pacheco, 2021). The capabilities facilitate proactive and not reactive traceability management, which builds system robustness and resilience.

6.5.5 User and stakeholder interaction layer

The last layer converts the results of the analytical outputs into role-specific information interfaces that are sensitive to the requirements of the farmers, processors, regulators, and consumers. It allows the involvement of the stakeholders in the access to the appropriate traceability information as per their role in supporting the operational decision-making, regulatory, and transparency to end users.

Well-designed user interfaces are already considered one of the main adoption determinants, which entails simplicity, analytical intent, and transparency in developing trust (Casino et al., 2021; Høyer et al., 2019). By closing the loop between data generation and value communication, this layer ensures that traceability information delivers tangible benefits across the dairy supply chain.

6.6 Case studies and industrial implementation scenarios

To illustrate the possibility of use and the value addition of the Traceability 4.0 Framework, the section presents the application of three Industry 4.0 technologies (i.e., IoT, Blockchain, and AI) in different geographical and socioeconomic settings. These case studies demonstrate the ability to adapt Industry 4.0 technologies (Table 6.3) to various dairy systems to obtain more efficient and effective traceability and tackle certain issues associated with food safety, product quality, operational effectiveness, and consumer confidence.

6.6.1 IoT-driven dairy supply chain traceability

IoT facilitates the exchange of data between and among devices using the internet, and in the dairy production ecosystem, it has robust potential in terms of building smart environments within agriculture and livestock systems (da Rosa Righi et al., 2020; Javadi et al., 2024).

In Saudi Arabia, data collected from 9 of the 12 firms operating in the dairy industry, comprising 4 large firms (over 500 employees), 3 medium-sized firms (50–499 employees), and 2 small firms (10–49 employees) reported that the positive motivation to invest in traceability technologies and to make their strategies aligned with the country's development objectives is provided by the alignment with Vision 2030. Technological cost and compatibility (83.74%,

TABLE 6.3 Industry 4.0 technologies used in dairy supply chain traceability.

S. no.	Technologies	Studies	References
1.	Blockchain	• Developed a secure, blockchain-based traceability system for the food supply chain. • Demonstrated feasibility with a dairy company case study, smart contracts, and private blockchain. • Highlighted managerial implications and future research opportunities.	Casino et al. (2021)
2.	IoT	• MooCare IoT: Automates and personalizes cow feeding. • Milk Yield Prediction: Uses ARIMA with IoT data for each cow. • Nutrition Management: Enables proactive feeding adjustments. • Validated: High accuracy and effective detection of nutritional issues.	da Rosa Righi et al. (2020)
3.	IoT and blockchain	• IoT + Blockchain Platform: Traceability system for Gioia del Colle Mozzarella PDO. • Full Milk Tracking: From production to delivery with geolocation verification. • Food Safety and Competitiveness: Ensures compliance and strengthens market position. • Efficiency and Security: Improves process control and data management for SMEs.	Giacalone et al. (2021)
4.	Blockchain	• Ecofriendly and efficient: Low energy consumption and reduced operational costs for the Fontina PDO cheese supply chain. • Algorand blockchain: Uses pure Proof-of-Stake for scalability and environmental sustainability. • Full traceability: Ensures data immutability and real-time access for stakeholders and consumers.	Varavallo et al. (2022)
5.	Blockchain	• 50-year-old Greek family dairy, products sold in 360+ points and exported to 12 countries. • Quality: ISO 22000:2005, HACCP, strict milk controls, partnerships with more than 100 breeders. • Traceability: Tracks all materials and batches from raw milk to delivery.	Casino et al. (2021)
6.	Blockchain	• Blockchain platform for safer, traceable dairy products. • Applied to milk, cheese, and butter across production and transport. • IoT, cloud, and AI enable real-time monitoring and removal of unsafe products.	Khanna et al. (2022)

(Continued)

TABLE 6.3 (Continued)

S. no.	Technologies	Studies	References
7.	AI	• Real-time system to proactively monitor raw milk quality. • Combines sensors, rule-based models, and MLg for early detection and alerts. • Tested on an Australian dairy farm with high accuracy.	Zakeri et al. (2018)
8.	AI	• An automated thermal imaging and ML system to predict dark-cutting meat in cattle. • Stress is detected via infrared-based face and eye temperature analysis. • High accuracy achieved.	Jaddoa et al. (2024)
9.	AI	• ML/DL models for milk production forecasting in Europe. • Deep learning outperformed traditional methods. • Milk production showed strong autoregressive behavior.	Cesarini et al. (2024)
10.	ML	• A simple, low-cost ssDNA–quencher sensor array was developed. • The system accurately identified eight milk-borne bacterial species. • High performance (94% accuracy) within 30 minutes of incubation.	Wang et al. (2025)
11.	ML	• Metabolomics and ML were used to study quality drivers of Parmigiano Reggiano. • Models accurately predicted rind inclusion and production altitude. • Ensures product authenticity and quality, mitigating fraudulent practices	Becchi et al. (2024)
12.	IoT and blockchain	• Food traceability ensures safety and consumer trust, especially in perishable supply chains. • Blockchain overcomes trust issues among multiple stakeholders. • BC-DRN integrates IoT and AI for efficient dairy product traceability.	Manisha and Jagadeeshwar (2023)
13.	Sensors, RFID, GPS	• Temperature-controlled transport is crucial for accurate milk sample analysis. • A smart cooler with sensors and USB data transfer tracks sample conditions. • System using RFID, GPS, and microcontrollers was successfully tested in vehicles.	de las Morenas et al. (2014)

frequency analysis) are the main factors that influence the adoption of milk traceability technologies based on IoT. Nevertheless, the significant obstacles are organizational culture and staff resistance, and the external pressures (99%, frequency analysis), especially the increased expectations of consumers after the COVID-19 pandemic, are an excellent opportunity to implement it to enhance transparency, operational effectiveness, and consumer confidence (Alessa, 2025).

IoT technologies are commonly incorporated into a wider range of solutions that would also include blockchain, AI, and other digital solutions. In this regard, da Rosa Righi et al. (2020) suggested MooCare, an IoT-based platform aimed at assisting dairy farmers with automated and personalized feeding control in southern Brazil. The model uses data that is received by connected devices to predict the yield of milk on the single cow scale based on the ARIMA-based forecasting model, so that the farmers can predict the change in production and modify nutritional plans accordingly. The system achieved a 94.3% prediction accuracy, with minimal differences between actual and predicted milk production. High performance was obtained by applying ARIMA to EMA-preprocessed lactation data rather than raw data. A 20% deviation threshold triggered targeted notifications, enabling farmers to take timely corrective actions when production diverged from expected values.

6.6.2 Blockchain for dairy supply chain traceability

Traceability, using blockchains, provides a strong solution to the improvement of transparency, integrity of data, and trust within complex food supply chains.

In this context, Giacalone et al. (2021) introduced a new traceability and certification platform that combines IoT technologies with blockchain to secure and certify production phases in a Protected Designation of Origin (PDO) food supply chain, namely, Gioia del Colle Mozzarella PDO. With the help of the system, it is possible to ensure that the product specifications are strictly followed as the geolocation data is verified, and each batch of milk in all its stages of the lifecycle, production, processing, disposal, and delivery can be traced. This transparent and regulation-compliant traceability model not only facilitates food safety and corporate social responsibility targets but also increases competitiveness in the food industry. The approach applied to Capurso Azienda Casearia S.R.L proves that blockchain, along with enhanced data gathering and big data handling, can enhance operational effectiveness, data safety, and process management to provide small and medium-sized dairy enterprises with a more accurate, streamlined, and smart control over their production chains.

Similarly, Varavallo et al. (2022) described a traceability system based on green blockchain with low energy use and low operational expenses, which is utilized in the supply chain of Fontina PDO cheese in the EU-funded project

"Typicalp." The suggested system is based on the Algorand blockchain on a Pure Proof-of-Stake consensus mechanism that consumes the fewest computational resources and has high scalability and environmental sustainability. In addition to the environmental and economic benefits, it allows complete digitalization of the production chain, guaranteeing the impossibility of data changes and the availability of information in real-time to the stakeholders and end consumers of the Fontina consortium. Moreover, according to Casino et al. (2021), Pagonis Sisters and Co. is a medium-sized Greek family-owned dairy company founded more than 50 years ago in Erythres, Attica. It manufactures traditional dairy products, which are sold in over 360 sales outlets in Greece and are also exported to 12 countries in Europe, Asia, and Africa. The company highly emphasizes the conservation of traditional production methods and the maintenance of high standards of food safety and quality. In this respect, it has long-term relationships with approximately 100 local and regional breeders, offering them daily technical help and having strict quality control on the quality of raw milk by daily physicochemical and microbiological control. The company manages its quality management system based on the ISO 22000:2005 certified solutions and implements the principles of HACCP management, the control of which is organized by the Quality Assurance Manager in collaboration with the CEO. Traceability in Pagonis Sisters and Co. involves the whole supply chain, starting with the reception of the raw material, all the way to the delivery of the product. Each material that comes into the facility is coded and cataloged properly with supplier details, amounts, transportation information, and dates. Milk can be identified by type and daily collection routes, whereas production batches are identified with different product and date codes. Comprehensive documentation is kept during the processing, such as Bills of Materials, control parameters, personnel in charge, and conformity checks. The final product codes are recorded in the delivery and invoicing records, and thus they can be completely traced back to the customers and help in compliance with regulatory provisions, operational efficiency, and consumer trust.

Khanna et al. (2022) suggested a blockchain-based supply chain system framework as a solution system to the dairy industry in India to improve the safety of products and end-to-end traceability, consequently decreasing the risk of counterfeit and adulterated products arriving at consumers. The framework developed is utilized for major dairy products, which include milk, cheese, and butter and in their production, storage, as well as transportation phases. The Indian dairy industry is, however, an experimental ground for the platform as the industry is big, diverse in the market, and has been reported to have challenges in areas concerning dairy contamination and adulteration. The paper illustrated the role that blockchain could play in combination with other complementary technologies, including IoT, cloud computing, and ML, to facilitate real-time tracking, smart detection, and successful elimination of hazardous dairy products on the supply chain.

6.6.3 AI-driven dairy supply chain traceability

The use of AI in recent years has become an important tool that improves the traceability of the dairy supply chain. Zakeri et al. (2018) presented an early detection system in Australia to have a proactive solution to monitor the milk quality using temperature and microbial indicators, which are major determinants of milk safety, and they can also have a significant economic effect at the farm level by minimizing losses. The system makes use of wireless sensor networks, nanosensors, and biosensors to monitor and early detect livestock diseases in real-time. The acquired data is applied to specify the best conditions of milk storage in the tank as per the set standards. This forecasted situation is then compared to the real conditions in the tank, and any discrepancies found will lead to alerts being given to the farmers on preventive actions to ensure that the milk remains of good quality. It is a hybrid system that integrates a rule-based system with ML, which is highly accurate in managing proactive raw milk quality control. The method was applied to a Queensland milk farm to establish its validity and to show how well it could be practiced.

Deep learning (DL) models have been used in European countries to effectively predict agricultural variables on a national scale, and they could be used with dairy insurance products and risk management practices. Cesarini et al. (2024) tested ten ML and DL models, such as SARIMA, Artificial Neural Networks (ANN), Convolutional Neural Networks (CNN), and Long Short-Term Memory (LSTM) networks, to predict milk production and manage resources in a more efficient way in France, Germany, and Italy. The models involved climatic and economic open datasets as inputs. Findings suggest that DL models are more effective than traditional statistical models with minimal complex model structures, giving strong predictions. More precisely, for Italy, SARIMA reaches an R^2 of 0.811 and an RMSE of 35.85 L, but CNN1D performs best, achieving the highest R^2 (0.843), lowest RMSE (32.45 L) and lowest MAPE (2.7%). LSTM models perform worse (R^2=0.741–0.7612, RMSE >40 L). For France, CNN2D and FCNN outperform SARIMA, reducing RMSE to around 66 L versus 76.08 L, and MAPE from 2.9% to 2.5%–2.6%. Parallel WaveNet shows lower RMSE and MAE despite a slightly lower R^2, indicating better robustness to extreme values.

An AI system called Nina was designed by the Buffalo Mozzarella Consortium in collaboration with Farzati Spa to fight against counterfeit. This system, capable of improving its detection abilities based on experience, employs seven different algorithms to analyze packaging (i.e., font verification, color scale analysis, CE stamp validation, translation accuracy, surface inspection, dimensional assessment, and anomaly detection) and detect cheese fraud. This system allowed 15,000 independent checks in 2023, making the Buffalo Mozzarella DOP one of the most controlled cheeses in Europe (Albertini, 2025). An equivalent approach was carried out by the Consortium of Parmigiano Reggiano (Brill, 2022). To protect this traditional Italian cheese, two companies, Kaasmerk Matec and p-Chip Corp., partnered to develop a high tracking product authentication system for ensuring consumer safety and consortium interests. For doing so, they proposed to integrate into the casein label (traditional identifier) a p-Chip device containing a micro transponder—a blockchain crypto-anchor that creates a digital twin for physical items (Brill, 2022).

6.7 Benefits, barriers, and strategic implications

6.7.1 Benefits of Traceability 4.0 in the dairy supply chain

6.7.1.1 Enhancing food safety, recall precision, and risk control

Traceability 4.0 is a significant move toward changing the dairy supply chain. It enables producers and regulators to operate more efficiently in terms of controlling food safety, recall operations, and risk prevention because it combines IoT, blockchain, and data analytics. The necessary variables like temperature, positioning, and handling of the product can be monitored in real-time with the assistance of the IoT-based sensors, which contribute to the discovery of the potentially contaminated areas in time and mitigate the risk of communicable diseases and harm to human health due to consuming dairy products (Nurgazina et al., 2021).

Availability of verifiable real-time data enhances the accuracy of tracing of the products and assists in a timely response to correct the situation in case of a safety issue. Not only does this digital transparency meet the food safety regulations, including the Food Safety Modernization Act (FSMA), but it also enhances consumer trust and confidence by offering information on product conditions, conditions of handling, and safety standards (Ayeni and Olagoke-Komolafe, 2024).

The other notable benefit is that it improves specific recall handling. The compromised batches can be isolated under better tracking in the shortest possible time, minimizing wastage and limiting the reputational impact (Jarh et al., 2025). The blockchain system is also ensuring no data will be changed; it can be checked, and this increases the responsibility of accounts and develops trust in the chain of supply (Anyibama et al., 2025).

6.7.1.2 Facilitating compliance and regulatory trust through digital integration

Traceability 4.0 also offers the ability to enhance regulatory compliance and automate significant operations, such as cold-chain monitoring, cleaning-in-place (CIP) operations, equipment maintenance, and laboratory tests (Hassoun et al., 2023a; Peres et al., 2025). Such computer-based solutions create verifiable and trustworthy documents that increase the level of accuracy and the alignment between the certification schemes, like HACCP, ISO 22000, halal (*halal food is a religious obligation based on Islamic law*), and other audit measures of the retailers (Vasileiou et al., 2025)

It is also supported by remote monitoring and digital documentation, which allow risk-based planning of the inspection, and less work on the administration side of producers and auditors (Hassoun et al., 2023b). Moreover, having the Industry 4.0 technologies, such as blockchain and immutable digital records, makes the whole process of developing a transparent system stronger, which increases regulatory trust and strengthens the validity of traceability systems. The larger compliance rates, which are enabled by the transparency in the distribution of information, do not merely enhance controls but also increase the confidence of the stakeholders in the quality and safety of the dairy products.

6.7.1.3 Operational efficiency, cost savings, and quality optimization

End-to-end traceability data has now become important in improving the efficiency of dairy logistics and processing. Traceability tools have revolutionized cold-chain control with real-time monitoring and exception control, whereby constant measurements on the major environmental variables in product quality and safety are taken (Lin et al., 2016).

These systems enhance transparency and enable active quality control through data collection and its reporting in real-time, which will help in failure recognition and simplify transport, storage, and handling operations (Protopappas et al., 2025). This interrelation between the wireless sensor network (WSNs) and the QR-based traceability tools will bring the environmental data and the product identifiers together, which will assist in the trace of the time-temperature exposure at a high level of precision and reduce the number of manual checks (Gowda and Raj, 2025; Nanda et al., 2022). The capabilities contribute to the visibility and operational decisions, such as shipment approvals, rerouting, and segregation (Sun et al., 2025).

As a way of addressing the risks of centralized control of data, blockchain traceability of supply chain data can ensure the integrity of data, dispute resolution, and offer safe and shared access to the records of the supply chain (Rahman and Hasan, 2025). The reduction of spoilage is possible with the help of better cold-chain visibility and the increase in shelf-life management. In comparison to the fixed shelf-life labels, the dynamic shelf-life estimation models use actual real-time condition data to better reflect the actual product status and minimize wastage (Corradini, 2018). Experimental research supports the idea that microbial growth, including the *Pseudomonas fluorescens*, is influenced by temperature changes during storage, and the addition of this data can be used to prolong shelf-life and minimize losses (Lin et al., 2016).

Enterprise Resource Planning (ERP) is an enhancement of traceability data integration to enhance the reliability and accessibility of the data in different operations (Rahman and Hasan, 2025). The innovations in AI and IoT also reinforce the optimization of the processes and enable intelligent decision-making and predictive maintenance. At the processing stage, AI and IoT-based pasteurization systems utilize less energy and less time, improving the quality of products and operational stability (Kusnadi et al., 2024).

6.7.1.4 Sustainability and circular-economy performance

Traceability 4.0 enhances sustainability, and circular economy results in converting fragmented supply chain actions, including production, handling, transport, and processing, into organized digital information exchanged among stakeholders. This digitization enables more environmental impacts to be seen and measured to enable organizations to track and enhance their sustainability performance (Freeman et al., 2022; Zhou et al., 2024). Enhanced transparency helps in busting food systems that are circular and helps policymakers to focus on ways to intervene in issues in certain areas, like dairy effluent and management of chemicals (Freeman et al., 2022).

At the enterprise level, digital traceability drives innovation and organizational learning, which allows the sustainability practices to be continuously improved (Zhou et al., 2024). Extended literature indicates that traceability and collaboration across firms are required to reuse, recycle, and remanufacture, which are part of circular activities, and to encourage the remodeling of operations to facilitate the closed-loop system (Annosi et al., 2021).

Minimizing food waste is one of the sustainability goals. Dairy supply chain losses are usually caused by a fluctuation in temperature or incorrect shelf-life data (Hertog et al., 2014; La Scalia et al., 2015). Real-time quality and shelf-life information in logistics, including first-expired-first-out (FIFO) policy, can lead to an improvement in the distribution efficiency and minimize spoilage (Hertog et al., 2014; La Scalia et al., 2015). Digital traceability data and monitoring technologies linked through smart logistics infrastructure are another way of bridging information gaps that restrict the ability to reduce waste (La Scalia et al., 2015).

Minimization of waste is also caused by innovation in quality control, logistics, and production planning because of digitalization. Such advances are in line with operational efficiency goals and sustainability goals that increase environmental and economic performance (Reyes et al., 2024). It is proven that inventory policy, shipping age, and packaging design have an impact on waste levels, and data-driven traceability is a solution that helps companies strike the right balance between cost, waste, and quality results (Akkas and Honhon, 2023; Li et al., 2017).

Moreover, Traceability 4.0 supports the shift to a circular economy by allowing real-time data on products and processes to move the process of recovery of materials and minimization of losses. Smart packaging and pallet sensing can be used to make decisions that prolong the life of the product and enhance its recyclability (Annosi et al., 2021; Fernandez et al., 2023). Another technological trend in the packaging industry that supports the principle of waste valorization and circular design is the use of milk-protein-based films as sustainable substitutes to traditional plastics (Gerna et al., 2023).

With intelligent packaging systems, sustainability is increased by integrating quality monitoring with traceability and communication with consumers. Such systems identify deterioration at an early stage, minimize unnecessary disposal, and directly relate to monitoring the condition of products to waste prevention and circular performance (Müller and Schmid, 2019; Poyatos-Racionero et al., 2018).

6.7.2 Barriers to implementing Traceability 4.0 in dairy

6.7.2.1 Financial and Economic Constraints

Financial and economic impediments are one of the most critical issues to Traceability 4.0 adoption in the dairy industry. The initial cost of establishing digital traceability infrastructure is very high, and many SMEs cannot afford it because of the lack of liquidity and scale economies (Shah et al., 2024; Vudugula, 2025). Research indicates that SMEs in the food supply chains are more challenged than bigger companies that usually enjoy greater financing power, professional workforce, and digital preparedness (Bourlakis et al., 2014).

Digital transformation in the process of dairy production will also be limited due to cost pressures associated with labor, feed, and production inputs (Bourlakis et al., 2014). In addition to installation, maintenance of the system and staff training, and cybersecurity continue to be prohibitive, particularly to small operators with small budgets (Omowole et al., 2024; Rui and Kaliani Sundram, 2024; Shah et al., 2024). These expenses are also complicated by the technical complexity of blockchain systems since this implies that additional costs will have to be introduced to guarantee interoperability, data security, and compliance with regulations (Olateju, 2025).

Lack of equal digital literacy and infrastructure among the supply chain participants is another issue of concern. Many SMEs cannot use or maintain Traceability 4.0 due to the absence of both the technological background and the necessary human resources to operate sophisticated digital platforms (Heussner and Hanf, 2025; Vuvor et al., 2025). The lack of finances is also a constant obstacle to the technological upgrade or the education of workers in the field of digital skills (Cunha et al., 2022; Esquivel et al., 2020; Tirichine et al., 2022).

Investment and value distribution within supply chain levels are often out of line at a structural level. Although full traceability involves both producers, processors, and distributors, the financial cost is usually placed on upstream participants, especially small manufacturers (Bourlakis et al., 2014; Ghadge et al., 2017). Such an imbalance does not promote its widespread use and reduces the ability of Traceability 4.0 to be rolled out throughout the dairy network (Tsolakis et al., 2021; Vuvor et al., 2025).

6.7.2.2 *Digital infrastructure and connectivity limitations*

Traceability 4.0 is also unfeasible in the dairy industry without a stable digital infrastructure. The efficient traceability will need to ensure the persistent information capturing and real-time coordination among farms, logistical systems, and common platforms, which will be impossible without stable and high-speed connections (Renzcherchen et al., 2025). Agriculture 4.0 research has indicated rural connectivity as a major challenge to digital transformation. Another limitation in African dairy value chains is the lack of internet infrastructure, which is crucial when it comes to using sensor-based systems, especially with small-scale producers (Vuvor et al., 2025). The same research has found similar results in other developing economies, showing that weak connectivity always erodes digital adoption, and it is a cross-sector constraint of rural innovation (Leochico et al., 2020; Sosa et al., 2023).

This problem is not limited only to access, but includes differences in the availability of devices, quality of networks, and digital interaction, which influence the capacity of the farms to gather and distribute traceability data (Pearce and Rice, 2013; Vuvor et al., 2025). Poor connectivity not only constrains data capture but also functions of greater value like automation, real-time analytics, and built-in decision-making.

In highly automated processes of dairy farm, sensor systems are often progressed beyond simple measurement to data interpretation and combination with other external sources, such as feed prices or weather data, which can then be used by humans or automated control systems to make informed decisions (Rutten et al., 2013). Nevertheless, in the absence of a trusted connection, the data integration and feedback loop are broken, which lowers the usefulness of digital technologies (Renzcherchen et al., 2025; Vuvor et al., 2025).

6.7.2.3 *Interoperability and data integration challenges*

The digital transformation of dairy supply chains relies heavily on the ability of multiple actors and technologies to exchange, interpret, and utilize data effectively. However, achieving interoperability remains one of the most significant challenges in Industry 4.0-enabled supply chains. Dairy value chains involve diverse stakeholders, including farmers, processors, logistics providers, retailers, certification bodies, and regulators, each often operating different information systems, data formats, and technological infrastructures (Lai et al., 2025). This fragmentation creates barriers to seamless information exchange and limits the effectiveness of digital traceability initiatives.

Data integration challenges stem from the diversity of technologies deployed across the supply chain. Traceability data are generated through multiple sources, including IoT devices, RFID systems, sensors, logistics platforms, and blockchain-enabled applications, requiring coordination across heterogeneous stakeholders and technological environments (Feng et al., 2020; Shahid et al., 2020). As a result, organizations may face difficulties in consolidating information from multiple sources into a unified and reliable traceability system. The lack of common standards and interoperable architectures can further hinder data consistency, accessibility, and information sharing across organizational boundaries (Feng et al., 2020).

Interoperability challenges are not limited to technical issues but also involve organizational and governance dimensions. Supply chain actors may be reluctant to share data because of concerns regarding confidentiality, competitive advantage, data ownership, and cybersecurity risks (Patelli and Mandrioli, 2020; Ding and Bai, 2022). Differences in digital maturity among stakeholders, particularly between large processors and smaller dairy farms, can further hinder the adoption of integrated information systems and create uneven participation in digital ecosystems.

Although emerging technologies such as blockchain have been proposed as mechanisms for improving data sharing and record synchronization, their successful implementation depends on the interoperability of the underlying systems that generate and upload data (Dey et al., 2021; Tran et al., 2024). Blockchain platforms cannot resolve inconsistencies in source data and may introduce additional integration challenges when they operate alongside existing enterprise systems. Consequently, the effectiveness of digital traceability depends not only on secure data storage but also on the establishment of common standards, protocols, and governance frameworks that enable reliable information exchange among stakeholders.

Recent studies emphasize the importance of standardized data architectures, application programming interfaces (APIs), and industry-wide data governance mechanisms to improve interoperability across agrifood supply chains (Lai et al., 2025; Malik et al., 2024). Enhancing compatibility between systems can facilitate real-time data sharing, improve traceability performance, and support more efficient decision-making throughout the dairy supply chain. Addressing these interoperability and integration challenges is therefore essential for realizing the full potential of Industry 4.0 technologies in creating transparent, resilient, and data-driven supply chain networks.

6.7.2.4 Data governance, privacy, and cybersecurity concerns

Issues of data governance, ownership, and privacy will continue to be the key factors in the implementation of Traceability 4.0 in the dairy system. Since these systems work based on continuous data collection and sharing among various actors, one would ask who is in control and reuses sensitive operation data (Amiri-Zarandi et al., 2022). In the context of smart farming and agridata ecosystems, cooperative data-sharing structures can hardly be achieved because of the technical and organizational complexity of getting various sources to interact. The governance will identify who is able to access and republish data, hence it is, in essence, a fundamental obstacle but not a secondary problem (Amiri-Zarandi et al., 2022).

Studies of secure agricultural supply chains indicate that stakeholder reluctance is likely due to ineffective regulation and inconsistent data management, which erode the trust in common digital infrastructure (Olateju, 2025). This is especially susceptible to centralized data architectures, since IoT-based systems are susceptible to breaches that disclose confidentiality and reliability, which are the two crucial factors of traceability credibility (Deep et al., 2024).

Traceability 4.0 is extremely interconnected, which enhances cybersecurity risks. The IoT sensors, cloud service providers, and distributed data exchanges form several vectors of vulnerability and make the system more vulnerable to attacks and manipulation (Hassija et al., 2019). With data flows becoming bigger and more complex, they are susceptible to sensor hijacking and privacy violations across distributed networks (Xiang et al., 2025; Zizic et al., 2022).

Some of the threats that have been identified include Distributed Denial of Service (DDoS) attacks, malware, ransomware, and botnet intrusions that increase when the environment is a multidevice traceability in farms, transport, and processing facilities (Musthafa et al., 2024). The technologies that facilitate the integration of the supply chain contribute to the sharp increase in exposure to cyber threats, thus necessitating proactive and adaptive efforts in defense (Sobb et al., 2020; Zizic et al., 2022).

Finally, Traceability 4.0 credibility is the most important factor of data integrity, availability, and confidentiality. They must be efficient, have access control, and antitampering and privacy measures that are higher against theft or manipulation (Deep et al., 2024; Husnoo et al., 2021; Sobb et al., 2020). The fact that the IoT platforms handle large amounts of sensitive data and that even their minor violations might prove to be both legal and reputational due to this fact is another reason to remind potential adopters (Hassija et al., 2019; Husnoo et al., 2021).

6.7.3 Strategic implications for dairy stakeholders

6.7.3.1 For dairy processors and brands

Traceability is no longer a necessity to the digital transformation and sustainability goals in the dairy processing business, but a strategic asset. Additional digital systems provide wide-ranging visibility of supply chains that increase product quality, safety, and regulatory compliance. Blockchain technology is aimed at ensuring transparent and unaffected records that create confidence among consumers and certify the products (Hastig and Sodhi, 2020; Rana et al., 2021).

Besides consumer trust, real-time data analytics enable the processors to avoid inefficiencies, manage recalls better, and establish collaborative relationships because of increased accountability and data sharing (Khanna et al., 2022). When operational data is effectively adopted, it may also turn into predictive information that can be used to improve production planning, inventory management, and efficiency in logistics (Ellahi et al., 2023; Lai et al., 2025). Additional automation tools also enhance the process of work, making it less expensive and less prone to human error.

Nevertheless, it is not enough to invest in technology. To achieve successful digital transformation, an organization needs to have a favorable organizational culture and long-term employee involvement (Yu et al., 2025). Digital capacity-building programs provide the staff with an opportunity to operate and adjust to the new system properly (Alves et al., 2022). Through the development of digital competence and the establishment of a culture of innovation, processors will be able to transform traceability as a compliance activity into a strategic asset and increase their sustainability, resilience, and long-term competitiveness (Zhang et al., 2022).

6.7.3.2 For farmers, cooperatives, and SMEs

The development of common online platforms provides farmers, cooperatives, and SMEs with an avenue toward higher inclusion and performance. The consolidation of services and data on these platforms facilitates cost management and collaboration, accessibility, and availability of more market insight, as well as the spread of educational resources. The studies associate the transfer of agricultural knowledge by digital means to the significant improvement of productivity among smallholder farmers (Fabregas et al., 2019). They further improve collective bargaining and group buying, reducing the operational costs and making them competitive.

Training methods that address the literacy levels of the rural population at different levels are required to bridge the digital divide. The programs must also have the capability of not only serving younger and digitally savvy farmers but also older farmers with little exposure or confidence in technologies (Hill et al., 2015; McCosker et al., 2023; Ok, 2025). The governments, the nongovernmental organizations, and the digital service providers must collaborate to provide sustainable digital training and technical services that are inclusive to everyone.

Furthermore, the involvement may be stimulated with the help of social media and interactive online applications to demonstrate the practical use and the culture of lifelong learning (Raina et al., 2025). The incorporation of digital literacy programs in the larger agricultural models would help to build an agricultural community that is more resilient and networked. In the long term, the strategy strengthens the rural agrifood systems through collaboration networks, resilience, and sustainability (Wang, 2025; Xie et al., 2021).

6.7.3.3 ***Recommendations for Regulators/Policymakers Implementing***

Implementing Traceability 4.0 within the dairy sector demands supportive regulatory frameworks and policies to encourage the adoption of digital tools and address unique challenges faced by dairy operations. A priority area should be the development of standardized traceability protocols and interoperable data systems that can facilitate information sharing across dairy farms, processors, distributors, retailers, and regulatory bodies (Aung & Chang, 2014; Casino et al., 2021). Policymakers need to acknowledge that the costs associated with sensors, blockchain platforms, and digital monitoring infrastructure may be prohibitive, particularly for smallholder dairy operations (Rotz et al., 2019; Wolfert et al., 2017). Additionally, unequal levels of digital literacy and technical expertise can create barriers to the effective utilization of Traceability 4.0 technologies throughout the dairy value chain (Eastwood et al., 2019; Klerkx et al., 2019). With increasing levels of digitalization, regulators should also prioritize policies surrounding data governance, cybersecurity, and information ownership to build stakeholder trust and safeguard Traceability 4.0 systems (Kamilaris et al., 2019; Rejeb et al., 2020). As such, policy coordination around standardization, financial incentives, capacity development, and data privacy will be critical to accelerate adoption of Traceability 4.0 to realize gains in transparency, food safety, and sustainability for dairy stakeholders.

6.7.3.4 *For technology providers and researchers*

It is important that agricultural innovations could generate real value in the rural economies by developing affordable and modular as well as user-friendly technologies. A successful design for the smallholder farmers must be based on the realities of the operations of the farmers, their resource constraints, and the environmental situations. The application of user-centered design (UCD) principles will improve adoption by using user feedback during development. Making tools and interfaces simple enough to be understood by the user is encouraging wider adoption (Lyon and Koerner, 2016), whereas the codesign approach at the iteration stage makes the technologies fit the farm routine, enhances their usability, and facilitates engagement in the long term (Lyon et al., 2020).

Agribusinesses have significant opportunities to improve their competitiveness and satisfy regulatory and sustainability requirements with the integration of Industry 4.0 technologies. Monitoring and data analytics based on sensors, as well as automation, can be used to enhance transparency, resource efficiency, and environmental performance (Lindblom et al., 2017). Companies that are consistent with consumer values, particularly as they relate to ethics and sustainability, have reputational and market benefits (Lindblom et al., 2017).

The success of digital transformation, however, is based on the ability to resolve socioorganizational barriers. Poor adoption is usually hindered by complex interfaces, lack of proper training, and coordination with current workflows (Lindblom et al., 2017). It is essential to understand that organizational culture, openness to change, and collaboration are all factors that contributed to the success of the implementation (Akhavan et al., 2014). It is also important to enhance digital literacy among smallholders. Training programs can be used to ensure that resistance is minimized by instilling technological confidence, empowering users, and keeping them digitally engaged.

6.8 Future directions and research opportunities

In the future, the processes of traceability will be increasingly connected with the concepts of sustainability, climate resilience, and environmental stewardship. With the increased scrutiny of the global dairy systems, which are subject to the assessment of their environmental impact, the use of digital product passports (DPPs), carbon tracking, and environmental monitoring tools will become essential. The new traceability systems in the future need to go beyond product identity and process tracking to encompass measures of greenhouse gas (GHG) emissions, land use intensity, biodiversity effects, and water footprint analytics. Implementation of these indicators within Traceability 4.0 architecture will enable producers, regulators, and consumers to assess the real-time performance of the product safety as well as sustainability performance (Freeman et al., 2022; Hassoun et al., 2023b).

There are also new opportunities introduced by emerging progress in generative AI. Natural language generation (NLG) as a form of AI may be used to convert complex, multidimensional traceability data into simple forms accessible to direct audiences, including consumers who need to know the origins of what they are purchasing, auditors who need to understand the compliance standards, or policymakers who need to know the risks posed by the sector in which they operate. This contextualized reporting may make traceability information more democratic and build better trust between supply chain actors in the dairy industry.

The creation of strong international standards is another area of concern in future research. Harmonization will be necessary as the traceability systems are becoming more digitized and globally connected. The Codex Alimentarius, Global Standards One (GS1), International Organization for Standardization (ISO), and regional authorities should work to find common ground in their data dictionaries, interoperable communication protocols, and consistent certification systems. Greater interoperability will help share data cross-border, enhance the ability to recover data globally, and make them more resilient in the face of disruption, whether it is a pandemic, geopolitical conflict, or a climate shock (Islam & Cullen, 2021).

Lastly, the sociotechnical aspects of adoption should be further researched. Research needs to be done on how the smallholder farmers and SMEs who form a large part of the world dairy industry can be empowered with capacity building, financing opportunities and policy incentives so that they can enjoy a healthy share in the digitally enabled traceability ecosystems. Knowledge of user behavior, confidence in digital systems, and the organizational preparedness needed to implement advanced technologies will be invaluable to the realization of massive and comprehensive implementation of Traceability 4.0 solutions (Eastwood et al., 2017).

6.9 Conclusion

The opportunity to redefine traceability as one of the pillars of the modern dairy supply chains is a unique opportunity offered by Industry 4.0. Instead of being an active measure that is used when any incidents happen, traceability can now be proactive, intelligent, and integrative and act as a continuous tool that ensures the quality of the produced products and specificity to regulations and enhances customer trust. Such transformation is needed and possible in an industry in which perishability, microbial hazards, and supply chain complexity continue to be important issues.

The dairy industry can build data-based, transparent, and connective traceability networks with the integration of IoT sensors, blockchain, analytics based on AI, digital twins, and cloud-edge infrastructures. These systems enable risks to be identified in time before they escalate, facilitate quick and accurate recalls, streamline workflow, and sustainability reporting. Traceability 4.0 Framework described in the chapter provides a systematic roadmap on how to facilitate this transition.

With the transition to the dairy industry of the Dairy 4.0 era, digital transformation is no longer a choice; it is also the key to resilience, competitiveness, and long-term sustainability. Researchers, policymakers, and industry leaders need to work together to enhance the speed of innovation, address barriers to adoption, and build homogeneous standards that can enable global integration. By working together, the dairy industry will be able to achieve the vision of an entirely transparent, smart, and climate-resilient supply chain.

AI disclosure

During the preparation of this work, the authors used ChatGPT-5.2 to improve readability, language, and illustrations. The authors reviewed and edited the content as needed after using this tool and take full responsibility for the content of the publication.

References

Abbas, M., Abbas, G., Jaffery, S., Chuanqiang, Z., Yunxia, L., Gaoping, Z., et al., 2025. Smart dairy farming: Enhanced efficiency, productivity and animal welfare through the Internet Of Things and cloud integration. The Journal of Animal and Plant Sciences (1), 18–35. Available at https://doi.org/10.36899/japs.2025.1.0002.

Agiomavriti, A.-A., Nikolopoulou, M.P., Bartzanas, T., Chorianopoulos, N., Demestichas, K., Gelasakis, A.I., 2024. Spectroscopy-based methods and supervised machine learning applications for milk chemical analysis in dairy ruminants. Chemosensors 12 (12), 263. Available at https://doi.org/10.3390/chemosensors12120263.

Agrawal, R., Wankhede, V.A., Kumar, A., Luthra, S., Huisingh, D., 2022. Progress and trends in integrating Industry 4.0 within Circular Economy: A comprehensive literature review and future research propositions. Business Strategy and the Environment 31 (1), 559–579. Available at https://doi.org/10.1002/bse.2910.

Akhavan, P., Ebrahim Sanjaghi, M., Rezaeenour, J., Ojaghi, H., 2014. Examining the relationships between organizational culture, knowledge management and environmental responsiveness capability. VINE 44 (2), 228–248. Available at https://doi.org/10.1108/VINE-07-2012-0026.

Akkas, A., Honhon, D., 2023. Determining maximum shipping age requirements for shelf life and food waste management. Production and Operations Management 32 (7), 2173–2188. Available at https://doi.org/10.1111/poms.13963.

Albertini, A., 2025. Italian AI system spots fake mozzarella, fights food counterfeiting. Retrieved from https://thetraceabilityhub.com/italian-ai-spots-fake-mozzarella-fights-fraud/.

Alessa, T., 2025. Anticipations and requirements of individuals with long-term health conditions regarding self-management and digital health support tools. Patient Preference and Adherence 19, 2901–2918. Available at https://doi.org/10.2147/PPA.S545496.

Alonso, R.S., Sittón-Candanedo, I., García, Ó., Prieto, J., Rodríguez-González, S., 2020. An intelligent Edge-IoT platform for monitoring livestock and crops in a dairy farming scenario. Ad Hoc Networks 98, 102047. Available at https://doi.org/10.1016/j.adhoc.2019.102047.

Alsaaod, M., Fadul, M., Steiner, A., 2019. Automatic lameness detection in cattle. The Veterinary Journal 246, 35–44. Available at https://doi.org/10.1016/j.tvjl.2019.01.005.

Alves, L., Ferreira Cruz, E., Lopes, S.I., Faria, P.M., Rosado da Cruz, A.M., 2022. Towards circular economy in the textiles and clothing value chain through blockchain technology and IoT: A review. Waste Management & Research: The Journal for a Sustainable Circular Economy 40 (1), 3–23. Available at https://doi.org/10.1177/0734242X211052858.

Alwadi, M., Alwadi, A., Chetty, G., Alnaimi, J., 2024. Smart dairy farming for predicting milk production yield based on deep machine learning. International Journal of Information Technology 16 (7), 4181–4190. Available at https://doi.org/10.1007/s41870-024-01998-5.

Amiri-Zarandi, M., Hazrati Fard, M., Yousefinaghani, S., Kaviani, M., Dara, R., 2022. A Platform approach to smart farm information processing. Agriculture 12 (6), 838. Available at https://doi.org/10.3390/agriculture12060838.

Amsaraj, R., Ambade, N.D., Mutturi, S., 2021. Variable selection coupled to PLS2, ANN and SVM for simultaneous detection of multiple adulterants in milk using spectral data. International Dairy Journal 123, 105172. Available at https://doi.org/10.1016/j.idairyj.2021.105172.

Anand, M.J., Sridhar, V., Ravi, R., 2019. Development of Prototype E nose for detection of adulterants in raw milk. COMPUSOFT, An International Journal of Advanced Computer Technology 8 (7), 3235–3241.

Annosi, M.C., Brunetta, F., Bimbo, F., Kostoula, M., 2021. Digitalization within food supply chains to prevent food waste. Drivers, barriers and collaboration practices. Industrial Marketing Management 93, 208–220. Available at https://doi.org/10.1016/j.indmarman.2021.01.005.

Anyibama, B.J., Orjinta, K.K., Onotole, E.F., Olalemi, A.A., Temitope Olayinka, O., Ogunwale, G.E., et al., 2025. Blockchain application in food supply chains: A critical review and agenda for future studies. International Journal of Innovative Science and Research Technology 2345–2352. Available at https://doi.org/10.38124/ijisrt/25feb1659.

Aqeel, M., Sohaib, A., Iqbal, M., Ullah, S.S., 2024. Hyperspectral identification of milk adulteration using advance deep learning. IEEE Access 12, 174965–174982. Available at https://doi.org/10.1109/ACCESS.2024.3504334.

Aqeel, M., Sohaib, A., Iqbal, M., Ullah, S.S., 2025. Milk adulteration identification using hyperspectral imaging and machine learning. Journal of Dairy Science 108 (2), 1301–1314. Available at https://doi.org/10.3168/jds.2024-25635.

Arago, N.M., Alvarez, C.I., Mabale, A.G., Legista, C.G., Repiso, N.E., Amado, T.M., et al., 2022. Smart dairy cattle farming and in-heat detection through the Internet of Things (IoT). International Journal of Integrated Engineering 14 (1), 157–172. Available at https://doi.org/10.30880/ijie.2022.14.01.014.

Aritua, B., Wagener, C., Wagener, N., Adamczak, M., 2021. Blockchain solutions for international logistics networks along the new silk road between Europe and Asia. Logistics 5 (3), 55. Available at https://doi.org/10.3390/logistics5030055.

Aung, M.M., Chang, Y.S., 2014. Traceability in a food supply chain: Safety and quality perspectives. Food Control 39, 172–184. Available at https://doi.org/10.1016/j.foodcont.2013.11.007.

Ayeni, O., Olagoke-Komolafe, O.E., 2024. Innovative risk management techniques in food safety: A review of best practices and emerging trends. International Journal of Applied Research in Social Sciences 6 (9), 2241–2257. Available at https://doi.org/10.51594/ijarss.v6i9.1589.

Bag, S., Pretorius, J.H.C., 2022. Relationships between industry 4.0, sustainable manufacturing and circular economy: Proposal of a research framework. International Journal of Organizational Analysis 30 (4), 864–898. Available at https://doi.org/10.1108/IJOA-04-2020-2120.

Bahri, A., Nawar, S., Selmi, H., Amraoui, M., Rouissi, H., Mouazen, A.M., 2019. Application of visible and near-infrared spectroscopy for evaluation of ewes milk with different feeds. Animal Production Science 59 (6), 1190–1200. Available at https://doi.org/10.1071/AN17240.

Baluprithviraj, K.N., Kalavathi, T.D., Madhan, M.M., Dharuhn, V.B., Madhusuthan, C., & Jaidharun, R., 2024. IoT integrated quality checking and stage classification for dairy products. In: 2024 5th International Conference on Data Intelligence and Cognitive Informatics (ICDICI) (pp. 267–272). https://doi.org/10.1109/ICDICI62993.2024.10810910.

Becchi, P.P., Rocchetti, G., García-Pérez, P., Michelini, S., Pizzamiglio, V., Lucini, L., 2024. Untargeted metabolomics and machine learning unveil quality and authenticity interactions in grated Parmigiano Reggiano PDO cheese. Food Chemistry 447, 138938. Available at https://doi.org/10.1016/j.foodchem.2024.138938.

Berman, B., Swani, K., 2010. Managing product safety of imported Chinese goods. Business Horizons 53 (1), 39–48. Available at https://doi.org/10.1016/j.bushor.2009.09.002.

Best, D., 2019. Arla Foods' Tomi Siren on how blockchain is helping the dairy giant's sustainability drive—Interview. Just Food.

Bittal, V., Alfatmi, K., Chaudhari, S., Shah, N., Patil, D., & Jain, M., 2025. Affordable IoT-based milk quality assessment for small-scale dairy farmers using machine learning. In: 2025 6th International Conference on Recent Advances in Information Technology (RAIT) (pp. 1–6). https://doi.org/10.1109/RAIT65068.2025.11088913.

Bosona, T., Gebresenbet, G., 2013. Food traceability as an integral part of logistics management in food and agricultural supply chain. Food Control 33 (1), 32–48. Available at https://doi.org/10.1016/j.foodcont.2013.02.004.

Bourlakis, M., Maglaras, G., Aktas, E., Gallear, D., Fotopoulos, C., 2014. Firm size and sustainable performance in food supply chains: Insights from Greek SMEs. International Journal of Production Economics 152, 112–130. Available at https://doi.org/10.1016/j.ijpe.2013.12.029.

Brill, J., 2022. Consorzio del Parmigiano Reggiano Partners on Digital Tracking Technology. Retrieved from https://www.qualityassurancemag.com/news/consorzio-del-parmigiano-reggiano-partners-with-others-on-digital-tracking-technology/.

Cabrera, V.E., Fadul-Pacheco, L., 2021. Future of dairy farming from the Dairy Brain perspective: Data integration, analytics, and applications. International Dairy Journal 121, 105069. Available at https://doi.org/10.1016/j.idairyj.2021.105069.

Cabrera, V.E., Barrientos-Blanco, J.A., Delgado, H., Fadul-Pacheco, L., 2020. Symposium review: Real-time continuous decision making using big data on dairy farms. Journal of Dairy Science 103 (4), 3856–3866. Available at https://doi.org/10.3168/jds.2019 17145.

Cabrera, V.E., Bewley, J., Breunig, M., Breunig, T., Cooley, W., De Vries, A., et al., 2025. Data integration and analytics in the dairy industry: Challenges and pathways forward. Animals 15 (3), 329. Available at https://doi.org/10.3390/ani15030329.

Caldarelli, G., Rossignoli, C., Zardini, A., 2020. Overcoming the Blockchain Oracle Problem in the Traceability of non-fungible products. Sustainability 12 (6), 2391. Available at https://doi.org/10.3390/su12062391.

Campuzano, S., Ruiz-Valdepeñas Montiel, V., Serafín, V., Yáñez-Sedeño, P., Pingarrón, J.M., 2020. Cutting-edge advances in electrochemical affinity biosensing at different molecular level of emerging food allergens and adulterants. Biosensors 10 (2), 10. Available at https://doi.org/10.3390/bios10020010.

Cannas, V.G., Ciccullo, F., Pero, M., Cigolini, R., 2020. Sustainable innovation in the dairy supply chain: enabling factors for intermodal transportation. International Journal of Production Research 58 (24), 7314–7333. Available at https://doi.org/10.1080/00207543.2020.1809731.

Casino, F., Kanakaris, V., Dasaklis, T.K., Moschuris, S., Stachtiaris, S., Pagoni, M., et al., 2021. Blockchain-based food supply chain traceability: A case study in the dairy sector. International Journal of Production Research 59 (19), 5758–5770. Available at https://doi.org/10.1080/00207543.2020.1789238.

Cesarini, L., Gonçalves, R., Martina, M., Romão, X., Monteleone, B., Lobo Pereira, F., et al., 2024. Comparison of deep learning models for milk production forecasting at national scale. Computers and Electronics in Agriculture 221, 108933. Available at https://doi.org/10.1016/j.compag.2024.108933.

Chakurkar, P., Shikalgar, S., & Mukhopadhyay, D., 2018. An Internet of Things (IOT) based monitoring system for efficient milk distribution. In: International Conference on Advances in Computing, Communication and Control 2017, ICAC3 2017 (pp. 1–5). https://doi.org/10.1109/ICAC3.2017.8318780

Chatterjee, S., Chaudhuri, R., Vrontis, D., Galati, A., 2024. Digital transformation using industry 4.0 technology by food and beverage companies in post COVID-19 period: From DCV and

IDT perspective. European Journal of Innovation Management 27 (5), 1475–1495. Available at https://doi.org/10.1108/EJIM-07-2022-0374.

Chryssochoidis, G., Karagiannaki, A., Pramatari, K., Kehagia, O., 2009. A cost-benefit evaluation framework of an electronic-based traceability system. British Food Journal 111 (6), 565–582. Available at https://doi.org/10.1108/00070700910966023.

Chung, H., Li, J., Kim, Y., Van Os, J.M.C., Brounts, S.H., Choi, C.Y., 2020. Using implantable biosensors and wearable scanners to monitor dairy cattle's core body temperature in real-time. Computers and Electronics in Agriculture 174, 105453. Available at https://doi.org/10.1016/j.compag.2020.105453.

Cockburn, M., 2020. Review: Application and prospective discussion of machine learning for the management of dairy farms. Animals 10 (9), 1690. Available at https://doi.org/10.3390/ani10091690.

Corradini, M.G., 2018. Shelf life of food products: From open labeling to real-time measurements. Annual Review of Food Science and Technology 9 (1), 251–269. Available at https://doi.org/10.1146/annurev-food-030117-012433.

Cunha, D.S., Coelho, M.C., Ribeiro, S.C., Silva, C.C.G., 2022. Application of *Enterococcus malodoratus* SJC25 for the manufacture of whey-based beverage naturally enriched with GABA. Foods 11 (3), 447. Available at https://doi.org/10.3390/foods11030447.

da Rosa Righi, R., Goldschmidt, G., Kunst, R., Deon, C., André da Costa, C., 2020. Towards combining data prediction and internet of things to manage milk production on dairy cows. Computers and Electronics in Agriculture 169. Available at https://doi.org/10.1016/j.compag.2019.105156.

Dasaklis, T.K., Casino, F., Patsakis, C., 2019. Defining granularity levels for supply chain traceability based on IoT and blockchain. Proceedings of the International Conference on Omni-Layer Intelligent Systems 184–190. Available at https://doi.org/10.1145/3312614.3312652.

Deep, A., Perrusquía, A., Aljaburi, L., Al-Rubaye, S., Guo, W., 2024. A novel distributed authentication of blockchain technology integration in IoT services. IEEE Access 12, 9550–9562. Available at https://doi.org/10.1109/ACCESS.2024.3349955.

de las Morenas, J., García, A., Blanco, J., 2014. Prototype traceability system for the dairy industry. Computers and Electronics in Agriculture 101, 34–41. Available at https://doi.org/10.1016/j.compag.2013.12.011.

Devi, P., Subburamu, K., Giridhari, V.A., Dananjeyan, B., Maruthamuthu, T., 2025. Integration of AI based tools in dairy quality control: Enhancing pathogen detection efficiency. Journal of Food Measurement and Characterization 19 (7), 4427–4438. Available at https://doi.org/10.1007/s11694-025-03269-8.

Dey, S., Saha, S., Singh, A.K., McDonald-Maier, K., 2021. FoodSQRBlock: Digitizing food production and the supply chain with blockchain and QR code in the cloud. Sustainability 13 (6), 3486. Available at https://doi.org/10.3390/su13063486.

Ding, J., Bai, S,, 2022. Optimal decision and coordination of organic food supply chain from the perspective of blockchain. Discrete Dynamics in Nature and Society 2022. Available at https://doi.org/10.1155/2022/6050185.

Drekalovic, N., 2021. Raw milk quality monitoring system. In *2021 10th Mediterranean Conference on Embedded Computing (MECO)* 1–4). Available at https://doi.org/10.1109/MECO52532.2021.9460257.

Durgun, M., 2024. Real-Time detection of milk adulteration with a portable multispectral analysis device: A multispectral sensor and optimized logistic regression approach. Journal of

Advanced Research in Natural and Applied Sciences 10 (4), 968–980. Available at https://doi.org/10.28979/jarnas.1569065.

Eastwood, C., Klerkx, L., Ayre, M., Dela Rue, B., 2017. Managing socio-ethical challenges in the development of smart farming: from a fragmented to a comprehensive approach for responsible research and innovation. Journal of Agricultural and Environmental Ethics 32, 741–768. Available at https://doi.org/10.1007/s10806-017-9704-5.

Ellahi, R.M., Wood, L.C., Bekhit, A.E.-D.A., 2023. Blockchain-based frameworks for food traceability: A systematic review. Foods 12 (16), 3026. Available at https://doi.org/10.3390/foods12163026.

Escobet, T., Puig, V., Quevedo, J., Palá-Schönwälder, P., Romera, J., Adelman, W., 2019. Optimal batch scheduling of a multiproduct dairy process using a combined optimization/constraint programming approach. Computers & Chemical Engineering 124, 228–237. Available at https://doi.org/10.1016/j.compchemeng.2018.08.040.

Esquivel, D.Y.A., Guo, Y., Brown, R.K., Müller, S., Schröder, U., Harnisch, F., 2020. Investigating community dynamics and performance during microbial electrochemical degradation of whey. ChemElectroChem 7 (4), 989–997. Available at https://doi.org/10.1002/celc.201902109.

European Commission, 2021. Brussels ANNEX to the communication from the commission to the European Parliament, the Council, the European Economic and Social Committee and the Committee of the Regions—2030 Digital Compass: The European way for the digital decade.

Fabregas, R., Kremer, M., Schilbach, F., 2019. Realizing the potential of digital development: The case of agricultural advice. Science (New York, N.Y.) 366 (6471). Available at https://doi.org/10.1126/science.aay3038.

Facchini, F., Mossa, G., Sassanelli, C., Digiesi, S., 2024. IoT-based milk-run routing for manufacturing system: An application case in an automotive company. International Journal of Production Research 62 (1–2), 536–555. Available at https://doi.org/10.1080/00207543.2023.2254408.

Fanelli, V., Mascio, I., Miazzi, M.M., Savoia, M.A., De Giovanni, C., Montemurro, C., 2021. Molecular approaches to agri-food traceability and authentication: An updated review. Foods 10 (7), 1644. Available at https://doi.org/10.3390/foods10071644.

Fang, C., & Stone, W.Z., 2021. An ecosystem for the dairy logistics supply chain with blockchain technology. In: International Conference on Electrical, Computer, Communications and Mechatronics Engineering, ICECCME 2021. https://doi.org/10.1109/ICECCME52200.2021.9591146.

FAO, 2023. FAO strategic priorities for food safety within the FAO strategic framework 2022-2031. FAO. https://doi.org/10.4060/cc4040en.

FAO. 2025. Traceability & Recalls. Food and Agriculture Organization of the United Nations. Available at: https://www.fao.org/food-safety/food-control-systems/supply-chains-and-consumers/traceability-and-recalls/en/.

Feng, H., Wang, X., Duan, Y., Zhang, J., Zhang, X., 2020. Applying blockchain technology to improve agri-food traceability: A review of development methods, benefits and challenges. Journal of Cleaner Production 260, 121031. Available at https://doi.org/10.1016/j.jclepro.2020.121031.

Fernandez, C.M., Alves, J., Gaspar, P.D., Lima, T.M., Silva, P.D., 2023. Innovative processes in smart packaging. A systematic review. Journal of the Science of Food and Agriculture 103 (3), 986–1003. Available at https://doi.org/10.1002/jsfa.11863.

Folinas, D., Manikas, I., Manos, B., 2006. Traceability data management for food chains. British Food Journal 108 (8), 622–633. Available at https://doi.org/10.1108/00070700610682319.

Fonseca, T.F. A., & Garrocho, C.T. B., 2024. Towards a milk quality recording and monitoring system based on blockchain and long-range Internet of Things. In: Anais Do VII Workshop Em Blockchain: Teoria, Tecnologias e Aplicações (WBlockchain 2024) (pp. 1–14). https://doi.org/10.5753/wblockchain.2024.2301.

Freeman, K.K., Valencia, V., Marzaroli, J., van Zanten, H.H., 2022. Digital traceability to enhance circular food systems and reach agriculture emissions targets. Outlook on Agriculture 51 (4), 414–422. Available at https://doi.org/10.1177/00307270221133854.

Galvez, J.F., Mejuto, J.C., Simal-Gandara, J., 2018. Future challenges on the use of blockchain for food traceability analysis. TrAC Trends in Analytical Chemistry 107, 222–232. Available at https://doi.org/10.1016/j.trac.2018.08.011.

Geethanjali, P., 2017. Application of technologies in Internet of Things (pp. 245–260). Springer. https://doi.org/10.1007/978-3-319-53472-5_12.

Gehlot, A., Malik, P.K., Singh, R., Akram, S.V., Alsuwian, T., 2022. Dairy 4.0: Intelligent Communication Ecosystem for the cattle animal welfare with blockchain and IoT enabled technologies. Applied Sciences 12 (14), 7316. Available at https://doi.org/10.3390/app12147316.

Gerna, S., D'Incecco, P., Limbo, S., Sindaco, M., Pellegrino, L., 2023. Strategies for exploiting milk protein properties in making films and coatings for food packaging: A review. Foods 12 (6), 1271. Available at https://doi.org/10.3390/foods12061271.

Ghadge, A., Kaklamanou, M., Choudhary, S., Bourlakis, M., 2017. Implementing environmental practices within the Greek dairy supply chain. Industrial Management & Data Systems 117 (9), 1995–2014. Available at https://doi.org/10.1108/IMDS-07-2016-0270.

Giacalone, M., Santarcangelo, V., Donvito, V., Schiavone, O., Massa, E., 2021. Big data for corporate social responsibility: Blockchain use in Gioia del Colle DOP. Quality & Quantity 55 (6), 1945–1971. Available at https://doi.org/10.1007/s11135-021-01095-w.

Gillaspie, E.A., Shipman, S.A., Timperley, J.B., Al-Refaie, W.B., Wong, S.L., 2025. Leveraging technology to enhance rural access to care. World Journal of Surgery 49 (8), 2074–2077. Available at https://doi.org/10.1002/wjs.12692.

Gomes, M.T.S.R., 2016. Electronic nose in dairy products. Electronic Noses and Tongues in Food Science. Elsevier, pp. 21–30. Available at https://doi.org/10.1016/B978-0-12-800243-8.00003-2.

Gossner, C.M.-E., Schlundt, J., Ben Embarek, P., Hird, S., Lo-Fo-Wong, D., Beltran, J.J.O., et al., 2009. The melamine incident: Implications for international food and feed safety. Environmental Health Perspectives 117 (12), 1803–1808. Available at https://doi.org/10.1289/ehp.0900949.

Gowda, S.D., Raj, K.T., 2025. An AI-integrated renewable-powered cold storage system with advanced environmental sensing for smallholder farmers. Engineering, Technology & Applied Science Research 15 (6), 29391–29396. Available at https://doi.org/10.48084/etasr.13983.

Group, P.I., Kasten, J., 2019. Blockchain application: The dairy supply chain. Journal of Supply Chain Management Systems 8 (1), 45–54.

Guldiken, B., Karliga, S., Capanoglu, E., Yolci-Omeroglu, P., Kamiloglu, S., 2021. Food traceability. Innovative food analysis. Elsevier, pp. 249–268. Available at https://doi.org/10.1016/B978-0-12-819493-5.00009-1.

Gutiérrez, P., Godoy, S.E., Torres, S., Oyarzún, P., Sanhueza, I., Díaz-García, V., et al., 2020. Improved antibiotic detection in raw milk using machine learning tools over the absorption spectra of a problem-specific nanobiosensor. Sensors 20 (16), 4552. Available at https://doi.org/10.3390/s20164552.

Habsari, W., Udin, F., Arkeman, Y., 2022. An analysis and design of fresh milk smart grading system based on internet of things. IOP Conference Series: Earth and Environmental Science 1063 (1). Available at https://doi.org/10.1088/1755-1315/1063/1/012059.

Hassija, V., Chamola, V., Saxena, V., Jain, D., Goyal, P., Sikdar, B., 2019. A survey on IoT security: Application areas, security threats, and solution architectures. IEEE Access 7, 82721–82743. Available at https://doi.org/10.1109/ACCESS.2019.2924045.

Hassoun, A., Garcia-Garcia, G., Trollman, H., Jagtap, S., Parra-López, C., Cropotova, J., et al., 2023a. Birth of dairy 4.0: Opportunities and challenges in adoption of fourth industrial revolution technologies in the production of milk and its derivatives. Current Research in Food Science 7, 100535. Available at https://doi.org/10.1016/j.crfs.2023.100535.

Hassoun, A., Kamiloglu, S., Garcia-Garcia, G., Parra-López, C., Trollman, H., Jagtap, S., et al., 2023b. Implementation of relevant fourth industrial revolution innovations across the supply chain of fruits and vegetables: A short update on Traceability 4.0. Food Chemistry 409, 135303. Available at https://doi.org/10.1016/j.foodchem.2022.135303.

Hassoun, A., Tarchi, I., Aït-Kaddour, A., 2024. Leveraging the potential of fourth industrial revolution technologies to reduce and valorize waste and by-products in the dairy sector. Current Opinion in Green and Sustainable Chemistry 47, 100927. Available at https://doi.org/10.1016/j.cogsc.2024.100927.

Hastig, G.M., Sodhi, M.S., 2020. Blockchain for supply chain traceability: Business requirements and critical success factors. Production and Operations Management 29 (4), 935–954. Available at https://doi.org/10.1111/poms.13147.

Hertog, M.L.A.T.M., Uysal, I., McCarthy, U., Verlinden, B.M., Nicolaï, B.M., 2014. Shelf life modelling for first-expired-first-out warehouse management. Philosophical Transactions of the Royal Society A: Mathematical, Physical and Engineering Sciences 372 (2017), 20130306. Available at https://doi.org/10.1098/rsta.2013.0306.

Heussner, J., Hanf, J., 2025. Digitalisation and sustainability in the Armenian wine sector: Challenges and opportunities. International Food and Agribusiness Management Review 28 (3), 621–641. Available at https://doi.org/10.22434/ifamr.1269.

Hill, R., Betts, L.R., Gardner, S.E., 2015. Older adults' experiences and perceptions of digital technology: (Dis)empowerment, wellbeing, and inclusion. Computers in Human Behavior 48, 415–423. Available at https://doi.org/10.1016/j.chb.2015.01.062.

Hosseini-Motlagh, S.-M., Samani, M.R.G., Rahmani, M., 2025. A digital twin framework integrated with a mixed proactive-reactive model for human milk supply chain planning. International Journal of Production Economics 287, 109683. Available at https://doi.org/10.1016/j.ijpe.2025.109683.

Høyer, M.R., Oluyisola, O.E., Strandhagen, J.O., Semini, M.G., 2019. Exploring the challenges with applying tracking and tracing technology in the dairy industry. Ifac-Papersonline 52 (13), 1727–1732.

Hu, Y., Feng, S., Gao, F., Li-Chan, E.C.Y., Grant, E., Lu, X., 2015. Detection of melamine in milk using molecularly imprinted polymers–surface enhanced Raman spectroscopy. Food Chemistry 176, 123–129. Available at https://doi.org/10.1016/j.foodchem.2014.12.051.

Husnoo, M.A., Anwar, A., Chakrabortty, R.K., Doss, R., Ryan, M.J., 2021. Differential privacy for IoT-enabled critical infrastructure: A comprehensive survey. IEEE Access 9, 153276–153304. Available at https://doi.org/10.1109/ACCESS.2021.3124309.

Hussain, S.A., Ramaiah, C.S., Prasad, M.N. G., & Hussain, S.M., 2016. Milk products monitoring system with arm processor for early detection of microbial activity. In: 2016 3rd MEC

International Conference on Big Data and Smart City (ICBDSC) (pp. 1–6). https://doi.org/10.1109/ICBDSC.2016.7460385.

Islam, S., Cullen, J.M., 2021. Food traceability: A generic theoretical framework. Food Control 123, 107848. Available at https://doi.org/10.1016/j.foodcont.2020.107848.

Iwasaki, W., Morita, N., Nagata, M.P.B., 2019. IoT sensors for smart livestock management. Chemical, gas, and biosensors for Internet of Things and related applications. Elsevier, pp. 207–221. Available at https://doi.org/10.1016/B978-0-12-815409-0.00015-2.

Jaddoa, M.A., Zaidan, A.A., Gonzalez, L.A., Deveci, M., Cuthbertson, H., Al-Jumaily, A., et al., 2024. An approach-based machine learning and automated thermal images to predict the dark-cutting incidence in cattle management of healthcare supply chain. Engineering Applications of Artificial Intelligence 135, 108804. Available at https://doi.org/10.1016/j.engappai.2024.108804.

Jaeger, M., Brügemann, K., Brandt, H., König, S., 2019. Associations between precision sensor data with productivity, health and welfare indicator traits in native black and white dual-purpose cattle under grazing conditions. Applied Animal Behaviour Science 212, 9–18. Available at https://doi.org/10.1016/j.applanim.2019.01.008.

Jarh, A., Lal, S.P., Gogoi, M., Sharma, M., Digra, M., 2025. Leveraging blockchain for food safety and traceability in the digital age. European Journal of Nutrition & Food Safety 17 (7), 366–382. Available at https://doi.org/10.9734/ejnfs/2025/v17i71791.

Javadi, B., Dadashi, N., Yazdi, F., Reza Abdali, M., 2024. Application of Internet of Things (IoT) to food supply chain under uncertainty-case: Traditional dairy products. IEEE Access 12, 102702–102717. Available at https://doi.org/10.1109/ACCESS.2024.3432325.

Ji, B., Banhazi, T., Phillips, C.J.C., Wang, C., Li, B., 2022. A machine learning framework to predict the next month's daily milk yield, milk composition and milking frequency for cows in a robotic dairy farm. Biosystems Engineering 216, 186–197. Available at https://doi.org/10.1016/j.biosystemseng.2022.02.013.

Kadam, P.R., & Shinde, K.P., 2018. Real time milk monitoring system. In: 2018 Fourth International Conference on Computing Communication Control and Automation (ICCUBEA) (pp. 1–5). https://doi.org/10.1109/ICCUBEA.2018.8697373.

Kamilaris, A., Fonts, A., Prenafeta-Boldύ, F.X., 2019. The rise of blockchain technology in agriculture and food supply chains. Trends in Food Science & Technology 91, 640–652. Available at https://doi.org/10.1016/j.tifs.2019.07.034.

Kang, N., Liu, W., Kim, S., Kim, H., & Lee, J., 2024. Innovations in real-time monitoring of cattle face recognition for feeding behaviors utilizing advanced deep learning techniques. In: 11th European Conference on Precision Livestock Farming (pp. 1641–1649).

Kaur, A., Singh, G., Kukreja, V., Sharma, S., Singh, S., Yoon, B., 2022. Adaptation of IoT with blockchain in food supply chain management: An analysis-based review in development, benefits and potential applications. Sensors 22 (21), 8174. Available at https://doi.org/10.3390/s22218174.

Khanna, A., Jain, S., Burgio, A., Bolshev, V., Panchenko, V., 2022. Blockchain-enabled supply chain platform for indian dairy industry: Safety and traceability. Foods 11 (17), 2716. Available at https://doi.org/10.3390/foods11172716.

Klerkx, L., Jakku, E., Labarthe, P., 2019. A review of social science on digital agriculture, smart farming and agriculture 4.0: New contributions and a future research agenda. NJAS: Wageningen Journal of Life Sciences 90-91, 1–16. Available at https://doi.org/10.1016/j.njas.2019.100315.

Konfo, T.R.C., Djouhou, F.M.C., Hounhouigan, M.H., Dahouenon-Ahoussi, E., Avlessi, F., Sohounhloue, C.K.D., 2023. Recent advances in the use of digital technologies in agri-food processing: A short review. Applied Food Research 3 (2), 100329. Available at https://doi.org/10.1016/j.afres.2023.100329.

Kumar, R., Kumar, D., 2025. Blockchain-based smart dairy supply chain: Catching the momentum for digital transformation. Journal of Agribusiness in Developing and Emerging Economies 15 (2), 225–248. Available at https://doi.org/10.1108/JADEE-07-2022-0141.

Kusnadi, A., Arkeman, Y., Syamsu, K., Hartono Wijaya, S., 2024. Assurance of halal beef products using a traceability system in enterprise resource planning applications. IOP Conference Series: Earth and Environmental Science 1358 (1), 012020. Available at https://doi.org/10.1088/1755-1315/1358/1/012020.

Lai, M.B., Vergamini, D., Brunori, G., 2025. Food supply chain: a framework for the governance of digital traceability. Foods 14 (12), 2032. Available at https://doi.org/10.3390/foods14122032.

Länge, K., 2022. Bulk and surface acoustic wave biosensors for milk analysis. Biosensors 12 (8), 602. Available at https://doi.org/10.3390/bios12080602.

Lanjewar, M.G., Parab, J.S., Kamat, R.K., 2024. Machine learning based technique to predict the water adulterant in milk using portable near infrared spectroscopy. Journal of Food Composition and Analysis 131, 106270. Available at https://doi.org/10.1016/j.jfca.2024.106270.

La Scalia, G., Nasca, A., Corona, O., Settanni, L., Micale, R., 2015. An innovative shelf life model based on smart logistic unit for an efficient management of the perishable food supply chain. Journal of Food Process Engineering 40 (1). Available at https://doi.org/10.1111/jfpe.12311.

Lasser, J., Matzhold, C., Egger-Danner, C., Fuerst-Waltl, B., Steininger, F., Wittek, T., et al., 2021. Integrating diverse data sources to predict disease risk in dairy cattle—A machine learning approach. Journal of Animal Science 99 (11). Available at https://doi.org/10.1093/jas/skab294.

Lavanya, R., Abuthakir, S., & Ahamed, D.A., 2023. IoT-enhanced livestock monitoring for animal health and productivity. In: 2023 International Conference on Sustainable Communication Networks and Application (ICSCNA) (pp. 504–509). https://doi.org/10.1109/ICSCNA58489.2023.10370264.

Lei, Z., Cai, S., Cui, L., Wu, L., Liu, Y., 2023. How do different Industry 4.0 technologies support certain Circular Economy practices? Industrial Management & Data Systems 123 (4), 1220–1251. Available at https://doi.org/10.1108/IMDS-05-2022-0270.

Leochico, C.F.D., Espiritu, A.I., Ignacio, S.D., Mojica, J.A.P., 2020. Challenges to the emergence of telerehabilitation in a developing country: A systematic review. Frontiers in Neurology 11. Available at https://doi.org/10.3389/fneur.2020.01007.

Li, Z., Liu, G., Liu, L., Lai, X., Xu, G., 2017. IoT-based tracking and tracing platform for prepackaged food supply chain. Industrial Management & Data Systems 117 (9), 1906–1916. Available at https://doi.org/10.1108/IMDS-11-2016-0489.

Lin, H., Shavezipur, M., Yousef, A., Maleky, F., 2016. Prediction of growth of *Pseudomonas fluorescens* in milk during storage under fluctuating temperature. Journal of Dairy Science 99 (3), 1822–1830. Available at https://doi.org/10.3168/jds.2015-10179.

Lin, J., Shen, Z., Zhang, A., & Chai, Y., 2018. Blockchain and IoT based food traceability for smart agriculture. In: Proceedings of the 3rd International Conference on Crowd Science and Engineering (pp. 1–6). https://doi.org/10.1145/3265689.3265692.

Lindblom, J., Lundström, C., Ljung, M., Jonsson, A., 2017. Promoting sustainable intensification in precision agriculture: Review of decision support systems development and strategies. Precision Agriculture 18 (3), 309–331. Available at https://doi.org/10.1007/s11119-016-9491-4.

Lopes, F.R., Silva, L.M., Zimpel, R., Munhoz, A.K., Vieira-Neto, A., Pereira, M.H.C., et al., 2020. Prostaglandin F2α influences pre-ovulatory follicle characteristics and pregnancy per AI in anovular dairy cows. Theriogenology 153, 122–132. Available at https://doi.org/10.1016/j.theriogenology.2020.04.038.

Lovarelli, D., Brandolese, C., Leliveld, L., Finzi, A., Riva, E., Grotto, M., et al., 2022. Development of a new wearable 3D Sensor node and innovative open classification system for dairy cows' behavior. Animals 12 (11), 1447. Available at https://doi.org/10.3390/ani12111447.

Lyon, A.R., Koerner, K., 2016. User-centered design for psychosocial intervention development and implementation. Clinical Psychology: Science and Practice 23 (2), 180–200. Available at https://doi.org/10.1111/cpsp.12154.

Lyon, A.R., Brewer, S.K., Areán, P.A., 2020. Leveraging human-centered design to implement modern psychological science: Return on an early investment. American Psychologist 75 (8), 1067–1079. Available at https://doi.org/10.1037/amp0000652.

Ma, X.Y., Liu, Y.X., Yu, W.J., 2013. Quality evaluation of dairy products based on electronic nose. Applied Mechanics and Materials 475–476, 524–527. Available at https://doi.org/10.4028/www.scientific.net/AMM.475-476.524.

Madhukara, S., Bhargavi, A.S., Chandushree, T.A., & Divya Reddy, P.R., 2024. Smart milk quality analysis and grading using IoT. In: 2024 International Conference on Knowledge Engineering and Communication Systems (ICKECS) (pp. 1–5). https://doi.org/10.1109/ICKECS61492.2024.10616698.

Malik, A.A., Brem, A., 2021. Digital twins for collaborative robots: A case study in human-robot interaction. Robotics and Computer-Integrated Manufacturing 68, 102092. Available at https://doi.org/10.1016/j.rcim.2020.102092.

Malik, T., Shahzadi, T., 2025. Fabrication of hybrid silver based nanostructured multi-sensor for cutting edge sensitive detection of dual adulterants from milk system: A statistically validated approach. Food Analytical Methods 18 (7), 1384–1397. Available at https://doi.org/10.1007/s12161-025-02799-7.

Malik, M., Gahlawat, V.K., Mor, R.S., Singh, M.K., 2024. Unlocking dairy traceability: Current trends, applications, and future opportunities. Future Foods 10, 100426. Available at https://doi.org/10.1016/j.fufo.2024.100426.

Malik, M., Dalal, M., Mor, A., Toor, K., Yadav, D.K., Gahlawat, V.K., 2025. Dairy traceability: Pathways to a sustainable and resilient supply chain. Trends in Animal-Based Foods. Springer Nature, Switzerland, pp. 137–148. Available at https://doi.org/10.1007/978-3-032-00575-5_7.

Mania, I., Delgado, A.M., Barone, C., Parisi, S., 2018. Traceability in the dairy industry in Europe. Springer International Publishing. Available at https://doi.org/10.1007/978-3-030-00446-0.

Manisha, N., Jagadeeshwar, M., 2023. BC driven IoT-based food quality traceability system for dairy product using deep learning model. High-Confidence Computing 3 (3), 100121. Available at https://doi.org/10.1016/j.hcc.2023.100121.

Marathe, A., Vaniya, P., Motale, V., Thokal, P., Mehare, Y., & Wayal, S., 2025. IoT-enabled real-time milk quality assessment system using multi-sensor fusion and XAI for microbial load prediction. In: 2025 6th International Conference on Electronics and Sustainable Communication Systems (ICESC) (pp. 1209–1215). https://doi.org/10.1109/ICESC65114.2025.11212238.

Marin, M.-P., Marin, I., & Vidu, L., 2019. Learning about the reduction of food waste using blockchain technology. In: INTED2019 Proceedings (pp. 3274–3277). https://doi.org/10.21125/inted.2019.0856.

McCosker, A., Critchley, C., Walshe, J., Tucker, J., Suchowerska, R., 2023. Accounting for diversity in older adults' digital inclusion and literacy: The impact of a national intervention. Ageing and Society 43 (11), 2629–2649. Available at https://doi.org/10.1017/S0144686X21001550.

McEntire, J.C., Arens, S., Bernstein, M., Bugusu, B., Busta, F.F., Cole, M., et al., 2010. Traceability (product tracing) in food systems: An IFT report submitted to the FDA, Volume 1: Technical aspects and recommendations. Comprehensive Reviews in Food Science and Food Safety 9 (1), 92–158. Available at https://doi.org/10.1111/j.1541-4337.2009.00097.x.

McMeekin, T.A., Baranyi, J., Bowman, J., Dalgaard, P., Kirk, M., Ross, T., et al., 2006. Information systems in food safety management. International Journal of Food Microbiology 112 (3), 181–194. Available at https://doi.org/10.1016/j.ijfoodmicro.2006.04.048.

Mehannaoui, R., Mouss, K.N., Aksa, K., Benchine, H., Ismail, S., 2025. A holochain-based IoT-enabled agri-food traceability system: application in the recombined milk supply chain. International Journal of Information Technology. Available at https://doi.org/10.1007/s41870-024-02396-7.

Mendonça, R.D., Gomes, O.S., Vieira, L.F. M., Vieira, A.B., Nacif, J.A. M., & Vieira, M.A. M., 2023. BlockColdChain: Vaccine cold chain blockchain. Authorea. https://doi.org/10.22541/au.167871951.17803821/v1.

Mhapsekar, R.U., Abraham, L., O'Shea, N., & Davy, S., 2022. Edge-AI Implementation for Milk Adulteration Detection. 2022 IEEE Global Conference on Artificial Intelligence and Internet of Things (GCAIoT), 108–113. https://doi.org/10.1109/GCAIoT57150.2022.10019173.

Mhapsekar, R.U., Abraham, L., Davy, S., Dey, I., 2025. Application adaptive light-weight deep learning (AppAdapt-LWDL) framework for enabling edge intelligence in dairy processing. IEEE Transactions on Mobile Computing 24 (2), 1105–1119. Available at https://doi.org/10.1109/TMC.2024.3475634.

Modi, P.M., Shah, M.M., Bhanushali, J.D., Dhulia, D.A., Barochiya, M.N., Jayswal, H., et al., 2025. Utilizing machine learning techniques and spectrometry for milk impurity detection (pp. 439–450). https://doi.org/10.1007/978-981-96-1206-2_34.

Moharkar, L.W., & Patnaik, S., 2019. Detection and quantification of milk adulteration by laser induced instumentation. In: 2019 IEEE 5th International Conference for Convergence in Technology (I2CT) (pp. 1–5). https://doi.org/10.1109/I2CT45611.2019.9033883.

Mottram, T., Rudnitskaya, A., Legin, A., Fitzpatrick, J.L., Eckersall, P.D., 2007. Evaluation of a novel chemical sensor system to detect clinical mastitis in bovine milk. Biosensors and Bioelectronics 22 (11), 2689–2693. Available at https://doi.org/10.1016/j.bios.2006.11.006.

Mu, F., Gu, Y., Zhang, J., Zhang, L., 2020. Milk source identification and milk quality estimation using an electronic nose and machine learning techniques. Sensors 20 (15), 4238. Available at https://doi.org/10.3390/s20154238.

Müller, P., Schmid, M., 2019. Intelligent packaging in the food sector: A brief overview. Foods 8 (1), 16. Available at https://doi.org/10.3390/foods8010016.

Musthafa, M.B., Huda, S., Kodera, Y., Ali, M.A., Araki, S., Mwaura, J., et al., 2024. Optimizing IoT intrusion detection using balanced class distribution, feature selection, and ensemble machine learning techniques. Sensors 24 (13), 4293. Available at https://doi.org/10.3390/s24134293.

Nanda, P.K., Bhattacharya, D., Das, J.K., Bandyopadhyay, S., Ekhlas, D., Lorenzo, J.M., et al., 2022. Emerging role of biosensors and chemical indicators to monitor the quality and safety of meat and meat products. Chemosensors 10 (8), 322. Available at https://doi.org/10.3390/chemosensors10080322.

Neethirajan, S., 2020. Transforming the adaptation physiology of farm animals through sensors. Animals 10 (9), 1512. Available at https://doi.org/10.3390/ani10091512.

Niaki, M.K., Nonino, F., Komijan, A.R., Dehghani, M., 2017. Food production in batch manufacturing systems with multiple shared-common resources: A scheduling model and its application in the yoghurt industry. International Journal of Services and Operations Management 27 (3), 345. Available at https://doi.org/10.1504/IJSOM.2017.084442.

Nishantha, G., Wanniarachchige, M., & Jehan, S. (2010). A pragmatic approach to traceability in food supply chains. In *The 12th International Conference on Advanced Communication Technology (ICACT).*

Niya, S.R., Dordevic, D., Hurschler, M., Grossenbacher, S., & Stiller, B. (2021). A blockchain-based supply chain tracing for the swiss dairy use case. In *2020 2nd International Conference on Societal Automation (SA)* (1–8). https://doi.org/10.1109/SA51175.2021.9507182

Nukapeyi, S., Bommala, R., Kondreddy, D.S., Amalakanti, C., Ravi, C., & Konuri, T.A. (2024). Milk chain with ML and blockchain for ensuring milk purity in dairy industry. In *2024 2nd World Conference on Communication & Computing (WCONF)* (pp. 1–7). https://doi.org/10.1109/WCONF61366.2024.10692146.

Nurgazina, J., Pakdeetrakulwong, U., Moser, T., Reiner, G., 2021. Distributed ledger technology applications in food supply chains: A review of challenges and future research directions. Sustainability 13 (8), 4206. Available at https://doi.org/10.3390/su13084206.

Ok, C., 2025. The dual impact of smartphone app usage diversity on quality of life: The moderating roles of age and digital literacy. European Journal of Investigation in Health, Psychology and Education 15 (11), 221. Available at https://doi.org/10.3390/ejihpe15110221.

Olateju, O.O., 2025. Blockchain and cybersecurity integration for secure and resilient agricultural supply chains in the United States. Asian Journal of Advances in Agricultural Research 25 (4), 126–138. Available at https://doi.org/10.9734/ajaar/2025/v25i4606.

Omowole, B.M., Olufemi-Phillips, A.Q., Ofodile, O.C., Eyo-Udo, N.L., Ewim, S.E., 2024. Conceptualizing green business practices in SMEs for sustainable development. International Journal of Management & Entrepreneurship Research 6 (11), 3778–3805. Available at https://doi.org/10.51594/ijmer.v6i11.1719.

Padma, T., Aruna, S., Sujipriya, J., Zainab, M., 2023. Blockchain based solutions for milk procurement management and adulteration detection. ITM Web of Conferences 56, 02008. Available at https://doi.org/10.1051/itmconf/20235602008.

Patelli, N., Mandrioli, M., 2020. Blockchain technology and traceability in the agrifood industry. Journal of Food Science 85 (11), 3670–3678. Available at https://doi.org/10.1111/1750-3841.15477.

Pearce, K.E., Rice, R.E., 2013. Digital divides from access to activities: Comparing mobile and personal computer internet users. Journal of Communication 63 (4), 721–744. Available at https://doi.org/10.1111/jcom.12045.

Peres, F.A.P., Bondarczuk, B.A., Gomes, L.de C., Jardim, L.de C., Corrêa, R.G.de F., Baierle, I.C., 2025. Advances in food quality management driven by Industry 4.0: A systematic review-based framework. Foods 14 (14), 2429. Available at https://doi.org/10.3390/foods14142429.

Pérez-Calabuig, A.M., Pradana-López, S., Lopez-Ortega, S., Beleño Sáenz, K.de J., Cancilla, J.C., Torrecilla, J.S., 2023. Application of residual neural networks to detect and quantify milk adulterations. Journal of Food Composition and Analysis 122, 105427. Available at https://doi.org/10.1016/j.jfca.2023.105427.

Phukkaphan, N., Eamsa-ard, T., Chairanit, C., & Kerdcharoen, T. (2021). The application of gas sensor array based electronic nose for milk spoilage detection. In *2021 7th International Conference on Engineering, Applied Sciences and Technology (ICEAST)* (pp. 273–276). https://doi.org/10.1109/ICEAST52143.2021.9426263.

Pigini, D., Conti, M., 2017. NFC-Based traceability in the food chain. Sustainability 9 (10), 1910. Available at https://doi.org/10.3390/su9101910.

Poghossian, A., Geissler, H., Schöning, M.J., 2019. Rapid methods and sensors for milk quality monitoring and spoilage detection. Biosensors and Bioelectronics 140, 111272. Available at https://doi.org/10.1016/j.bios.2019.04.040.

Poyatos-Racionero, E., Ros-Lis, J.V., Vivancos, J.-L., Martínez-Máñez, R., 2018. Recent advances on intelligent packaging as tools to reduce food waste. Journal of Cleaner Production 172, 3398–3409. Available at https://doi.org/10.1016/j.jclepro.2017.11.075.

Priyadharshini, N., Kiruthika, S., Radha, J., Vaishya, R., Pranavi, Sasidharan, P., et al. (2024). IoT - based quality monitoring and control automation in dairy industry. In *2024 15th International Conference on Computing Communication and Networking Technologies (ICCCNT)* (pp. 1–5). IEEE. https://doi.org/10.1109/ICCCNT61001.2024.10724285.

Protopappas, L., Bechtsis, D., Tsotsolas, N., 2025. IoT services for monitoring food supply chains. Applied Sciences 15 (13), 7602. Available at https://doi.org/10.3390/app15137602.

Punzalan, J.K., Punzalan, M.G., 2025. Integrating digital health competencies in community engaged medical education: A scoping review for developing teaching and learning strategies for digital health. Journal of Education and Health Promotion 14 (1). Available at https://doi.org/10.4103/jehp.jehp_1240_24.

Putra, M.A., Rivai, M., & Arifin, A., 2018. Milk Assessment using potentiometric and gas sensors in conjunction with neural network. In: 2018 International Seminar on Intelligent Technology and Its Applications (ISITIA) (pp. 409–412). https://doi.org/10.1109/ISITIA.2018.8710944.

Qian, J., Ruiz-Garcia, L., Fan, B., Robla Villalba, J.I., McCarthy, U., Zhang, B., et al., 2020. Food traceability system from governmental, corporate, and consumer perspectives in the European Union and China: A comparative review. Trends in Food Science & Technology 99, 402–412. Available at https://doi.org/10.1016/j.tifs.2020.03.025.

Rahman, M., Hasan, S., 2025. Improving transparency and compliance in agricultural supply chains through IT and traceability systems using blockchain and ERP integration. International Journal For Multidisciplinary Research 7 (4). Available at https://doi.org/10.36948/ijfmr.2025.v07i04.54505.

Raina, P., Darji, R.K., Mittal, R., 2025. Impact of social media campaigns on farmers awareness of environmental conservation practices. Journal of Biodiversity and Environmental Sciences (JBES) 1–8. Available at https://doi.org/10.12692/jbes/27.4.1-8.

Rajakumar, G., Kumar, T.A., Samuel, T.A., Kumaran, E.M., 2018. IoT based milk monitoring system for detection of milk adulteration. International Journal of Pure and Applied Mathematics 118 (9), 21–32.

Rajput, S., Singh, S.P., 2020. Industry 4.0 Model for circular economy and cleaner production. Journal of Cleaner Production 277, 123853. Available at https://doi.org/10.1016/j.jclepro.2020.123853.

Rambim, D., & Awuor, F.M., 2020. Blockchain based milk delivery platform for stallholder dairy farmers in Kenya: Enforcing transparency and fair payment. In: 2020 IST-Africa Conference (IST-Africa) (pp. 1–6).

Rana, R.L., Tricase, C., De Cesare, L., 2021. Blockchain technology for a sustainable agri-food supply chain. British Food Journal 123 (11), 3471–3485. Available at https://doi.org/10.1108/BFJ-09-2020-0832.

Rejeb, A., Keogh, J.G., Zailani, S., Treiblmaier, H., Rejeb, K., 2020. Blockchain technology in the food industry: A review of potentials, challenges and future research directions. Logistics 4, 27. Available at https://doi.org/10.3390/logistics4040027.

Renny, E.F., Daniel, D.K., Krastanov, A.I., Zachariah, C.A., Elizabeth, R., 2005. Enzyme based sensor for detection of urea in milk. Biotechnology & Biotechnological Equipment 19 (2), 198–201. Available at https://doi.org/10.1080/13102818.2005.10817216.

Renzcherchen, S.K., Gomes, M.A.S., Siatkowski, A., Góes, H., 2025. IoT in Agriculture 4.0 and the challenge of connectivity in rural areas: A bibliometric study. International Journal of Innovation 13 (2), e27573. Available at https://doi.org/10.5585/2025.27573.

Reyes, V., Cahill, E., Mis Solval, K.E., 2024. The potential for reducing food waste through shelf-life extension: Actionable insights from data digitization. Sustainability 16 (7), 2986. Available at https://doi.org/10.3390/su16072986.

Romanello, R., Veglio, V., 2022. Industry 4.0 in food processing: Drivers, challenges and outcomes. British Food Journal 124 (13), 375–390. Available at https://doi.org/10.1108/BFJ-09-2021-1056.

Rossi, S., Gemma, S., Borghini, F., Perini, M., Butini, S., Carullo, G., et al., 2025. Agri-food traceability today: Advancing innovation towards efficiency, sustainability, ethical sourcing, and safety in food supply chains. Trends in Food Science & Technology 163, 105154. Available at https://doi.org/10.1016/j.tifs.2025.105154.

Rotz, S., Gravely, E., Mosby, I., Duncan, E., Finnis, E., Horgan, M., LeBlanc, J., Martin, R., Neufeld, H.T., Nixon, A., Pant, L., Shalla, V., Fraser, E., 2019. Automated pastures and the digital divide: How agricultural technologies are shaping labour and rural communities. Journal of Rural Studies 68, 112–122. Available at https://doi.org/10.1016/j.jrurstud.2019.01.023.

Rui, F., Kaliani Sundram, V.P., 2024. Technological innovation for sustainable supply chain management in the food industry. Information Management and Business Review 16 (3S(I)a), 892–903. Available at https://doi.org/10.22610/imbr.v16i3S(I)a.4173.

Rutten, C.J., Velthuis, A.G.J., Steeneveld, W., Hogeveen, H., 2013. Invited review: Sensors to support health management on dairy farms. Journal of Dairy Science 96 (4), 1928–1952. Available at https://doi.org/10.3168/jds.2012-6107.

Ryan, T., 2022. Carrefour uses blockchain to offer consumers greater supply chain transparency. RetailWire. Available at https://retailwire.com/discussion/carrefour-uses-blockchain-to-offer-consumers-greater-supply-chain-transparency/.

Sahu, A., Agrawal, S., Kumar, G., 2022. Integrating Industry 4.0 and circular economy: A review. Journal of Enterprise Information Management 35 (3), 885–917. Available at https://doi.org/10.1108/JEIM-11-2020-0465.

Senturk, S., Senturk, F., Karaca, H., 2023. Industry 4.0 technologies in agri-food sector and their integration in the global value chain: A review. Journal of Cleaner Production 408, 137096. Available at https://doi.org/10.1016/j.jclepro.2023.137096.

Sezer, B., Durna, S., Bilge, G., Berkkan, A., Yetisemiyen, A., 2018. Identification of milk fraud using laser-induced breakdown spectroscopy (LIBS). International Dairy Journal 81, 1–7. Available at https://doi.org/10.1016/j.idairyj.2017.12.005.

Shah, S., Hussain Madni, S.H., Hashim, S.Z.B.M., Ali, J., Faheem, M., 2024. Factors influencing the adoption of industrial internet of things for the manufacturing and production small and medium enterprises in developing countries. IET Collaborative Intelligent Manufacturing 6 (1). Available at https://doi.org/10.1049/cim2.12093.

Shahid, A., Almogren, A., Javaid, N., Al-Zahrani, F.A., Zuair, M., Alam, M., 2020. Blockchain-based agri-food supply chain: A complete solution. IEEE Access 8, 69230–69243. Available at https://doi.org/10.1109/ACCESS.2020.2986257.

Shen, W., Sun, Y., Zhang, Y., Fu, X., Hou, H., Kou, S., et al., 2021. Automatic recognition method of cow ruminating behaviour based on edge computing. Computers and Electronics in Agriculture 191, 106495. Available at https://doi.org/10.1016/j.compag.2021.106495.

Sivalingam, D., Rayappan, J., 2012. Development of e-nose prototype for raw milk quality discrimination. Milchwissenschaft 67 (4), 381–385.

Sobb, T., Turnbull, B., Moustafa, N., 2020. Supply Chain 4.0: A survey of cyber security challenges, solutions and future directions. Electronics 9 (11), 1864. Available at https://doi.org/10.3390/electronics9111864.

Sosa, K.E.R., Leochico, C.F.D., Rimando, C.R.D., 2023. A web-based survey on the telerehabilitation knowledge, attitude, and practice of physical therapists in a developing country during the COVID-19 pandemic: An analytical cross-sectional study. Acta Medica Philippina 58 (2), 54–62. Available at https://doi.org/10.47895/amp.vi0.6664.

Sowmya, N., Ponnusamy, V., 2021. Development of spectroscopic sensor system for an IoT application of adulteration identification on milk using machine learning. IEEE Access 9, 53979–53995. Available at https://doi.org/10.1109/ACCESS.2021.3070558.

Sun, H., Jiang, G., Kong, Q., Chen, Z., Li, X., 2016. Design of real-time monitoring system on raw milk transport process. International Journal of Multimedia and Ubiquitous Engineering 11 (4), 335–342. Available at https://doi.org/10.14257/ijmue.2016.11.4.33.

Sun, Q., Yuan, Y., Xu, B., Gao, S., Zhai, X., Xu, F., et al., 2025. Innovative technologies reshaping meat industrialization: Challenges and opportunities in the intelligent era. Foods 14 (13), 2230. Available at https://doi.org/10.3390/foods14132230.

Suresh Babu, G., Rama Krishna, V., Dharma Teja, V., Balraj Yadav, S., 2018. Milk quality testing using intelligent inference performance evaluation system integrated with IoT. International Journal of Engineering & Technology 7 (2.20), 74. Available at https://doi.org/10.14419/ijet.v7i2.20.11757.

Tan, A., Ngan, P.T., 2020. A proposed framework model for dairy supply chain traceability. Sustainable Futures 2, 100034. Available at https://doi.org/10.1016/j.sftr.2020.100034.

Taneja, M., Byabazaire, J., Jalodia, N., Davy, A., Olariu, C., Malone, P., 2020. Machine learning based fog computing assisted data-driven approach for early lameness detection in dairy cattle. Computers and Electronics in Agriculture 171, 105286. Available at https://doi.org/10.1016/j.compag.2020.105286.

Tangorra, F.M., Lopez, A., Ighina, E., Bellagamba, F., Moretti, V.M., 2024. Handheld NIR spectroscopy combined with a hybrid LDA-SVM model for fast classification of retail milk. Foods 13 (22), 3577. Available at https://doi.org/10.3390/foods13223577.

Tirichine, N., Saber, M., Khitous, M., Lounici, H., Bouarab, R., 2022. Mesophilic anaerobic co-digestion of cheese whey with cow manure in batch reactor. ENP Engineering Science Journal 2 (2), 26–30. Available at https://doi.org/10.53907/enpesj.v2i2.171.

Tohidi, M., Ghasemi-Varnamkhasti, M., Ghafarinia, V., Bonyadian, M., Mohtasebi, S.S., 2018. Development of a metal oxide semiconductor-based artificial nose as a fast, reliable and non-expensive analytical technique for aroma profiling of milk adulteration. International Dairy Journal 77, 38–46. Available at https://doi.org/10.1016/j.idairyj.2017.09.003.

Tran, D., De Steur, H., Gellynck, X., Papadakis, A., Schouteten, J.J., 2024. Consumers' valuation of blockchain-based food traceability: role of consumer ethnocentrism and communication via QR codes. British Food Journal 126 (13), 72–93. Available at https://doi.org/10.1108/BFJ-09-2023-0812.

Tsolakis, N., Niedenzu, D., Simonetto, M., Dora, M., Kumar, M., 2021. Supply network design to address United Nations Sustainable Development Goals: A case study of blockchain implementation in Thai fish industry. Journal of Business Research 131, 495–519. Available at https://doi.org/10.1016/j.jbusres.2020.08.003.

Umapathi, N., Jegadeesan, R., Karthick, G., Venkateswaran, N., & Ramana, R., 2025. IoT based Smart Feeding System for Livestock. In: 2025 4th International Conference on Innovative Mechanisms for Industry Applications (ICIMIA) (pp. 214–219). https://doi.org/10.1109/ICIMIA67127.2025.11200970.

Usman, A., 2022. The future of marketing: Nestle's digital transformation. Medium.

Vara Martínez, J.A.de la, García Higuera, A., Román Esteban, M., Romero Asensio, J., Carmona Delgado, M., Berruga, I., et al., 2018. Monitoring bulk milk quality by an integral traceability system of milk. Journal of Applied Animal Research 46 (1), 784–790. Available at https://doi.org/10.1080/09712119.2017.1403327.

Varavallo, G., Caragnano, G., Bertone, F., Vernetti-Prot, L., Terzo, O., 2022. Traceability platform based on green blockchain: An application case study in dairy supply chain. Sustainability (Switzerland) 14 (6). Available at https://doi.org/10.3390/su14063321.

Vasafi, P.S., Paquet-Durand, O., Brettschneider, K., Hinrichs, J., Hitzmann, B., 2021. Anomaly detection during milk processing by autoencoder neural network based on near-infrared spectroscopy. Journal of Food Engineering 299, 110510. Available at https://doi.org/10.1016/j.jfoodeng.2021.110510.

Vasanthraj, V., Potdar, V., Agrawal, H., Kaur, A., 2024. Transforming milk supply chains with blockchain: enhancing visibility and cost reduction. Benchmarking: An International Journal. Available at https://doi.org/10.1108/BIJ-10-2023-0702.

Vasileiou, M., Kyrgiakos, L.S., Kleisiari, C., Lappas, P.Z., Tsinopoulos, C., Kleftodimos, G., et al., 2025. Digital transformation of food supply chain management using blockchain: A systematic literature review towards food safety and traceability. Business & Information Systems Engineering. Available at https://doi.org/10.1007/s12599-025-00948-0.

Vidhya, D.S., Jaividhya, S., Devanand, K., Karthicksoundar, C., & Keerthivaasan, D., 2023. Automatic Industrial Fault Detection and IoT based Remote Monitoring. In: 2023 Second International Conference on Electronics and Renewable Systems (ICEARS) (pp. 552–555).

Vishnu, K., & Kumar, S.P., 2024. Enhancing the quality of milk prediction using random forest algorithm and logistic regression algorithm. AIP Conference Proceedings, 3193, 020180. https://doi.org/10.1063/5.0238130.

Vudugula, S., 2025. Sustainable smart supply chains: A review of green technologies and their impact on logistics. American Journal of Scholarly Research and Innovation 4 (01), 1–32. Available at https://doi.org/10.63125/1fehce37.

Vuvor, C.M., Epifani, S., Pulina, G., 2025. Exploring opportunities in Africa's dairy production: digital transition as a catalyst for marketing innovation and value chain synergies—A systematic review laying foundations for future research. Frontiers in Animal Science 6. Available at https://doi.org/10.3389/fanim.2025.1632532.

Wang, J., 2025. Opportunities, challenges, and strategic responses for international agricultural trade in the digital era. European Journal of Business, Economics & Management 1 (4), 21–29. Available at https://doi.org/10.71222/ycvm1g76.

Wang, Y., Feng, Y., Xiao, Z., Luo, Y., 2025. Machine learning supported single-stranded DNA sensor array for multiple foodborne pathogenic and spoilage bacteria identification in milk. Food Chemistry 463, 141115. Available at https://doi.org/10.1016/j.foodchem.2024.141115.

Wolfert, S., Ge, L., Verdouw, C., Bogaardt, M.-J., 2017. Big data in smart farming – a review. Agricultural Systems 153, 69–80. Available at https://doi.org/10.1016/j.agsy.2017.01.023.

World Health Organization (2022). WHO global strategy for food safety 2022-2030: towards stronger food safety systems and global cooperation. World Health Organization. https://iris.who.int/handle/10665/363475. License: CC BY-NC-SA 3.0 IGO.

Wu, D., Wu, Q., Yin, X., Jiang, B., Wang, H., He, D., et al., 2020. Lameness detection of dairy cows based on the YOLOv3 deep learning algorithm and a relative step size characteristic vector. Biosystems Engineering 189, 150–163. Available at https://doi.org/10.1016/j.biosystemseng.2019.11.017.

Xiang, B., Zheng, R., Zhang, K., Li, C., Zheng, J., 2025. FFT-RDNet: A time–frequency-domain-based intrusion detection model for IoT security. Sensors 25 (15), 4584. Available at https://doi.org/10.3390/s25154584.

Xie, L., Luo, B., Zhong, W., 2021. How are smallholder farmers involved in digital agriculture in developing countries: A case study from China. Land 10 (3), 245. Available at https://doi.org/10.3390/land10030245.

Xue, D., Zhao, H., 2024. Machine learning and electrochemistry techniques for detecting adulteration of goat milk with cow milk. Journal of Food Measurement and Characterization 18 (7), 6012–6019. Available at https://doi.org/10.1007/s11694-024-02625-1.

Yang, B., Huang, X., Yan, X., Zhu, X., Guo, W., 2020. A cost-effective on-site milk analyzer based on multispectral sensor. Computers and Electronics in Agriculture 179, 105823. Available at https://doi.org/10.1016/j.compag.2020.105823.

Yu, Z., Liu, Y., Yu, S., Wang, R., Song, Z., Yan, Y., et al., 2022. Automatic detection method of dairy cow feeding behaviour based on yolo improved model and edge computing. Sensors 22 (9), 3271. Available at https://doi.org/10.3390/s22093271.

Yu, R., Yang, W., Zhang, L., 2025. A privacy protection method for trustworthy traceability of rice supply chain based on blockchain and multilayer encryption. Concurrency and Computation: Practice and Experience 37 (27–28). Available at https://doi.org/10.1002/cpe.70413.

Zakeri, A., Saberi, M., Hussain, O.K., Chang, E., 2018. An early detection system for proactive management of raw milk quality: An Australian case study. IEEE Access 6, 64333–64349. Available at https://doi.org/10.1109/ACCESS.2018.2877970.

Zhang, J., Bhatt, T., 2014. A guidance document on the best practices in food traceability. Comprehensive Reviews in Food Science and Food Safety 13 (5), 1074–1103. Available at https://doi.org/10.1111/1541-4337.12103.

Zhang, A., Zhong, R.Y., Farooque, M., Kang, K., Venkatesh, V.G., 2020. Blockchain-based life cycle assessment: An implementation framework and system architecture. Resources, Conservation and Recycling 152, 104512. Available at https://doi.org/10.1016/j.resconrec.2019.104512.

Zhang, X., Sun, Y., Sun, Y., 2022. Research on cold chain logistics traceability system of fresh agricultural products based on blockchain. Computational Intelligence and Neuroscience 2022, 1–13. Available at https://doi.org/10.1155/2022/1957957.

Zhang, Y., Zhang, Y., Gao, M., Dai, B., Kou, S., Wang, X., et al., 2023. Digital twin perception and modeling method for feeding behavior of dairy cows. Computers and Electronics in Agriculture 214, 108181. Available at https://doi.org/10.1016/j.compag.2023.108181.

Zhou, H., Nanseki, T., 2023. Traceability system of dairy products and its impacts on consumer behavior in China: An application of multinominal logit model. Agricultural Innovation in Asia. Springer Nature, Singapore, pp. 149–157. Available at https://doi.org/10.1007/978-981-19-9086-1_9.

Zhou, X., Lu, H., Kumar Mangla, S., 2024. The impact of digital traceability on sustainability performance: Investigating the roles of sustainability-oriented innovation and supply chain learning. Supply Chain Management: An International Journal 29 (3), 497–522. Available at https://doi.org/10.1108/SCM-01-2023-0047.

Zizic, M.C., Mladineo, M., Gjeldum, N., Celent, L., 2022. From Industry 4.0 towards Industry 5.0: A Review and analysis of paradigm shift for the people, organization and technology. Energies 15 (14), 5221. Available at https://doi.org/10.3390/en15145221.

Chapter 7

Dairy waste management using Industry 4.0 technologies

Wessam Galia[1], Françoise Leriche[2], Caroline Wybraniec[1], Cécile Moussard[2], Geneviève Gagne[2], and Benoit Cournoyer[1]
[1]*Laboratoire d'Ecologie Microbienne, Université Claude Bernard Lyon 1, INRAE, VetAgro Sup, Marcy l'étoile, France,* [2]*UMRF, Université Clermont Auvergne, INRAE, VetAgro Sup, Aurillac, France*

7.1 Introduction

Global waste generation is a growing environmental and economic concern, driven by population growth, urbanization, industrial expansion, and increasing resource consumption (Ahmed et al., 2021; Shemer et al., 2023). Among industrial sectors, the dairy industry is particularly resource-intensive, producing substantial volumes of both liquid and solid waste throughout its supply chains. Waste streams from dairy operations include wastewater, whey residues, sludge, packaging materials, and food losses, all of which contain nutrients, organic matter, or energy potential that can impact environmental sustainability if not properly managed.

Freshwater scarcity is a growing global concern, driven by population growth, urbanization, industrial expansion, and the impacts of climate change (Ahmed et al., 2021; Shemer et al., 2023). Wastewater generation further exacerbates this problem by degrading both the quality and quantity of freshwater resources. In Europe, industries contribute approximately 54% of total wastewater, highlighting the critical role of industrial water management in freshwater conservation strategies (Stasinakis et al., 2022). Wastewater recycling offers a promising solution to mitigate industrial impacts on water scarcity. It allows for the reuse of water within plants, conserving freshwater resources and reducing overall demand. Additionally, it promotes environmental sustainability by minimizing wastewater discharge into natural water bodies (Stasinakis et al., 2022).

The dairy industry is one of the most water-intensive industrial sectors. A significant amount of water used in dairy plants is discharged as wastewater. According to the Best Available Techniques (BAT) document for the milk industry, about 0.3–10 L of wastewater are generated per liter of milk produced. Dairy plants in the EU release large volumes of wastewater, mainly from cleaning and processing operations. Frequent cleaning activities, which include washing

Dairy 4.0. DOI: https://doi.org/10.1016/B978-0-443-44730-3.00009-2

floors, tanks, silos, and equipment with various sanitizers and detergents such as sodium hydroxide, sodium carbonate, nitric acid, and phosphoric acid, significantly contribute to wastewater pollution (Bella and Rao, 2023; Danalewich et al., 1998).

Dairy wastewater (DWW) typically exhibits a high organic load, with chemical oxygen demand (COD) reaching up to 13,000 mg O_2/L and biochemical oxygen demand (BOD) around 11,100 mg O_2/L. In addition to milk constituents (proteins, lipids, and carbohydrates), DWW contains suspended solids, chlorides, sulfates, trace organics, and nutrients such as nitrogen and phosphorus. The volume and composition of DWW vary daily and seasonally, reflecting batch processing and diverse product streams, which complicates treatment and recycling efforts (Bella and Rao, 2023; Horňák et al., 2025; Stasinakis et al., 2022).

While wastewater represents the largest liquid waste stream, dairy processing also generates diverse solid and by-product wastes, including whey residues, sludge, packaging materials, and food losses. Many of these streams are rich in nutrients, organic matter, or energy potential, and can be valorized through techniques such as whey fractionation, anaerobic digestion, composting, or bioconversion. Effective waste management therefore requires an integrated, multistream approach addressing prevention, monitoring, treatment, and resource recovery across both liquid and solid fractions. Traditional treatment and disposal approaches are increasingly constrained by stricter regulations, rising energy costs, and consumer expectations for sustainability, underscoring the need for innovative and circular strategies (Almeida et al., 2023; Granato et al., 2022).

The Fourth Industrial Revolution (Industry 4.0) offers a comprehensive technological framework to support this transition. The convergence of pervasive sensing (Internet of Things, IoT), advanced analytics (machine learning and artificial intelligence), model-based simulation (digital twins), automation and robotics, and distributed ledger technologies (blockchain) enables continuous process visibility, predictive control, and valorization of diverse waste streams as products or energy vectors. When effectively deployed, these technologies allow a shift from reactive compliance to proactive optimization and from isolated treatment units to networked, circular resource systems (Andrade Cruz et al., 2022; Kegl et al., 2025).

This chapter examines how Industry 4.0 components can reduce waste at its source, optimize treatment and valorization outcomes (e.g., whey fractionation, anaerobic digestion, packaging reuse), and support regulatory compliance across both liquid and solid waste streams. It also addresses adoption barriers and outlines future research directions, providing a roadmap toward sustainable, efficient, and resilient dairy waste management in a circular economy framework.

7.2 Typology, characteristics and conventional management of dairy waste streams

The load and composition of dairy waste streams can vary depending on production schedules and product types (such as cheese, yogurt, and milk powder). Understanding the physicochemical and temporal properties of these streams is essential for selecting appropriate digital and process interventions.

7.2.1 Liquid effluents

Dairy process wastewater primarily originates from cleaning-in-place (CIP) operations, product handling, pasteurization, and equipment rinsing. They are typically characterized by high COD, BOD, suspended solids, fats, proteins, and sugars (especially lactose). The high organic loads and variable hydraulic regimes present significant challenges for conventional biological treatment plants, highlighting the need for real-time monitoring and management strategies (Cvetković et al., 2025).

To address these challenges, recent studies have explored Minimal Liquid Discharge (MLD) systems, which can recover up to 89.8% of water from dairy effluents. Although complex and energy-intensive, these systems offer promising sustainable solutions. Experimental MLD setups, combining microfiltration, reverse osmosis, and vacuum evaporation, produced treated effluents with low COD, BOD, and total dissolved solids, suitable for reuse in nonfood applications. Techno-economic assessments further indicated payback periods as short as 2.8 years, suggesting that MLD approaches can be both environmentally beneficial and economically feasible (Brião et al., 2019; Horňák et al., 2025).

7.2.2 Whey and permeates

Whey, generated in large quantities during cheese manufacturing, presents both a significant environmental challenge and a valuable resource. Its dual nature is well documented: if discharged untreated, whey's high COD can severely impact receiving water bodies. Conversely, controlled processing of whey enables the recovery of proteins, lactose, and bioactive peptides for applications in both food and nonfood markets. These recovery pathways, including membrane concentration, enzymatic hydrolysis, fermentation, and drying, require precise process control and benefit from digital monitoring, ensuring product quality, process efficiency, and reduced environmental impact (Buchanan et al., 2023; Chourasia et al., 2022; Hafeez et al., 2015).

7.2.3 Solid residues and sludges

In addition to liquid effluents and whey streams, solid residues and sludge form a significant component of dairy waste streams. Screening solids, sludge from primary and secondary treatment, and fat scums are biodegradable and can be valorized through energy recovery (e.g., anaerobic digestion) or nutrient recycling following appropriate conditioning (Usmani et al., 2022). However, their heterogeneity and variable moisture content pose challenges for handling and operational logistics. Recent studies analyzing digestates from processes fed with a mix of dairy and other agricultural wastes, as well as urban organic wastes (including wastewater treatment sludge), have identified the occurrence of multiple sanitary indicators. This highlights the challenges of controlling microbial and chemical contaminants when valorizing organic residues, including those from dairy waste streams (Aigle et al., 2021; Wybraniec et al., 2024).

7.2.4 Packaging and inorganic waste

In dairy processing, packaging materials such as multilayer films, cartons, and plastics can become contaminated and require careful segregation and traceability upstream. This ensures efficient recycling and prevents cross-contamination of organic waste streams. Implementing digital tracking of packaging flows within dairy facilities can significantly improve capture rates and enhance recyclability, contributing to overall waste management efficiency (Malik et al., 2024; Matthes and Schmid, 2024).

7.2.5 Gaseous emissions

Dairy waste streams also contribute to gaseous emissions, including methane (CH4), nitrous oxide (N2O), carbon dioxide (CO2), and volatile organic compounds (VOCs), which can arise during storage, anaerobic digestion (AD), or heating processes. Continuous monitoring and greenhouse gas accounting are essential for evaluating the environmental footprint of dairy waste treatment. Integrating sensing technologies and predictive analytics enables real-time control, helping to mitigate emissions while optimizing operational efficiency (Singaravadivelan et al., 2023; Udourioh et al., 2025).

7.2.6 Additional dairy by-products and nonedible residues

Beyond liquid effluents, whey, and sludge, the dairy industry generates additional organic residues that are relevant for sustainability and valorization strategies. These include spoiled or outdated dairy products from retail and distribution, clarified butter residues from ghee production, sweet buttermilk powders, cream residue powders, and secondary whey streams from cheese and butter processing. These residues are rich in proteins, fats, and bioactive compounds and can be valorized for energy production, feed, or food ingredient applications (Hassoun et al., 2024).

Taken together, the management of solid, liquid, packaging, and gaseous components of dairy waste necessitates multimodal interventions. Combining advanced sensing, process control, predictive analytics, and networked resource allocation supports circularity objectives and enhances the sustainability of dairy waste valorization systems (Granato et al., 2022; Wang et al., 2020). Moreover, the presence of trace contaminants in organic and inorganic waste streams highlights gaps between current methodologies and environmental knowledge, further underscoring the potential for Industry 4.0 technologies to improve detection, monitoring, and control across dairy waste management (Buchanan et al., 2023; Kazemi et al., 2021; Wang et al., 2020; Avadi et al., 2022).

This integrative perspective emphasizes that a holistic approach, addressing all components of dairy waste, is essential to advance environmental performance and operational efficiency in dairy processing. Table 7.1 provides a comprehensive synthesis of these waste streams, associated challenges, and potential valorization pathways supported by Industry 4.0 technologies.

7.3 Core Industry 4.0 technologies applicable to dairy waste management

Industry 4.0 provides a technological framework that transforms waste management from a traditionally reactive operation into a predictive, data-driven system. These technologies constitute the digital backbone of modern dairy waste management. The convergence of advanced sensing infrastructures, artificial intelligence (AI), digital modeling, automation, and distributed data systems enables continuous process supervision, sanitary risk mitigation, resource optimization, and strengthened regulatory compliance. In the context of the food sector, Industry 4.0 can be applied across digital, physical, and biological domains. Digital innovations include AI, big data analytics, and digital twins; physical technologies encompass robotics and smart sensors; and biological approaches involve precision fermentation and 3D printing (Konstantinidis et al., 2023a,b; Hassoun et al., 2024). The dairy industry, in particular, has emerged as a fast adopter of such technologies due to the need for tighter control and enhanced management across the entire supply chain. Reviews of recent literature highlight key Industry 4.0 enablers for Dairy 4.0, including robotics, the IoT, blockchain, big data, 3D printing, and AI. Additional tools such as digital twins, machine vision, machine learning (ML), and advanced sensor networks are increasingly recognized as accelerating the transition from conventional operations to fully integrated Dairy 4.0 systems (Hassoun et al., 2024, 2023). Rather than focusing on specific treatment processes, this section presents the core digital and cyber–physical technologies that enable smart, connected, and adaptive waste management systems.

7.3.1 Internet of Things and advanced sensing

At the foundation of Industry 4.0 systems lies the ability to observe processes continuously and in real time. IoT technologies provide this observational layer by enabling high-frequency, distributed measurements across dairy waste management infrastructures. Specifically, IoT allows continuous monitoring of critical operational parameters in dairy effluent treatment systems, including flow rate, temperature, pH, conductivity, turbidity, and redox potential (Jagtap et al., 2021).

TABLE 7.1 Overview of dairy waste streams, environmental challenges, and conventional and emerging valorization pathways.

Waste stream	Main source	Key characteristics	Risks/constraints	Potential valorization	Emerging digital tools
Liquid effluents	Cleaning-in-place (CIP), equipment rinsing	High COD/BOD, suspended solids, proteins, lactose	Water pollution, daily and seasonal variability	Water recovery, fermentation, biogas production	IoT sensors, real-time monitoring, AI for process optimization
Whey and permeates	Cheese manufacturing	Proteins, lactose, bioactive peptides	Organic pollution if discharged untreated	Protein/lactose fractionation, bioactive compounds	Sensors, ML for process control and monitoring
Solids and sludge	Screening, fat skimming, primary/secondary sludge	Heterogeneous, variable moisture	Difficult handling, microbial contamination	Anaerobic digestion, composting, biogas	Digital twins, AI for digestion optimization
Packaging and inorganic waste	Films, cartons, plastics	Possible contamination	Complex segregation	Recycling	Blockchain for traceability
Gaseous emissions (CH_4, N_2O, CO_2, VOCs)	Storage, digestion, heating	Greenhouse gases, odors	Environmental impact	Methane capture, energy recovery	IoT, predictive analytics
Additional organic residues	Spoiled/outdated products, ghee residues, secondary whey, cream powders	Proteins, fats, bioactive compounds	Microbial contamination, high variability	Energy recovery, feed, food ingredients	AI for quality monitoring, IoT for tracking, blockchain for traceability

Recent advances in biosensors and sensor fusion, which combine spectroscopic proxies with conductivity and turbidity measurements, further enhance real-time estimation of key targets such as protein and lactose concentrations, as well as volatile fatty acids. By integrating these hybrid sensing approaches, predictive capabilities and operational responsiveness are significantly improved (Hemdan et al., 2025).

The deployment of modular wireless sensor networks at multiple process points provides high spatial and temporal resolution, enabling the detection of short-term spikes in organic load, fat content, or hydraulic disturbances (Zhou et al., 2025; Di Puglia Pugliese et al., 2025). Successful practical implementation, however, requires careful attention to calibration, fouling mitigation, telemetry reliability, and robust data governance to ensure both data quality and security.

7.3.2 Data analytics, machine learning, and artificial intelligence

While sensor networks generate large volumes of data, their value depends on the ability to transform raw measurements into actionable operational knowledge. Data analytics, ML, and AI constitute the cognitive layer of Industry 4.0, providing predictive, diagnostic, and decision-support capabilities that bridge process monitoring, regulatory compliance, and public health surveillance (Neethirajan, 2023; Shandilya et al., 2024).

Supervised learning techniques are widely used to predict influent quality indicators, such as COD, BOD, and nutrient loads, from upstream process variables. In microbiological surveillance, ML methods can predict antimicrobial resistance (AMR) patterns from genomic and metagenomic inputs by identifying key markers and classifying resistant phenotypes (Galia et al., 2017, 2015; Valavarasu et al., 2025). Commonly employed models include Random Forest, Support Vector Machines, and XGBoost (eXtreme Gradient Boosting) (Idrissi Khaldi et al., 2025; Sheth et al., 2022). XGBoost is specifically employed for its capacity to handle high-dimensional tabular datasets, capture nonlinear relationships, and provide interpretable feature importance metrics, complementing time-series models such as long short-term memory neural network that forecast dynamic reactor conditions (Ching et al., 2022). These approaches form the core of the predictive layer, linking sensor measurements to actionable insights.

At a broader system level, large-scale data platforms and deep learning infrastructures further enhance these capabilities by enabling real-time monitoring, automated reporting, and integration with external databases dedicated to AMR and microbial risk assessment (e.g., PATRIC, CARD, MEGARes), thereby strengthening links between industrial process control and public health surveillance (Chen et al., 2025; Dikmen et al., 2025).

Beyond prediction, unsupervised anomaly detection methods are used to flag unexpected operational events, such as rapid pump degradation, sudden surges in fat content, or foaming episodes, enabling timely corrective actions and improving system robustness. Reinforcement learning approaches further support the optimization of multivariable operational regimes, including aeration rates, feeding schedules, and adaptive control policies, by continuously learning optimal strategies from historical and real-time process feedback.

Beyond process optimization, machine learning models trained on physico-chemical, molecular, and operational datasets can also provide near real-time predictions of both sanitary performance and environmental impacts of treatment processes. In an industrial context, recent advances in structure-informed and data-driven ML approaches demonstrate their potential to rapidly estimate life-cycle environmental impacts, offering a powerful complement to conventional life-cycle assessment (LCA) frameworks, which are often limited by data availability, cost, and temporal resolution (Zhao et al., 2025). When coupled with process monitoring data, such models can support operational decision-making by recommending optimized treatment cycles or retention times that jointly improve hygienization efficiency while reducing environmental footprints.

To ensure trust, robustness, and regulatory acceptance, feature engineering strategies, such as resistome profiling or the integration of molecular descriptors, combined with interpretability techniques (e.g., SHAP values), are increasingly used to improve model transparency and accountability. Explainable AI approaches are essential to facilitate the adoption of ML-based recommendations by operators, regulators, and stakeholders, particularly in contexts involving sanitary risk management (Bilal et al., 2025).

Finally, the predictive outputs generated by data analytics and AI models provide the foundation for their integration into digital twins and operator dashboards (see next section). These cyber–physical systems enable closed-loop architectures in which sensor data (*sense*) are transformed into model-based decisions (*decide*), supporting automated or operator-assisted interventions (*act*) within a traceable and auditable Industry 4.0 framework.

7.3.3 Digital twins and cyber–physical integration

Beyond point-wise prediction, Industry 4.0 introduces a system-level understanding of complex treatment processes through dynamic simulation environments known as digital twins. Digital twins are dynamic, physics-aware virtual replicas of individual treatment units or entire processing plants that continuously integrate mechanistic models, data-driven empirical modules, and real-time sensor streams (Konstantinidis et al., 2023a,b; Udugama et al., 2021). By synchronizing physical assets with their virtual counterparts, digital twins enable continuous state estimation, scenario exploration, and predictive diagnostics across multiple operational scales.

Unlike static models, digital twins evolve in parallel with the physical system, allowing virtual experimentation under realistic constraints. This capability significantly reduces operational risks by enabling operators to test alternative strategies, boundary conditions, or failure scenarios in silico before implementation. As a result, learning cycles are accelerated, decision-making becomes more robust, and costly trial-and-error approaches at full scale are minimized.

For biologically driven treatment processes, digital twins can incorporate modules describing microbial community dynamics, biokinetic reactions, and the fate of AMR indicators. Such integrated models enable the assessment of scenarios including changes in substrate composition, modification or cessation of cotreatment strategies, shifts in microbial community structure, or the introduction of functional amendments aimed at improving hygienization performance. These simulations support the evaluation of operational envelopes, control strategies, and mitigation measures before deployment on live systems, thereby strengthening both process stability and sanitary risk management.

More broadly, the coupling of digital twins with real-time data streams forms the backbone of cyber–physical systems, where physical processes and digital intelligence continuously interact. This integration establishes a foundation for adaptive control, predictive maintenance, and closed-loop optimization, enabling treatment systems to respond proactively to evolving operational and sanitary conditions.

7.3.4 Automation, robotics, and advanced actuation

The translation of digital intelligence into physical action is achieved through automation, robotics, and advanced actuation systems, which together form the execution layer of Industry 4.0. Automation plays a critical role in reducing process variability, optimizing resource use, and ensuring consistent operational performance. Automated control of CIP cycles, sludge extraction, membrane cleaning, and valve sequencing ensures repeatable and effective cleaning procedures while minimizing water consumption, chemical usage, and downtime.

Advanced actuation systems allow precise, real-time adjustments of key operational parameters, such as agitation intensity, feeding rates, temperature profiles, and hydraulic retention times. These capabilities are particularly important in biologically driven processes, where small deviations can have significant impacts on stability or sanitary outcomes. When coupled with AI-driven control algorithms, automated actuation enables dynamic responses to predicted disturbances, facilitating proactive rather than reactive process management.

Robotic systems further enhance operational reliability by performing hazardous, labor-intensive, or repetitive tasks, including inspection of confined spaces, handling of contaminated components, and routine maintenance operations. The use of robotics improves worker safety, reduces human error, and ensures consistent execution of standardized procedures. Collectively, automation, robotics, and intelligent actuation contribute to improved process resilience and sustained sanitary performance, aligning operational efficiency with environmental and health objectives (Riboli et al., 2023; Aly et al., 2023).

7.3.5 Blockchain and traceability platforms

In parallel with advances in operational intelligence, Industry 4.0 also transforms governance, accountability, and trust through secure digital infrastructures. Blockchain technology provides decentralized, immutable ledgers that record provenance, chain-of-custody information, and transactional data related to waste streams and derived by-products. Such features are particularly relevant in waste management systems, where transparency and traceability are critical for regulatory compliance and stakeholder confidence (Casino et al., 2021; Khanna et al., 2022).

In the case of dairy waste, which may contain pathogenic microorganisms or AMR determinants, blockchain-enabled traceability supports transparent documentation of treatment conditions, process histories, and sanitary assessments (Mangla et al., 2021). Distributed ledgers allow critical information such as operational parameters, analytical results, and certification records to be time-stamped and securely shared among stakeholders, including operators, regulators, and downstream users (Stasinakis et al., 2022).

When paired with AI-derived sanitary scores and predictive indicators, blockchain platforms can certify by-product quality in a tamper-proof manner, providing verifiable evidence of compliance with sanitary and environmental regulations. This integration facilitates trusted valorization pathways, supports auditing and reporting requirements, and enhances public and regulatory confidence in circular economy practices (Mangla et al., 2021). Ultimately, blockchain and traceability platforms complement sensing, analytics, and automation layers by embedding trust and accountability into Industry 4.0–enabled waste management systems. Table 7.2 provides a structured

TABLE 7.2 Industry 4.0 technologies applied to dairy waste management.

Industry 4.0 technology	Primary function	Example applications in dairy waste management
IoT and sensors	Real-time monitoring	Measure COD, BOD, flow rates, temperature, pH in effluents
AI/machine learning	Prediction and optimization	Adjust retention times, nutrient dosing, predict digester acidification
Digital twins	Simulation and diagnostics	Optimize anaerobic digesters, evaluate codigestion scenarios
Automation/ robotics	Physical execution	Automated CIP, sludge handling, equipment cleaning
Blockchain	Traceability and compliance	Record treatment history of digestates, certify quality and compliance

synthesis of the key Industry 4.0 technologies, their operating principles, and their functional roles in dairy waste management.

7.4 Waste treatment and valorization: smart control, optimization, and industrial-scale strategies

7.4.1 Valorization pathways for dairy waste

Industry 4.0 technologies can substantially improve the efficiency and outputs of treatment and valorization pathways in dairy waste management. Whey valorization, for instance, involves high-value fractionation of proteins and lactose using advanced membrane controls and fouling management. Deploying sensor-assisted membrane operations together with AI-guided cleaning schedules can maximize yields and ensure process stability (Buchanan et al., 2023). Similarly, Smart CIP operations, which account for a large portion of water and chemical use in dairy plants, can be optimized through real-time cleanliness sensors and automated control. Dynamic adjustment of cycle duration, temperature, and chemical dosing minimizes effluent volumes and resource consumption while supporting process validation, traceability, and compliance with HACCP systems (Jin et al., 2025).

Emerging valorization pathways, such as fermentative conversion to bioethanol or bioplastic precursors, increasingly rely on integrated analytics for process control. These approaches leverage real-time monitoring and predictive algorithms to maintain optimal operating conditions, minimizing variability and enhancing yield. Such strategies exemplify how Industry 4.0 frameworks facilitate scalable, data-driven optimization across both core valorization and ancillary operations.

7.4.2 Anaerobic digestion as a central valorization platform

Among available pathways, AD plays a central role, converting organic residues into biogas and digestate. AD performance is influenced by feed composition, loading rate, pH, temperature, and the presence of inhibitory compounds (Aigle et al., 2021). By incorporating IoT sensor arrays alongside AI-based controllers, it is possible to dynamically adjust feed blends and rates. This optimization allows for increased methane production while ensuring stability in the face of variable dairy feedstocks (Andrade Cruz et al., 2022).

At industrial scale, AD can be positioned as a central step within cascade biorefinery systems, followed by downstream extraction and valorization processes to maximize nutrient and energy recovery. Reported biomethane production can reach 0.60 L CH_4/g volatile solids, with organic matter removal exceeding 83%, reflecting significant solid valorization potential (Escalante et al., 2018). Two-stage fluidized bed reactor systems codigesting whey and glycerol further enhance methane yields. Mesophilic reactors demonstrate better performance under heavy loads compared to thermophilic systems, while combining thermophilic acidogenesis with mesophilic methanogenesis leads to higher energetic yield and organic matter removal relative to single-stage processes (Almeida et al., 2023).

Such performance metrics demonstrate the technical maturity of AD as an industrial-scale valorization platform when supported by advanced monitoring and control systems.

7.4.3 Smart control and process optimization

Intelligent wastewater treatment plants utilize AI to optimize aeration, dosing, and retention times, resulting in effluent compliance with reduced energy and chemical inputs. Coupling digital twin models with ML enhances predictive maintenance and process control, further supporting sustainability objectives.

Complementing classical mathematical models, real-time process control integrates biosensors, edge analytics, and ML to adaptively tune operational parameters (loading rate, temperature, nutrient dosing) in situ. These responsive frameworks safeguard process stability, increase methane output, and reduce disturbances such as acidification or foaming. This approach aligns with Industry 4.0 principles that promote circular bioenergy production and resource-efficient industrial operations (Kazemi et al., 2021; Kegl et al., 2025; Borisov et al., 2016).

The use of AI-driven controllers and predictive models facilitates decision support by linking feedstock composition, operational parameters, and expected outputs, enabling proactive interventions before process deviations occur.

7.4.4 Sanitary risk and digestate quality

Recent studies on French digestates underscore the importance of monitoring sanitary indicators and microbial consortia to ensure safe digestate valorization. Variability in feedstock origin and digestion conditions can result in significant differences in pathogen prevalence and antimicrobial resistance genes (ARGs) (Wybraniec et al., 2024).

Wybraniec et al. (2024) identified multiple sanitary indicators in reactors fed with a mix of dairy and other agricultural wastes, as well as urban organic wastes (including wastewater treatment sludge). This highlights the challenges of controlling microbial and chemical contaminants when valorizing organic residues, including those from dairy waste streams (Aigle et al., 2021; Wybraniec et al., 2024).

Empirical studies revealed that reactors fed solely with activated sludge and manure exhibited high triclosan concentrations correlated with increased ARG abundance (*bla*$_{TEM}$, *intI1/2*). In contrast, codigestion with wheat straw and zeolite reduced micropollutant concentrations and achieved up to 70% reduction in ARG prevalence (Wybraniec et al., 2024). These contrasted scenarios demonstrate the need for integrated monitoring and predictive analytics to safeguard digestate quality while maintaining process efficiency.

7.4.5 Conceptual integration of Industry 4.0 components

To synthesize the technological architecture discussed, Fig. 7.1 provides a non–site-specific conceptual overview of how Industry 4.0 components interact within a digitally enabled AD system. The schematic illustrates the functional integration of sensing, analytics, modeling, actuation, and data governance as a general cyber–physical architecture applicable to dairy waste management. The figure emphasizes a layered logic comprising data acquisition, intelligence generation, process control, human oversight, and regulatory traceability. Together, these layers form the foundation for smart, integrated waste management systems, demonstrating how Industry 4.0 technologies enable adaptive, optimized, and safe valorization strategies at industrial scale.

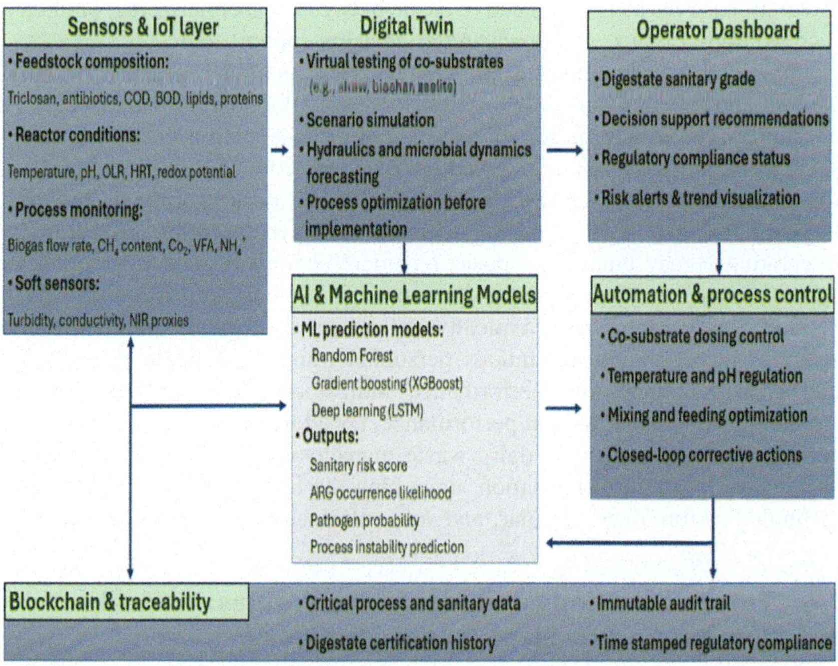

FIGURE 7.1 Industry 4.0–enabled digital architecture for monitoring, control, and sanitary risk management in anaerobic digestion of dairy waste. The framework integrates real-time sensing, machine learning models, automation, digital twin simulation, and blockchain-based traceability to support process optimization, digestate safety, and regulatory compliance. *AI*, Artificial intelligence; *ARG*, antimicrobial resistance genes; *CH_4*, methane; *COD*, chemical oxygen demand; *BOD*, biochemical oxygen demand; *HRT*, hydraulic retention time; *IoT*, Internet of Things; *LSTM*, long short-term memory neural network; *XGBoost*, eXtreme Gradient Boosting; *ML*, machine learning; *NH_4^+*, ammonium; *OLR*, organic loading rate; *VFA*, volatile fatty acids.

7.5 Data integration, traceability, compliance, and adoption challenges

While Industry 4.0 technologies can significantly enhance process efficiency, valorization outcomes, and sanitary risk management in dairy waste systems (as illustrated in Section 7.4), their successful implementation requires a robust data infrastructure and careful consideration of sociotechnical constraints. Within an Industry 4.0 framework, SCADA systems, coupled with cloud analytics, edge computing, and secure APIs, provide the backbone for real-time data acquisition, storage, and processing. Blockchain-enabled traceability frameworks ensure transparent, immutable, and auditable records, supporting regulatory compliance and enabling circular business models (Khanna et al., 2022).

These infrastructures complement smart control and optimization workflows by facilitating secure data sharing, performance benchmarking, and integration of predictive analytics across treatment units. In this way, data integration serves as the connective tissue between sensor networks, predictive models, digital twins, and downstream decision-support tools.

Despite these advantages, significant barriers can hinder adoption. The high capital and operational costs of sensor networks, digital twins, and AI controllers necessitate thorough cost-benefit analyses that account for energy savings, reduced waste, and potential revenue from valorization pathways (Xiao et al., 2024). Addressing workforce skill gaps is crucial, requiring targeted training in digital maintenance and cybersecurity. Additionally, integrating legacy equipment poses technical challenges, and interconnected systems increase cybersecurity risks that must be managed carefully.

Successful deployment typically relies on a stepwise implementation strategy, incorporating continuous personnel training, strategic partnerships, and clearly defined key performance indicators (KPIs) to monitor both technological and operational performance. By addressing these sociotechnical and regulatory dimensions, dairy waste management facilities can achieve a sustainable digital transformation, unlocking the full potential of Industry 4.0 technologies for smart, circular, and safe valorization of organic residues.

7.6 Future directions and research priorities: toward autonomous and resilient dairy biorefineries

The evolution of dairy biorefineries will integrate biochemical innovations, advanced digital platforms, and resilient operational strategies to achieve autonomous, efficient, and safe valorization of waste streams. Building on the smart control, optimization, and traceability frameworks discussed in Sections 7.4 and 7.5, future developments will focus on enhancing process resilience, operational autonomy, and circularity.

Whey valorization is moving beyond conventional uses, emphasizing the recovery of bioactive peptides through enzymatic hydrolysis. These peptides, including lactokines, exhibit multifunctional properties, such as antimicrobial activity that can stabilize methanogenic consortia by inhibiting pathogenic bacteria. This directly reduces operational risks, including digester acidification and foaming, thereby enhancing process resilience and reliability (Granato et al., 2022). Further resilience can be achieved through biofilm reactor technologies, which improve microbial retention and shock resistance, particularly under variable whey loading scenarios (Murshid et al., 2023).

Looking forward, the next generation of dairy biorefineries will rely on autonomous, AI-driven platforms embedded within an Industry 4.0 framework. These self-optimizing systems integrate digital twins, real-time sensors, and smart actuators to dynamically adapt to variable feedstock qualities and operational conditions, requiring minimal human intervention. By continuously learning from historical and real-time process data, these platforms can optimize bioprocess parameters, predict potential disturbances, and implement corrective actions autonomously, aligning with Industry 4.0 principles to maximize efficiency while maintaining stability and compliance.

In parallel, multiproduct valorization pathways combining AD with fermentation, bioplastic production, and protein extraction are emerging. These integrated cascade processes rely heavily on continuous process analytics and predictive algorithms to ensure efficient, stable, and flexible operation while meeting sustainability and circular economy goals. Advanced process control enables selective valorization of different dairy waste fractions, optimizing energy recovery, nutrient recycling, and production of high-value bioproducts, thereby enhancing both economic and environmental performance. Although the application of machine learning to the discovery of bioactive peptides from dairy waste streams remains largely unexplored, recent advances in ML-assisted peptide identification from alternative protein-rich matrices provide a strong methodological foundation for future developments. For instance, machine learning approaches combining multivariate analysis and structure–activity relationship models have been successfully applied to identify novel bioactive peptides from complex plant-based protein hydrolysates, enabling rapid screening and experimental validation of multifunctional peptides (Cournoyer et al., 2025). Transferring such data-driven discovery frameworks to dairy-derived substrates, such as whey protein hydrolysates, represents a promising research direction to accelerate the identification of bioactive compounds while integrating peptide valorization into autonomous and digitally controlled dairy biorefineries.

Robust data governance remains a critical enabler for these developments. Blockchain-based traceability, together with harmonized open standards, ensures transparent, auditable, and circular business models. Such frameworks support regulatory compliance, foster stakeholder confidence, and allow verification of both process performance and product quality. Integrated digital infrastructures, encompassing sensors, predictive models, and digital twins, provide the backbone for autonomous decision-support and traceable operations, ensuring that technological advances translate into safe, reliable, and circular valorization practices.

By combining biochemical innovations, autonomous AI-driven platforms, multiproduct valorization, and robust data governance, dairy biorefineries can transition toward more resilient, efficient, and sustainable operations. Future research priorities include optimizing peptide-based and microbial strategies to enhance process stability, expanding digital twin capabilities for fully autonomous process management, integrating modular and scalable multi-product pathways, developing harmonized data standards and blockchain-enabled traceability, and addressing sociotechnical barriers such as workforce training, cybersecurity, and legacy equipment integration. Together, these directions outline a clear roadmap toward next-generation dairy biorefineries, where Industry 4.0 technologies enable fully integrated, adaptive, and safe valorization of dairy waste streams.

7.7 Conclusion

Industry 4.0 is transforming dairy waste management, turning waste streams into valuable resources within a circular bioeconomy. By integrating real-time sensing, AI-driven control, and predictive analytics, dairy operations can enhance environmental sustainability, operational efficiency, and economic resilience.

Despite these opportunities, adoption challenges remain. High initial costs, skills gaps, legacy infrastructure, and cybersecurity risks require careful planning, workforce training, and scalable deployment strategies. Addressing these sociotechnical barriers is essential to ensure reliable and safe implementation.

Future research should focus on autonomous biorefineries, resilient biofilm reactors, and traceable valorization pathways, supported by robust data governance frameworks such as blockchain. These innovations will enable adaptive, efficient, and safe valorization of dairy waste streams, contributing to a low-carbon, circular, and resilient dairy industry.

Ultimately, the successful integration of Industry 4.0 technologies will unlock the full potential of dairy waste, drive innovation, and support the transition toward sustainable, circular, and resilient bio-based economies.

AI disclosure

During the preparation of this work, the authors used ChatGPT in order to reformulate sentences. After using this tool/service, the authors reviewed and edited the content as needed and take full responsibility for the content of the publication.

References

Ahmed, J., Thakur, A., & Goyal, A. (2021). *Industrial wastewater and its toxic effects*. The Royal Society of Chemistry. Available at https://doi.org/10.1039/9781839165399-00001.

Aigle, A., Bourgeois, E., Marjolet, L., Houot, S., Patureau, D., Doelsch, E., et al., 2021. Relative weight of organic waste origin on compost and digestate 16S rRNA gene bacterial profilings and related functional inferences. Frontiers in Microbiology 12. Available at https://doi.org/10.3389/fmicb.2021.667043.

Almeida, P.D.S., de Menezes, C.A., Camargo, F.P., Sakamoto, I.K., Lovato, G., Rodrigues, J.A.D., et al., 2023. Biomethane recovery through co-digestion of cheese whey and glycerol in a two-stage anaerobic fluidized bed reactor: Effect of temperature and organic loading rate on methanogenesis. Journal of Environmental Management 330. Available at https://doi.org/10.1016/j.jenvman.2022.117117.

Aly, B.A., Low, T., Long, D., Baillie, C., Brett, P., 2023. Robotics and sensing technologies in red meat processing: A review. Trends in Food Science and Technology 137, 142–155. Available at https://doi.org/10.1016/j.tifs.2023.05.015.

Andrade Cruz, I., Chuenchart, W., Long, F., Surendra, K.C., Renata Santos Andrade, L., Bilal, M., et al., 2022. Application of machine learning in anaerobic digestion: Perspectives and challenges. Bioresource Technology 345, 126433. Available at https://doi.org/10.1016/j.biortech.2021.126433.

Avadi, A., Benoit, P., Bravin, M., Cournoyer, B., Feder, F., Galia, W., et al., 2022. Trace contaminants in the environmental assessment of organic waste recycling in agriculture: Gaps between methods and knowledge. Advances in Agronomy. Available at https://doi.org/10.1016/bs.agron.2022.03.002.

Bella, K., Rao, P.V., 2023. Anaerobic digestion of dairy wastewater: Effect of different parameters and co-digestion options—a review. Biomass Conversion and Biorefinery 13 (4), 2527–2552. Available at https://doi.org/10.1007/s13399-020-01247-2.

Bilal, H., Khan, M.N., Khan, S., Shafiq, M., Fang, W., Khan, R.U., et al., 2025. The role of artificial intelligence and machine learning in predicting and combating antimicrobial resistance. Computational and Structural Biotechnology Journal 27, 423–439. Available at https://doi.org/10.1016/j.csbj.2025.01.006.

Borisov, M., Dimitrova, N., Simeonov, I., 2016. Mathematical modelling of anaerobic digestion with hydrogen and methane production* IFAC-PapersOnLine, Foundations of Systems Biology in Engineering - FOSBE 2016 (49), 231–238. Available at https://doi.org/10.1016/j.ifacol.2016.12.131.

Brião, V.B., Vieira Salla, A.C., Miorando, T., Hemkemeier, M., Cadore Favaretto, D.P., 2019. Water recovery from dairy rinse water by reverse osmosis: Giving value to water and milk solids. Resources, Conservation and Recycling 140, 313–323. Available at https://doi.org/10.1016/j.resconrec.2018.10.007.

Buchanan, D., Martindale, W., Romeih, E., Hebishy, E., 2023. Recent advances in whey processing and valorisation: Technological and environmental perspectives. International Journal of Dairy Technology 76 (2), 291–312. Available at https://doi.org/10.1111/1471-0307.12935.

Casino, F., Kanakaris, V., Dasaklis, T.K., Moschuris, S., Stachtiaris, S., Pagoni, M., et al., 2021. Blockchain-based food supply chain traceability: A case study in the dairy sector. International Journal of Production Research 59 (19), 5758–5770. Available at https://doi.org/10.1080/00207543.2020.1789238.

Chen, Y., Zhang, H., You, Y., Zhang, J., Tang, L., 2025. A hybrid deep learning model based on signal decomposition and dynamic feature selection for forecasting the influent parameters of wastewater treatment plants. Environmental Research 266, 120615. Available at https://doi.org/10.1016/j.envres.2024.120615.

Ching, P.M.L., Zou, X., Wu, D., So, R.H.Y., Chen, G.H., 2022. Development of a wide-range soft sensor for predicting wastewater BOD5 using an eXtreme gradient boosting (XGBoost) machine. Environmental Research 210, 112953. Available at https://doi.org/10.1016/j.envres.2022.112953.

Chourasia, R., Phukon, L.C., Abedin, M.M., Padhi, S., Singh, S.P., Rai, A.K., 2022. Whey valorization by microbial and enzymatic bioprocesses for the production of nutraceuticals and value-added products. Bioresource Technology Reports 19. Available at https://doi.org/10.1016/j.biteb.2022.101144.

Cournoyer, A., Bernier, M.È., Aboubacar, H., de Toro-Martín, J., Vohl, M.C., Ravallec, R., et al., 2025. Machine learning-driven discovery of bioactive peptides from duckweed (Lemnaceae) protein hydrolysates: Identification and experimental validation of 20 novel antihypertensive, antidiabetic, and/or antioxidant peptides. Food Chemistry 482. Available at https://doi.org/10.1016/j.foodchem.2025.144029.

Cvetković, B.R., Memišin, N., Nježić, Z., Pezo, L., Juodeikiene, G., Vitas, J., et al., 2025. Addressing wastewater challenges in the dairy industry: A focused case study. Journal of Dairy Research 92 (2), 135–145. Available at https://doi.org/10.1017/S0022029925101131.

Danalewich, J.R., Papagiannis, T.G., Belyea, R.L., Tumbleson, M.E., Raskin, L., 1998. Characterization of dairy waste streams, current treatment practices, and potential for biological nutrient removal. Water Research 32 (12), 3555–3568. Available at https://doi.org/10.1016/s0043-1354(98)00160-2.

Dikmen, F., Demir, A., Özkaya, B., Raza, M.O., Rasheed, J., Asuroglu, T., et al., 2025. AI-driven wastewater management through comparative analysis of feature selection techniques and predictive models. Scientific Reports 15 (1). Available at https://doi.org/10.1038/s41598-025-07124-0.

Escalante, H., Castro, L., Amaya, M.P., Jaimes, L., Jaimes-Estévez, J., 2018. Anaerobic digestion of cheese whey: Energetic and nutritional potential for the dairy sector in developing countries. Waste Management 71, 711–718. Available at https://doi.org/10.1016/j.wasman.2017.09.026.

Galia, W., Mariani-Kurkdjian, P., Bastien, S., Loukiadis, E., Blanquet-Diot, S., Leriche, F., et al., 2015. Correction for Galia et al., Genome sequence and annotation of a human infection isolate of *Escherichia coli* O26:H11involved in a raw milk cheese outbreak. Genome Announcements 3, e00880-15. Available at https://doi.org/10.1128/genomeA.00880-15.

Galia, W., Leriche, F., Cruveiller, S., Garnier, C., Navratil, V., Dubost, A., et al., 2017. Strand-specific transcriptomes of Enterohemorrhagic *Escherichia coli* in response to interactions with ground beef microbiota: Interactions between microorganisms in raw meat. BMC Genomics 18 (1). Available at https://doi.org/10.1186/s12864-017-3957-2.

Granato, D., Carocho, M., Barros, L., Zabetakis, I., Mocan, A., Tsoupras, A., et al., 2022. Implementation of Sustainable Development Goals in the dairy sector: Perspectives on the use of agro-industrial side-streams to design functional foods. Trends in Food Science and Technology 124, 128–139. Available at https://doi.org/10.1016/j.tifs.2022.04.009.

Hafeez, Z., Cakir-Kiefer, C., Girardet, J.M., Lecomte, X., Paris, C., Galia, W., et al., 2015. New insights into the proteolytic system of *Streptococcus thermophilus*: Use of isracidin to characterize cell-associated extracellular peptidase activities. Journal of Agricultural and Food Chemistry 63 (34), 7522–7531. Available at https://doi.org/10.1021/acs.jafc.5b01647.

Hassoun, A., Garcia-Garcia, G., Trollman, H., Jagtap, S., Parra-López, C., Cropotova, J., et al., 2023. Birth of dairy 4.0: Opportunities and challenges in adoption of fourth industrial revolution technologies in the production of milk and its derivatives. Current Research in Food Science 7, 100535. Available at https://doi.org/10.1016/j.crfs.2023.100535.

Hassoun, A., Tarchi, I., Aït-Kaddour, A., 2024. Leveraging the potential of fourth industrial revolution technologies to reduce and valorize waste and by-products in the dairy sector. Current Opinion in Green and Sustainable Chemistry 47, 100927. Available at https://doi.org/10.1016/j.cogsc.2024.100927.

Hemdan, M., Abuelhaded, K., Shaker, A.A.S., Ashour, M.M., Abdelaziz, M.M., Dahab, M.I., et al., 2025. Recent advances in nano-enhanced biosensors: Innovations in design, applications in healthcare, environmental monitoring, and food safety, and emerging research challenges. Sensing and Bio-Sensing Research 48. Available at https://doi.org/10.1016/j.sbsr.2025.100783.

Horňák, D., Procházková, M., Miklas, V., Vondra, M., Touš, M., Kollmann, M., et al., 2025. Minimal Liquid Discharge system for sustainable dairy wastewater management. Separation and Purification Technology 373, 133556. Available at https://doi.org/10.1016/j.seppur.2025.133556.

Idrissi Khaldi, M., Erraissi, A., Hain, M., Banane, M., 2025. Comparative analysis of supervised machine learning classification models. In: Farhaoui, Y., Herawan, T., Imoize, A.L., Allaoui, A.E. (Eds.), Intersection of artificial intelligence, data science, and cutting-edge technologies: From concepts to applications in smart environment. Springer Nature Switzerland, pp. 321–326. Available at https://doi.org/10.1007/978-3-031-88304-0_44.

Jagtap, S., Garcia-Garcia, G., Rahimifard, S., 2021. Optimisation of the resource efficiency of food manufacturing via the Internet of Things. Computers in Industry 127, 103397. Available at https://doi.org/10.1016/j.compind.2021.103397.

Jin, G., Jiang, Z., Sun, Y., Liu, Z., Liu, S., Wu, F., 2025. Intelligent clean-in-place (CIP) system in beverage (healthy water) cleaner production. Food Control 168, 110877. Available at https://doi.org/10.1016/j.foodcont.2024.110877.

Kazemi, P., Bengoa, C., Steyer, J.P., Giralt, J., 2021. Data-driven techniques for fault detection in anaerobic digestion process. Process Safety and Environmental Protection 146, 905–915. Available at https://doi.org/10.1016/j.psep.2020.12.016.

Kegl, T., Torres Jiménez, E., Kegl, B., Kovač Kralj, A., Kegl, M., 2025. Modeling and optimization of anaerobic digestion technology: Current status and future outlook. Progress in Energy and Combustion Science 106, 101199. Available at https://doi.org/10.1016/j.pecs.2024.101199.

Khanna, A., Jain, S., Burgio, A., Bolshev, V., Panchenko, V., 2022. Blockchain-enabled supply chain platform for indian dairy industry: Safety and traceability. Foods 11 (17), 2716. Available at https://doi.org/10.3390/foods11172716.

Konstantinidis, F. K., Balaska, V., Symeonidis, S., Psarommatis, F., Psomoulis, A., Giakos, G. et al. (2023a). Achieving zero defected products in Diary 4.0 using digital twin and machine vision. In *ACM International Conference Proceeding Series*.(pp. 528-534). Association for Computing Machinery Greece. https://doi.org/10.1145/3594806.3596554.

Konstantinidis, F.K., Balaska, V., Symeonidis, S., Tsilis, D., Mouroutsos, S.G., Bampis, L., et al., 2023b. Automating dairy production lines with the yoghurt cups recognition and detection process in the Industry 4.0 era. Procedia Computer Science, 4th International Conference on Industry 4.0 and Smart Manufacturing 217, 918–927. Available at https://doi.org/10.1016/j.procs.2022.12.289.

Malik, M., Gahlawat, V.K., Mor, R.S., Singh, M.K., 2024. Unlocking dairy traceability: Current trends, applications, and future opportunities. Future Foods 10. Available at https://doi.org/10.1016/j.fufo.2024.100426.

Mangla, S.K., Kazancoglu, Y., Ekinci, E., Liu, M., Özbiltekin, M., Sezer, M.D., 2021. Using system dynamics to analyze the societal impacts of blockchain technology in milk supply chainsrefer. Transportation Research Part E: Logistics and Transportation Review 149. Available at https://doi.org/10.1016/j.tre.2021.102289.

Matthes, J., Schmid, M., 2024. Biogenic raw materials from food waste and by-products for smart packaging applications. Current Opinion in Green and Sustainable Chemistry 46, 100894. Available at https://doi.org/10.1016/j.cogsc.2024.100894.

Murshid, S., Antonysamy, A.J., Dhakshinamoorthy, G.P., Jayaseelan, A., Pugazhendhi, A., 2023. A review on biofilm-based reactors for wastewater treatment: Recent advancements in biofilm carriers, kinetics, reactors, economics, and future perspectives. Science of The Total Environment 892, 164796. Available at https://doi.org/10.1016/j.scitotenv.2023.164796.

Neethirajan, S., 2023. Artificial intelligence and sensor technologies in dairy livestock export: Charting a digital transformation. Sensors 23 (16), 7045. Available at https://doi.org/10.3390/s23167045.

Di Puglia Pugliese, L., Guerriero, F., Mitton, N., 2025. Optimizing wireless sensor networks deployment with coverage and connectivity requirements. Annals of Operations Research 346, 1997–2008. Available at https://doi.org/10.1007/s10479-025-06487-x.

Riboli, M., Jaccard, M., Silvestri, M., Aimi, A., Malara, C., 2023. Collision-free and smooth motion planning of dual-arm Cartesian robot based on B-spline representation. Robotics and Autonomous Systems 170, 104534. Available at https://doi.org/10.1016/j.robot.2023.104534.

Shandilya, S.K., Datta, A., Kartik, Y., Nagar, A., 2024. Role of artificial intelligence and machine learning. In: Shandilya, S.K., Datta, A., Kartik, Y., Nagar, A. (Eds.), Digital resilience: Navigating disruption and safeguarding data privacy. Springer Nature Switzerland, pp. 313–399. Available at https://doi.org/10.1007/978-3-031-53290-0_6.

Shemer, H., Wald, S., Semiat, R., 2023. Challenges and solutions for global water scarcity. Membranes 13, 612. Available at https://doi.org/10.3390/membranes13060612.

Sheth, V., Tripathi, U., Sharma, A., 2022. A comparative analysis of machine learning algorithms for classification purpose. Procedia Computer Science, 4th International Conference on Innovative Data Communication Technology and Application 215, 422–431. Available at https://doi.org/10.1016/j.procs.2022.12.044.

Singaravadivelan, A., Sachin, P.B., Harikumar, S., Vijayakumar, P., Vindhya, M.V., Farhana, F.M.B., et al., 2023. Life cycle assessment of greenhouse gas emission from the dairy production system—Review. Tropical Animal Health and Production 55 (5). Available at https://doi.org/10.1007/s11250-023-03748-4.

Stasinakis, A.S., Charalambous, P., Vyrides, I., 2022. Dairy wastewater management in EU: Produced amounts, existing legislation, applied treatment processes and future challenges. Journal of Environmental Management 303, 114152. Available at https://doi.org/10.1016/j.jenvman.2021.114152.

Udourioh, G.A., Chizoma, A.N., Okolie, J.A., Udourioh, M.G., Solomon, M.M., 2025. Greenhouse gases emission from dairy industries and promising abatement processes and technologies. International Journal of Dairy Technology 78 (3). Available at https://doi.org/10.1111/1471-0307.70040.

Udugama, I.A., Lopez, P.C., Gargalo, C.L., Li, X., Bayer, C., Gernaey, K.V., 2021. Digital twin in biomanufacturing: Challenges and opportunities towards its implementation. Systems Microbiology and Biomanufacturing 1 (3), 257–274. Available at https://doi.org/10.1007/s43393-021-00024-0.

Usmani, Z., Sharma, M., Gaffey, J., Sharma, M., Dewhurst, R.J., Moreau, B., et al., 2022. Valorization of dairy waste and by-products through microbial bioprocesses. Bioresource Technology 346, 126444. Available at https://doi.org/10.1016/j.biortech.2021.126444.

Valavarasu, S., Sangu, Y., Mahapatra, T., 2025. Prediction of antibiotic resistance from antibiotic susceptibility testing results from surveillance data using machine learning. Scientific Reports 15, 30509. Available at https://doi.org/10.1038/s41598-025-14078-w.

Wang, L., Long, F., Liao, W., Liu, H., 2020. Prediction of anaerobic digestion performance and identification of critical operational parameters using machine learning algorithms. Bioresource Technology 298, 122495. Available at https://doi.org/10.1016/j.biortech.2019.122495.

Wybraniec, C., Cournoyer, B., Moussard, C., Beaupère, M., Lusurier, L., Leriche, F., et al., 2024. Occurrence of 40 sanitary indicators in French digestates derived from different anaerobic digestion processes and raw organic wastes from agricultural and urban origin. Frontiers in Microbiology 15. Available at https://doi.org/10.3389/fmicb.2024.1346715.

Xiao, J., Xu, Z., Xiao, A., Wang, X., Skare, M., 2024. Overcoming barriers and seizing opportunities in the innovative adoption of next-generation digital technologies. Journal of Innovation & Knowledge 9 (4), 100622. Available at https://doi.org/10.1016/j.jik.2024.100622.

Zhao, K., Peng, X., Hu, S., Wang, X., 2025. Machine learning models for rapid prediction of chemicals' life-cycle environmental impacts: Current status, challenges, and future directions. Green Carbon 3 (4), 433–442. Available at https://doi.org/10.1016/j.greenca.2025.05.004.

Zhou, P., Kan, M., Chen, W., Wang, Y., Cao, B., 2025. An adaptive coverage method for dynamic wireless sensor network deployment using deep reinforcement learning. Scientific Reports 15 (1). Available at https://doi.org/10.1038/s41598-025-16031-3.

Chapter 8

Packaging 4.0 in the dairy industry

Yana Cahyana[1], Qurratu Aini Alya Adzkia[2], Rivan Satria Jaluwardhana[1], Atiek Rostika Noviyanti[2], and Abderrahmane Aït-Kaddour[3]
[1]*Department of Food Technology, Universitas Padjadjaran, Bandung, Indonesia, France,* [2]*Department of Chemistry, Universitas Padjadjaran, Bandung, Indonesia, France,* [3]*VetAgro Sup, Departement of Sustainable Foods, Management and Innovation/Clermont-Auvergne, INRAE, VetAgro Sup, Cheese Research Unit, Lempdes, France*

8.1 Introduction

Packaging plays a critical role in the dairy industry by protecting products from physical, chemical, and biological hazards while preserving quality and extending shelf life. Dairy products are particularly sensitive to environmental factors such as temperature fluctuations, oxygen exposure, and light, making packaging a key determinant of product stability during distribution and storage. Beyond its protective function, packaging also serves as a communication interface, conveying product information, branding, and increasingly, interactive content to consumers. As dairy markets become more diversified and sustainability expectations intensify, conventional packaging approaches are no longer sufficient to address emerging industry challenges (Davidescu et al., 2025).

The concept of Packaging 4.0 has emerged in response to the broader transformation associated with Industry 4.0, which emphasizes digitalization, connectivity, and intelligent systems across industrial operations. Packaging 4.0 refers to the integration of digital technologies and other Industry 4.0 innovations into packaging design and systems, enabling packaging to become adaptive, responsive, and value-generating (Hassoun et al., 2023a). In the dairy sector, this transition represents a shift from static packaging solutions toward intelligent systems that actively support quality preservation, operational flexibility, consumer engagement, and sustainability objectives (Hassoun et al., 2023b).

Several factors drive the adoption of Packaging 4.0 in the dairy industry. Increasing product diversification such as lactose free, functional, and premium dairy products has led to greater complexity in packaging requirements and shorter product life cycles. At the same time, heightened environmental awareness and

Dairy 4.0. DOI: https://doi.org/10.1016/B978-0-443-44730-3.00006-7

regulatory pressure have intensified the need for sustainable packaging solutions that minimize material use and environmental impact (Hussain et al., 2024). Technological advancements, including the Internet of Things (IoT), automation and artificial intelligence (AI), among others, have further enabled the redesign of packaging systems to meet these evolving demands. Together, these drivers underscore the need for packaging systems that are both technologically advanced and strategically aligned with market and sustainability goals (Palazzo et al., 2023).

Packaging 4.0 in the dairy industry can be understood through three interrelated functional dimensions: smart and intelligent packaging, flexible and customized packaging systems, and sustainable packaging design. Active and intelligent packaging incorporates active and intelligent elements, such as sensors and indicators, that enhance product–package interaction and information availability (Jafarzadeh et al., 2024). Flexible and customized packaging systems emphasize modularity, automation, and digital printing technologies that allow rapid adaptation to changing product specifications and market needs (Vărzaru and Bocean, 2024). Sustainable packaging design focuses on the use of digital tools to optimize material selection, reduce environmental footprints, and support ecoefficient packaging solutions from the early design stage (Georgakoudis et al., 2022).

Importantly, Packaging 4.0 should be viewed as a system level transformation rather than the mere automation of packaging lines. It involves the integration of packaging into cyber–physical systems, where digital tools support design decisions, enable system responsiveness, and enhance human–-machine collaboration. While data and automation play a central role, human expertise remains essential in managing, interpreting, and governing intelligent packaging systems. This perspective aligns with the emerging vision of human-centric and sustainable industrial systems (Islam et al., 2025).

Fig. 8.1 presents a conceptual framework of Packaging 4.0 in the dairy industry, illustrating the relationships between core packaging functions, enabling digital technologies, and expected outcomes. The framework highlights how smart packaging, flexible systems, and sustainable design are supported by Industry 4.0 technologies and collectively contribute to extended shelf life, enhanced consumer engagement, improved sustainability, and greater operational responsiveness. This framework provides a foundation for the subsequent sections of the chapter, which explore these dimensions in greater detail.

The scope of this chapter is deliberately focused on Packaging 4.0 as a distinct domain within the dairy value chain.

8.2 Smart packaging for dairy products

8.2.1 General overview

Dairy products are particularly susceptible to deterioration because of their high moisture content, abundant nutrients, and vulnerability to temperature variations during storage and distribution, which collectively shorten shelf life

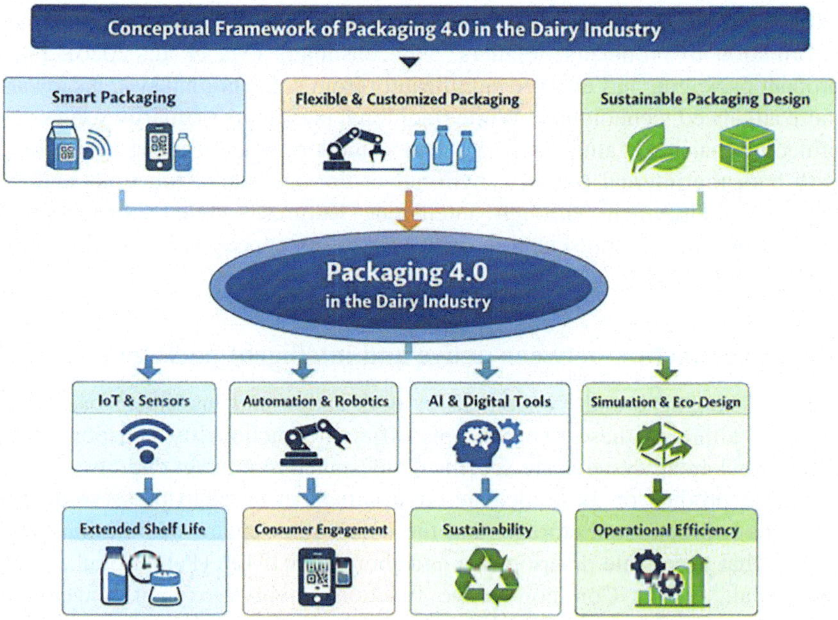

FIGURE 8.1 Conceptual framework of Packaging 4.0 in the dairy industry.

and complicate handling. These characteristics also make dairy foods an ideal environment for microbial proliferation, including pathogenic bacteria and spoilage yeasts, leading to product degradation and potential health risks (Eissa et al., 2018; Hawashin et al., 2025). Packaging, therefore, plays a critical role in preserving product quality, minimizing food wastes, and safeguarding consumers' health.

Conventional packaging systems mainly serve as passive barriers against humidity, oxygen, enzymatic degradation, microbial contamination, pressure, spurious odors, uncontrolled temperature, and other storage conditions; however, they remain inadequate for managing dynamic spoilage processes across the dairy supply chain that shorten product shelf life (Adzkia et al., 2026; Gouvêa et al., 2015; Mirza Alizadeh et al., 2022). Spoiled food products exhibit adverse changes in visual, textural, and sensory characteristics that compromise their safety for consumption (Karanth et al., 2023). In this context, active and intelligent packaging technologies have emerged as promising tools for food preservation, enhancing food safety, extending shelf life, and enabling real-time quality monitoring for various dairy products (Palanisamy et al., 2025).

Packaging systems are no longer designed solely for product protection but also incorporate additional functional properties, allowing them to interact with internal or external environments and communicate quality-related information to producers, retailers, and consumers (Ma et al., 2026). Food product packaging has evolved significantly from conventional systems toward more advanced technologies. While traditional packaging primarily serves as a protective barrier against external factors, modern smart packaging interacts with both the internal product environment through active functions and the external environment through intelligent features, while simultaneously enhancing product presentation and consumer appeal (Du et al., 2023; Mkhari et al., 2025).

8.2.2 Distinction between active and intelligent packaging

Smart packaging broadly encompasses both active and intelligent packaging systems, although these two concepts differ in functionality and objectives. Active packaging systems are specifically designed to interact directly with the packaged product or its immediate environment in order to preserve quality throughout production, storage, and the distribution chain, while minimizing factors that accelerate deterioration and shorten shelf life (Perera et al., 2023; Qin et al., 2024). Common active functions involve oxygen scavenging, moisture control, antimicrobial release, and ethylene management (Flórez et al., 2022). These systems operate by either releasing or absorbing specific compounds within the package or food matrix, thereby slowing spoilage reactions (Palanisamy et al., 2025). In dairy applications, active materials, such as antimicrobial films, oxygen scavengers, and moisture regulators, are widely employed to suppress microbial growth, limit oxidative degradation, and enhance product stability (Foltynowicz, 2020).

In contrast, intelligent packaging focuses on monitoring and communicating information about the packaged product's condition without directly modifying it from the point of manufacture to the customer. The packaging systems rely on sensors, indicators, or data carriers that respond to environmental changes such as temperature, gas composition, or pH, providing visual or digital signals related to product freshness, integrity, or safety (Mirza Alizadeh et al., 2022; Mkhari et al., 2025). Examples include time–temperature indicators (TTIs) (Ma et al., 2026), freshness indicators (Becerril et al., 2021; Deng et al., 2026), gas sensors (Roy and Rhim, 2021), and radio-frequency identification (RFID)-based tracking systems (Liao et al., 2019; Zuo et al., 2022). Therefore, customers will have the opportunity to conduct in-house quality control to minimize unnecessary food waste and prevent them from potential food poisoning.

While active packaging improves product stability through functional interactions, intelligent packaging enhances transparency and traceability by providing real-time or near-real-time information. The integration of both systems into a single packaging platform represents a key trend toward fully smart packaging solutions for dairy products.

8.2.3 Smart indicators for dairy products

8.2.3.1 Freshness indicators

Freshness indicators are designed to reflect biochemical changes associated with microbial growth and metabolic activity in dairy products. These indicators typically respond to volatile compounds, such as organic acids, sulfur compounds, and amines, produced during spoilage (Palanisamy et al., 2025). In dairy matrices, especially milk and fresh cheeses, freshness indicators based on colorimetric dyes embedded in polymer matrices or labels can visually signal product deterioration by changing color as spoilage metabolites accumulate (Becerril et al., 2021; Deng et al., 2026).

8.2.3.2 Gas- and pH-sensitive labels

Gas-sensitive indicators respond to changes in oxygen or carbon dioxide concentrations within the package headspace, which are closely related to microbial respiration and package integrity. For example, CO_2-sensitive labels are particularly relevant for cheese packaging, where microbial fermentation can increase CO_2 levels during storage (Roy and Rhim, 2021; Siciliano et al., 2025). Similarly, pH-sensitive labels respond to acidification processes, which are critical quality parameters in fermented dairy products such as yogurt and cultured milk (Barleany et al., 2025; Guo et al., 2023; Romero et al., 2021). These systems provide low-cost, noninvasive monitoring tools that enhance quality control throughout the distribution chain.

8.2.3.3 Temperature abuse indicators

Temperature abuse remains one of the most significant contributors to dairy spoilage during transportation and storage (Ma et al., 2026). TTIs record cumulative thermal exposure and visually reflect whether a product has been subjected to temperatures beyond its recommended storage conditions. TTIs are particularly useful for chilled dairy products such as milk, cream, and fresh cheeses, where even short-term temperature deviations can accelerate microbial growth and compromise safety (Liu et al., 2023; Ma et al., 2026).

8.2.4 Integration of sensors with digital interfaces

Recent advances in smart packaging have enabled the integration of physical sensors with digital communication technologies such as QR codes, near-field communication (NFC), and RFID. These interfaces allow consumers, retailers, and logistics operators to access real-time or stored information regarding product history, storage conditions, and freshness status using smartphones or scanning devices (Liao et al., 2019; Zuo et al., 2022).

In dairy packaging, QR codes linked to cloud-based databases can provide dynamic shelf-life predictions, traceability information, and quality alerts.

NFC-enabled labels further enhance interactivity by enabling contactless data exchange and authentication. The combination of sensor-generated data with digital interfaces supports improved transparency, supply chain monitoring, and consumer trust, while also facilitating data-driven decision-making for inventory management and waste reduction (Kumar et al., 2025).

8.2.5 Role of Industry 4.0 enablers in smart packaging

The adoption of active and intelligent packaging systems increasingly aligns with the principles of Industry 4.0, which emphasize digitalization, connectivity, and automation across manufacturing and supply chains (Kumar et al., 2025). IoT technologies enable continuous data capture at the packaging level, enabling monitoring of temperature, humidity, and gas composition throughout transportation and storage (Javaid et al., 2022; Ma et al., 2026). These data streams support real-time quality assurance and early detection of deviations from optimal storage conditions.

AI and machine learning tools can further enhance the value of smart packaging data by enabling pattern recognition, spoilage prediction, and shelf-life modeling. Although most current applications in the dairy sector remain at the conceptual or pilot stage, AI-driven analytics offer significant potential for predictive quality management, dynamic shelf-life labeling, and intelligent logistics optimization (Kumar et al., 2025; Prajapati et al., 2025).

8.2.6 Applications across dairy categories

Active and intelligent packaging technologies have been increasingly adapted to address the specific quality and safety challenges associated with different dairy product categories. Variations in moisture content, microbial activity, gas production, and storage requirements among milk, cheese, yogurt, and fermented dairy products necessitate tailored packaging strategies (Adzkia et al., 2026; Karanth et al., 2023; Mirza Alizadeh et al., 2022). Consequently, both active and intelligent systems have been developed to optimize product stability, extend shelf life, and provide real-time quality information across diverse dairy matrices, as summarized in.

8.2.6.1 Fluid milk

In fluid milk packaging, active and intelligent technologies are primarily focused on temperature monitoring, freshness indication, and microbial spoilage detection (Deng et al., 2026). TTIs and freshness indicators integrated into milk cartons or bottles can provide visual signals of cold-chain integrity and remaining shelf life (Barleany et al., 2025; Piao et al., 2026). Oxygen scavengers and antimicrobial coatings in active packaging systems further extend shelf life by limiting oxidation and microbial growth.

8.2.6.2 Cheese

Cheese represents a highly suitable application for smart packaging systems owing to its widespread consumption and susceptibility to spoilage, especially after package opening. Its physicochemical and microbiological properties vary considerably across different types and stages of ripening (Roozbahani et al., 2025). Throughout cheese ripening, proteolytic breakdown of casein releases free amino acids, which are subsequently transformed into diverse flavor-and aroma-active compounds by the microbial populations involved (Lima et al., 2026). Active packaging systems incorporating CO_2 absorbers, moisture regulators, and antimicrobial agents have been widely investigated to control mold growth and surface contamination (Falih et al., 2024). Intelligent packaging approaches, including gas sensors and freshness indicators, enable monitoring of fermentation activity and spoilage progression, particularly in soft and semi-soft cheeses (Kontogianni et al., 2023).

8.2.6.3 Yogurt and fermented dairy products

In yogurt and fermented dairy products, packaging must support continuous microbial metabolism and progressive acidification. Once opened, exposure to uncontrolled conditions accelerates microbial growth, metabolite formation, and texture deterioration, while conventional date labeling cannot reflect real-time product quality, increasing the risk of safety concerns and economic losses. Consequently, pH-responsive indicators and TTIs are particularly valuable for tracking fermentation behavior and storage integrity (Guo et al., 2023; Ma et al., 2026). Smart labels integrated with QR or NFC technologies also enable real-time access to quality information and intuitive quality guidance, supporting improved product transparency and shelf-life management (Liu et al., 2023; Ma et al., 2026).

8.3 Flexible and customized dairy packaging systems

Personalization and customization are expected to emerge as key drivers of future change, supported by technological advancements and influencing all industries (Pollet, 2025). One of the industrial sectors being driven toward product customization is the dairy industry, where global milk production and the number of milking cows are expected to continue increasing in response to sustained demand for dairy products (Bell and Jauernik, 2023; FAO, 2018). This section explains why the dairy industry requires a high degree of packaging flexibility. Several topics demonstrate how Packaging 4.0 enables the industrial sector to respond to the market quickly and adaptively.

8.3.1 Market drivers for customization in dairy

Market drivers for customization in dairy refer to factors arising from market dynamics, such as consumer preferences, retail requirements, regulations, and industry competition, that encourage dairy producers to adopt product and packaging customization rather than rely on a uniform mass-production approach. The following are several sources of market pressure that compel the dairy industry to undertake product and packaging customization, such as SKU (Stock Keeping Unit) proliferation, private labels, regional, and functional dairy products.

8.3.1.1 SKU proliferation

SKU is an internal identifier (Fig. 8.2) used by retailers and e-commerce sellers to distinguish individual products and their variants, typically reflecting attributes such as style, size, color, collection, or packaging format (Kowalczyk and Sarwas, 2024). Manufacturers view regarded SKU proliferation as a strategy to expand market presence and share, while retailers have been concerned that reducing product variety may negatively affect consumers' perceptions of assortment and reduce store traffic (Zhang and Krishna, 2007). SKU proliferation refers to the rapid increase in the number of product variants in the dairy market, where each variation in size, formulation, or labeling constitutes a distinct stock-keeping unit, thereby necessitating flexible and customizable packaging systems (Silva et al., 2025).

FIGURE 8.2 Digital prototyping and evaluation tools for customized biobased dairy packaging configurations.

Several recommended examples of SKU implementation in the dairy industry include packaging volume, formulation, flavor profile and targets, consumer segments, labelling, claims, and regulatory requirements.

8.3.1.1.1 Differences in packaging volume

Variations in packaging volume are closely linked to consumer preferences, pricing strategies, and distribution channels, such as modern retail, food service, and regional markets. Within the context of Packaging 4.0, the integration of production data, modular design, and automation enables dairy producers to efficiently manage differences in volume-based SKUs while maintaining packaging quality consistency and sustainability. There are studies that analyze variations in packaging design based on differences in packaging volume and shape, such as those by Klein et al. (2024) and (Musso et al., 2022). Volume-based SKUs in the dairy industry range from small single-serving packages to large food service formats, each serving distinct market needs. Packaging volumes of 100–200 mL are commonly used for flavored milk, yogurt drinks, and children's milk, while medium-sized formats of 250–330 mL are typical for standard Ultra High Temperature (UHT) milk and high-protein dairy beverages. Pasteurized milk is frequently marketed in 500 mL packages, whereas larger household consumption is addressed through family-size formats of 750 mL to 1 L. Economy-size milk products are generally offered in 1.5–2 L packages, while institutional and food service applications rely on bulk packaging volumes ranging from 5 to 10 L.

8.3.1.1.2 Differences in formulation

Differences in formulation represent a major driver of SKU proliferation in the dairy industry, as products derived from the same base ingredients are reformulated to meet diverse nutritional, functional, and health-related consumer demands. Variations such as low-fat, lactose-free, high-protein, fortified, and probiotic dairy products exhibit distinct physicochemical properties that influence stability, shelf life, and sensitivity to oxygen and light. Consequently, packaging systems must be customized to provide appropriate barrier performance, material specifications, and labeling requirements for each formulation (UNCTAD, 2025). These formulation-driven differences necessitate flexible and adaptive packaging solutions capable of efficiently managing multiple SKUs without increasing production complexity or operational costs.

8.3.1.1.3 Differences in flavor profiles and target consumer segments

Differences in flavor profiles constitute an important source of SKU differentiation in the dairy industry, as products are tailored to meet the preferences of distinct consumer segments such as children, adults, and health-conscious consumers. Variations in flavors, including chocolate, fruit-based, or plain formulations, are often accompanied by differences in sweetness level and sensory attributes, which influence product positioning and branding (Pedersen et al., 2023). These flavor-driven distinctions require customized packaging designs and product identification to clearly communicate the intended consumer segment, thereby increasing the demand for flexible packaging systems capable of handling multiple SKUs efficiently.

8.3.1.1.4 Differences in labeling language, claims, and regulatory requirements

Differences in labeling language, product claims, and regulatory requirements further contribute to SKU proliferation in the dairy industry, particularly in regional and international markets. Packaging must be adapted to accommodate multiple languages (Wagner & Charinsarn, 2020), culturally appropriate terminology, and region-specific regulations (Pettigrew et al., 2022, UNCTAD, 2025) related to nutritional information (Afroza et al., 2024), health claims (Venkatesh et al., 2025), and allergen declarations. As a result, even identical dairy formulations may require distinct packaging and labeling configurations, underscoring the need for digitally enabled and customizable packaging solutions that ensure regulatory compliance while maintaining production efficiency.

8.3.1.2 Private labels

Private label (PL) refers to a situation in which a seller or retailer, including online platforms, markets products under its own brand while outsourcing production to third-party manufacturers (Sgroi and Salamone, 2022). The implementation of PL represents a strategic approach adopted by food distributors to enhance profitability. Within the framework of Packaging 4.0, the adoption of digital printing and flexible packaging systems enables efficient small-scale production without high setup costs, enabling dairy manufacturers to meet PL requirements while maintaining operational efficiency, build a purchase intention, and loyalty (Calvo Porral and Levy-Mangin, 2016; Musso et al., 2022).

8.3.1.3 Regional and functional dairy products

This aspect emphasizes the need for the dairy industry to adapt to consumer preferences, including local tastes, language, and regulatory requirements. Dairy product packaging can be modified or customized to reflect local language or commonly used terms and the community's cultural characteristics. The consumer or community perspective represents a central consideration in the development of food packaging (Jagoda et al., 2023; Wikström et al., 2014). Although consumer expectations vary by factors such as ethnicity, age, and food category, specific "must-be" attributes are essential.

In contrast, "one-dimensional" attributes are desirable and should be fulfilled when possible.

Kano's theory categorizes "must-be" attributes as compulsory features, where failure to meet them results in total consumer dissatisfaction (Löfgren and Witell, 2005; Maja Brozović Dorotea Kovačević, 2021; Williams et al., 2008). Meanwhile, "one-dimensional" attributes influence satisfaction in a linear manner, increasing satisfaction when present and causing dissatisfaction when absent. As a result, these attributes are valuable indicators for evaluating consumer preferences, given the proportional relationship between functional performance and consumer satisfaction (Dash, 2021). Addressing these functional requirements is crucial, as they directly affect consumer preferences and purchasing decisions. In the context of functional dairy products, such as high protein, lactose free, or probiotic formulations, packaging plays a significant role in clearly communicating nutritional benefits, functional claims, and usage information to specific consumer segments. For functional dairy products, "*must-be*" attributes typically include clear nutritional labeling, safety information, and regulatory compliance, whereas "*one-dimensional*" attributes relate to the visibility and credibility of functional claims, convenience features, and package usability (Kim et al., 2013).

8.3.2 Modular and reconfigurable packaging lines

Dairy packaging development based on Industry 4.0 promotes the use of modular packaging systems, in which dairy product packaging lines are constructed from interchangeable modules (such as filling, sealing, and labeling) that can be reconfigured as needed. The advantages gained from the implementation of this technological innovation include (1) greater ease in adapting to new packaging formats and (2) more efficient long-term investment. Several examples of the application of this technology in Packaging 4.0 include the use of sensors, IoT systems, Digital Twin (DT), Augmented Reality/Virtual Reality (AR/VR), Additive Manufacturing (AM), among others that enable self-adjusting capabilities based on predefined SKUs (Ching et al., 2022; Shaban et al., 2026).

8.3.3 Digital printing and short production runs

One of the most transformative developments is the emergence of digital printing technologies. In contrast to conventional analog printing methods, digital printing enables data-driven, on-demand production without the use of physical plates, thereby providing greater flexibility for short-run production and customization (Denise, 2022). The major types of digital printing technology include digital laser technology, electrostatic imaging, inkjet imaging technology, digital surface technology, and digital light solidification technology (Wang, 2025). Additionally, the range of compatible substrates is more limited compared with traditional printing technologies (Ryynänen et al., 2012). Examples of digital printing applications in dairy packaging include: Inkjet digital printing (Denise, 2022), electrophotographic (digital toner) printing (Denise, 2022), Variable Data Printing, and Digital roll-to-roll label printing. The capability of digital printing directly supports short production runs, enabling small batches of customized dairy packaging to be produced at a competitive cost.

8.3.4 Automation and robotics as flexibility enablers

One key enabler of flexibility in dairy packaging systems is the use of automation and robotics, including industrial robotic arms for pick-and-place operations, collaborative robots that can be quickly reprogrammed for multiple SKU formats, and machine vision systems that guide robots in real time. These technologies, integrated with Industry 4.0 data streams and AI-based design insights from big data analytics (Gan et al., 2022), allow dairy producers to respond rapidly to SKU variations and custom label designs without extensive downtime or manual intervention.

Within the framework of Packaging 4.0, robotic packaging systems are no longer operated as isolated units but are digitally connected to production control systems and Enterprise Resource Planning (ERP) platforms (Gan et al., 2022). This integration enables packaging equipment to automatically receive production schedules, SKU specifications, and material requirements directly from ERP systems, ensuring alignment between production planning, inventory management, and packaging execution. As a result, adjustments in packaging format, volume, or labeling can be implemented seamlessly without extensive manual intervention.

8.3.5 Benefits of flexible and customized dairy packaging

The implementation of flexible, customized dairy packaging systems supported by Packaging 4.0 technologies delivers tangible operational and market benefits. By integrating modular equipment design, digital printing, and automated robotic systems, dairy producers are able to translate technological flexibility into measurable performance improvements. These benefits are particularly evident in the performance indicators, such as reduced changeover time (achieved through modular equipment, automation, and digital printing, enabling rapid transitions between different SKUs with minimal downtime), improved responsiveness (allows dairy producers to adapt quickly to changing market demands, private label requirements, and regional preferences), and enhanced branding based on study by Hussain et al. (2024).

8.4 Sustainable packaging design enabled by Industry 4.0

Industry 4.0 technologies provide the necessary tools to support dairy industry transformation. For instance, IoT-enabled sensors, advanced data analytics, artificial intelligence algorithms, automation, and cloud computing allow manufacturers to continuously monitor operations, identify inefficiencies in real time, and promptly implement corrective measures (Shaban et al., 2026).

AI, machine learning (ML), and deep learning algorithms are optimizing business processes and strengthening organizational capabilities for intelligent decision-making. In recent years, these technologies have been adopted across nearly all major industries, including food, pharmaceuticals, and cosmetics (Kumar et al., 2025). As most products in these sectors require packaging solutions, AI-driven tools are increasingly transforming packaging operations to improve efficiency and responsiveness to consumer needs (Kampakis, 2022).

8.4.1 Sustainability challenges specific to dairy packaging

Food waste at the household level has been extensively examined and is driven by various factors. Packaging, alongside consumer behavior, plays a crucial role, particularly through excessive package sizes, multipack offerings, price-based promotions, and the interpretation of "best before" dates (Klein et al., 2024; WRAP, 2007). There are several effects of nonsustainable food packaging on dairy packaging (Hussain et al., 2024) as shown in Table 8.1.

8.4.2 Digital tools for ecodesign

Digital tools for ecodesign play a critical role in enabling sustainable packaging development in the dairy industry by shifting environmental considerations to the early stages of the design process. There are two key digital approaches to ecodesign in dairy packaging:

8.4.2.1 Simulation-based material reduction

In the dairy industry, where packaging must provide adequate mechanical strength and effective barrier protection against oxygen and moisture, simulation-based material reduction enables the optimization of packaging design without compromising product safety or quality. Digital simulations are used to evaluate stress resistance, deformation, and barrier performance under conditions typical of dairy processing, storage, and distribution, enabling controlled reductions in material thickness. This approach minimizes the need for physical prototyping while ensuring that packaging maintains its functional integrity. As a result, plastic consumption can be significantly reduced, leading to lower material costs and decreased greenhouse gas emissions associated

TABLE 8.1 Environmental impacts associated with nonsustainable food packaging materials.

Effects	Examples	Solutions
High Energy consumption and resource depletion	Plastic packaging production relies on fossil fuel extraction, which is a major source of greenhouse gas emissions.	Plastic packaging reduction for Dairy Industries or change the type of biodegradable plastic (not fossil fuel based)
Waste generation	Materials such as plastic, polystyrene, and aluminum may require hundreds of years to decompose, leading to the accumulation of waste in landfills and marine environments.	Nonbiodegradable packaging from fossil fuel-based preprocessing before it is released to nature
Chemical pollution	Contain hazardous chemicals, such as bisphenol A (BPA) or phthalates, which may migrate into food and pose risks to human health.	Reduce the use of certain synthetic chemicals through new packaging developed from natural resources. Moreover, the strict or comprehensive research should be held.
Hazardous chemicals	Concentrations of per- and polyfluoroalkyl substances (PFAS), chloropropanols, and pesticide residues	

with packaging production and transportation (Haba et al., 2025; Houssini et al., 2025). Jagoda et al. (2023) in their study used SolidWorks 2016 as the software platform for modeling design alternatives of the food packaging (FP) components.

8.4.2.2 Virtual prototyping

Virtual prototyping (VP) has been increasingly studied and applied in engineering design in recent years (Ambaw et al., 2022). Compared with conventional simulation methods, VP places greater emphasis on human–production interaction. Using this approach, Wu et al. (2019) were among the first to demonstrate how package design, pallet positioning, stacking configurations, and cold-chain conditions influence cooling kinetics and the evolution of fruit quality. The approach integrates computer-aided graphics, computational structural dynamics, computational fluid dynamics (CFD),

JavaCAD (Hirt et al., 2001), and computational heat transfer techniques. Virtual prototyping required only a few hours to generate data on carton compression strength, airflow resistance, produce cooling rates, spatiotemporal temperature distribution, and quality mapping throughout the cold chain.

8.4.3 Lightweighting and material optimization

Packaging redesign (Georgakoudis et al., 2022) may provide additional benefits, including reduced material consumption, which consequently lowers the volume of waste directed to landfills and helps mitigate environmental impacts on air, soil, and water (Craighill and Powell, 1996; da Cruz et al., 2012). According to a study by Georgakoudis and Pechlivanidou (2025), two alternative bottle designs with the same capacity were evaluated, with emphasis on differences in geometry, material consumption, and space efficiency. A comprehensive quantitative comparison reveals the respective strengths and limitations of each design. The results indicate that an optimized bottle configuration can achieve significant reductions in material use and waste generation.

This comparative evaluation of two alternative bottle designs with identical capacity highlights the critical role of lightweighting and material optimization in the dairy industry. Differences in bottle geometry, material consumption, and space efficiency directly influence the environmental and economic performance of dairy packaging, particularly given the high production volumes and distribution intensity typical of dairy products. The quantitative comparison demonstrates that an optimized bottle configuration can substantially reduce material usage and waste generation without compromising functional requirements, such as structural integrity and product protection. These findings underscore the importance of geometry-driven design optimization, supported by digital design and assessment tools, as a key strategy for achieving sustainable packaging solutions in the dairy sector.

The concept of packaging redesign aimed at reducing material use without compromising product protection is of critical importance and should be a key concern for both industry stakeholders and policymakers (Georgakoudis and Pechlivanidou, 2025).

8.4.4 Bio-based, recyclable, and monomaterial packaging

Bio-based packaging materials are recognized as promising renewable alternatives to conventional packaging materials. These materials can be derived from biomass, synthesized through chemical processes, or produced by microorganisms (Donkor et al., 2023). Furthermore, monomaterial multilayer packaging represents a practical approach to mitigating carbon emissions, as it delivers required functionalities, such as mechanical strength and barrier performance, through layered structures while remaining compatible with recycling processes (Wei et al., 2024).

Bioplastics are increasingly regarded as a favorable alternative to petrochemical-based plastics due to the greater environmental and health concerns associated with conventional plastics (Hussain et al., 2024). Advantages of bioplastics include a lower carbon footprint, reliance on renewable resources, and contributions to rural economic development. In contrast, petrochemical-based plastics are less attractive because of their nonbiodegradability, limited renewability, and potential health risks associated with the release of hazardous emissions (Hussain et al., 2024). There are several evidences of industries that implement bio-based packaging (European Dairy Association (EDA), 2020), as presented in Table 8.2.

8.4.5 Role of Industry 4.0 in sustainable packaging

Industry 4.0 enables data-driven dairy packaging development by supporting informed design decision-making and early-stage sustainability assessment through integrated digital tools and analytics.

8.4.5.1 Design decision support

The integration of robotics with Enterprise Resource Planning (ERP) and Customer Relationship Management (CRM) systems (Gan et al., 2022) enables real-time, reinforcing flexibility, data-driven decision-making, operational efficiency, and responsiveness across dairy packaging operations. This connectivity allows dairy producers to dynamically adjust packaging configurations in response to customer-specific or regional requirements. Moreover, design decision support tools in dairy packaging include Life Cycle Assessment (LCA) for environmental evaluation (Cappiello et al., 2022), Finite Element Analysis for assessing mechanical strength, CFD (Hirt et al., 2001) for analyzing thermal and flow behavior, and Digital Twin supported by material databases to enable data-driven selection of optimal packaging designs based on performance, sustainability, and cost.

8.4.5.2 Early-stage sustainability assessment

Early-stage sustainability assessment in dairy packaging design can be effectively conducted using a LCA approach to evaluate potential environmental impacts from the initial concept phase (Cappiello et al., 2022; Jagoda et al., 2023; Kan and Miller, 2022). By applying simplified or screening-level LCA, designers can identify key environmental hotspots related to material choice, packaging weight, and processing requirements before physical prototyping or large-scale manufacturing. In the dairy industry, where packaging must meet strict functional and barrier requirements, integrating LCA with Industry 4.0 tools, such as digital twins and material databases,

TABLE 8.2 The bio-based and recyclable implementations in several brands (European Dairy Association (EDA), 2020).

Implementations	Industries/brands
It achieved a 40% reduction in plastic use and a 30% decrease in CO_2 emissions by eliminating 16 g of plastic and incorporating cardboard to ensure sufficient container rigidity.	Arla's Global Skyr brand in Denmark April, 2020
Introducing a new renewable packaging format for its fresh organic white milk range in March 2020, utilizing unbleached cardboard and bio-based plastics to reduce reliance on fossil-derived materials.	Arla Germany's Fresh White Organic Milk March, 2020
The new packaging format for Kærgården spreadable butter, branded as the "Natural Tub," achieves full recyclability by adopting a monomaterial polypropylene (PP) container, complemented by a removable cardboard sleeve to ensure structural rigidity.	Arla, launched in 2020–2021
All gable-top cartons for milk, buttermilk, cream, and yoghurt are produced entirely from plant-based materials, including fully plant-based caps.	Valio
Implemented a range of bio-based and sustainable packaging initiatives, including recyclable cheese packaging, further reductions in plastic usage, the adoption of paper straws, and the use of plant-based packaging for Chocomel products. In addition, sustainability performance is evaluated using a dedicated packaging assessment tool developed by Quantis, which applies a life cycle assessment (LCA)-based approach incorporating indicators for recyclability and food waste reduction.	Friesland Campina
The black Actimel bottle incorporates specialized pigments that are detectable in the near-infrared range, enabling efficient sorting and recycling.	Danone

supports data-driven decision-making despite limited data availability. This approach ensures that sustainability considerations are embedded in packaging design from the outset, rather than addressed at later stages of development.

8.4.5.3 *Reduced packaging and product waste*

Industry 4.0 technologies can substantially reduce waste in dairy packaging systems by enabling precise material use, minimizing defects, and improving packaging line efficiency. Smart sensors, machine vision, and AI-based quality control can detect sealing failures, incorrect filling levels, label misplacement, and microleaks in real time, allowing immediate corrective actions and reducing the volume of rejected packs (Cruz et al., 2021; Garbacz et al., 2026; Sagar and Rani, 2026). In parallel, digital twins and predictive analytics support optimized packaging design and process parameters, limiting over-packaging and material excess while maintaining barrier and safety performance (Huang et al., 2024; Madhu, 2025; Qin et al., 2024; Shrivastava et al., 2023). Furthermore, enhanced traceability through IoT, blockchain, and data integration across the supply chain improves inventory planning and cold chain monitoring, reducing product spoilage and packaging waste linked to expired or damaged dairy products (Dzreke, 2025; Han et al., 2026; Rahman et al., 2026). Generally, Industry 4.0 supports a shift from reactive waste handling to proactive waste prevention in dairy packaging operations.

8.5 Challenges, adoption barriers, and future outlook of Packaging 4.0

Although Packaging 4.0 offers substantial potential to enhance functionality, flexibility, and sustainability in the dairy industry, its adoption remains uneven and faces multiple challenges. These challenges extend beyond technological limitations and include economic, organizational, and regulatory factors. Addressing these barriers is essential to ensure that Packaging 4.0 contributes effectively to the broader transformation of dairy systems.

8.5.1 Technological and integration challenges

A major challenge in adopting Packaging 4.0 is the integration of advanced digital technologies into existing dairy packaging infrastructures. Many dairy facilities rely on legacy packaging lines that were designed for standardized, high-throughput production and lack compatibility with smart sensors, digital communication systems, and real-time data platforms. Retrofitting these systems can be technically complex and costly, often requiring significant changes to hardware, software, and control architectures.

Interoperability issues further hinder system integration. Packaging 4.0 relies on seamless data exchange among machines, sensors, and digital platforms; however, the absence of harmonized communication standards may result in fragmented systems with limited functionality. In addition, increased connectivity raises cybersecurity concerns, as packaging systems become potential targets for data breaches and operational disruptions. Ensuring system robustness and data security therefore represents an important technical consideration.

8.5.2 Economic and investment barriers

Economic constraints represent a significant barrier to the widespread adoption of Packaging 4.0, particularly for small- and medium-sized dairy enterprises. The initial investment required for smart packaging materials, advanced equipment, and digital infrastructure can be substantial, while the economic benefits are often realized gradually. This imbalance creates uncertainty regarding return on investment and may discourage adoption in cost-sensitive dairy markets (Taramuel-Taramuel et al., 2025).

Moreover, many benefits of Packaging 4.0, such as improved brand value, enhanced consumer engagement, or sustainability performance, are difficult to quantify in monetary terms. To address these challenges, scalable and modular Packaging 4.0 solutions are increasingly important, enabling incremental adoption and reducing financial risk. Collaborative initiatives involving industry, technology providers, and research institutions may also help lower adoption costs and facilitate knowledge transfer.

8.5.3 Human and organizational challenges

The successful implementation of Packaging 4.0 depends heavily on human and organizational readiness. Advanced packaging systems require personnel with digital competencies and the ability to interact effectively with intelligent machines. In many dairy operations, skill gaps limit the effective use of Packaging 4.0 technologies and reduce their potential benefits (Turcan and Pojar, 2024).

Organizational resistance to change may also impede adoption, particularly when digitalization is perceived as a threat to employment or established work practices. Effective change management, workforce training, and clear communication are therefore critical to fostering acceptance. Rather than replacing human labor, Packaging 4.0 should be viewed as a means of enhancing human–machine collaboration, supporting decision-making, and improving operational performance.

8.5.4 Regulatory and standardization considerations

Regulatory and standardization issues play a crucial role in shaping the adoption of Packaging 4.0 in the dairy sector. Smart packaging solutions must comply with food contact material regulations, labeling requirements, and data protection laws. The introduction of novel materials and digital features may raise regulatory uncertainties, particularly where existing frameworks have not yet adapted to technological innovation (Nichifor et al., 2025).

In addition, the lack of standardized protocols for smart packaging components and data exchange can limit interoperability and scalability.

Harmonized standards and clear regulatory guidance are therefore essential to support innovation while ensuring consumer safety and system reliability.

8.5.5 Future outlook: toward Packaging 5.0

Looking forward, Packaging 4.0 is expected to evolve toward Packaging 5.0, which emphasizes human-centricity, sustainability, and resilience. In the dairy industry, this transition involves integrating advanced digital technologies with societal values, environmental responsibility, and inclusive development. Future packaging systems are likely to enhance interaction with consumers, support sustainable consumption, and enable greater transparency (Herrmann et al., 2022).

In conclusion, while Packaging 4.0 presents significant opportunities for innovation in the dairy industry, its successful adoption depends on addressing technological, economic, human, and regulatory challenges in an integrated manner. By adopting a holistic and human-centered approach, Packaging 4.0 can serve as a key enabler of sustainable and resilient dairy systems and contribute to the broader transition toward future Dairy 5.0.

8.6 Conclusion

Packaging 4.0 is reshaping dairy packaging from a primarily protective barrier into a connected, data-enabled system that supports product quality, operational agility, and sustainability. Across the chapter, smart active and intelligent solutions demonstrate how packaging can either interact with the pack environment to extend shelf life or monitor conditions and communicate freshness and handling information to stakeholders, strengthening transparency and consumer trust. At the same time, flexible and customized packaging systems, enabled by digital printing, modular lines, automation, and interoperable data flows, help dairy processors manage SKU proliferation and shorter product cycles with reduced changeover burdens. These advances align with the Industry 4.0 vision of cyber–physical integration and connected manufacturing.

Looking ahead to the future of food packaging, wider adoption will depend on resolving integration, investment, skills, and regulatory challenges while ensuring cybersecurity and standardization. With responsible governance and workforce upskilling, Packaging 4.0 can serve as a practical pathway toward more resilient and human-centric Packaging 5.0 strategies in the dairy sector.

AI disclosure

During the preparation of this work, the authors used ChatGPT in order to rephrase some sentences and Canva to edit Fig. 8.2. After using this tool/service, the authors reviewed and edited the content as needed and take full responsibility for the content of the publication.

References

Adzkia, Q.A.A., Noviyanti, A.R., Cahyana, Y., Rukayadi, Y., Ogino, C., Putra, F.J.N., 2026. Sustainable antibacterial films from cellulose acetate and hydroxyapatite: A green approach to food packaging. Results in Chemistry 20, 102973. Available at https://doi.org/10.1016/j.rechem.2025.102973.

Afroza, U., Abrar, A.K., Nowar, A., Sobhan, S.M.M., Ide, N., Choudhury, S.R., 2024. Global overview of government-endorsed nutrition labeling policies of packaged foods: A document review. Frontiers in Public Health 12. Available at https://www.frontiersin.org/journals/public-health/articles/10.3389/fpubh.2024.1426639.

Ambaw, A., Mukama, M., Fadiji, T., Opara, U.L., 2022. Fresh fruit packaging design verification through virtual prototyping technique. Food Packaging and Shelf Life 32, 100858. Available at https://doi.org/10.1016/j.fpsl.2022.100858.

Barleany, D.R., Yulvianti, M., Jayanuddin, J., Setiadi, H.R., Prasetyo, R.M., 2025. Green coconut (*Cocos nucifera* L) husk extract as milk freshness indicator in chitosan/ polyvinyl alcohol-based smart packaging films. Jurnal Riset Kimia 16 (1), 1–13. Available at https://doi.org/10.25077/jrk.v16i1.738.

Becerril, R., Nerín, C., Silva, F., 2021. Bring some colour to your package: Freshness indicators based on anthocyanin extracts. Trends in Food Science and Technology 111. Elsevier Ltd, pp. 495–505. Available at https://doi.org/10.1016/j.tifs.2021.02.042.

Bell, M.J., Jauernik, G.-M., 2023. Selecting the 'Sustainable'' cow using a customized breeding index: Case study on a commercial UK dairy herd.' In Agriculture 13 (2), 423. Available at https://doi.org/10.3390/agriculture13020423.

Calvo Porral, C., Levy-Mangin, J.-P., 2016. Food private label brands: The role of consumer trust on loyalty and purchase intention. British Food Journal 118 (3), 679–696. Available at https://doi.org/10.1108/BFJ-08-2015-0299.

Cappiello, G., Aversa, r, Genovesi, A., Barletta, M., 2022. Life cycle assessment (LCA) of bio-based packaging solutions for extended shelf-life (ESL) milk. Environmental Science and Pollution Research International 29 (13), 18617–18628. Available at https://doi.org/10.1007/s11356-021-17094-1.

Ching, N.T., Ghobakhloo, M., Iranmanesh, M., Maroufkhani, P., Asadi, S., 2022. Industry 4.0 applications for sustainable manufacturing: A systematic literature review and a roadmap to sustainable development. Journal of Cleaner Production 334, 130133. Available at https://doi.org/10.1016/j.jclepro.2021.130133.

Craighill, A.L., Powell, J.C., 1996. Lifecycle assessment and economic evaluation of recycling: A case study. Resources, Conservation and Recycling 17 (2), 75–96. Available at https://doi.org/10.1016/0921-3449(96)01105-6.

Cruz, S., Paulino, A., Duraes, J., Mendes, M., 2021. Real-time quality control of heat sealed bottles using thermal images and artificial neural network. Journal of Imaging 7 (2), 24. Available at https://doi.org/10.3390/jimaging7020024.

da Cruz, N.F., Sim\~{o}es, P., Marques, R.C., 2012. Economic cost recovery in the recycling of packaging waste: The case of Portugal. Journal of Cleaner Production 37, 8–18. Available at https://doi.org/10.1016/j.jclepro.2012.05.043.

Dash, S., 2021. Identifying and classifying attributes of packaging for customer satisfaction—A Kano Model Approach. International Journal of Production Management and Engineering 9, 57. Available at https://doi.org/10.4995/ijpme.2021.13683.

Davidescu, M.A., Pânzaru, C., Mădescu, B.M., Poroșnicu, I., Simeanu, C., Usturoi, A., et al., 2025. Advances and challenges in smart packaging technologies for the food industry: Trends, applications, and sustainability considerations. Foods 14 (24), 4347. Available at https://doi.org/10.3390/foods14244347.

Deng, L., Liu, Q., Weng, W., Zhang, Y., 2026. Dragon fruit peel pigment extraction residue based intelligent and biodegradable film for indicating milk freshness. Innovative Food Science & Emerging Technologies 108, 104418. Available at https://doi.org/10.1016/j.ifset.2025.104418.

Denise, B., 2022. Influence of digital printing innovations on the efficiency and quality of packaging production. International Journal of Printing and Packaging Technology (IJPPT 3 (1), 1–7. Available at https://iaeme.com/Home/article_id/IJPPT_03_01_001.

Donkor, L., Kontoh, G., Yaya, A., Bediako, J.K., Apalangya, V., 2023. Bio-based and sustainable food packaging systems: Relevance, challenges, and prospects. Applied Food Research 3 (2), 100356. Available at https://doi.org/10.1016/j.afres.2023.100356.

Du, H., Sun, X., Chong, X., Yang, M., Zhu, Z., Wen, Y., 2023. A review on smart active packaging systems for food preservation: Applications and future trends. Trends in Food Science and Technology 141 Elsevier Ltd. Available at https://doi.org/10.1016/j.tifs.2023.104200.

Dzreke, S.S., 2025. Systematic Analysis of IoT, AI, active packaging, and blockchain for food waste reduction across the farm-to-fork supply chain. International Journal of Management Science and Application 4 (2), 95–122. Available at https://doi.org/10.58291/ijmsa.v4i2.441.

Eissa, M.A., Nasralla, N.N., Gomah, N.H., Osman, D.M., El-Derwy, Y.M., 2018. Evaluation of natural fertilizer extracted from expired dairy products as a soil amendment 695 Evaluation of natural fertilizer extracted from expired dairy. In Journal of Soil Science and Plant Nutrition 18 (3).

European Dairy Association (EDA). (2020). *Packaging in the dairy industry*. Retrieved from https://share.google/n6UzukQoFlnraV0zI.

Falih, M.A., Altemimi, A.B., Alkaisy, Q.H., Awlqadr, F.H., Abedelmaksoud, T.G., Amjadi, S., et al., 2024. Enhancing safety and quality in the global cheese industry: A review of innovative preservation techniques. Heliyon 10 Elsevier Ltd. Available at https://doi.org/10.1016/j.heliyon.2024.e40459 23.

FAO, G.D. P. (2018). *Climate change and the global dairy cattle sector—The role of the dairy sector in a low-carbon future*. Rome: Food and Agriculture Organization of the United Nations and Global Dairy Platform Inc.

Flórez, M., Guerra-Rodríguez, E., Cazón, P., Vázquez, M., 2022. Chitosan for food packaging: Recent advances in active and intelligent films. Food Hydrocolloids Vol. 124 Elsevier B.V. Available at https://doi.org/10.1016/j.foodhyd.2021.107328.

Foltynowicz, Z., 2020. Oxygen scavengers applications in the dairy industry. Journal of Dairy Research & Technology 3 (1), 1–6. Available at https://doi.org/10.24966/DRT-9315/100016.

Gan, F., Romainoor, N.H., Guo, Z., 2022. Research on innovative design of product packaging based on big data technology. Scientific Programming 2022 (1), 4973875. Available at https://doi.org/10.1155/2022/4973875.

Garbacz, P., Burghardt, A., Czajka, P., Mężyk, J., Mizak, W., 2026. Predictive modeling of packaging seal strength: A hybrid vision and process data approach for non-destructive quality assurance. Applied Sciences 16 (2), 923. Available at https://doi.org/10.3390/app16020923.

Georgakoudis, E.D., & Pechlivanidou, G.G., 2025. Sustainable packaging design: Packaging optimization and material reduction for environmental protection and economic benefits to industry and society. Applied Science, 15, 8289.

Georgakoudis, E.D., Pechlivanidou, G.G., Tipi, N.S., 2022. The packaging redesign issue—Space exploitation and environmental benefits. Management of Environmental Quality: An International Journal 34 (1), 1–20. Available at https://doi.org/10.1108/MEQ-08-2021-0195.

Gouvêa, D.M., Mendonça, R.C.S., Soto, M.L., Cruz, R.S., 2015. Acetate cellulose film with bacteriophages for potential antimicrobial use in food packaging. LWT - Food Science and Technology 63 (1), 85–91. Available at https://doi.org/10.1016/j.lwt.2015.03.014.

Guo, J., Zhang, H., Linguo, C., Ding, L., Hu, F., Pei, L., et al., 2023. A visual detection films based on waterborne polyurethane for real-time monitoring of yogurt freshness. Dyes and Pigments 210. Available at https://doi.org/10.1016/j.dyepig.2022.110995.

Haba, B., Djellali, S., Abdelouahed, Y., Boudjelida, S., Faleschini, F., Carraro, M., 2025. Transforming plastic waste into value: A review of management strategies and innovative applications in sustainable construction. Polymers 17 (7), 881. Available at https://doi.org/10.3390/polym17070881.

Han, J., Sun, C., Ji, Z., Yang, X., 2026. Smart cold chain logistics for fresh agricultural products: Key technologies, challenges, and future trends. Trends in Food Science & Technology 167, 105421. Available at https://doi.org/10.1016/j.tifs.2025.105421.

Hassoun, A., Boukid, F., Ozogul, F., Aït-Kaddour, A., Soriano, J.M., Lorenzo, J.M., et al., 2023a. Creating new opportunities for sustainable food packaging through dimensions of industry 4.0: New insights into the food waste perspective. Trends in Food Science & Technology 142, 104238. Available at https://doi.org/10.1016/j.tifs.2023.104238.

Hassoun, A., Garcia-Garcia, G., Trollman, H., Jagtap, S., Parra-López, C., Cropotova, J., et al., 2023b. Birth of dairy 4.0: Opportunities and challenges in adoption of fourth industrial revolution technologies in the production of milk and its derivatives. Current Research in Food Science 7, 100535. Available at https://doi.org/10.1016/j.crfs.2023.100535.

Hawashin, D., Salah, K., Jayaraman, R., Ahmad, R.W., Kanhere, S., Damiani, E., 2025. Enhancing sustainability in dairy industry: Blockchain-based waste reduction. Sustainable Futures 9. Available at https://doi.org/10.1016/j.sftr.2025.100529.

Herrmann, C., Rhein, S., Sträter, K.F., 2022. Consumers' sustainability-related perception of and willingness-to-pay for food packaging alternatives. Resources, Conservation and Recycling 181, 106219. Available at https://doi.org/10.1016/j.resconrec.2022.106219.

Hirt, E., Tröster, G., & Scheffler, M., 2001. Virtual prototyping for high density packaging systems. IEEE Transactions on Advanced Packaging, 24, 392–400.

Houssini, K., Li, J., Tan, Q., 2025. Complexities of the global plastics supply chain revealed in a trade-linked material flow analysis. Communications Earth & Environment 6 (1), 257. Available at https://doi.org/10.1038/s43247-025-02169-5.

Huang, Y., Ghadge, A., Yates, N., 2024. Implementation of digital twins in the food supply chain: A review and conceptual framework. International Journal of Production Research 62 (17), 6400–6426. Available at https://doi.org/10.1080/00207543.2024.2305804.

Hussain, S., Akhter, R., Maktedar, S.S., 2024. Advancements in sustainable food packaging: From eco-friendly materials to innovative technologies. Sustainable Food Technology 2 (5), 1297–1364. Available at https://doi.org/10.1039/D4FB00084F.

Islam, M.T., Sepanloo, K., Woo, S., Woo, S.H., Son, Y.-J., 2025. A Review of the Industry 4.0 to 5.0 transition: Exploring the intersection, challenges, and opportunities of technology and human–machine collaboration. Machines 13 (4), 267. Available at https://doi.org/10.3390/machines13040267.

Jafarzadeh, S., Yildiz, Z., Yildiz, P., Strachowski, P., Forough, M., Esmaeili, Y., Naebe, M., Abdollahi, M., 2024. Advanced technologies in biodegradable packaging using intelligent sensing to fight food waste. International Journal of Biological Macromolecules 261, 129647. Available at https://doi.org/10.1016/j.ijbiomac.2024.129647.

Jagoda, S.U.M., Gamage, J.R., Karunathilake, H.P., 2023. Environmentally sustainable plastic food packaging: A holistic life cycle thinking approach for design decisions. Journal of

Cleaner Production 400, 136680. Available at https://doi.org/10.1016/j.jclepro.2023.136680 February.

Javaid, M., Haleem, A., Singh, R.P., Suman, R., Gonzalez, E.S., 2022. Understanding the adoption of Industry 4.0 technologies in improving environmental sustainability. Sustainable Operations and Computers 3, 203–217. Available at https://doi.org/10.1016/j.susoc.2022.01.008.

Kampakis, S. (2022). How artificial intelligence is revolutionizing the packaging industry? Retrieved from https://thedatascientist.com/how-artificial-intelligence-is-revolutionizing-the-packaging-industry/.

Kan, M., Miller, S.A., 2022. Environmental impacts of plastic packaging of food products. Resources, Conservation and Recycling 180, 106156. Available at https://doi.org/10.1016/j.resconrec.2022.106156.

Karanth, S., Feng, S., Patra, D., Pradhan, A.K., 2023. Linking microbial contamination to food spoilage and food waste: The role of smart packaging, spoilage risk assessments, and date labeling. Frontiers in Microbiology 14 Frontiers Media SA. Available at https://doi.org/10.3389/fmicb.2023.1198124.

Kim, M.K., Lopetcharat, K., Drake, M.A., 2013. Influence of packaging information on consumer liking of chocolate milk. Journal of Dairy Science 96 (8), 4843–4856. Available at https://doi.org/10.3168/jds.2012-6399.

Klein, M., Werner, C., Tacker, M., Apprich, S., 2024. Influence of packaging design on technical emptiability of dairy products and implications on sustainability through food waste reduction. In Sustainability 16 (15), 6335. Available at https://doi.org/10.3390/su16156335.

Kontogianni, V.G., Kosma, I., Mataragas, M., Pappa, E., Badeka, A.V., Bosnea, L., 2023. Innovative intelligent cheese packaging with whey protein-based edible films containing Spirulina. Sustainability (Switzerland) 15 (18). Available at https://doi.org/10.3390/su151813909.

Kowalczyk, A., Sarwas, G., 2024. One-Shot Learning from Prototype Stock Keeping Unit Images. Information 15 (9), 526. Available at https://doi.org/10.3390/info15090526.

Kumar, L., Dutt, R., Gaikwad, K.K., 2025. Packaging 4.0: Artificial intelligence and machine learning applications in the food packaging industry. Current Food Science and Technology Reports 3 (1), 1–14. Available at https://doi.org/10.1007/s43555-025-00064-w.

Liao, Y., Zhang, R., Qian, J., 2019. Printed electronics based on inorganic conductive nanomaterials and their applications in intelligent food packaging. RSC Advances 9. Royal Society of Chemistry, pp. 29154–29172. Available at https://doi.org/10.1039/c9ra05954g 50.

Lima, S.P., Da Silva, D.D.M., Sobral, D., De Paula, J.C.J., De, M.M., Moreira, G., et al., 2026. Effect of packaging type on the formation of biogenic amines, physicochemical characteristics and texture profile of blue cheese. International Dairy Journal 174, 106510. Available at https://doi.org/10.1016/j.idairyj.2025.106510.

Liu, Y., Li, L., Yu, Z., Ye, C., Pan, L., Song, Y., 2023. Principle, development and application of time–temperature indicators for packaging. Packaging Technology and Science 36 (10), 833–853. Available at https://doi.org/10.1002/pts.2761.

Löfgren, M., Witell, L., 2005. Kano's theory of attractive quality and packaging. Quality Management Journal 12, 7. Available at https://doi.org/10.1080/10686967.2005.11919257.

Ma, L., Liu, Y., Xu, S., 2026. Enzyme-responsive glucoamylase microcapsule time-temperature indicator: Intelligent monitoring of yogurt freshness dynamics. Food Control 181, 111760. Available at https://doi.org/10.1016/j.foodcont.2025.111760.

Madhu, B., 2025. AI-driven food packaging systems: A new frontier in intelligent food safety and shelf-life management. Journal of Food Science 90 (12). Available at https://doi.org/10.1111/1750-3841.70716.

Maja Brozović Dorotea Kovačević, J.B., 2021. Consumer satisfaction with packaging materials: Kano model analysis approach. Tehnički Vjesnik 28 (4), 1203–1210. Available at https://doi.org/10.17559/TV-20200831093518.

Mirza Alizadeh, A., Masoomian, M., Shakooie, M., Zabihzadeh Khajavi, M., Farhoodi, M., 2022. Trends and applications of intelligent packaging in dairy products: A review. Critical Reviews in Food Science and Nutrition 62 (2), 383–397. Available at https://doi.org/10.1080/10408398.2020.1817847.

Mkhari, T., Adeyemi, J.O., Fawole, O.A., 2025. Recent advances in the fabrication of intelligent packaging for food preservation: A review. Processes 13 (2), 539. Available at https://doi.org/10.3390/pr13020539.

Musso, F., Colamatteo, A., Bravi, L., Pagnanelli, M.A., Murmura, F., Sansone, M., 2022. Analysis of factors affecting the purchase of private label products by different age consumers. British Food Journal 124 (13), 619–636. Available at https://doi.org/10.1108/BFJ-03-2022-0256.

Nichifor, B., Zait, L., Timiras, L., 2025. Drivers, barriers, and innovations in sustainable food consumption: A systematic literature review. Sustainability 17 (5), 2233. Available at https://doi.org/10.3390/su17052233.

Palanisamy, Y., Kadirvel, V., Ganesan, N.D., 2025. Recent technological advances in food packaging: Sensors, automation, and application. Sustainable Food Technology 3 (1), 161–180. Available at https://doi.org/10.1039/D4FB00296B.

Palazzo, M., Gigauri, I., & Ferri, M.A. (2023). Intelligent packaging: A strategy for boosting sustainable marketing and contributing to the sustainable development goals. In I. Gigauri, M. Palazzo, & M. Ferri (Eds.), *Handbook of research on achieving sustainable development goals with sustainable marketing* (pp. 1-17). IGI Global Scientific Publishing . https://doi.org/10.4018/978-1-6684-8681-8.ch001.

Pedersen, L., Bertelsen, A.S., Byrne, D.V., Kidmose, U., 2023. Sensory interactions between sweetness and fat in a chocolate milk beverage. Foods 12 (14), 2711. Available at https://doi.org/10.3390/foods12142711.

Perera, K.Y., Jaiswal, A.K., Jaiswal, S., 2023. Extending cheese shelf-life using eco-friendly sodium alginate-gelatin films reinforced with nanoclay. Food Bioscience 56. Available at https://doi.org/10.1016/j.fbio.2023.103304.

Pettigrew, S., Coyle, D., McKenzie, B., Vu, D., Lim, S.C., Berasi, K., et al., 2022. A review of front-of-pack nutrition labelling in Southeast Asia: Industry interference, lessons learned, and future directions. The Lancet Regional Health - Southeast Asia 3, 100017. Available at https://doi.org/10.1016/j.lansea.2022.05.006.

Piao, Z., Qiao, J., Liu, X., Ying, Y., Pan, P., Liu, Z., et al., 2026. A multifunctional nano-intelligent film based on self-assembled amphiphilic H-HPC/d-MWCNTs coatings grafted via H-HPC for goat milk shelf-life extension and freshness monitoring. Progress in Organic Coatings 211, 109811. Available at https://doi.org/10.1016/j.porgcoat.2025.109811.

Pollet, E., 2025. The future of milk in 2050: Dairy in the age of personalization. JDS Communications 6, S3–S8. Available at https://doi.org/10.3168/jdsc.2025-0859.

Prajapati, C., Tigga, A., Prajapati, R.J., Minz, P.S., & Scholar, P.D. (2025). *Revolutionizing Dairy with AI and Machine Learning*. Retrieved from https://www.researchgate.net/publication/394015794_Revolutionizing_Dairy_with_AI_and_Machine_Learning.

Qin, Z., Fang, W., Jiang, Q., Li, J., Zhang, H., 2024. The urchin-like gold nanoparticles/poly(e-caprolactone)/chitosan electrospun nanofibers for antibacterial active packaging. International

Journal of Biological Macromolecules 274, 133287. Available at https://doi.org/10.1016/j.ijbiomac.2024.133287.

Rahman, M., Sobhan, A., Sadak, O., 2026. Emerging trends in intelligent packaging for tackling food waste in the modern food supply chain. Trends in Food Science and Technology 167 Elsevier Ltd. Available at https://doi.org/10.1016/j.tifs.2025.105436.

Romero, A., Sharp, J.L., Dawson, P.L., Darby, D., Cooksey, K., 2021. Evaluation of two intelligent packaging prototypes with a pH indicator to determine spoilage of cow milk. Food Packaging and Shelf Life 30, 100720. Available at https://doi.org/10.1016/j.fpsl.2021.100720.

Roozbahani, S., Parsaeimehr, M., Javan, A.J., Hesarinejad, M.A., 2025. New application of smart packaging by using sumac extract for detecting lactic cheese aging. Applied Food Research 5 (2). Available at https://doi.org/10.1016/j.afres.2025.101163.

Roy, S., Rhim, J.W., 2021. Fabrication of cellulose nanofiber-based functional color indicator film incorporated with shikonin extracted from Lithospermum erythrorhizon root. Food Hydrocolloids 114. Available at https://doi.org/10.1016/j.foodhyd.2020.106566.

Ryynänen, M., Sirviö, P., Tanninen, P., Lindell, H., 2012. A productivity study of digital printing in the packaging industry. Packaging Technology and Science 25 (2), 119–124. Available at https://doi.org/10.1002/pts.960.

Sagar, N.A., Rani, N., 2026. Recent trends and innovations in smart and AI-based food packaging: A review. Frontiers in Food Science and Technology 5. Available at https://doi.org/10.3389/frfst.2025.1665055.

Sgroi, F., Salamone, P.A., 2022. Private label food products: Consumer perception and distribution strategies. Journal of Agriculture and Food Research 8, 100287. Available at https://doi.org/10.1016/j.jafr.2022.100287.

Shaban, I.A., Ajaj, R., Elshimy, H., 2026. Industry 4.0 enabled sustainable manufacturing. Sustainability 18 (156), 1–22 https://doi.org/Sustainabi/10.3390/su18010156.

Shrivastava, C., Schudel, S., Shoji, K., Onwude, D., da Silva, F.P., Turan, D., 2023. Digital twins for selecting the optimal ventilated strawberry packaging based on the unique hygrothermal conditions of a shipment from farm to retailer. Postharvest Biology and Technology 199, 112283. Available at https://doi.org/10.1016/j.postharvbio.2023.112283.

Siciliano, S., Lopresto, C.G., Carnì, D.L., Lamonaca, F., 2025. Optical gas sensors in smart food bio-packaging: Innovation for monitoring the product freshness and safety. Measurement: Food 19 Elsevier Ltd. Available at https://doi.org/10.1016/j.meafoo.2025.100245.

Silva, C., Santelli, R.E., Braz, B.F., Silva, M.M.N., Melo, L., Lemes, A.C., Taramuel-Taramuel, J.P., Aza-Fuelantala, O.E., Ader, D., Mayorga, A., Barrios, D., 2025. Technology adoption in smallholder dairy farms in Indigenous Pastos communities of Colombia. Journal of Agriculture and Food Research 23, 102191. Available at https://doi.org/10.1016/j.jafr.2025.102191.

Taramuel-Taramuel, J.P., Aza-Fuelantala, O.E., Ader, D., Mayorga, A., Barrios, D., 2025. Technology adoption in smallholder dairy farms in Indigenous Pastos communities of Colombia. Journal of Agriculture and Food Research 23, 102191. Available at https://doi.org/10.1016/j.jafr.2025.102191.

Turcan, R., Pojar, D., 2024. Impact of Industry 4.0 technologies on skill requirements and workforce availability in key sectors. Competitiveness and Sustainable Development 203–212. Available at https://doi.org/10.52326/csd2024.31.

UNCAD, United Nations Conference on Trade and Development. (2025). A review of parameters and requirements for an effective consumer label on plastics and plastics alternatives. Geneva,

Switzerland. UN trade & Development. Retrieved from https://unctad.org/system/files/official-document/tcsditcinf2025d7_en.pdf.

Vărzaru, A.A., Bocean, C.G., 2024. Digital transformation and innovation: The influence of digital technologies on turnover from innovation activities and types of innovation. Systems 12 (9), 359. Available at https://doi.org/10.3390/systems12090359.

Venkatesh, S., Steinberg, J., Ryan, C., 2025. Do food packaging health claims depict healthiness or mislead patients? PRiMER (Leawood, Kan.) 9, 41. Available at https://doi.org/10.22454/PRiMER.2025.586640.

Wagner, U., Charinsarn, A., 2020. What language should be displayed on product packaging? How Unconventional lettering influences packaging and product evaluation. Journal of International Consumer Marketing 33, 1–18. Available at https://doi.org/10.1080/08961530.2020.1741483.

Wang, J., 2025. Innovative practice of digital printing technology and computer image fusion in personalized packaging. Journal of Combinatorial Mathematics and Combinatorial Computing 1285–1302. Available at https://doi.org/10.61091/jcmcc127a-075 July 2024.

Wei, L., Zhan, S., Zhou, M., Xu, X., You, F., Zheng, H., 2024. A powerful strategy for carbon reduction: Recyclable mono-material polyethylene functional film. In Polymers 16 (15), 2196. Available at https://doi.org/10.3390/polym16152196.

Wikström, F., Williams, H., Verghese, K., Clune, S., 2014. The influence of packaging attributes on consumer behaviour in food-packaging life cycle assessment studies—A neglected topic. Journal of Cleaner Production 73, 100–108. Available at https://doi.org/10.1016/j.jclepro.2013.10.042.

Williams, H., Wikström, F., Löfgren, M., 2008. A life cycle perspective on environmental effects of customer focused packaging development. Journal of Cleaner Production 16 (7), 853–859. Available at https://doi.org/10.1016/j.jclepro.2007.05.006.

WRAP. (2007). Retail programme—Food waste: Final report, food behaviour consumer research: quantitative phase. WRAP.

Wu, W., Cronjé, P., Verboven, P., Defraeye, T., 2019. Unveiling how ventilated packaging design and cold chain scenarios affect the cooling kinetics and fruit quality for each single citrus fruit in an entire pallet. Food Packaging and Shelf Life 21, 100369. Available at https://doi.org/10.1016/j.fpsl.2019.100369.

Zhang, J., Krishna, A., 2007. Brand-level effects of stockkeeping unit reductions. Journal of Marketing Research 44 (4), 545–559. Available at http://www.jstor.org/stable/30162502.

Zuo, J., Feng, J., Gameiro, M.G., Tian, Y., Liang, J., Wang, Y., et al., 2022. RFID-based sensing in smart packaging for food applications: A review. Future Foods 6 Elsevier B.V. Available at https://doi.org/10.1016/j.fufo.2022.100198.

Part III

Current development and future landscape of dairy technology

Chapter 9

Precision dairy farming from sensor-based monitoring to socio-technical transformation

Moutaz Alhamada[1,2]
[1]*Istom, Higher School of International Agro-Development, Angers, France,* [2]*Department of Animal Husbandry, Faculty of Agronomy, Al-Furat University, Deir ez-Zur, Syria*

9.1 Introduction

The dairy sector is undergoing a profound transformation driven by advances in digital technologies, automation, and data-driven decision-making. Within this context, precision dairy farming (PDF) has emerged as a central component of the broader Industry 4.0 paradigm applied to animals' productions, aiming to improve productivity, animal health and welfare, labor efficiency, and sustainability through continuous monitoring and intelligent data use (Rutten et al., 2013; Barkema et al., 2015; Neethirajan and Kemp, 2021). These developments respond to increasing pressures on dairy systems to produce more efficiently while addressing societal expectations related to environmental performance, animal welfare, transparency, and food safety (Barkema et al., 2015; Hassoun et al., 2023; Lokhorst et al., 2019; Von Keyserlingk et al., 2024). PDF has evolved progressively over several decades rather than emerging as a single disruptive innovation. Early developments focused on mechanization and electronic recording of production parameters, followed by the introduction of sensor-based monitoring and automated systems such as activity sensors, milk quality sensors, and automatic milking systems (AMS) (Rutten et al., 2013; Hogenboom et al., 2019). More recently, the convergence of big data analytics, artificial intelligence (AI), cloud computing, and connected devices has positioned PDF within the framework of Dairy 4.0, emphasizing automation, connectivity, and optimization of farm processes (Hassoun et al., 2023; Kaur et al., 2023).

Despite rapid technological progress, the adoption and impact of PDF remain highly heterogeneous across farms and regions. Numerous studies report that the availability of large volumes of sensors data does not necessarily translate into improved management outcomes or sustained technology use

Dairy 4.0. DOI: https://doi.org/10.1016/B978-0-443-44730-3.00005-5

(Borchers and Bewley, 2015; Rojo-Gimeno et al., 2019). This discrepancy highlights a persistent gap between data generation and actionable decision-making, underscoring the importance of decision support systems (DSS) as the interface between digital analytics and on-farm action. Reviews consistently emphasize that many precision technologies underperform in practice because information is insufficiently contextualized, poorly prioritized, or inadequately aligned with farmer's decision processes (Rutten et al., 2013; Rojo-Gimeno et al., 2019). Furthermore, the development of PDF has been historically shaped by intensive indoor dairy systems, where stable connectivity, controlled environments, and regular data flows facilitate automation and digital monitoring (Marques et al., 2023; Lokhorst et al., 2019). In contrast, extensive and pasture-based dairy systems challenge these foundational assumptions due to spatial dispersion, environmental variability, and infrastructural constraints and, consequently, remain underrepresented in both research and commercial applications (Aquilani et al., 2022; Bocquier et al., 2014). This imbalance raises important questions regarding the transferability of existing precision technologies and the need for context-specific approaches.

In parallel, social and ethical dimensions of PDF have gained increasing attention. Farmer acceptance, changes in human–animal relationships, data governance, and societal trust are now recognized as critical determinants of successful and legitimate digital transformation in livestock systems (Borchers and Bewley, 2015; Vaarst et al., 2024; Von Keyserlingk et al., 2024). These perspectives emphasize that the challenges of PDF are not solely technological, but deeply sociotechnical, requiring alignment between digital systems, human expertise, and animal-centered management. Against this background, this chapter provides a comprehensive synthesis of PDF within the context of Industry 4.0. It first examines the historical evolution of PDF, highlighting key technological milestones and structural biases. It then reviews the core technological components of PDF, including sensors and IoT systems, robotics and automation, AI and machine learning (ML), and big data and cloud platforms. Particular attention is given to the role of DSS in translating data into practical value. The chapter further discusses the benefits and barriers of PDF adoption, addresses issues of traceability and trust through blockchain-based approaches, and situates PDF within an emerging transition from Dairy 4.0 toward Dairy 5.0, characterized by a shift toward human-centered, ethical, resilient, and inclusive digital systems (Hassoun et al., 2023; Von Keyserlingk et al., 2024). By integrating technological, socioeconomic, and conceptual perspectives, this chapter positions PDF not merely as a set of digital tools, but as a sociotechnical framework whose future development will depend as much on governance, human engagement, and contextual adaptation as on continued technological innovation.

9.2 Historical evolution of precision dairy farming

The evolution of PDF reflects broader trends in agricultural innovation, from early mechanization and basic electronic monitoring to contemporary digital ecosystems enabled by Industry 4.0 technologies. These historical overviews situate PDF within a continuum of technological and conceptual advancement, highlighting both its origins in intensive systems and its gradual relevance to extensive. The foundations of modern PDF emerged in the latter half of the 20th century, driven by a transition from manual, labor-intensive practices to mechanization and electronic data collection (Rutten et al., 2013; Barkema et al., 2015). Mechanization in dairy farming, including vacuum milking machines and motorized feeding, provided the first steps toward standardized, repeatable processes. These innovations laid the groundwork for individual animal monitoring by freeing labor and introducing standardized measurements (Rutten et al., 2013; Barkema et al., 2015; Bewley et al., 2017).

With the advent of electronic identification (EID) and RFID technologies, it became possible to link individual animals to digital records, enabling longitudinal tracking of production and health parameters. Initially developed for traceability and regulatory compliance, these systems soon became integral elements of herd management databases, enabling individual-level data collection across herds for the first time. Conceptually, this represented a shift from herd averages to individual animals monitoring, a foundational principle of PDF (Trapanese et al., 2025; Lokhorst et al., 2019). In the 1990s, the first sensor applications in dairy involved simple proxies such as electrical conductivity (EC) of milk, an early indicator of udder health, and basic behavioral sensors like pedometers for estrus detection. This early technologies, while limited in sophistication, introduced the idea of continuous, automated monitoring as an alternative to episodic human observation (Bernabucci et al., 2025; Trapanese et al., 2025; Borchers and Bewley, 2015). However, these innovations were predominantly implemented in intensive indoor systems, where infrastructure supported regular cow traffic through parlors or barns and facilitated power supply and data collection (Bewley et al., 2017). Pasture-based systems, with extensive grazing areas and limited connectivity, remained largely outside the ambit of early precision technologies (Alhamada et al., 2021; Aquilani et al., 2022; Bocquier et al., 2014).

The early 2000s witnessed the formalization of precision livestock farming (PLF) as a conceptual framework, emphasizing the continuous, automated monitoring of livestock to improve decision-making and management outcomes. This period coincided with the rise of digital data storage and the early spread of Internet-based farm management systems, enabling greater data accessibility and record keeping (Groher et al., 2020). Technological advancements in microprocessors, battery technology, and wireless communication enabled the widespread deployment of wearable accelerometers, rumination sensors, and automated milk meters. These devices provided time-series data on activity, feeding, and production patterns, which, although often analyzed with simple threshold-based methods, began to support more systematic decision support compared with traditional practices (Hussein et al., 2025; Vidal et al., 2023). Bibliometric analyses show that even in this early phase, the literature on PLF began growing steadily, with increasing attention to sensor integration, automated monitoring, and physiological parameter tracking (Vidal et al., 2023; Lokhorst et al., 2019). Nonetheless, adoption remained uneven, largely confined to contexts where technical infrastructure and economic incentives were favorable.

Perhaps the most transformative technological milestone in dairy farming was the development and commercialization of AMS. First introduced experimentally in the Netherlands in 1992, AMS technologies evolved throughout the 1990s and became commercially viable in the early 2000s, integrating robotics, sensors, and data management in a single platform (Hogenboom et al., 2019; Rebuli et al., 2015). AMS not only automated the milking process but also served as central data hubs, recording milking frequency, yield, milk composition, and cow activity. The integration of AMS with herd management software marked a qualitative shift: it enabled continuous, high-frequency data streams and informed decisions that extended beyond milking into broader aspects of herd health and productivity (Trapanese et al., 2025; Rebuli et al., 2015). Reviews of robotic milking systems (Marques et al., 2023) document significant adoption growth worldwide, driven by labor shortages, rising production costs, and the promise of enhanced management efficiency. The technology was adopted most rapidly in intensive, free-stall and compost-barn systems, largely in Europe, North America, and parts of Oceania, further entrenching intensive production as the early epicenter of PDF innovation (Marques et al., 2023).

The past decade has seen the convergence of multiple digital, physical, and biological technologies under the umbrella of Industry 4.0, prompting scholars to define the concept of Dairy 4.0, un digitally enabled, data-intensive paradigm of dairy production that leverages robotics, AI, the Internet of Things (IoT), cloud computing, and blockchain for enhanced management and traceability (Hassoun et al., 2023). Big data analytics and ML have transformed raw sensor data into predictive insights, supporting early detection of disease, nuanced reproductive management, and optimization of feeding strategies (Alhamada et al., 2016; Chi et al., 2016; Hussein et al., 2025). Our literatures review on PDF confirms the bibliometric reviews (Kamilaris et al., 2017; Lokhorst et al., 2019), which reveal a rapid increase in research outputs on PLF since the early 2000s, with a notable acceleration over the past decade, reflecting the growing integration of AI, IoT, and cloud computing in livestock systems. In parallel, the dairy industry has embraced data integration across domains, linking on-farm monitoring with genetic evaluations, supply-chain traceability, and sustainability assessment (Sajid et al., 2021; Hawashin et al., 2025). Recent reviews emphasize the potential of centralized data infrastructures, transparent data-sharing agreements, and global standards to support effective data use and cross-system interoperability, indicating maturation of the field beyond isolated sensor deployments (Brito et al., 2025).

Although precision technologies initially proliferated in intensive dairy systems, there is a nascent but growing body of research exploring their applicability to extensive and pasture-based production (González-García et al., 2018; Aquilani et al., 2022; Alhamada et al., 2017). Reviews of PLF in grazing systems highlight both opportunities and limitations: tools such as wearable sensors, global positioning system (GPS) tracking, remote sensing, and geospatial analytics can provide real-time behavioral and environmental information, while challenges persist in battery life, connectivity, and algorithm robustness in heterogeneous environments (Alhamada et al., 2021; Bocquier et al., 2014; Heins et al., 2023; Lovarelli et al., 2020). In extensive systems, technologies like virtual fencing and satellite monitoring are being trialed to optimize grazing management and reduce labor costs, opening new avenues for precision approaches in contexts where conventional automation is less feasible (González-García et al., 2018; Odintsov Vaintrub et al., 2021). Nonetheless, empirical adoption remains limited compared with indoor dairy contexts, underscoring a research gap and an opportunity for future innovation.

The historical evolution of PDF highlights a trajectory from mechanization and basic monitoring toward integrated, data-intensive digital ecosystems. As the field matures, current research emphasizes not only technological performance but also ethical, social, and sustainability dimensions, aligning PDF with broader goals of resilient and responsible food systems.

Emerging themes include the integration of sustainability metrics into decision support, the use of explainable AI to improve trust and transparency, and policy frameworks that support equitable technology diffusion across farms of different scales and contexts (Rojo-Gimeno et al., 2019; Lokhorst et al., 2019; Von Keyserlingk et al., 2024). Together, these developments points toward what some scholars describe as Dairy 5.0 (Fig. 9.1), a paradigm that integrates intelligent technologies with human expertise, animal welfare considerations, and societal expectations, fostering dairy systems that are not only productive but also sustainable and socially legitimate (Khanashyam et al., 2025; Vaarst et al., 2024; Hassoun et al., 2023).

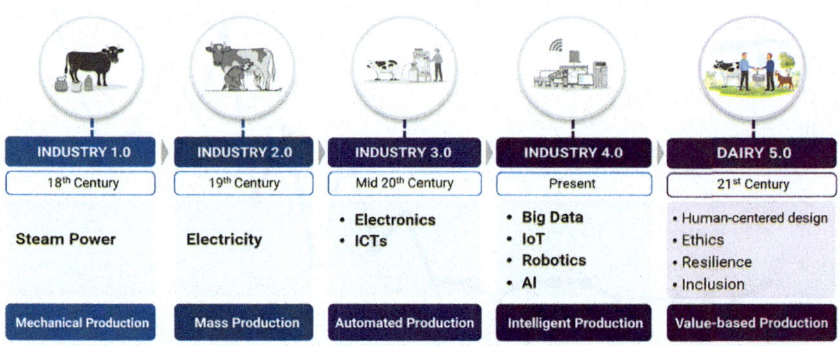

FIGURE 9.1 Evolution of dairy production systems from Industry 1.0 to Dairy 5.0, highlighting the transition from mechanized and automated production toward intelligent, human-centered, and value-based dairy systems.

To further illustrate the temporal dynamics of research activity in PDF, a complementary bibliographic screening was conducted using the ScienceDirect database with the search term "Precision Dairy Farming" applied to (Title, abstract or author-specified keywords). This screening identified 133 peer-reviewed articles published between 2001 and December 2025, revealing a clear growth in scientific output over time (Fig. 9.2). During the early 2000s, publications remained sparse, reflecting the exploratory and technology-driven nature of early research. From approximately 2010–2012 onward, the number of publications increased markedly, coinciding with the widespread adoption of AMS, the maturation of sensor technologies, and the emergence of data integration and analytics frameworks. The sustained acceleration observed in recent years highlights the consolidation of PDF as a distinct and rapidly evolving research field, increasingly aligned with the broader paradigms of Dairy 4.0 and digital agriculture. While this screening is limited to a single bibliographic database and should be interpreted as illustrative rather than exhaustive, it provides a clear indication of the growing scientific and technological interest in PDF over the past two decades.

The geographical distribution of affiliations associated with first and last authors further highlights a strong regional concentration of research activity. Prior to 2010, most publications originated from European countries and Japan, with recurring affiliations in Western and Northern Europe. This pattern reflects early adoption of AMS, strong research–industry linkages, and supportive policy environments for digital agriculture in these regions. After 2011, research activity expanded to a broader range of countries, including Brazil (5 articles), India (4 articles), China (8 articles), Turkey (2 articles), and

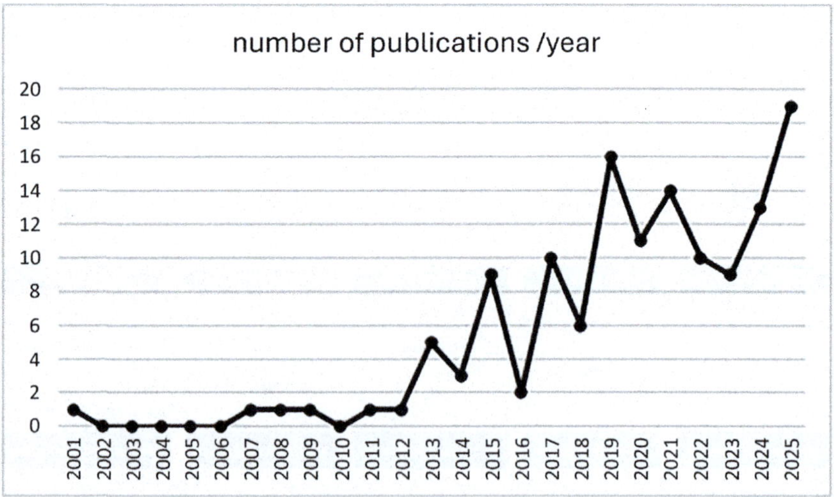

FIGURE 9.2 Temporal evolution of scientific publications on precision dairy farming indexed in ScienceDirect from 2001 to December 2025.

Iran (1 article), indicating a gradual geographical diversification of the field. Nevertheless, the majority of publications continue to originate from developed countries, and studies explicitly addressing pasture-based dairy systems, extensive production systems, or smallholder contexts remain comparatively limited.

Overall, this geographical imbalance mirrors the thematic distribution of the literature and underscores the dominant role of intensive indoor production systems in shaping the historical development of PDF research. Despite recent diversification, extensive livestock systems continue to be underrepresented, highlighting persistent research gaps that remain relevant for the future development of PDF within global and heterogeneous production contexts.

The distribution of main research topics further illustrates the structural orientation of the PDF literature (Fig. 9.3). A large proportion of studies fall into multicategory approaches, indicating a progressive shift toward integrative frameworks that combine multiple application domains rather than addressing isolated technologies or single management objectives. Among domain-specific studies, reproductive management and feeding and nutrition clearly dominate, reflecting their historical role as early and tangible entry points for PDF through activity-based monitoring, automated estrus detection, and feeding optimization systems.

Management and societal aspects also represent a substantial share of the literature, highlighting increasing attention to farm organization, decision-making processes, and human factors associated with digital technology adoption. In contrast, comparatively fewer studies focus on health and well-being, adoption, and economics, and particularly on environmental impacts and genetics and breeding, despite their recognized importance for long-term

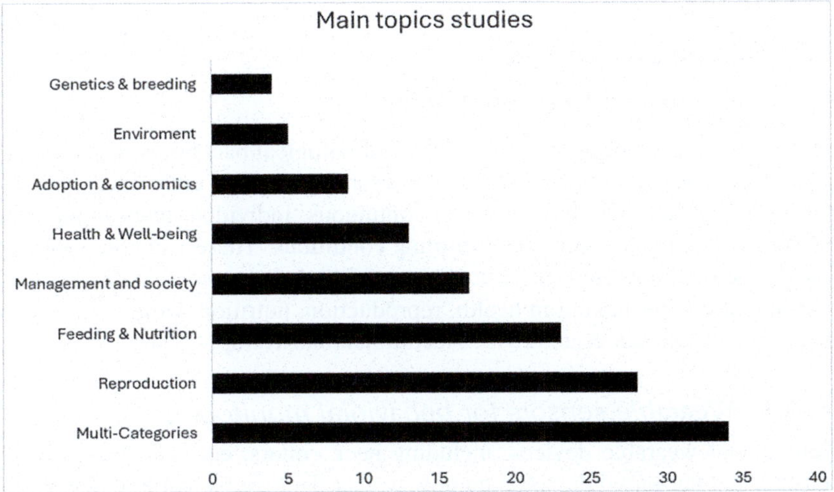

FIGURE 9.3 Distribution of main research topics in scientific publications on precision dairy farming.

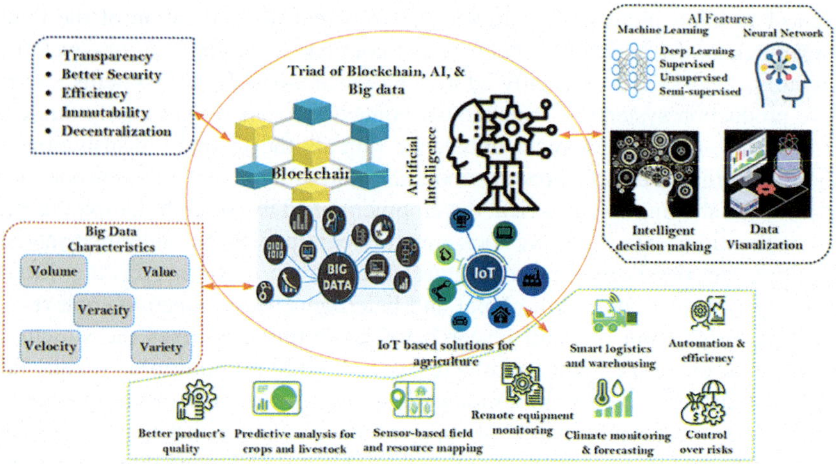

FIGURE 9.4 The figure synthesizes the integrated digital ecosystem of PDF, illustrating how IoT-based sensing, big data analytics, AI, and decision support systems interact to transform animal-level data into system-level intelligence, while also establishing the foundations for secure and transparent data exchange addressed in the following section.

sustainability and resilience of dairy systems. Overall, this thematic distribution confirms that PDF research has been primarily driven by operational and productivity-oriented objectives, while the integration of environmental, genetic, and economic dimensions has progressed more gradually, revealing persistent gaps that remain critical for future research and system-level innovation (Fig. 9.4).

9.3 Enabling technologies

9.3.1 Sensors and Internet of Things

Sensor-based monitoring constitutes the technological backbone of PDF. Over the past two decades, rapid advances in wearable sensors, embedded systems, and IoT infrastructures have enabled continuous, individual-level monitoring of dairy cattle under commercial farming conditions. These technologies have transformed animals into dynamic sources of high-frequency data, supporting real-time decision-making in health, reproduction, nutrition, welfare, and farm management (Rutten et al., 2013; Neethirajan and Kemp, 2021).

9.3.1.1 Wearable sensors for behavioral monitoring

Commercial wearable devices, including neck collars, ear tags, leg-mounted sensors, and tail-based devices, are now widely deployed to capture behavioral patterns such as grazing, rumination, activity, lying, standing, and locomotion. These systems typically integrate accelerometers and gyroscopes, sometimes complemented by magnetometers, to infer behavioral states through pattern recognition algorithms (Higaki et al., 2020; Odintsov Vaintrub et al., 2021; Bouchon et al. 2025).

Recent validation studies indicate that the sensitivity, specificity, and overall reliability of wearable behavioral sensors vary considerably across devices, algorithms, and production contexts, with performance strongly influenced by sensor placement, calibration procedures, and the environmental conditions under which data are collected (Rutten et al., 2013; Dolecheck et al., 2015; Bouchon et al., 2025). These findings are consistent with others work showing that dynamic behaviors with distinct movement signatures (e.g., grazing, estrus-related activity) are more reliably detected than static postures such as lying or standing (Dolecheck et al., 2015). Multisensor studies further demonstrate that behavioral changes detected by wearables are biologically meaningful. Dolecheck et al. (2015) showed that estrus events are characterized by a 200%–300% increase in activity, accompanied by marked reductions in rumination and lying time, patterns consistently captured across several sensor platforms (Dolecheck et al., 2015; Alhamada et al., 2016). Similarly, long-term behavioral trajectories derived from accelerometers have been shown to precede clinical disease events, reinforcing the role of wearables as early-warning systems rather than simple activity trackers (Mardhati et al., 2021).

Beyond cattle, studies in small ruminants demonstrate the broader relevance of behavioral sensing. Automated detection of mating and estrus-related behaviors in sheep, validated against continuous visual observation, illustrates that complex reproductive behaviors can be quantified reliably under real farming conditions, foreshadowing applications transferable to indoor dairy systems (Alhamada et al., 2016; Alhamada et al., 2017).

9.3.1.2 *Milk sensors and in-line monitoring systems*

Milking systems, particularly AMS, serve as central hubs for IoT-based data acquisition. In-line milk sensors routinely measure milk yield, flow rate, EC, somatic cell count (SCC), fat, protein, lactose, and milk temperature, generating rich longitudinal datasets at each milking event.

Numerous studies confirm the value of these sensors for early detection of udder health disorders. Changes in EC, milk flow asymmetry, and SCC dynamics can signal subclinical mastitis days before clinical signs or laboratory confirmation, enabling earlier and more targeted interventions (Hogenboom et al., 2019; Rutten et al., 2013). However, sensor performance varies widely depending on algorithms, gold standards, and farm conditions, and no single indicator currently meets stringent diagnostic thresholds on its own, highlighting the need for multiparameter integration (Rutten et al., 2014). From a technological perspective, the foundations of modern milk sensing date back more than two decades. Early work using near-infrared spectroscopy (NIRS) demonstrated that milk spectra could reliably estimate SCC, milk urea nitrogen (MUN), and metabolic indicators at the individual cow level, anticipating today's embedded milk analyzers in AMS (Tsenkova and Toyoda, 2001). These early studies already framed milk as a noninvasive integrative biomarker of whole-animal physiology, a concept that remains central to contemporary PDF.

9.3.1.3 Physiological sensing: temperature, weight, and internal status

Physiological sensors complement behavioral and milk-based measurements by capturing internal states that are otherwise difficult to observe. Endo-ruminal boluses continuously measure internal temperature, enabling detection of fever, calving, heat stress, and metabolic disturbances. Studies have shown that ruminal temperature increases of as little as 0.4°C–0.5°C can be associated with estrus or disease events, although environmental and drinking-related confounders necessitate careful signal processing (Bumbálek et al., 2026; Higaki et al., 2020; Dolecheck et al., 2015).

Similarly, automated weighing systems, including stationary and walk-over scales, provide frequent live weight measurements. However, short-term weight fluctuations are largely driven by gut fill and water balance, rather than changes in body tissue, limiting their direct interpretability without advanced filtering and contextual data integration (Mardhati et al., 2021). These findings underscore a recurring theme in PDF: sensor data are biologically meaningful only when interpreted through appropriate models and temporal frameworks.

9.3.1.4 Internet of Things infrastructure and data integration

Collectively, wearable sensors, milk analyzers, and physiological monitoring devices generate dense, high-velocity data streams, which are transmitted via IoT infrastructures (e.g., Bluetooth, LoRa, Wi-Fi, cellular networks) to on-farm servers or cloud-based platforms. These systems increasingly rely on edge and fog computing to preprocess data locally, reduce latency, and ensure robustness under variable connectivity conditions (Kaur et al., 2023; Neethirajan and Kemp, 2021). Despite this technological maturity, several reviews emphasize that most commercial systems remain confined to data collection and basic alert generation, corresponding to early maturity levels of sensor-based decision support (Rutten et al., 2013). Interoperability between devices, ownership of data, and meaningful integration across behavioral, physiological, and production domains remain major challenges, limiting the full realization of IoT-enabled PDF.

Overall, from sensing to value creation, sensor and IoT technologies have enabled an unprecedented level of individual animal monitoring in dairy systems. They provide the raw material for early disease detection, reproductive management, welfare assessment, and productivity optimization.

However, evidence across multiple studies consistently shows that the mere availability of sensor data does not guarantee improved outcomes. The value of these technologies depends critically on data quality, preprocessing, integration across sensors, and translation into actionable, user-centered decision support (Rojo-Gimeno et al., 2019; Borchers and Bewley, 2015). This growing body of literature highlights that sensors and IoT form a necessary, but not sufficient, foundation for PDF, setting the stage for advanced analytics, AI, and decision-support systems addressed in subsequent sections.

9.3.2 Robotics and automation

Robotics and automation represent one of the most mature and visible dimensions of PDF, with AMS constituting the flagship application. Over the last two decades, AMS have transitioned from experimental technologies to fully commercialized systems that fundamentally reshape labor organization, animal–human interactions, and data generation in dairy farms.

9.3.2.1 Automatic milking systems as sociotechnical systems

A recent global scoping review of AMS research highlighted a strong geographical concentration of scientific output, with Europe accounting for approximately 73% of published studies, reflecting both early adoption and dense research–industry linkages in this region (Marques et al., 2023), with 16% for North America and 10% for Asia and Australia/Oceania.

Beyond automation of milking itself, AMS function as integrated cyber-physical systems, combining robotics, sensors, and software platforms. Each milking event generates continuous streams of data on milk yield, flow rate, electrical conductivity, SCC, and milk composition, transforming the milking process into a central node for real-time monitoring of animal health and productivity (Tse et al., 2018; Marques et al., 2023). This high-frequency data acquisition enables detection of deviations at the individual cow level that would be difficult or impossible to observe in conventional milking systems.

9.3.2.2 Labor efficiency, work organization, and adoption drivers

One of the most consistently documented benefits of AMS adoption is the reduction and reorganization of labor, rather than its elimination. Longitudinal analyses of technology adoption show that robotic milking is primarily adopted as a labor-saving and lifestyle-enhancing innovation, especially in contexts characterized by increasing herd sizes and limited availability of skilled labor (Yang et al., 2021; Gargiulo et al., 2018).

Generally, AMS adoption has been associated with increased flexibility in daily schedules, and reduced physical workload, in the same time improved farmer well-being and job satisfaction, while maintaining or increasing milk production per cow (Yang et al., 2021; Borchers and Bewley, 2015).

Importantly, adoption studies demonstrate that AMS are rarely implemented in isolation. Farms equipped with AMS tend to adopt clusters of complementary technologies, including electronic identification, automated sorting gates, and herd management software, reinforcing the role of AMS as a gateway technology within broader PDF ecosystems (Marques et al., 2023; Yang et al., 2021; Gargiulo et al., 2018).

On the far side of milking, a growing range of robotic peripherals contributes to automation in dairy farms. Robotic calf feeders allow individualized milk allowance and feeding frequency, generating detailed data on intake patterns that can be linked to health and growth trajectories. Genetic and phenotypic studies have shown that such data can be used to characterize feeding behavior and disease susceptibility, particularly for respiratory disorders in calves (Jin et al., 2024; Mahato and Neethirajan, 2025). Similarly, robotic floor cleaners and manure scrapers reduce manual labor and contribute to improved barn hygiene, indirectly affecting animal welfare and udder health. These systems are particularly relevant in AMS-equipped barns, where cow traffic is continuous and hygiene conditions can change rapidly. (Jin et al., 2024; Vidal et al., 2023).

9.3.2.3 *Milk quality, animal welfare, and technological trade-offs*

While AMS offer clear advantages, their impacts are not uniformly positive and depend strongly on management practices. Comprehensive reviews of milk produced under robotic milking indicate that milk composition and hygienic quality are generally comparable to conventional systems, provided that hygiene protocols and equipment maintenance are well managed (Hogenboom et al., 2019; Bianchi et al., 2022). However, AMS are often associated with an increased variability in milking intervals, with slightly elevated free fatty acid (FFA) levels, in the same time of transient increases in SCC during the adaptation phase. These effects are attributed not to the technology itself but to changes in milking frequency, cow behavior, and management routines, reinforcing the notion that AMS are management-intensive systems rather than "plug-and-play" solutions (Hogenboom et al., 2019; Tse et al., 2018). From an animal welfare perspective, AMS can enable more animal-driven milking behavior, but they also require careful design of barn layout and traffic flow to avoid competition, uneven milking frequency, and social stress, particularly in large herds (Barkema et al., 2015; Trapanese et al., 2025).

9.3.2.4 *Economic and energetic constraints*

Despite their benefits, AMS remain associated with substantial capital investment, maintenance costs, and energy consumption. High upfront costs represent a major barrier to adoption, This may explain its spread in developing countries, as shown in a study of (Marques et al., 2023), particularly for small- and medium-sized farms, contributing to structurally uneven diffusion of robotic technologies across the dairy sector (Gargiulo et al., 2018; Borchers and Bewley, 2015). Energy use is another emerging concern. AMS and associated robotic systems operate continuously and rely on electrically powered actuators, sensors, and data infrastructure. Rising energy prices and sustainability objectives are therefore pushing the sector toward more energy-efficient robotic designs and smarter control strategies, including data-driven optimization of milking frequency and robot utilization (Marques et al., 2023).

Overall, AMS and related robotic technologies function as data-generating and data-structuring platforms, enabling continuous monitoring at a level of granularity that underpins advanced analytics, decision support, and long-term sustainability strategies. However, as emphasized across multiple studies, the value of robotic systems depends critically on their integration with sensors, analytics, and human decision-making, rather than on automation alone (Rojo-Gimeno et al., 2019; Rutten et al., 2013).

9.3.3 Artificial intelligence and machine learning

AI and ML have become central components of PDF, enabling the transformation of large, heterogeneous sensor datasets into actionable information. As sensor density and data velocity increase, AI methods play a critical role in detecting complex, nonlinear patterns that precede health disorders, characterize individual animals, and support long-term genetic and management decisions (Khanashyam et al., 2025).

9.3.3.1 Disease detection and early warning systems

One of the most advanced and widely studied applications of AI in PDF is early disease detection. Supervised ML models have been developed for a range of health conditions, including mastitis, metritis, lameness, and infectious diseases, using behavioral, physiological, and milk-derived data streams.

For udder health, ML models trained on milk flow dynamics and in-line sensor data have demonstrated superior performance compared with traditional laboratory indicators. Jin et al. (2024) reported that gradient boosting and random forest models applied to milk flow profiles achieved an area under the curve (AUC) of 0.793 for subclinical mastitis detection, outperforming single-indicator approaches based on SCC alone (Hogenboom et al., 2019; Bumbálek et al., 2026; Borchers and Bewley, 2015). These findings are consistent with earlier reviews showing that no single sensor meets diagnostic gold standards, whereas multivariate ML approaches substantially improve sensitivity–specificity trade-offs (Rutten et al., 2013).

Similarly, Vidal et al. (2023) demonstrated that ML performance in detecting metritis depends as much on data preprocessing and temporal feature extraction as on algorithm choice. Across multiple algorithms (random forest, gradient boosting, support vector machines), predictive accuracy varied widely depending on how behavioral trajectories were aligned relative to calving, highlighting that AI performance in PDF is critically linked to biological context and data engineering, not only model complexity (Graham et al., 2024; Mardhati et al., 2021).

Beyond high-input systems, AI-based disease detection has also been explored in resource-limited contexts. ML frameworks for infectious disease detection in smallholder systems demonstrate that AI can support animal health monitoring even where veterinary infrastructure is weak, although challenges related to data quality, validation, and scalability remain substantial (Rojo-Gimeno et al., 2019; Borchers and Bewley, 2015). Together, these studies indicate that AI-driven disease detection is one of the most mature PDF applications, yet remains highly sensitive to data quality, labeling strategies, and farm-specific conditions.

9.3.3.2 *Behavioral modeling and event prediction*

AI and ML are particularly well suited to modeling temporal behavioral dynamics, enabling prediction of discrete biological events rather than simple state classification. In reproductive management, supervised ML models integrating behavioral and physiological time series have achieved high accuracy in predicting estrus and calving events. For example, ML models combining accelerometer-derived activity, rumination patterns, and temperature data have consistently outperformed visual observation for estrus detection, achieving accuracies exceeding 90% in controlled validation studies (Dolecheck et al., 2015; Higaki et al., 2020). Importantly, these models capture multisignal biological signatures, reinforcing that estrus and calving are not single-variable phenomena but emerge from coordinated behavioral and physiological changes. These findings align with broader methodological insights showing that trajectory-based ML approaches, which model deviations from individual baselines, are more robust than static threshold-based systems, particularly in large herds with high inter-animal variability (Mardhati et al., 2021; Rutten et al., 2013).

9.3.3.3 *Computer vision and biometric identification*

Another rapidly expanding AI application in PDF is computer vision for individual identification and phenotyping. Deep learning approaches, particularly convolutional neural networks (CNNs), have enabled noninvasive, automated identification of cows using facial, body, or 3D morphological features.

Recent studies demonstrate that biometric facial recognition systems can identify individual cows with accuracies exceeding 90%, even under variable lighting and positioning conditions, providing a humane alternative to physical identification methods (Mahato and Neethirajan, 2025; O' Mahony et al., 2019). Advances in meta-learning and few-shot learning further reduce data requirements: models trained on a limited number of images per animal can generalize effectively, making deployment feasible under commercial farm conditions where labeled datasets are scarce (Neethirajan and Kemp, 2021; Mahato and Neethirajan, 2025).

Beyond identification, AI-driven vision systems support automated body condition scoring, lameness detection, and morphological monitoring, reinforcing the role of computer vision as a core sensing modality within PDF ecosystems (O' Mahony et al., 2019; Rutten et al., 2013).

9.3.3.4 AI-enabled genetic prediction and long-term resilience

AI methods are increasingly applied to genetic evaluation and resilience phenotyping, leveraging high-frequency behavioral data that were previously inaccessible. Data from automated milk feeders and activity sensors provide detailed records of feeding behavior, intake dynamics, and growth trajectories, which have been shown to exhibit moderate heritability. Studies using these data demonstrate strong genetic correlations between feeding behavior traits and health and resilience outcomes, including resistance to bovine respiratory disease in calves (Hassoun et al., 2023; Mahato and Neethirajan, 2025). These results support the integration of AI-derived phenotypes into genomic selection programs, expanding breeding objectives beyond production traits toward robustness, welfare, and longevity (Barkema et al., 2015; Mahato and Neethirajan, 2025).

9.3.3.5 From data to decisions: the central role of decision support systems

While sensors, (AI), and big data analytics constitute the technological backbone of PDF, their practical value ultimately depends on the effectiveness of decision support systems (DSS). In this context, AI and ML do not replace DSS, but rather function as analytical engines within DSS, transforming raw, high-frequency data into predictions, classifications, and risk scores. DSS represent the critical interface that contextualizes these AI-generated outputs and translates them into recommendations that can be understood, trusted, and acted upon by farmers. A persistent challenge in PDF is the gap between data availability and actionable decision-making. Early reviews already emphasized that many sensor-based technologies fail to deliver sustained value because they generate information without providing clear guidance on how, when, or whether to intervene (Rutten et al., 2013). More recent analyses confirm that increasing data volume and algorithmic sophistication, often driven by advanced ML models, does not automatically improve decision quality, particularly when alerts are poorly prioritized, insufficiently contextualized, or misaligned with farm management objectives (Rojo-Gimeno et al., 2019). Within DSS, AI and ML play a supportive and enabling role, contributing to pattern recognition, anomaly detection, and prediction of health, reproduction, or production events. However, effective DSS must extend beyond predictive accuracy and address key functional requirements, including timeliness, interpretability, transparency, and integration into existing management workflows. Studies on technology adoption consistently show that farmers value systems that enhance decision confidence and situational awareness rather than replacing human judgment, reinforcing the importance of human-centered DSS design (Borchers and Bewley, 2015).

In this perspective, decision support systems (DSS) act as mediators between automated analytics and experiential knowledge, enabling hybrid decision-making processes in which AI-generated insights are interpreted and combined with farmer expertise and contextual understanding (Rutten et al., 2013; Rojo-Gimeno et al., 2019). For example, predictive models may identify an increased risk of mastitis or metabolic disorders based on sensor-derived indicators, but effective DSS contextualize these predictions by integrating lactation stage, production objectives, and farm-specific management constraints before recommending targeted interventions. Such mediation is essential to ensure that algorithmic outputs support, rather than override, human judgment. Evidence from adoption studies suggests that the limited uptake or abandonment of some precision dairy technologies is often not attributable to deficiencies in sensing or ML algorithms, but rather to the absence of robust, interpretable, and user-oriented DSS architectures (Borchers and Bewley, 2015; Rojo-Gimeno et al., 2019). Systems that generate frequent, poorly prioritized alerts or provide limited explanation of underlying risk factors tend to erode user trust and decision confidence, even when predictive performance is high. Conversely, DSS that emphasize clarity, transparency, and relevance to daily management routines are more likely to be perceived as valuable and sustainably adopted. This perspective reframes DSS not as auxiliary tools, but as central components of PDF systems, essential for converting technological capability into measurable economic, animal welfare, and sustainability outcomes (Rutten et al., 2013; Lokhorst et al., 2019). As PDF evolves toward more integrated and value-oriented paradigms within Dairy 4.0 and beyond, the explicit development of transparent, adaptive, and human-centered DSS—including explainable analytics and participatory design approaches—will remain a key determinant of successful and legitimate implementation (Vaarst et al., 2024; Von Keyserlingk et al., 2024).

9.3.3.6 Limitations and the need for integrative AI frameworks

Despite their promise, AI and ML applications in PDF face well-documented limitations. Model performance often declines when transferred across farms, breeds, or management systems, reflecting context dependency and limited external validation (Graham et al., 2024; Rutten et al., 2013). Moreover, high predictive accuracy does not automatically translate into management impact if alerts are poorly timed, insufficiently interpretable, or ignored by users (Rojo-Gimeno et al., 2019; Borchers and Bewley, 2015).

Consequently, recent literature emphasizes the need for hybrid approaches combining AI with mechanistic models, domain knowledge, and explicit decision rules. Such integrative frameworks aim to move AI applications in PDF from pattern recognition tools toward robust, transparent, and value-generating decision-support systems (Graham et al., 2024; Barkema et al., 2015). This emerging application illustrates a key strength of AI in PDF: its ability to bridge short-term decision support and long-term system sustainability, linking real-time monitoring with genetic improvement strategies.

9.3.4 Big data and cloud platforms

Big data analytics is widely regarded as the computational backbone of Dairy 4.0, enabling the transition from sensor-rich environments to integrated, data-driven decision-making systems. In PDF, big data emerges from the continuous aggregation of heterogeneous data streams generated by sensors, robotic systems, genomic tools, remote sensing, and farm management software, increasingly processed and stored through cloud-based platforms.

A seminal review of 142 studies (Lokhorst et al., 2019) highlighted that 83% of big data applications in livestock farming focus on animal-level analytics, particularly health and reproduction, while farm-, network-, and value-chain–level integration remains marginal (Groher et al., 2020; Hussein et al., 2025; Trapanese et al., 2025). This structural imbalance reflects both the maturity of individual-animal monitoring and the unresolved challenges associated with interoperability, governance, and system-level analytics.

9.3.4.1 The five "V"s of big data in precision dairy farming

Big data in agriculture and livestock systems is commonly described through five core dimensions: Volume, Velocity, Variety, Veracity, and Value. While these dimensions are well established conceptually, empirical evidence shows that they are unevenly addressed in PDF.

Volume: Modern dairy systems generate massive data volumes. AMS alone produce thousands of records per cow per year, including milk yield curves, electrical conductivity, milking frequency, and behavioral metadata (Hogenboom et al., 2019). When combined with high-frequency accelerometer data, feeding records, genetic information, and historical health events, data volumes quickly exceed the processing capacity of traditional on-farm software. Similar volume-related challenges have been documented across agriculture, where the scale of data generation necessitates distributed storage and computing infrastructures (Kamilaris et al., 2017; Chi, et al., 2016).

Variety: Data variety is one of the most actively exploited dimensions in PDF. Dairy big data integrates multisource, multitemporal, and multiresolution datasets, including animal-mounted sensors, in-line milk analyzers, environmental sensors, genomic data, and increasingly, remote sensing and contextual farm-level information (Dolecheck et al., 2015; Kamilaris et al., 2017). Reviews of agricultural big data consistently emphasize that value arises not from individual data streams but from their cross-domain integration, which allows complex biological and management processes to be modeled more realistically (Hussein et al., 2025).

9.3.4.1.1 Velocity

Despite the real-time nature of many dairy sensors, velocity remains the least exploited dimension. Most analytics pipelines operate in batch or near-batch modes, delivering delayed alerts rather than real-time decision support. This gap mirrors broader findings in agriculture and Industry 4.0, where data are generated at high speed but processed too slowly to support time-critical interventions (Sajid et al., 2021; Rojo-Gimeno et al., 2019). In dairy systems, this limitation reduces the effectiveness of early-warning applications for health and welfare.

9.3.4.1.2 Veracity

Data veracity represents a major bottleneck for dairy big data. Sensor noise, calibration drift, missing values, and farm-specific biases can substantially degrade analytical performance. Across both livestock and broader agricultural applications, studies consistently show that data preprocessing, validation, and contextualization have a larger impact on outcomes than algorithm choice (González-García et al., 2018; Alhamada et al., 2017; Kamilaris et al., 2017). Low veracity undermines trust in analytics and contributes to alert fatigue among farmers (Trapanese et al., 2025; Brito et al., 2025).

9.3.4.1.3 Value

Finally, value creation remains the most elusive dimension. While large volumes of data are collected and stored, their translation into measurable economic, welfare, or environmental benefits is inconsistent (Pardo et al., 2022). Conceptual frameworks emphasize that data only generate value when embedded into decision-making processes that are timely, interpretable, and aligned with user needs (Rojo-Gimeno et al., 2019; Hussein et al., 2025).

9.3.4.2 Cloud platforms as core enablers of Dairy 4.0

Cloud computing has become the dominant infrastructure for managing dairy big data, offering scalable storage, high-performance computing, and remote accessibility. Cloud platforms support several support like aggregation of heterogeneous data streams, in addition to centralized model deployment and updating, and in the same time the cross-farm benchmarking and advisory services. However, as highlighted in both agricultural and Industry 4.0 literature (Sajid et al., 2021; Neethirajan and Kemp, 2021), most existing platforms are vertically integrated and vendor-specific, limiting interoperability and data portability. Similar challenges have been observed in predictive maintenance systems in Industry 4.0, where siloed architectures constrain system-wide optimization despite abundant data availability (Sajid et al., 2021; Hussein et al., 2025).

However, Interoperability is now widely recognized as a critical bottleneck for big data in agriculture and livestock systems. a few project provides a concrete response to this challenge by proposing federated, open architectures that allow heterogeneous systems to exchange data while preserving farmer control and data sovereignty (Yalçin., 2025). These principles are consistent with broader agricultural big data reviews, which emphasize the need for standardized data models and APIs, with semantic interoperability, in the same time to transparent governance frameworks defining data ownership and access rights (Kamilaris et al., 2017; Neethirajan and Kemp, 2021), Without this three frameworks, cloud-based dairy platforms risk reinforcing data monopolies rather than enabling collaborative innovation.

9.3.4.3 *From animal-level data to system-level intelligence*

Despite substantial technological advances, most applications of big data in dairy farming remain largely confined to zootechnical and individual–animal optimization, such as disease detection, health monitoring, or reproductive management (Alhamada et al., 2016; Graham et al., 2024; Magrin et al., 2026). In contrast, relatively few studies address whole-farm optimization, including labor organization, energy use, or greenhouse gas emissions, or explore multifarm learning, benchmarking, and collective intelligence across production systems (Kamilaris et al., 2017; Hussein et al., 2025). This pattern mirrors broader findings across agriculture, where big data adoption often remains fragmented and underutilizes key dimensions such as velocity, veracity, and value creation. In theory, cloud platforms provide the infrastructure required to scale data integration from the animal level to the farm, regional, or network level. However, empirical implementations of such system-level intelligence in dairy farming remain limited, with few operational examples demonstrating real-time coordination across multiple data streams and decision domains. As a result, the transformative potential of cloud-based analytics in dairy systems has yet to be fully realized.

Overall, big data and cloud platforms constitute the technical foundation of Dairy 4.0, enabling large-scale data aggregation, storage, and advanced analytics. Nevertheless, evidence from dairy farming, agriculture, and Industry 4.0 consistently indicates that big data alone does not guarantee improved decision-making or sustainability outcomes. Key challenges increasingly relate to data quality, contextualization, interoperability, and real-time analytical capacity, rather than data volume per se (Brito et al., 2025; Hussein et al., 2025). These limitations position big data not as an end in itself, but as a necessary yet insufficient enabler of truly intelligent, resilient, and sustainable dairy systems.

9.3.5 Blockchain and traceability

Blockchain technologies are increasingly explored in dairy supply chains as a means to enhance traceability, transparency, accountability, and consumer trust in increasingly complex and globalized production networks. Unlike conventional centralized databases, blockchain provides an immutable, distributed ledger in which transactions and events are cryptographically secured and shared among authorized stakeholders, reducing information asymmetry and the risk of fraud.

In the dairy sector, blockchain applications primarily target product traceability and certification assurance, including the verification of organic labels or geographical indications, the tracing of milk batches back to individual farms, and the documentation of compliance with quality and food safety standards. Reviews of agrifood blockchain applications consistently identify the dairy sector as a particularly relevant use case, due to the perishability of milk, the multiplicity of actors along the value chain, and the high sensitivity of consumers to food safety, origin, and authenticity claims (Hassoun et al., 2023; Malik et al., 2026; Tran et al., 2025). Concrete use cases include end-to-end traceability systems linking on-farm production data with processing, distribution, and retail stages, as well as tamper-proof recording of animal health treatments, milk quality indicators, and processing operations. Blockchain-based architectures have also been proposed for cold-chain monitoring, enabling secure and immutable logging of temperature data for dairy products that are particularly sensitive to storage and transport conditions (Magrin et al., 2026; Hassoun et al., 2023). Beyond technical traceability, such systems aim to strengthen trust and coordination among stakeholders, not only between producers and consumers, but also among cooperatives, processors, distributors, and regulatory authorities. This trust-enhancing function is especially critical in premium, certified, or protected markets, where transparency, auditability, and data integrity are central to market access and value creation (Malik et al., 2026; Tran et al., 2025). Beyond traceability, recent blockchain research emphasizes accountability as a central mechanism for improving sustainability outcomes. A detailed blockchain-based framework for dairy supply chains demonstrated how smart contracts can enforce commitments at each stage of production, distribution, and retail, enabling precise identification of overproduction, losses during transport, and under-consumption at the retail level (Hawashin et al., 2025).

By recording all actions as immutable events and linking them to specific actors, blockchain systems can quantify where and when dairy waste occurs, and assign responsibility for waste generation, in the same time to incentivize better coordination between supply and demand.

This represents a conceptual shift from traceability as a passive documentation tool toward active governance of supply-chain behavior, directly aligned with environmental and economic sustainability objectives (Malik et al., 2026; Hawashin et al., 2025).

While early blockchain applications focused on simple transaction logging, recent developments emphasize hybrid architectures combining blockchain with IoT, cloud platforms, and AI. Proof-of-concept studies in the dairy industry show that blockchain can act as a trusted backbone, while AI models process sensor and operational data off-chain to generate predictions and alerts, which are then securely recorded and shared through the ledger

Such hybrid blockchain–AI systems address key limitations of standalone blockchain solutions, notably scalability constraints, high transaction costs, and limited analytical capabilities. By using off-chain storage and computation (e.g., cloud platforms, IPFS), these architectures maintain data integrity and provenance while enabling advanced analytics, making them more compatible with data-intensive Dairy 4.0 environments (Rojo-Gimeno et al., 2019).

Despite their potential, blockchain applications in dairy remain largely confined to pilot projects and proof-of-concept implementations. Large-scale adoption faces significant challenges related to (i) interoperability with existing farm management and cloud platforms; (ii) alignment of incentives among farmers, cooperatives, processors, and retailers; (iii) regulatory acceptance and standardization; and (iv) technical complexity and energy consumption as mentioned by Tran et al. (2025) and Malik et al. (2026). Holistic assessments of blockchain-based traceability systems, including protected designation of origin (PDO) dairy products, emphasize that technological feasibility alone is insufficient. Successful implementation requires institutional coordination, clear governance frameworks, and shared standards for data access and ownership. Overall, blockchain should be viewed not as a standalone solution, but as an enabling layer within broader Dairy 4.0 ecosystems. When integrated with sensors, big data platforms, and AI-driven DSS, blockchain can enhance trust, accountability, and transparency across the dairy value chain. However, its added value depends critically on interoperability, governance, and user acceptance, rather than on the technology itself.

As such, blockchain represents a promising, put still emerging, tool for addressing traceability, sustainability, and trust challenges in modern dairy systems, complementing rather than replacing existing digital farming technologies.

9.4 Benefits of precision dairy farming

Beyond the technological components detailed in previous sections, PDF delivers a broad set of benefits that extend from farm-level performance to societal trust and long-term sustainability. Rather than representing a collection of isolated digital tools, PDF fundamentally reshapes how decisions are made, how animals are managed, and how dairy systems interact with their economic, environmental, and social environments. These benefits are multidimensional and interdependent, reflecting the systemic nature of Industry 4.0 transformations.

9.4.1 Improved productivity through decision efficiency

One of the most immediate benefits of PDF lies not simply in increased milk yield, but in enhanced decision efficiency. By providing continuous, individualized information on animal behavior, production, and health status, PDF enables a shift from reactive management toward anticipatory and adaptive decision-making. This transition allows producers to better align feeding strategies, milking frequency, and health interventions with the physiological and productive status of each animal, thereby reducing inefficiencies and variability within the herd (Rutten et al., 2013; Rojo-Gimeno et al., 2019; Borchers and Bewley, 2015).

Empirical studies demonstrate that precision feeding and decision-support tools can improve income over feed costs by optimizing nutrient allocation at the individual or group level. Precision-fed diets have been shown to generate tangible economic gains, with reported savings exceeding one dollar per cow per day while maintaining or improving milk production (Campos et al., 2023). Importantly, these gains arise not from intensification alone, but from improved targeting and timing of inputs, highlighting the role of PDF in enhancing efficiency by doing better with the same or fewer resources rather than simply increasing production intensity (Gargiulo et al., 2018; Rutten et al., 2013).

9.4.2 Health management and animal welfare: from treatment to prevention

A major benefit of PDF lies in the transformation of animal health management paradigms. Traditional herd health approaches rely largely on visual observation and periodic clinical assessments, often resulting in disease detection only after overt clinical signs have appeared. In contrast, PDF enables the early identification of deviations from normal behavioral, physiological, or production patterns through continuous monitoring, thereby supporting a preventive rather than curative approach to health management (Rutten et al., 2013; Barkema et al., 2015; Dolecheck et al., 2015). Early detection of health disorders such as mastitis, lameness, metabolic disturbances, or respiratory diseases allows for more timely and targeted interventions, which can reduce disease severity, shorten treatment duration, and lower associated economic costs (Jin et al., 2024; Vidal et al., 2023). From a welfare perspective, continuous behavioral monitoring provides a more objective and longitudinal assessment of animal well-being, capturing subtle changes that may be missed by sporadic observation. These data streams complement integrated welfare assessment frameworks, such as the Herd Status Index, which combine multiple indicators to evaluate welfare status at both individual and herd levels (Dallago et al., 2022). Importantly, PDF does not replace human observation but rather augments it by extending farmers' perceptual capacity across time, space, and herd size. By providing consistent and continuous information, PDF supports more systematic welfare monitoring, particularly in large or highly automated herds where continuous individual observation is no longer feasible. In this way, precision technologies help reinforce the alignment between productivity, animal health, and animal-cantered management, contributing to more resilient and welfare-oriented dairy systems (Barkema et al., 2015; Von Keyserlingk et al., 2024).

9.4.3 Environmental sustainability and system resilience

PDF contributes to sustainability by improving resource-use efficiency and supporting longer-term system resilience. By optimizing feed utilization, health management, and reproductive performance, PDF can reduce the environmental footprint of milk production on a per unit basis. Improved feed efficiency and reduced disease incidence translate into lower greenhouse gas emissions and nutrient losses per kilogram of milk produced Lovarelli et al., 2020; (Gargiulo et al., 2018).

Beyond short-term efficiency gains, PDF enables the integration of high-frequency phenotypic data into breeding and selection strategies, facilitating selection for traits related to robustness, resilience, and longevity (Graham et al., 2024). This shift supports a move away from narrow production-focused objectives toward more balanced breeding goals that incorporate health, welfare, and environmental performance (Brito et al., 2025; Lovarelli et al., 2020).

Preventive health management enabled by PDF also contributes to sustainability by reducing reliance on antimicrobials and hormonal treatments, aligning dairy production with public health objectives and societal expectations regarding responsible medicine use (Barkema et al., 2015; Alhamada et al., 2016). In this sense, PDF supports a One Health perspective, linking animal health, environmental stewardship, and human well-being.

9.4.4 Labor efficiency and quality of life

Labor-related benefits are among the most frequently cited motivations for PDF adoption. Automation and decision-support tools do not eliminate labor but fundamentally reconfigure work organization. Technologies such as AMS reduce repetitive and physically demanding tasks, while digital monitoring tools decrease the need for constant manual supervision (Marques et al., 2023; Hogenboom et al., 2019; Borchers and Bewley, 2015).

As a result, farmers often report increased flexibility in daily schedules, improved work–life balance, and reduced physical strain (Marques et al., 2023; Von Keyserlingk et al., 2024). Importantly, PDF also changes the nature of required skills, shifting emphasis from routine manual tasks toward interpretation, planning, and strategic decision-making. Smartphone applications and digital dashboards further enhance accessibility and convenience, enabling farmers to engage with herd information remotely and more efficiently (Michels et al., 2019).

These changes have implications beyond individual farms, influencing labor attractiveness, generational renewal, and the long-term viability of dairy farming as a profession.

9.4.5 Transparency, trust, and societal legitimacy

At the level of the dairy value chain and society, PDF contributes to enhanced transparency and trust. Digital traceability systems, particularly when combined with blockchain-based architectures, enable reliable documentation of production practices, animal welfare conditions, and product origin. This supports verification of quality labels, organic certification, and sustainability claims, addressing growing consumer concerns about food safety, fraud, and ethical production (Hassoun et al., 2023; Malik et al., 2026; Tran et al., 2025). By increasing the visibility and credibility of on-farm practices, PDF helps strengthen the social license to operate of the dairy sector. Transparent data flows not only benefit consumers, but also facilitate communication among farmers, cooperatives, processors, and regulatory authorities, particularly in premium or protected markets where trust and compliance are central to value creation (Vaarst et al., 2024; Lokhorst et al., 2019) In an increasingly scrutinized agrifood landscape, such transparency is becoming a strategic asset rather than an optional feature.

9.4.6 From technological capability to system-level value

Taken together, these benefits illustrate that the value of PDF lies not in individual technologies, but in their integration into coherent sociotechnical systems. By combining sensor-based monitoring, advanced analytics, and decision support within operational farm contexts, PDF supports improvements in productivity, animal health and welfare, environmental sustainability, labor efficiency, and societal trust, enabling better-informed decisions at multiple scales, from individual animals to entire farms and dairy value chains (Rutten et al., 2013; Rojo-Gimeno et al., 2019; Lokhorst et al., 2019). However, realizing these benefits consistently requires alignment between technological capabilities, management practices, and human factors. Evidence from adoption and social science studies indicates that when precision technologies are embedded within user-centered designs and supportive institutional frameworks, their potential value is more likely to be translated into sustained on-farm use and measurable outcomes (Borchers and Bewley, 2015; Von Keyserlingk et al., 2024). In this sense, PDF represents not merely a collection of digital tools, but a key enabler of the broader Industry 4.0 transformation of the dairy sector, provided that technological innovation is accompanied by appropriate governance, skills development, and human-centered implementation strategies.

9.5 Barriers and challenges to the adoption of precision dairy farming

Despite the well-documented technical capabilities and potential benefits of PDF, its adoption remains heterogeneous and far from universal. Evidence across regions and production systems indicates that PDF diffusion is constrained by a combination of economic, technical, social, infrastructural, and ethical barriers that interact and reinforce one another. Importantly, these constraints are not merely transitional obstacles linked to technological immaturity, but reflect deeper structural and sociotechnical challenges within contemporary dairy systems.

9.5.1 Economic barriers

Economic factors remain the most frequently cited barriers to PDF adoption. High upfront investment costs associated with technologies such as automatic milking systems (AMS), robotic feeders, and integrated sensor platforms represent a substantial financial risk, particularly for small- and medium-sized farms. Comparative analyses indicate that AMS investments can exceed those of conventional milking systems by more than 50%, while energy consumption may account for up to 40% of annual operational costs, increasing exposure to volatile energy markets (Marques et al., 2023; Gargiulo et al., 2018).

Although subscription-based services, leasing models, and trial periods have been proposed to lower entry barriers, financial uncertainty persists. Producers consistently consider not only expected profitability but also risk exposure, debt tolerance, and payback time, which often discourages adoption in regions with narrow profit margins or limited access to credit (Borchers and Bewley, 2015). As a result, PDF technologies tend to diffuse preferentially among larger herds, reinforcing structural inequalities within the dairy sector.

9.5.2 Technical and data-related barriers

Technical limitations constitute a second major category of barriers, particularly when technologies move from experimental validation to routine farm use. While sensor-based systems may achieve high accuracy under controlled conditions, their performance often declines in commercial environments due to sensor noise, calibration drift, missing data, and farm-specific variability. Behavioral sensors may misclassify resting or transitional activities, while disease detection algorithms can suffer from overfitting and limited generalizability when trained on homogeneous datasets (Graham et al., 2024; Hassoun et al., 2023).

Beyond individual device performance, lack of interoperability remains a critical bottleneck. Farmers frequently operate multiple vendor-specific platforms that do not communicate seamlessly, resulting in fragmented data silos and duplicated effort. Even when cloud-based infrastructures exist, the absence of standardized data models and interfaces limits cross-system integration and constrains the development of holistic decision support systems (Rojo-Gimeno et al., 2019). These technical constraints undermine user confidence and contribute to alert fatigue, ultimately reducing perceived system reliability.

9.5.3 Social and cultural barriers

Adoption of PDF technologies is deeply embedded in social and cultural contexts. Empirical studies show that adoption decisions are influenced by farmer age, education, professional identity, risk perception, and values, rather than by technical performance alone. Younger and more formally educated farmers generally exhibit greater openness to digital tools, whereas older farmers may express skepticism toward data-driven decision-making and prioritize experiential knowledge (Borchers and Bewley, 2015; Yang et al., 2021). Qualitative social science research further highlights concerns related to changing human–animal relationships, particularly in highly automated systems. Some farmers perceive that technologies such as AMS reduce direct contact with animals, potentially weakening observational skills and emotional bonds central to traditional husbandry practices (Barkema et al., 2015). These perceptions reflect broader concerns about control, responsibility, and professional meaning in increasingly automated farming systems.

9.5.4 Infrastructural barriers

Infrastructure-related constraints remain particularly salient in rural and remote regions. Reliable broadband connectivity, mobile network coverage, and stable energy supply are prerequisites for cloud-based analytics, real-time data transfer, and remote technical support. However, such infrastructure remains unevenly distributed, especially in developing regions and extensive production systems, limiting the feasibility of advanced PDF applications (Malik et al., 2026). These infrastructural disparities exacerbate the digital divide within the dairy sector, restricting access to PDF technologies precisely in contexts where productivity gains and improved animal health monitoring could yield substantial benefits. Consequently, infrastructural barriers extend beyond farm-level decisions and represent a policy-relevant constraint requiring coordinated public investment.

9.5.5 Ethical and governance challenges

Ethical and societal considerations are emerging as significant challenges for PDF adoption. Automation and continuous monitoring raise questions related to animal agency, privacy, and acceptability, particularly when technologies such as biometric identification or advanced genetic tools are perceived as intrusive or unnatural (Barkema et al., 2015; O' Mahony et al., 2019). Public acceptance of digital livestock farming is shaped not only by outcomes but also by perceptions of transparency, intent, and respect for animal integrity. In parallel, concerns related to data ownership, surveillance, and power asymmetries between farmers, technology providers, and downstream actors have gained prominence. Without clear governance frameworks, data-intensive systems risk eroding trust rather than enhancing it. Ethical design, transparent data governance, and stakeholder engagement therefore represent essential conditions for maintaining social license and long-term legitimacy.

9.5.6 Human-centered design and coconstruction

Importantly, these barriers do not operate in isolation. Economic risk amplifies resistance to technically imperfect systems; technical failures undermine trust and reinforce social skepticism; infrastructural deficits magnify economic inequalities; and ethical concerns influence both consumer acceptance and regulatory responses. Growing evidence highlights that the success of PDF critically depends on human-centered design and coconstruction processes. Social science research consistently shows that limited adoption or abandonment of digital tools often results from a mismatch between system design and farmers' real-world practices, values, and decision-making routines, rather than from technological inadequacy (Borchers and Bewley, 2015; Vaarst et al., 2024; Von Keyserlingk et al., 2024). Participatory and codesign approaches, in which farmers are actively involved in the development and refinement of technologies, are increasingly recognized as essential for sustainable adoption. From this perspective, effective PDF relies on the hybridization of digital intelligence and embodied human knowledge, fostering decision support systems that enhance situational awareness while preserving farmer autonomy and established human–animal relationships.

9.6 Conclusion

PDF has become a defining element of the digital transformation of the dairy sector, reflecting broader shifts associated with Industry 4.0 while simultaneously exposing its limits. This chapter has shown that PDF is not merely a technological upgrade of existing production systems, but a profound transformation in how information is generated, interpreted, and mobilized to support decision-making across dairy farms and value chains. Its significance lies less in individual tools or algorithms than in the way digital technologies reshape relationships between animals, farmers, technologies, and society. The synthesis presented in this chapter highlights that the primary contribution of PDF is its capacity to enable more informed, timely, and adaptive management. Through continuous, individualized monitoring, PDF supports a transition from reactive interventions toward anticipatory and preventive strategies in animal health, nutrition, reproduction, and welfare. When effectively integrated into management workflows, these capabilities contribute to improvements in productivity, resource-use efficiency, labor organization, and transparency, while also supporting broader sustainability and societal objectives. However, these benefits do not arise automatically from data availability or technological sophistication. They depend critically on the presence of robust decision support systems that translate complex data streams into actionable, interpretable, and context-sensitive guidance.

A key conclusion of this chapter is that many of the limitations currently observed in PDF adoption are not technological in nature, but sociotechnical. Persistent challenges related to investment costs, interoperability, data governance, and infrastructure are compounded by issues of trust, usability, professional identity, and changing human–animal relationships. These findings underscore that successful implementation of PDF requires alignment between technology design, farm management practices, and human factors. PDF must therefore be understood as a sociotechnical system, in which technological performance and human judgment are inseparable. The analysis also reveals a structural orientation of PDF toward intensive indoor dairy systems, where assumptions of continuous connectivity, controlled environments, and regular data flows are more easily satisfied. While recent research demonstrates growing interest in adapting precision technologies to pasture-based and extensive systems, such contexts challenge foundational design assumptions and remain comparatively underexplored. Addressing this imbalance will require not only technical adaptation, but also conceptual rethinking of what precision means in systems characterized by variability, autonomy, and limited infrastructural control. Looking forward, this chapter positions PDF at a critical juncture between the established Dairy 4.0 paradigm and emerging visions of Dairy 5.0. While Dairy 4.0 has emphasized automation, data integration, and efficiency, Dairy 5.0 extends this trajectory by foregrounding human-centered design, ethical governance, resilience, and inclusiveness. Within this framework, digital technologies are no longer viewed solely as instruments of optimization, but as tools that must align with farmer expertise, animal welfare, environmental stewardship, and societal expectations. Such a shift calls for greater emphasis on transparency, explainability, participatory design, and institutional support structures.

Ultimately, the future of PDF will depend on its ability to move beyond technological promise toward meaningful, context-sensitive, and socially legitimate implementation. By integrating data, decision-making, and values, PDF has the potential to contribute to more resilient and sustainable dairy systems. Realizing this potential will require continued interdisciplinary research, thoughtful system design, and governance approaches that place humans, animals, and society at the center of digital innovation in the dairy sector.

AI disclosure

During the preparation of this work, the author used ChatGpt in order to translate and correct sentences. After using this tool/service, the authors reviewed and edited the content as needed and takes full responsibility for the content of the publication.

References

Alhamada, M., Debus, N., Bocquier, F., 2017. An automated method for the evaluation of ram libido in real mating conditions. Animal 11, 2036–2044. Available at https://doi.org/10.1017/S1751731117000611, https://doi.org/10.13140/RG.2.2.35543.44963.

Alhamada, M., Debus, N., Lurette, A., Bocquier, F., 2016. Validation of automated electronic oestrus detection in sheep as an alternative to visual observation. Small Ruminant Research 134, 97–104. Available at https://doi.org/10.1016/j.smallrumres.2015.12.032.

Alhamada, M., González-García, E., Debus, N., Lurette, A., Menassol, J.B., & Bocquier, F. (2021). Using an accelerometer based on inertial measurement unit to evaluate the sexual activity of the ram. In *72th Annual meeting of the european federation of animal science (EAAP 2022)*. Davos, Switzerland. P437.

Aquilani, C., Confessore, A., Bozzi, R., Sirtori, F., Pugliese, C., 2022. Review: Precision Livestock Farming technologies in pasture-based livestock systems. Animal 16 (1), 100429. Available at https://doi.org/10.1016/j.animal.2021.100429.

Barkema, H.W., von Keyserlingk, M.A.G., Kastelic, J.P., Lam, T.J.G.M., Luby, C., Roy, J.-P., et al., 2015. Invited review: Changes in the dairy industry affecting dairy cattle health and welfare. Journal of Dairy Science 98 (11), 7426–7445. Available at https://doi.org/10.3168/jds.2015-9377.

Bernabucci, G., Evangelista, C., Girotti, P., Viola, P., Spina, R., Ronchi, B., et al., 2025. Precision livestock farming: An overview on the application in extensive systems. Italian Journal of Animal Science 24 (1), 859–884. Available at https://doi.org/10.1080/1828051x.2025.2480821.

Bewley, J.M., Robertson, L.M., Eckelkamp, E.A., 2017. A 100-year review: Lactating dairy cattle housing management. Journal of Dairy Science 100, 10418–10431. Available at https://doi.org/10.3168/jds.2017-13251.

Bianchi, M.C., Bava, L., Sandrucci, A., Tangorra, F.M., Tamburini, A., Gislon, G., Zucali, M., 2022. Diffusion of precision livestock farming technologies in dairy cattle farms. animal 16, 100650. Available at https://doi.org/10.1016/j.animal.2022.100650.

Bocquier, F., Debus, N., Lurette, A., Maton, C., Viudes, G., Moulin, C.H., et al., 2014. Elevage de précision en systèmes d'élevage peu intensifiés. INRA Prod. Anim. 27, 101–112. Available at https://doi.org/10.20870/productions-animales.2014.27.2.3058.

Borchers, M.R., Bewley, J.M., 2015. An assessment of producer precision dairy farming technology use, prepurchase considerations, and usefulness. Journal of Dairy Science 98 (6), 4198–4205. Available at https://doi.org/10.3168/jds.2014-8963.

Bouchon, M., Chanel, H., Rouchez, L., Martin, B., Coppa, M., 2025. Method: Using a commercial precision livestock farming activity collar to automatically record and classify dairy cow activity at pasture. Animal - Open Space 4, 100099. Available at 10.1016/j.anopes.2025.100099.

Brito, L.F., Heringstad, B., Klaas, I.C., Schodl, K., Cabrera, V.E., Stygar, A., et al., 2025. Invited review: Using data from sensors and other precision farming technologies to enhance the sustainability of dairy cattle breeding programs. Journal of Dairy Science 108 (10), 10447–10474. Available at https://doi.org/10.3168/jds.2025-26554.

Bumbálek, R., Stehlík, R., Zoubek, T., Ufitikirezi, J.de D.M., Umurungi, S.N., Kuneš, R., et al., 2026. Automated detection and localization of hoof diseases in dairy cattle using an integrated computer vision and infrared thermography system. Computers and Electronics in Agriculture 243, 111404. Available at https://doi.org/10.1016/j.compag.2025.111404.

Campos, L.M., Ringer, H., Chung, M., Hanigan, M.D., 2023. Application of a mathematical framework for the optimization of precision-fed dairy cattle diets. animal 17, 101001. Available at https://doi.org/10.1016/j.animal.2023.101001.

Chi, M., Plaza, A., Benediktsson, J.A., Sun, Z., Shen, J., Zhu, Y., 2016. Big data for remote sensing: challenges and opportunities. Proceedings of the IEEE 104, 2207–2219.Available at https://doi.org/10.1109/jproc.2016.2598228.

Dallago, G.M., Mauyenova, N., Warner, D., Cue, R.I., Vasseur, E., 2022. Using the Herd Status Index to remotely assess the welfare status of dairy herds based on prerecorded data. animal 16 (10), 100641. Available at https://doi.org/10.1016/j.animal.2022.100641.

Dolecheck, K.A., Silvia, W.J., Heersche, G., Chang, Y.M., Ray, D.L., Stone, A.E., et al., 2015. Behavioral and physiological changes around estrus events identified using multiple automated monitoring technologies. Journal of Dairy Science 98 (12), 8723–8731. Available at https://doi.org/10.3168/jds.2015-9645.

Gargiulo, J.I., Eastwood, C.R., Garcia, S.C., Lyons, N.A., 2018. Dairy farmers with larger herd sizes adopt more precision dairy technologies. Journal of Dairy Science 101 (6), 5466–5473. Available at https://doi.org/10.3168/jds.2017-13324.

González-García, E., Alhamada, M., Pradel, J., Douls, S., Parisot, S., Bocquier, F., et al., 2018. A mobile and automated walk-over-weighing system for a close and remote monitoring of liveweight in sheep. Computers and Electronics in Agriculture 153, 226–238. Available at https://doi.org/10.1016/j.compag.2018.08.022.

Graham, J.R., Montes, M.E., Pedrosa, V.B., Doucette, J., Taghipoor, M., Araujo, A.C., et al., 2024. Genetic parameters for calf feeding traits derived from automated milk feeding machines and number of bovine respiratory disease treatments in North American Holstein calves. Journal of Dairy Science 107 (4), 2175–2193. Available at https://doi.org/10.3168/jds.2023-23794.

Groher, T., Heitkämper, K., Umstätter, C., 2020. Digital technology adoption in livestock production with a special focus on ruminant farming. Animal 14, 2404–2413. Available at https://doi.org/10.1017/s1751731120001391.

Hassoun, A., Garcia-Garcia, G., Trollman, H., Jagtap, S., Parra-López, C., Cropotova, J., et al., 2023. Birth of dairy 4.0: Opportunities and challenges in adoption of fourth industrial revolution technologies in the production of milk and its derivatives. Current Research in Food Science 7, 100535. Available at https://doi.org/10.1016/j.crfs.2023.100535.

Hawashin, D., Salah, K., Jayaraman, R., Ahmad, R.W., Kanhere, S., Damiani, E., 2025. Enhancing sustainability in dairy industry: Blockchain-based waste reduction. Sustainable Futures 9, 100529. Available at https://doi.org/10.1016/j.sftr.2025.100529.

Heins, B.J., Pereira, G.M., Sharpe, K.T., 2023. Precision technologies to improve dairy grazing systems. JDS Communications 4, 318–323. Available at https://doi.org/10.3168/jdsc.2022-0308.

Higaki, S., Koyama, K., Sasaki, Y., Abe, K., Honkawa, K., Horii, Y., et al., 2020. Technical note: Calving prediction in dairy cattle based on continuous measurements of ventral tail base skin temperature using supervised machine learning. Journal of Dairy Science 103 (9), 8535–8540. Available at https://doi.org/10.3168/jds.2019-17689.

Hogenboom, J.A., Pellegrino, L., Sandrucci, A., Rosi, V., D'Incecco, P., 2019. Invited review: Hygienic quality, composition, and technological performance of raw milk obtained by robotic milking of cows. Journal of Dairy Science 102 (9), 7640–7654. Available at https://doi.org/10.3168/jds.2018-16013.

Hussein, J.B., Workneh, T.S., Kassim, A., Ntsowe, K., Melesse, S.F., El-Mesery, H.S., 2025. A review on the impact of big data analytics in transforming agricultural practices, food processing, and preservation strategies. Applied Food Research 5, 101234. Available at https://doi.org/10.1016/j.afres.2025.101234.

Jin, C., Upton, J., Bazargani, M.H.Z., Mac Namee, B., 2024. Using milk flow profiles for subclinical mastitis detection. Smart Agricultural Technology 9, 100537. Available at https://doi.org/10.1016/j.atech.2024.100537.

Kamilaris, A., Kartakoullis, A., Prenafeta-Boldú, F.X., 2017. A review on the practice of big data analysis in agriculture. Computers and Electronics in Agriculture 143, 23–37. Available at https://doi.org/10.1016/j.compag.2017.09.037.

Kaur, U., Malacco, V.M.R., Bai, H., Price, T.P., Datta, A., Xin, L., et al., 2023. Invited review: Integration of technologies and systems for precision animal agriculture–A case study on precision dairy farming. Journal of Animal Science 101. Available at https://doi.org/10.1093/jas/skad206, https://academic.oup.com/jas.

Von Keyserlingk, M.A.G., Mills, K.E., Weary, D.M., 2024. Attitudes of western Canadian dairy farmers toward technology. Journal of Dairy Science 107, 933–943. Available at https://doi.org/10.3168/jds.2023-23279.

Khanashyam, A.C., Jagtap, S., Agrawal, T.K., Thorakkattu, P., Malav, O.P., Trollman, H., et al., 2025. Applications of artificial intelligence in the dairy Industry: From farm to product development. Computers and Electronics in Agriculture 238. Available at https://doi.org/10.1016/j.compag.2025.110879.

Lokhorst, C., De Mol, R.M., Kamphuis, C., 2019. Invited review: Big Data in precision dairy farming. Animal 13, 1519–1528. Available at https://doi.org/10.1017/S1751731118003439.

Lovarelli, D., Bacenetti, J., Guarino, M., 2020. A review on dairy cattle farming: Is precision livestock farming the compromise for an environmental, economic and social sustainable production? Journal of Cleaner Production 262, 121409. Available at https://doi.org/10.1016/j.jclepro.2020.121409.

Magrin, L., Dante, S., Contiero, B., Serva, L., Gottardo, F., Cozzi, G., 2026. When beef cattle drink less: Automated water intake monitoring as an early warning for respiratory disease during the adaptation period at the fattening unit. Preventive Veterinary Medicine 247, 106772. Available at https://doi.org/10.1016/j.prevetmed.2025.106772.

Mahato, S., Neethirajan, S., 2025. Integrating Artificial Intelligence in dairy farm management – biometric facial recognition for cows. Information Processing in Agriculture 12, 312–325. Available at https://doi.org/10.1016/j.inpa.2024.10.001.

Malik, M., Mor, R.S., Gahlawat, V.K., Kumar, V., 2026. Unlocking the potential: Hybrid blockchain and AI-enabled traceability model development and implementation in the dairy industry – proof-of-concept. Transportation Research Part E: Logistics and Transportation Review 205. Available at https://doi.org/10.1016/j.tre.2025.104552, https://www.sciencedirect.com/science/journal/13665545.

Mardhati, M., González, L.A., Thomson, P.C., Clark, C.E.F., García, S.C., 2021. Short-term liveweight changes of dairy cows measured by stationary and walk-over weighing scales. Journal of Dairy Science 104 (7), 8202–8213. Available at https://doi.org/10.3168/jds.2020-19912.

Marques, T.C., Lage, C.F.A., Bruno, D.R., Fausak, E.D., Endres, M.I., Ferreira, F.C., et al., 2023. Geographical trends for automatic milking systems research in non-pasture-based dairy farms: A scoping review. Journal of Dairy Science 106 (11), 7725–7736. Available at https://doi.org/10.3168/jds.2023-23313.

Michels, M., Bonke, V., Musshoff, O., 2019. Understanding the adoption of smartphone apps in dairy herd management. Journal of Dairy Science 102, 9422–9434. Available at https://doi.org/10.3168/jds.2019-16489.

Neethirajan, S., Kemp, B., 2021. Digital livestock farming. Sensing and Bio-Sensing Research 32, 100408. Available at https://doi.org/10.1016/j.sbsr.2021.100408.

O' Mahony, N., Campbell, S., Carvalho, A., Krpalkova, L., Riordan, D., Walsh, J., 2019. 3D vision for precision dairy farming. IFAC-PapersOnLine 52, 312–317. Available at https://doi.org/10.1016/j.ifacol.2019.12.555.

Odintsov Vaintrub, M., Levit, H., Chincarini, M., Fusaro, I., Giammarco, M., Vignola, G., 2021. Review: Precision livestock farming, automats and new technologies: Possible applications in extensive dairy sheep farming. Animal 15, 100143. Available at https://doi.org/10.1016/j.animal.2020.100143.

Pardo, G., del Prado, A., Fernández-Álvarez, J., Yáñez-Ruiz, D.R., Belanche, A., 2022. Influence of precision livestock farming on the environmental performance of intensive dairy goat farms. Journal of Cleaner Production 351, 131518. Available at https://doi.org/10.1016/j.jclepro.2022.131518.

Rebuli, K.B., Ozella, L., Masía, F., Vrieze, E., Giacobini, M., 2025. Assessing the stability of herd productivity groups across lactation periods in automatic milking systems using multi-algorithm clustering. Computers and Electronics in Agriculture 235, 110295. Available at https://doi.org/10.1016/j.compag.2025.110295.

Rojo-Gimeno, C., van der Voort, M., Niemi, J.K., Lauwers, L., Kristensen, A.R., Wauters, E., 2019. Assessment of the value of information of precision livestock farming: A conceptual framework. Wageningen Journal of Life Sciences 90-91. Available at https://doi.org/10.1016/j.njas.2019.100311.

Rutten, C.J., Velthuis, A.G.J., Steeneveld, W., Hogeveen, H., 2013. Invited review: Sensors to support health management on dairy farms. Journal of Dairy Science 96 (4), 1928–1952. Available at https://doi.org/10.3168/jds.2012-6107.

Sajid, S., Haleem, A., Bahl, S., Javaid, M., Goyal, T., Mittal, M., 2021. Data science applications for predictive maintenance and materials science in context to Industry 4.0. Materials Today: Proceedings 45, 4898–4905. Available at https://doi.org/10.1016/j.matpr.2021.01.357.

Tran, D., Melissari, F., Le Feon, S., Papadakis, A., Zahariadis, T., Chatzitheodorou, D., et al., 2025. A holistic assessment of blockchain-based traceability systems: The case of protected designation of origin feta cheese production. Computers and Electronics in Agriculture 230, 109863. Available at https://doi.org/10.1016/j.compag.2024.109863.

Trapanese, L., Bifulco, G., Calanni Macchio, A., Aragona, F., Purrone, S., Campanile, G., et al., 2025. Precision livestock farming applied to the dairy sector: 50 years of history with a text mining and topic analysis approach. Smart Agricultural Technology 10, 100827. Available at https://doi.org/10.1016/j.atech.2025.100827.

Tse, C., Barkema, H.W., DeVries, T.J., Rushen, J., Pajor, E.A., 2018. Impact of automatic milking systems on dairy cattle producers' reports of milking labour management, milk production and milk quality. Animal 12 (12), 2649–2656. Available at https://doi.org/10.1017/s1751731118000654.

Tsenkova, R., Toyoda, K., 2001. Artificial Inteligence in Dairy Farming : Near Infrared Approach. IFAC Proceedings Volumes 34, 167–172. Available at https://doi.org/10.1016/s1474-6670(17)32843-4.

Vaarst, M., Ritter, C., Saraceni, J., Roche, S., Wynands, E., Kelton, D., et al., 2024. Invited review: Qualitative social and human science research focusing on actors in and around dairy farming. Journal of Dairy Science 107 (12), 10050–10065. Available at https://doi.org/10.3168/jds.2024-25329.

Vidal, G., Sharpnack, J., Pinedo, P., Tsai, I.C., Lee, A.R., Martínez-López, B., 2023. Impact of sensor data pre-processing strategies and selection of machine learning algorithm on the prediction of metritis events in dairy cattle. Preventive Veterinary Medicine 215, 105903. Available at https://doi.org/10.1016/j.prevetmed.2023.105903.

Yang, W., Edwards, J.P., Eastwood, C.R., Dela Rue, B.T., Renwick, A., 2021. Analysis of adoption trends of in-parlor technologies over a 10-year period for labor saving and data capture on pasture-based dairy farms. Journal of Dairy Science 104 (1), 431–442. Available at https://doi.org/10.3168/jds.2020-18726.

Yalçin, H., 2025. A sustainable partial least squares structural equation modeling framework for integrating milk quality and yield dynamics in Awassi ewes. Journal of Dairy Science 108, 9222–9237. Available at https://doi.org/10.3168/jds.2025-26896.

Chapter 10

Precision dairy processing: integrating Industry 4.0 technologies for smart, data-driven dairy manufacturing

Anum Ishaq[1], Hafiz Ubaid ur Rahman[1,2], Sadia Aslam[1], Amna Sahar[3], and Abderrahmane Aït-Kaddour[4]

[1]*School of Food and Agricultural Sciences, University of Management and Technology, Lahore, Pakistan,* [2]*Department of Animal Sciences, Purdue University, West Lafayette, IN, United States,* [3]*National Institute of Food Science and Technology, University of Agriculture, Faisalabad, Pakistan,* [4]*VetAgro Sup, Departement of Sustainable Foods, Management and Innovation/ Clermont-Auvergne, INRAE, VetAgro Sup, Cheese Research Unit, Lempdes, France*

10.1 Introduction

Traditionally, dairy processing relied on periodic sampling, operator adjustments, and reactive, batch-based control strategies to ensure product quality and safety. Decisions were largely based on offline laboratory analyses and the experience of processors. Although these approaches enabled the production of safe dairy products, they were labor-intensive and inherently reactive, often leading to variability caused by fluctuations in raw milk composition and processing conditions (Chen et al., 2026). In contrast, modern precision dairy processing (PDP) represents a shift toward automated, sensor-supported, and data-driven systems. This paradigm enables continuous monitoring and control of processing conditions at both the component and operational levels. Similar transformations have occurred in other industries, particularly in pharmaceutical manufacturing, where the production of monoclonal antibodies, insulin, and vaccines increasingly relies on real-time data to enhance reliability, efficiency, and process control (Maruthamuthu et al., 2020). By integrating various sensing technologies with automation, process analytical technology (PAT), and advanced strategies, PDP achieves closed-loop decision making across dairy supply and value chains. Unlike traditional processing approaches, which often use retrospective quality indicators (like changes in fats, proteins, and micro-organisms) measured after processing, precision processing relies on dynamic quality indicators that allow continuous

Dairy 4.0. DOI: https://doi.org/10.1016/B978-0-443-44730-3.00007-9

monitoring and real-time adjustment of processing conditions. Continuous quality monitoring and data collection from near-infrared (NIR), Fourier Transform Infrared (FTIR), and Raman spectroscopies, along with other microbiological and physical sensors, facilitate these changes. Dynamic processing quality indicators help encourage improvements with critical quality attributes (composition, functionality, and safety) and help to minimize overprocessing, yield losses, and waste (Mohammadi et al., 2024; Zhang et al., 2025).

The factory floor is not the only place where PDP takes place. The whole scope includes milk reception and raw material grading, processing automation, intelligent packaging, end-to-end traceability, and the crafting of innovative dairy and dairy substitute ingredients. During reception, the dynamic channeling of milk streams to process lines is achieved through rapid screening of composition and quality. In processing, pasteurization, fermentation, concentration, and drying are achieved through process control and sensor automation. In packaging and distribution, digital traceability integrates process data to aid in quality and sustainability claims, targeted recalls, and process control. These features tackle industry problems such as cross-batch quality, resource efficiency, rapid product innovations, and constant food safety regulations (Hussain, 2021; Boulis and Boskou, 2025).

To create aligned precision dairy processing for Industry 4.0 and farm-to-fork digitization, we need to develop a flexible, interconnected food system through cyber-physical systems, the Industrial Internet of Things (IIoT), cloud computing, and artificial intelligence (AI). Integrated data capture at the farm level (animal health and milk yield sensors), processing, and logistics optimize the dairy value chain. This integration balances operational sustainability with energy and water efficiency, reduced food loss, and greater integration of fermented dairy constituents using less environmentally damaging precision. As dairy systems evolve to become more complex and more transparent, precision dairy processing becomes the key framework for complex, transparent, resilient, and innovative manufacturing (Wolfert et al., 2017; Sauer et al., 2021).

This chapter seeks to create a simple and plain understanding of PDP through the prism of data-based, sensor-integrated, and automated solutions along the dairy value chain. It attempts to address the most relevant questions pertaining to the recent advancements in sensing technologies, advanced process control, digital twins, and AI analytics, and their role in enhancing consistency, safety, sustainability, and adherence to safety and quality standards in dairy systems. The chapter also aims to explain the position of PDP in the context of Industry 4.0 and digital farm-to-fork initiatives and discusses the challenges, regulations, and potential research areas, including precision fermentation and innovative dairy ingredients.

10.2 Conceptual framework

The conceptual framework of PDP focuses on the integration of new technologies such as advanced analytics, automation, and systems control

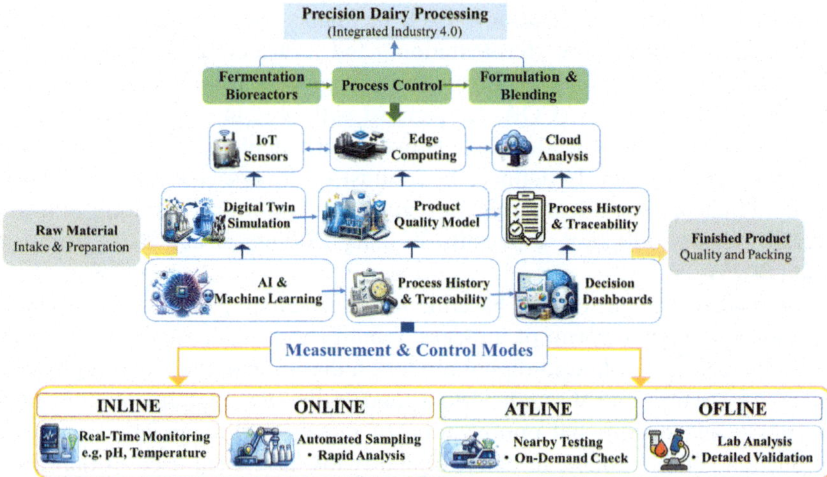

FIGURE 10.1 Framework of precision dairy processing integrating Industry 4.0 technologies and multilevel measurement and control strategies (inline, online, at line, and offline). This figure described how precision can be reached in dairy processing by using industry 4.0 tools (e.g., digital twin, AI, Machine learning, and sensors). These tools can be used in both measurements and control modes during processing from the raw material analysis to the finished product.

that allow a departure from experience and end-point quality control toward predictive real-time adaptive process control. The "smart" data from inline, online, at line, and offline sensor systems is the basis of this framework. Fig. 10.1 shows an example of a fully integrated framework of PDP combined with the technologies of Industry 4.0 and multitier measurement and control to enhance process effectiveness, product quality, and product traceability. Of the various types of sensors (e.g., NIR, FTIR, and Raman probes) that provide continuous and nondestructive measurements within the process streams, the real-time feedback control is solely in the hands of the inline sensor controls. Alternatively, the online and at line sensors (e.g., online NIR) provide rapid verification for and redundancy to a few of the key quality attributes. While offline lab analyses that are time-consuming are still required for reference corrections, compliance, and methodology approvals. Sensor drift, variability in the milk matrix, and challenges related to seasonal and cross-plant calibration transfer are among the most critical issues reported for sensor technologies across different stages of dairy processing. These factors highlight the need for robust chemometric models and regular calibration updates to ensure reliable measurements (van Nuenen et al., 2024).

PAT offers a cohesive philosophy that integrates various sensing technologies with multivariate data analysis and control. Though initially applied to the pharmaceutical industry, its real-time quality assurance (as opposed to end-process testing) has made it increasingly applicable to dairy processing. In dairy applications, PAT facilitates the automated control of individual pasteurization, fermentation, and concentration processes by correlating and dynamically adjusting critical process parameters (CPPs) and critical quality attributes (CQAs) associated with protein denaturation, microbial inactivation, and functionality. However, the deployment of PAT in the food sector is still poorly defined. In contrast to the pharmaceutical industry, the food industry has a less stringent regulatory framework, resulting in lower overall industry adoption, especially in dairy (Sibono et al., 2025; Tarapata et al., 2025).

Advanced PAT provides a baseline for understanding digital twins, which represents a higher form of process intelligence. More than mechanistic models, analytics, or hybrid digital models, a digital twin replicates a physical process for dairy, and through virtual models, fosters experimentation, fault diagnosis, and operational optimization of processes without interference to production (Purlis, 2024). In dairy, digital twins have been used in optimization of heat treatment, membrane filtration, and fermentation kinetics. However promising, digital twins in dairy remain unit operation specific rather than plant wide. The main challenges are model synthesis and integration, extensive computing requirements, and real time data synchronization that must be reliable. This reinforces the need for research to be conducted in the areas of scalable and interoperable architecture for digital twins specific to dairy operations (Tao et al., 2018; Negri et al., 2017).

The decision-making engine in this context is Model Predictive Control (MPC), which converts sensor data and model predictions into optimized control actions in real time. While classical Proportional Integral Derivative (PID) control cannot handle multivariable interactions, constraints, and time delays, which are characteristic of processes such as coupled heat–mass transfer and fermentation in dairy systems, MPC does. MPC is increasingly able to outperform conventional PID control in minimizing energy consumption while maintaining consistency and quality of the end product, but in dairy systems industry use is still restricted because of the challenges associated with model building, the need to train the operators, and the requirement of quality real time data (Thostrup et al., 2022; Indumathy et al., 2025).

The overarching framework for this is Industry 4.0, which offers the digital building blocks for connectivity and interoperability, as well as the real-time data driven decision making across the dairy value chain. Industry 4.0 facilitates the integration of farm level data (e.g., data on milk composition, animal health) with the processing, packaging, and distribution systems, via IIoT, cloud and edge computing, and the data stream protection system that incorporates fundamental principles of cybersecurity. However, the data streams protection systems that incorporate fundamental principles of cybersecurity, coupled with the digital divide that exists between small and medium processors and large-scale dairy, pose the greatest challenges (Wolfert et al., 2017; Sauer et al., 2021).

Lastly, precision fermentation broadens the milk processing paradigm by adding biologically designed systems that can generate milk proteins, enzymes, and functional lipids. Coupled with precision processing and digital control, precision fermentation provides a powerful tool to reimagine the dairy ecosystem by separating the production of certain functional food components from livestock. Although there is considerable innovation and sustainability potential, its alignment with current dairy processing infrastructures and regulatory classifications remains controversial and problematic, needing a greater degree of integration regarding safety assessment, labeling, and consumer communication (Stephens et al., 2019; Tubb and Seba, 2021).

10.3 Precision in raw milk intake

Raw milk characterization and control at the point of intake is critical to the practice of PDP, as any deviations at this point will carry through to all subsequent processes and will negatively impact the quality, consistency, and efficiency of the end products (cheese, yogurt, other fermented dairy products). The intake procedures currently employed, which rely on composite sampling and delayed laboratory results, are increasingly inadequate for future dairy systems that depend on high levels of automation and real-time data analytics for decision-making (Liu et al., 2025).

To overcome the aforementioned issues, modern precision dairy intake systems utilize various sensing technologies in combination with the assessment of the composition, physical properties, and/or risk parameters of the milk. These modern intake systems enable not only enhanced characterization of milk quality at intake but also create conditions for adaptive control of processing parameters. These sensors, by analyzing and continuously tracking the processing-related constituents of milk (fat, protein, and lactose), facilitate the adjustment of specific processing parameters (pasteurization, sterilization, fermentation, and coagulation) in accordance with the quality of milk received, thereby maximizing yield, quality of the products, and efficiency of the entire process (Liu et al., 2025).

Portable and handheld NIR spectroscopy devices also bolster this adaptive framework. While these devices likely differ in how precise and accurate they are compared to benchtop Mid-Infrared (MIR/FTIR) systems, they can provide rapid and nondestructive measurements of major milk components with the appropriate calibration and a chemometric model (Pu et al., 2021). This offers a means to make decisions about how to adapt a process in real time, allowing for the maintenance of production standards to produce high-quality dairy products while accounting for variations in the composition of raw milk.

Precision tools are also used to enhance safety and prohibit food fraud. The combination of NIR and FTIR portable systems and machine learning models has proven to be an effective fraud detection mechanism for the intentional adulteration of unprocessed milk. Liang et al. (2025) and Lanjewar et al. (2024)

describe models that can predict the dilution of milk in real time, while also identifying economically important contaminants. These tools reinforce milk intake analytics and provide support for fraud detection, quality assurance, and adaptive processing. Methods such as MIR analyzers, cryoscopy, and autosampling methods remain important for verification of intake quality, and for calibration transfer. The autosampling and autosorting capabilities of milk collection systems allow for real-time, quality-based separation of milk and therefore further support the optimization of quality-specific milk processing (Liu et al., 2023). While direct microbiological surveillance at intake remains largely indirect, methods based on spectroscopy and machine learning have improved risk surveillance, although the ability to detect pathogens at the individual level still has a lot of room for improvement (Ahranjani et al., 2025).

In addition, a well-planned sampling strategy will be important to guarantee that rapid analytics at the point of intake provide credible data. Predictive spectroscopy is very sensitive to sampling error, variability of milk, and problems of calibration. Therefore, statistically sound sampling and a strong validation of the model are necessary to prevent the occurrence of false positives and negatives, which are detrimental to the quality of both the decision-making processes and the quality of the final products (Pu et al., 2021). The integration of accurate analytics at the point of intake, adaptive control of the processing steps, and a strong data governance structure sets the foundation of PDP, as it facilitates the automated process control to enhance operational efficiency, optimize quantities, and ensure superior dairy products quality.

10.4 Inline and real-time sensing for process control

Recent advancements in dairy processing are utilizing real-time continuous monitoring of milk quality to enhance traditional quality control methods, which rely on time-lagged sampling and analyses conducted in laboratories. Inline sensor technology has the capacity to enhance process control, standardization of milk composition, optimization of resource consumption, and prevention of quality and safety issues, thereby improving the prevention of production loss (Fernández González et al., 2024). Inline sensing in the dairy industry can be divided into three categories: traditional (or classical) physical and chemical sensors, spectral (or spectroscopic) sensors, and biosensors. Despite offering different functionalities, each of the above categories faces unique design-, functional-, and application-related challenges (Fernández González et al., 2024).

In dairy automation, classical physical and chemical sensors are used to continuously monitor and control temperature, pH, pressure, flow, electrical conductivity, and turbidity. These sensors are critical in the operation of systems used in pasteurization, fermentation, membrane filtration, evaporation, and cleaning-in-place (CIP) processes. For example, flow and temperature sensors ensure that all pasteurization requirements are satisfied, and the control of pH is critical for the fermentation of yogurt and cheese. In addition, conductivity and turbidity sensors are vital in the monitoring of processes, changes in the flow of the product, and the validation of cleaning (Kaveti, 2023; Aydogdu et al., 2023).

These sensors are easy to incorporate into industrial control systems and are reliable and inexpensive. However, they measure the physical variables indirectly and do not measure compositional variables. Calibration is frequently required due to sensor drift, and the stability of the signal is compromised due to the fouling of sensors caused by proteins and fats. Traditional sensors cannot measure and detect structural changes in the individual components of the dairy matrix such as fat, protein, or lactose, which limits their applicability for advanced compositional optimization (Kunes et al., 2021). Spectral techniques, as part of the PAT framework, provide rapid, nondestructive and often multicomponent analysis of dairy streams (Fernández González et al., 2024). For the analysis of fat, protein, lactose, and total solids in milk, NIR and FTIR are the most used techniques. They are also used for real-time control during the processes of cream separation and milk standardization, but their performance is affected by variability of temperature and other physical factors, as well as optical fouling, and seasonal or feed variability of milk can cause chemometric calibration drift (van Nuenen et al., 2024). Some of the more advanced properties of Raman spectroscopy are its specificity at the molecular level, the ability to detect changes in the structure of a protein, and the ability to determine the composition of a sample. However, its application in dairy products is hindered by poor signal detection, the interference of fluorescence, the high costs of the equipment, and a decrease in the strength of the equipment in typical industrial conditions (He et al., 2019). Besides Raman spectroscopy, fluorescence spectroscopy is useful in detecting specific aromatic amino acids and vitamins. However, its application in the dairy industry is limited due to interference from the matrix and the quenching of fluorescence in the complex streams of dairy products. Hyperspectral Imaging allows detection of heterogeneities and adulterations in a sample and provides a compositional analysis of the sample. For this purpose, the system needs to have a large capacity (for storage and processing of the image) to regulate the large number of variables (light scattering and the positioning of the sample), which can complicate routine industrial deployment (Fernández González et al., 2024).

Although rapid real-time spectroscopic methods give information about the composition and quality of milk, the addition of rapid microbial detection techniques-especially quantitative polymerase chain reaction quantitative Polymerase Chain Reaction (qPCR) and bioluminescence detection allows for effective monitoring of both compositional and microbiological criteria at optimum process control during the milk intake. Recent advances in rapid microbiological detection, including the various forms of biosensors,

ATP (Adenosine Triphosphate) bioluminescence, and rapid qPCR offer solutions to the challenges posed by the classical methods, due to their slow and retrospective nature. In dairy settings, ATP bioluminescence is commonly used for quick and easy verification of cleanliness, although it is only a proxy for total biological residue and cannot distinguish between live and dead cells (Yang et al., 2024). Biosensor-based methods and rapid qPCR present greater target specificity and sensitivity, as well as greater sensitivity for pathogens and spoilage organisms. However, issues with sample preparation, robustness, and fully inline configurations have yet to be addressed (Singh & Sharanagat, 2024).

Temperature, pH, flow, turbidity, and total solids physical sensors are still central to dairy precision engineering. These sensors facilitate control over-heat, fermentation, and fluid operations and are often combined with composite data for multivariate control. Because of their reliability, systems such as these are important and invaluable due to controlling the operations previously stated. However, the long-term precision of these systems is still affected by maintenance, drift, and fouling (Mhapsekar et al., 2025; Tarapata et al., 2025). Trust in real-time systems is mostly dependent on the validation, transfer, and drift compensation of the most reliable sensors. For chemometric calibration models of spectroscopic sensors, there is a requirement for maintenance, reliability, and a drift compensation of the model to be validated in the production line as well as on the instrument. This need for constant validation increases the necessity for the standardization of models and the repetition of updating the model (Tarapoulouzi et al., 2026).

In dairy systems, the combination of compositional, microbiological, and physical sensing creates a shift away from end-point quality testing to proactive and predictive process control. Though automated inline sensing, combined with sophisticated analytics, reduces variability, optimizes resource efficiency, and improves the quality and granularity of the information, it also improves high-quality data resource efficiency. Such datasets help build digital twins and AI-driven optimization for PDP (Demartini et al., 2018).

10.5 Automation, robotics, and equipment innovations

Automation in the dairy sector has resulted in flexible systems that integrate robotics across the entire dairy value chain to improve the process control, ensuring the safety and quality. In the context of PDP, this automation is specifically aimed at achieving high-resolution control of critical process variables. Adopting automation and robotics for greater reliability and safety relieves workers from performing injury-prone manual tasks and, instead, lets them supervise automated systems and focus on higher-level tasks. This shift supports data-driven supervision and real-time decision making, which are core principles of PDP. Panda et al. (2023) noted that cleaning, packaging and material handling are the tasks performed by automation and robotics in dairy plants today. With the automation of those different tasks, workers can be exposed to fewer work-related manual handling injuries. Additionally, the introduction of automation improved production throughput and operational quality consistency across shifts and product types (Sandey et al., 2017; Loria, 2025). Such consistency is fundamental to reducing batch-to-batch variability in PDP. Many types of machinery can be reconfigured, in a short downtime, to work with a variety of dairy products, and with different types of packaging and production environments. This reconfiguration enables precise adaptation to product specifications and market demands, a key attribute of PDP. Automated packaging lines can support rapid changeover to work and process varying types of packaging materials, including bottles, cartons, and pouches. Additionally, highly advanced robotics can produce handling with very little (or even no) human coding. This increased flexibility allows for more efficient responses to changes in consumer markets and allows for a greater range of products to be produced (Gibeson, 2022;).

In the Programmable Logic Controllers (PLC) and Supervisory Control and Data Acquisition (SCADA) controlled systems for dairy, automation helps adhere to microbial safety and quality requirements through temperature and time control. Within PDP, these systems form the digital control layer that enables continuous monitoring and closed loop corrective actions. For example, the absence of direct human interventions in systems that are already used for robotic case packing, thermoforming, and automated packaging, as described by Sandey et al. (2017), creates less risk of contamination, and the quality of packaging remains consistent and unchanged.

Distributed control systems, in conjunction with smart valves and variable frequency drives, provide greater control over rate, pressure, and pumping speed. The integration of control networks and devices improves energy efficiency and responsiveness to sensor feedback, thereby reducing variability compared to fixed speed, as manual controls, and other controls are used within traditional plants (Bordeasu et al., 2023). Electromechanical actuators are critical for repeatable filling and transfer tasks that meet the required physical and hygienic standards and provide control of precise motion for valves and conveyors. This interconnectivity creates cyber-physical environment that underpins PDP architecture.

Robotic systems beyond the control of machinery are an additional enhancement of dairy processing. These include robotic milking systems that autonomously control the handling and milking of the animals. These systems improve the consistency of yields and the welfare of the animals. Other systems include autonomous guided vehicles (AGVs) and multi axis systems that control the packaging, palletizing, and the warehousing logistics with accuracy and incredibly low downtime (Sandey et al., 2017). These robotic systems, in conjunction with satisfying the requirements of repetitive and hygiene sensitive tasks, often integrate with vision systems and programmable logic controllers to provide inspection and decision-making support, thus reducing the need for human involvement (Heema et al., 2022).

IoT has made it possible for robots, sensors, and control systems to be interconnected at various points throughout a dairy processing plant. IoT-enabled devices transmit and receive data in real time, allowing machines to communicate with control and analytics systems to carry out predictive maintenance, optimize task assignments, and provide real-time responses to process disruptions. All these factors contribute to improved visibility and the ability to process data to increase the efficiency of decision-making and operating functions of the systems integrate to form a cyber-physical system (Hassoun et al., 2023; Michie et al., 2020).

Overall, transition toward automation and robotics in PDP is beneficial due to improved control of process variables, reduced batch-to-batch variability, minimized downtime through predictive maintenance, and enhanced compliance with food safety standards (Heema et al., 2022). From a PDP perspective, the primarily value lies in the systematic reduction of process variability through data integration, predictive maintenance, and automated response mechanisms.

10.6 Process control strategies and digital twins

Advanced Process Control (APC) in precision dairy involves the use of advanced control methods and real-time data analysis to track, forecast, and autonomously regulate specific dairy processing activities. The integration of APC strategies and digital twin technologies exemplifies this trend. Process control in agrifood settings relies primarily on classical control strategies, and particularly the use of PID controllers, which remain popular for managing feedback control due to their effectiveness and ease of use for temperature, flow, and pressure control in the various unit operations of dairy process engineering. PID control tends to be ineffective when process control variables remain uncontrolled, especially in processing systems with time delays and nonlinear behaviors, such as dairy processes involving pasteurization and fermentation (Agachi et al., 2023). This has created the need and opportunity for the application of more sophisticated processes, such as Model Predictive Control (MPC) and adaptive control. Compared to other predictive control methods, model MPC has been the most successful and preferred because of how it intelligently utilizes a process model to forecast the process and control multiple tunable variables of key quality attributes, simultaneously minimizing energy and variability (Wu et al., 2025).

In conjunction with real-time sensor data, predictive models, and control algorithms, digital twins strengthen these abilities by providing a virtual replica of the actual dairy process. Through the controlled optimization of process parameters (e.g., heat treatment profiles in pasteurization) with simulation and scenario analysis, digital twins support the achievement of microbial safety and the preservation of denaturable whey proteins and bioactive peptides (Rao & Neethirajan, 2025; Szpicer et al., 2025). Optimally, digital twin platforms for high-capacity milk pasteurization lines use predictive modeling to adjust and balance holding-time and temperature processes for microbial inactivation and nutrient retention (Kannapinn et al., 2025).

When sensor networks, predictive models, and control loops are integrated within the digital twin framework, users can track processes and continuously adjust optimally to detect and repair faults. For PDP, a digital twin consists of (i) real-time data collection via mechanistic, microbiological, and physical sensors; (ii) model updates via cloud and edge analytics; (iii) model updates suited to Adaptive Predictive Control (MPC); and (iv) feedback to the user or automated control system to perform (Verboven et al., 2020). The combination of digital twin technology, precision process control, and precision farming is significant progress toward highly resilient and sustainable dairy operations of Industry 4.0.

10.7 Data infrastructure, edge, and cloud analytics

The digital infrastructure of PDP includes technology for data collection, storage, and management, via remote collection centers, for processing facilities. Some remote technologies include milk analyzers, automatic pasteurizers, homogenizers, fermentation sensors, and environmental sensors. These technologies create data streams on parameters of milk composition, microbial count, temperatures, pH, processing durations, and other fundamental factors. Data is of higher quality if context parameters are completely and accurately described (Serrano-Torres et al., 2025). These context parameters include sensor calibration, sampling frequency, and the presence of metadata for data items that describe, for example, the location of data collection.

Edge computing creates the opportunity to analyze data on-site and in real-time. Edge devices can be programmed to make automatic adjustments to the processing variables, predictive maintenance of processing equipment, and immediate quality control. These on-site adjustments lessen the need for urgent operational decisions to be made in the cloud. Edge devices can be coupled with the appropriate segmentation of operational technology (OT) and information technology (IT) to create a secure environment for automated dairy processing (Wolfert et al., 2017; Bhat & Huang, 2021; Rose et al., 2015).

Cloud systems enhance edge computing by providing an additional level of detail in storage and analytics that can be tailored to the temporal and volumetric needs of large-scale historical processing data. Time-series analytics clouds use databases that can be segmented to allow comparative analytics to be achieved for discerning trends, optimizing processes, and benchmarking performance. Cloud analytics in conjunction with digital twins, machine learning, and advanced analytics of the relevant processed data can facilitate predictive control of the processes, optimize yields, and maintain traceability from raw milk to end products. Standards, metadata, data compression, and system interoperability by means of controlled coupling facilitate interoperability and reliability of analyses in competitive digital systems (Wolfert et al., 2017; Bhat & Huang, 2021; Kahraman et al., 2023). Firewalls, access control, and other encryption protocols secure the safe integration of OT and IT to support data-driven decision-making (Rose et al., 2015).

The integration of edge and cloud computing and data architectures is critical to the successful precision processing of dairy. Integrated data and computing systems that are secure, scalable, and structured from sensor data collection to edge computational analytics and cloud storage provide the basis for accurate, real-time control of dairy processing, predictive maintenance of dairy processing equipment, assurance of the quality of processed dairy products, and complete traceability of all processed dairy products, all of which is critical to the sustainability of contemporary dairy farming.

10.8 Machine learning and AI applications

Emerging concepts in machine learning (ML) and AI offer data driven solutions for managing farm and plant level operations in precision dairy farming. ML models can analyze large data sets from sensors, images, and processes to identify complex and nonlinear relationships, unlike traditional models that work based on fixed rules. This unique capability improves predictive, operational, and automation parameters within the dairy value chain (Wolfert et al., 2017; Jox et al., 2025).

One of the main uses of AI in precision dairy systems is predictive maintenance. Algorithms examine time series data from equipment sensors (e.g., vibration, temperature, pressure, motor load) to predict potential breakdowns. Data-based maintenance methods have proven in food and dairy processing settings to lessen maintenance costs and unplanned downtimes compared to reactive and calendar-based maintenance and to better hygienic design compliance due to avoiding emergency meetings during production (Carvalho et al., 2019; Lee et al., 2018).

AI and ML have also been used to make quality predictions, especially those relating to shelf-life and sensory evaluation resulting from fusion of the senses. ML and multivariate models, which integrate and analyze spectroscopic data (NIR/MIR) with other variables, such as process parameters and storage conditions, have demonstrated the ability to predict quality deterioration better than traditional univariate models, which allows for dynamic shelf-life assignment and decreased product loss. These models are particularly important in dairy systems due to the greater compositional variability of raw milk and its downstream impacts on stability and organoleptic quality (Grassi and Casiraghi, 2022). The ability to use computer vision to inspect packaging integrity, the presence of labels, and surface erosion of dairy products (especially cheese and fermented products) has also been enhanced by AI. Vision-based systems can perform inline defect removal and maintain compliance to quality standards with better efficiency and consistency than human inspection. In cheese manufacturing, the combination of image analysis and Convolutional Neural Networks (CNNs) has been used to detect surface defects and rind irregularities and to assess contamination risks that are difficult to identify using human inspection (Ganesh et al., 2024).

The use of AI in precision dairy processing is overcoming challenges of model explain-ability, validation, and lifecycle management. When models lack explainability, users may lack trust in the models and regulators may ignore them, especially when the model's predictions relate to food safety and animal welfare. As a result, models that incorporate AI represent a risk in the dairy farming industry and must be accompanied by XAI (explainable AI) techniques to increase transparency and confidence among the users. Also, systematic validation is essential, along with retraining from the users to ensure the continued reliability of the AI-assisted models (Arrieta et al., 2019). The use of AI in precision dairy processing has the potential to be very beneficial, especially from the perspective of efficiency, quality consistency, and sustainability. For AI to be successfully implemented, a quality data infrastructure that integrates many different fields of knowledge is required.

10.9 Traceability, blockchain, and digital quality assurance

Advancements in digital technology, sensor networks, and data analytics in dairy processing plants enable more precise, real-time control over products quality, safety, and composition. Digital traceability within processing facilitated real-time tracking of milk and dairy products throughout the production chain, from intake, pasteurization, homogenization, to final product packaging, ensuring full visibility of quality parameters and process conditions (Wolfert et al., 2017). Inline sensors, PAT, and automated data collection systems continuously provide and integrate information on temperature, pH, fat and protein content, flow rates, and microbial counts. Such data is meant to integrate with downstream processing and logistics data for real-time decision making and actions (Siddiqui et al., 2024).

Blockchain can enhance data integrity, security, and traceability at the processing stage. Integrating blockchain with production data allows batch-level records of milk attributes, pathogen test results, processing parameters, and packaging history, which supports regulatory requirements and GS1 (Global Standards 1) traceability standards (Malik et al., 2024). The integration of this technology can also improve consumer confidence and facilitate targeted product recalls based on batch-level traceability (Treiblmaier, 2018; Kamilaris et al., 2019).

By combining automated data capture, digital quality assurance, and blockchain-based traceability, dairy processors can make possible rapid detection of fraud, immediate evaluation of quality, and reporting on sustainability of production processes. Although, the high costs of the adoption of early innovations in dairy, the integration of new technologies, and the adoption of new processes are all significant challenges. The preliminary case studies are indicative of the "sustainable dairy precursor" to the full adoption of Industry 4.0 in the dairy sector, especially in the areas of safety and the effectiveness of recalls (Vargas-Bello-Pérez et al., 2025).

10.10 Precision fermentation and novel dairy ingredients

Precision fermentation is the process of biotechnologically engineering of certain microorganisms, such as yeasts, bacteria, or fungi, to produce specific molecules like proteins or enzymes through the control of specific metabolic pathways and fermentation conditions to predictably high-purity yield (Hilgendorf et al., 2024). While traditional fermentation produces a variety of unrefined end-products of fermentation, precision fermentation produces a singular defined component using synthetic biology and engineered DNA constructs in bioreactors. This process is positive in the optimization of microorganisms and the reduction of waste and of agrifoods and provides an alternative sustainable method of food production (Verma et al., 2025). In the case of dairy applications, precision fermentation allows the production of animal-free dairy-identical proteins, such as casein, whey, lactoferrin milk-identical proteins, and functional dairy-derived lipids and enzymes. These proteins produced via precision fermentation replicate the nutrition, functionality, and processing of traditional dairy proteins; therefore, they can be used in milk analogues, cheeses, yogurts, and creams, as well as in fortified formulations, all while decreasing the carbon footprint (Dairy Global, 2024).

Knychala et al. (2024) and Rout and Srivastav (2025) explain how precision fermentation helps create new dairy products and ingredients, such as the production of animal-free milk proteins, the production of lactoferrin and other bioactive compounds, and the production of tailored components that enhance the texture, taste, or functionality of high-value formulations. Microbial platforms have proven the commercial viability of precisely bioengineering functional dairy proteins (casein and whey) that are nutritionally and functionally identical to those derived from cows. Additionally, the industrial collaboration between Bel Group and Standing Ovation companies resulted in the production of casein from acid whey by using microbial fermentation under tightly controlled process conditions, contributing to the valorization of dairy by-products. This precision biomanufacturing approach reduces environmental impact by decreasing greenhouse gas emissions, land, and water use compared to traditional dairy farming (Augustin et al., 2024).

Dairy process line integration of precision-fermented ingredients entails maintaining functional equivalence through solids handling, emulsification, and coagulation. The challenges are mostly casein micelle phosphorylation replicative to authentic texture and structuring of cheese and yogurt (Knychala et al., 2024). The optimization of sensory attributes, cost, and nutrition with hybrid models of traditional and precision-fermented proteins is also being explored (Knychala et al., 2024).

In the United States, precision-fermented dairy proteins enjoy GRAS status (or "no questions" letters), but absence of global regulatory guidance persists. Transparent labeling is important for consumer trust, such as allergenicity and "animal-free" (Ronchetti et al., 2024; Augustin et al., 2024). The market potential and the sustainability of precision-fermented dairy proteins are quite high. Technology provides the ability to produce more high-quality dairy proteins and with fewer environmental impacts. Compared to traditional livestock farming, life cycle assessments show significant reductions in the negative environmental impacts, from greenhouse gas emissions and the water and land used, with precision fermentation (AgriFutures Grow-AG, 2025; Manzoki et al., 2025). These factors make precision fermentation one of the most promising technologies to provide high quality dairy products in an environmentally sustainable way.

The impact of precision fermentation is further accentuated by the tools of Industry 4.0, such as real-time analytics, AI, and ML. When integrated with IoT and real-time analytics, AI and ML provide predictive control, automated optimization of fermentation, and process streams. By increasing the ability to monitor and control fermentation processes with a high degree of automation, these processes can be dynamically and automatically adjusted to provide improved yields, consistency, and scalability, as well as improved process efficiencies with respect to fermentation and bioconversion process streams (Verma et al., 2025).

10.11 Food quality and safety in dairy processing—from HACCP meets process analytical technology in precision dairy processing

The HACCP system provides a comprehensive approach to ensure the safety of food products. However, the conventional HACCP approach has limitations particularly in terms of process precision and data control, as it mainly relies on manual checks to verify food safety. Within the framework of PDP, the integration of PAT with the HACCP transforms dairy processing from reactive to proactive and highly precise process control (Fig. 10.2). PAT provides in-line, real-time data on quality and safety parameters, enabling immediate corrective actions and continuous monitoring of CCPs (critical control points) such as pasteurization or UHT temperatures, moisture levels, and determination of fat and/or protein contents of different dairy products (He et al., 2019).

Several PATs tools, such as NIR spectroscopy, Raman spectroscopy, and acoustic sensors, can be integrated with the conventional Food Safety and Quality Management systems to strengthen control over variability. This integration enables precision monitoring of acidification, crystallization,

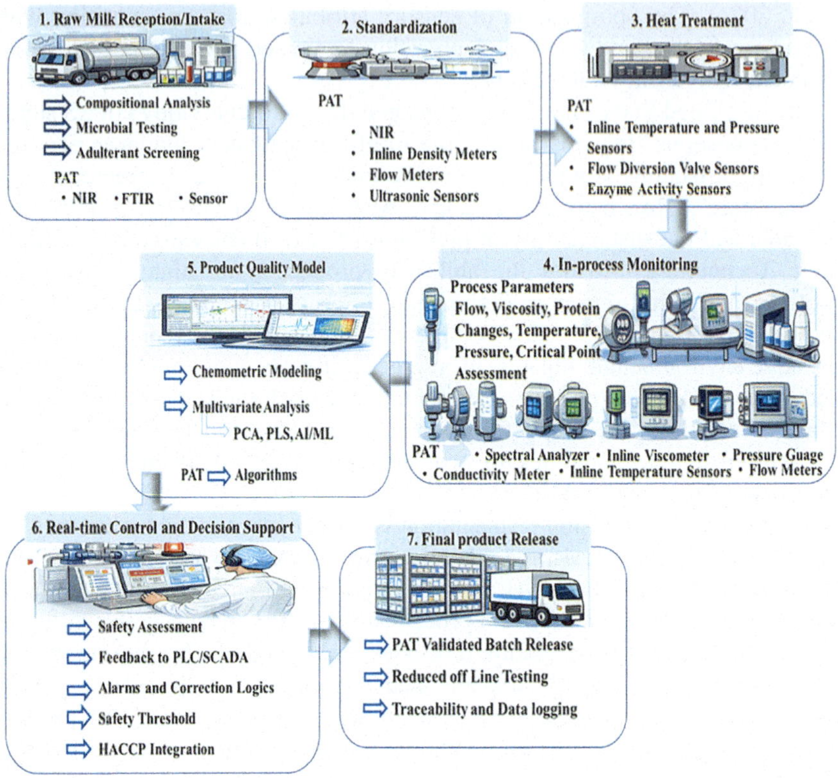

FIGURE 10.2 Improving quality and safety in dairy processing through integrating PATs. This figure illustrates how quality and safety in dairy plants can be improved by the integration of PATs (e.g., NIR spectroscopy, Raman spectroscopy, and acoustic sensors) to improve control over variability. This integration enables precision monitoring of unit operations (e.g., homogenization, concentration, drying, and sterilization operations) to avoid delays compared to traditional lab-based techniques. *PAT*, Process analytical technology.

standardization, homogenization, concentration, and drying operations using online sensors, avoiding delays inherent to traditional lab-based techniques. PAT also ensures online monitoring of critical limits for key quality and safety parameters such as temperature, pH acidity, flow rate, and time-temperature combinations during processing. This allows rapid detection of process deviations and timely decision about the required corrective measures. Additionally, this integration would be helpful in ensuring blending uniformity and compositional consistency of dairy products, reducing rework burden, waste production, and variability between batches. Furthermore, PAT facilitates the implementation of preventive quality and safety frameworks. Automated and continuous record-keeping ensures HACCP compliance, with PAT systems support GMPs (Good Manufacturing Practices) and

SSOPs (Sanitization Standard Operating Procedures) by monitoring CIP (cleaning-in-place) efficiency and detecting residues of cleaning agents. PAT also strengthens the validation step of HACCP that the processes operate consistency with predefined limits (Moran et al., 2017). In the PDP context, this approach not only ensures safety and quality but also maximizes process precision, reduces variability, and enables real-time decision-making, key features of Dairy 4.0.

10.12 Challenges, risks, and regulatory considerations

Transitioning from conventional to PDP creates various technological, operational, and regulatory challenges. PAT along with AI, automation, and precision fermentation have all demonstrated positive impact in improving quality, efficiency, and sustainability, but challenges such as operational and regulatory restrictions still exist for large-scale application of the technology (Strani, 2021; Srivastava & Atri, 2025).

The greatest challenges to adopting precision dairy technologies are the high costs. The investment in new inline sensors (NIR, FTIR, Raman), automated systems for pasteurization, robotic filling lines, digital twins, and cloud analytics platforms is substantially high in cost, and for many small and medium dairy processors in developing countries, the ability to achieve a return on investment is often out of reach. The addition of sensor and data integrated systems to existing retrofitted legacy systems is another barrier to implementation. In addition to the financial barriers, organizational and human behavioral barriers are significant. PDP relies on a diverse range of skills, spanning dairy science, microbiology, automation engineering, and data analytics. Most processing establishments lack staff with competencies and expertise in advanced controls, cybersecurity, and machine learning system (model) validation. It is important to note that during the transition to new technologies, the risk associated with lowering operational productivity and increasing operational risks is significant. A primary technical issue is the reliability of sensors. The processing environment includes high moisture content, extreme temperatures, and chemical cleaning cycles, all of which are harmful to the inline volumetric and microbiological sensors, as well as the other processing equipment. Variable composition and drifting calibration all negatively impact the sensors' accuracy. In AI, model drift is a major issue. When the raw materials' composition and processing conditions (temperature, and humidity) change, the predictive models fail to work as intended. Without any retraining or validity measures, predictive models drift and result in poor estimation of a product's shelf-life, microbial load, and other attributes. A positive inaccuracy means a product goes waste without any compensation. A negative inaccuracy means endangering consumers' safety compromise the regulation compliance (Pu et al., 2020; Rathod and Dixit, 2020).

IoT technology, cloud computing, and digital traceability are all essential components of PDP. With the combination of OT and IT, the risk of cyberattacks increases. Malicious attempts to access a process control system, ransomware attacks on process control systems, and traceability system attacks can disrupt production and undermine consumer confidence. The necessity for data governance is compulsory. Milk data, supplier data, sustainability claims, and data related to proprietary processes and their parameters need to be collected and stored so the security of the data is not compromised. There is still a great deal of risk despite the existence of clear legislative and organizational data protection policies, such as the implementation of cyberpolicies, encryption, firewalls, and even real-time monitoring. Most dairy plants remain unprotected at the cyber level regardless of the measures (Jachimczyk et al., 2021).

There is a great deal of uncertainty surrounding the production of milk proteins (casein and whey) via precise fermentation of engineered microorganisms. The regulation regarding the production of milk proteins (casein and whey) through precise fermentation of engineered microorganisms is highly variable and, therefore, poses a great deal of uncertainty. The regulation of these proteins declares them as novel foods in some jurisdictions and as biotechnological products in others creating a significant challenge for international trade and product labeling. To fulfill consumer demand and regulatory compliance, manufacturers are compelled to refine certification processes, transparently articulate production methodologies, and implement standardized procedures for safety claims (Mirsalami & Mirsalami, 2025).

10.13 Future directions and research needs

An emerging transformative trend is the on-site, continuous biomanufacturing of dairy ingredients. Rather than relying solely on centralized production systems, future dairy plants may integrate modular precision fermentation units capable of producing targeted functional dairy proteins, enzymes, or other bioactive ingredients directly within processing facilities. Such integration could significantly reshape traditional dairy supply chains by decentralizing the production of key functional components and reducing dependence on conventional biomanufacturing infrastructures. Another important emerging trend is the development of comprehensive digital twins of the entire dairy value chain, spanning precision dairy farming, milk intake, processing, packaging, and distribution. These are dynamic simulation models that use predictive algorithms, in addition to real-time sensor data, to control and optimize heating, fermentation, membrane filtration, and energy management. Enhanced MPC is expected to enable dairy processors to optimize the balance between microbial safety and the retention of sensory nutritional quality (Alamirew et al., 2017).

In future studies, online detection methods that are less expensive, quicker, and have lower limits of detection would be valuable. Creating a final product that integrates advanced biosensors, microfluidics, and rapid molecular diagnostics within a continuous flow environment will create a more robust preventative food safety system that complies with HACCP and Food Safety Modernization Act (FSMA) directives. The development of simple and robust sensing systems for small- and medium-scale dairy plants is also a major goal. To expand the range of potential target systems, there is a need for edge computing systems, modular sensing, and additional low-cost precision technologies. Corresponding inexpensive and adequate IT systems will also be needed. Improved interoperability will be critical for the integration of systems and the smooth flow of information within a supply chain. Meeting interoperability will be achievable with the proposed blockchain system for supply chain traceability. The internationally accepted blockchain traceability system will ensure the appropriate safety measures and labeling for international trade facilitation. Thus, the precision processing technology for dairy products will be enhanced (Malik et al., 2024). Numerous challenges, such as technology, economy, and regulations, exist for the PDP. Nevertheless, if innovations and partnerships continue, more intelligent and adaptive systems will be developed.

10.14 Conclusion

PDP represents a transformative approach to modern dairy manufacturing by integrating advanced sensing technologies, automation, AI, and digital infrastructures within the framework of Industry 4.0. Through real-time monitoring, predictive control, and data-driven decision-making, these technologies enhance process efficiency, product consistency, food safety, and sustainability across the dairy value chain. Innovations such as digital twins, machine learning, and precision fermentation further expand the possibilities for optimizing dairy processing and developing novel dairy ingredients. Despite these advancements, challenges related to technological integration, investment costs, data governance, and regulatory harmonization remain significant barriers to large-scale adoption. Continued interdisciplinary research, technological standardization, and collaborative innovation will be essential to fully realize the potential of PDP and to support the transition toward more resilient, transparent, and sustainable dairy systems.

AI disclosure

During the preparation of this work, the authors used ChatGPT in order to correct English language and reformulate some sentences throughout the chapter. Some tool graphics in the figures were also created with minor AI assistance. After using this tool/service, the authors reviewed and edited the content as needed and take full responsibility for the content of the publication.

References

Agachi, P.S., Cristea, M.V., Csavdari, A.A., Szilagyi, B., 2023. Advanced process engineering control. Walter de Gruyter GmbH & Co KG.

AgriFutures (Grow-AG)., 2025. Precision fermentation explained: how it works & why it matters. https://www.growag.com/article/precision-fermentation-explained-how-it-works-why-it-matters.

Ahranjani, P.J., Dehghan, K., Esfandiari, Z., Ahranjani, P.J., 2025. A systematic review of spectroscopic techniques for detecting milk adulteration. Critical Reviews in Analytical Chemistry 1–32. Available at https://doi.org/10.1080/10408347.2025.2477535.

Alamirew, T., Balaji, V., Gabbeye, N., 2017. Comparison of PID controller with model predictive controller for milk pasteurization process. Bulletin of Electrical Engineering and Informatics 6 (1), 24–35.

Arrieta, A.B., Díaz-Rodríguez, N., Ser, J.D., Bennetot, A., Tabik, S., Barbado, A., et al., 2019. Explainable Artificial Intelligence (XAI): Concepts, taxonomies, opportunities and challenges toward responsible AI. Information Fusion 58, 82–115.

Augustin, M.A., Hartley, C.J., Maloney, G., Tyndall, S., 2024. Innovation in precision fermentation for food ingredients. Critical Reviews in Food Science and Nutrition 64 (18), 6218–6238.

Aydogdu, T., O'Mahony, J.A., McCarthy, N.A., 2023. pH, the fundamentals for milk and dairy processing: A review. Dairy 4 (3), 395–409.

Bhat, S.A., Huang, N.-F., 2021. Big data and AI revolution in precision agriculture: Survey and challenges. Institute of Electrical and Electronics Engineers Inc., IEEE Access 9, 110209–110222.

Bordeasu, D., Prostean, O., Filip, I., Vasar, C., 2023. Adaptive control strategy for a pumping system using a variable frequency drive. Machines 11 (7), 688.

Boulis, A., Boskou, G., 2025. Analysis of food safety data and its importance in risk assessment. Journal of Risk Research 1–18.

Carvalho, T.P., Soares, F.A., Vita, R., Francisco, R.D.P., Basto, J.P., Alcalá, S.G., 2019. A systematic literature review of machine learning methods applied to predictive maintenance. Computers & Industrial Engineering 137, 106024.

Chen, J., Chen, Z., Lai, Y., Lai, X., Xiao, X., Peng, J., et al., 2026. Recent advances in rapid analytical technologies for the detection of illicit pharmaceutical adulteration in food: A review. TrAC Trends in Analytical Chemistry 199, 118785.

Cimini, A., Moresi, M., 2016. Assessment of the optimal operating conditions for pale lager clarification using novel ceramic hollow-fiber membranes. Journal of Food Engineering 173, 132–142.

Dairy Global, 2024. Precision fermentation: Disrupting dairy. Available at https://www.dairyglobal.net/industry-and-markets/market-trends/precision-fermentation-disrupting-dairy.

Demartini, M., Pinna, C., Tonelli, F., Terzi, S., Sansone, C., Testa, C., 2018. Food industry digitalization: From challenges and trends to opportunities and solutions. IFAC-Papers OnLine 51 (11), 1371–1378.

Fernández González, A., Badía Laíño, R., Costa-Fernández, J.M., Soldado, A., 2024. Progress and challenge of sensors for dairy food safety monitoring. Sensors 24 (5), 1383. Available at https://doi.org/10.3390/s24051383.

Ganesh, N., Shankar, R., Mahdal, M., Murugan, J., Chohan, J., Kalita, K., 2024. Exploring deep learning methods for computer vision applications across multiple sectors: Challenges and future trends. Computer Modeling in Engineering & Sciences 139 (1), 103.

Gibeson, A., 2022. Dairy industry turning to automation for packaging solutions. https://www.dairyprocessing.com/articles/1272-dairy-industry-turning-to-automation-for-packaging-solutions.

Grassi, S., Casiraghi, E., 2022. Advances in NIR spectroscopy analytical technology in food industries. Foods 11 (9), 1250.

Hassoun, A., Garcia-Garcia, G., Trollman, H., Jagtap, S., Parra-López, C., Cropotova, J., et al., 2023. Birth of dairy 4.0: Opportunities and challenges in adoption of fourth industrial revolution technologies in the production of milk and its derivatives. Current Research in Food Science 7, 100535.

He, H., Sun, D.W., Pu, H., Chen, L., Lin, L., 2019. Applications of Raman spectroscopic techniques for quality and safety evaluation of milk: A review of recent developments. Critical Reviews in Food Science and Nutrition 59 (5), 770–793.

Heema, R., Sivaranjani, S., Gnanalakshmi, K.S., 2022. An insight in to the automation of the dairy industry: A review. Asian Journal of Dairy and Food Research 41 (2), 125–131.

Hilgendorf, K., Wang, Y., Miller, M.J., Jin, Y.S., 2024. Precision fermentation for improving the quality, flavor, safety, and sustainability of foods. Current Opinion in Biotechnology 86, 103084. Available at https://doi.org/10.1016/j.copbio.2024.103084.

Hussain, A.S., 2021. Process Analytical Technology (PAT): Understanding validity of pharmaceutical quality control and assurance handbook of validation in pharmaceutical processes: Fourth Edition. CRC Press, United States, pp. 883–915.

Indumathy, M., Subramani, S., Panda, R.C., Panda, A., 2025. Quadratic dynamic predictive control framework for continuous high-temperature-short-time pasteurization process. International Journal of Food Engineering 21 (6), 399–421.

Jachimczyk, B., Tkaczyk, R., Piotrowski, T., Johansson, S., Kulesza, W., 2021. IoT-based dairy supply chain-an ontological approach. Elektronika ir Elektrotechnika 27 (1), 71–83.

Jox, D., Borsum, C., Hummel, D., Hinrichs, J., Krupitzer, C., 2025. Enhancing dairy processing with machine learning and domain knowledge: A combined analysis of offline and time series data. Journal of Food Engineering 391, 112423.

Kahraman, I., Köse, A., Koca, M., Anarim, E., 2023. Age of information in Internet of Things: A survey. IEEE Internet of Things Journal 11 (6), 9896–9914.

Kamilaris, A., Fonts, A., Prenafeta-Boldú, F.X., 2019. The rise of blockchain technology in agriculture and food supply chains. Trends in Food Science & Technology 91, 640–652.

Kannapinn, M., Dorer, D., Schäfer, M., Weeger, O., 2025. Digital twins for autonomous thermal food processing: A model predictive control study with reduced-order models of augmented neural ordinary differential equation type. Journal of Food Engineering 410, 112918.

Kaveti, B., 2023. Sensors in dairy processing: Ensuring milk quality and safety. AZoSensors. Available at https://www.azosensors.com/article.aspx?ArticleID=2919.

Knychala, M.M., Boing, L.A., Ienczak, J.L., Trichez, D., Stambuk, B.U., 2024. Precision fermentation as an alternative to animal protein, A review. Fermentation 10 (6), 315.

Kunes, R., Bartos, P., Iwasaka, G.K., Lang, A., Hankovec, T., Smutny, L., et al., 2021. In-line technologies for the analysis of important milk parameters during the milking process: A review. Agriculture 11 (3), 239.

Lanjewar, M.G., Parab, J.S., Kamat, R.K., 2024. Machine learning based technique to predict the water adulterant in milk using portable near infrared spectroscopy. Journal of Food Composition and Analysis 131, 106270.

Lee, J., Davari, H., Singh, J., Pandhare, V., 2018. Industrial artificial Intelligence for industry 4.0-based manufacturing systems. Manufacturing letters 18, 20–23.

Liang, Q., Xia, Y., Che, J., Liu, Y., Zhang, H., Guo, J., et al., 2025. Detection of water adulteration levels in milk using near-infrared spectroscopy combined with chemometrics. Journal of Dairy Science 108 (7), 6852–6866.

Liu, N., Qi, J., An, X., Wang, Y., 2023. A review on information technologies applicable to precision dairy farming: Focus on behavior, health monitoring, and the precise feeding of dairy cows. Agriculture 13 (10), 1858.

Liu, Y., Wen, Y., Hao, X., Xu, L., Zhao, X., Ping, J., 2025. Recent advances in analytical technologies and devices for raw milk quality. TrAC Trends in Analytical Chemistry 118510.

Loria, K., 2025. Robotics, AI enhancing supply chain efficiencies. DairyProcessing.com. Available at https://www.dairyprocessing.com/articles/3016-robotics-ai-enhancing-supply-chain-efficiencies.

Malik, M., Gahlawat, V.K., Mor, R.S., Hosseinian-Far, A., 2024. Towards white revolution 2.0: Challenges and opportunities for the industry 4.0 technologies in Indian dairy industry. Operations Management Research 17 (3), 811–832.

Malik, M., Gahlawat, V.K., Mor, R.S., Singh, M.K. (2024). Unlocking dairy traceability: Current trends, applications, and future opportunities. Future Foods, 10, 100426. Available at https://doi.org/10.1016/j.fufo.2024.100426.

Manzoki, M.C., Negreiros Piazenski, I., Mânica Candelário, J.P., Soccol, V.T., de Melo Pereira, G.V., Grace Karp, S., et al., 2025. Advances in bovine cultivated meat and precision fermentation-derived milk proteins: Technological, regulatory, and sensory perspectives. Food Reviews International 1–34.

Maruthamuthu, M.K., Rudge, S.R., Ardekani, A.M., Ladisch, M.R., Verma, M.S., 2020. Process analytical technologies and data analytics for the manufacture of monoclonal antibodies. Trends in Biotechnology 38 (10), 1169–1186.

Mhapsekar, R., Kilbane, D., Davy, S., Abraham, L., Fenelon, M., O'Shea, N., 2025. A systematic review of the internet of things and artificial intelligence applications in milk quality monitoring and analysis. International Journal of Dairy Technology 78 (3), e70049.

Michie, C., Andonovic, I., Davison, C., Hamilton, A., Tachtatzis, C., Jonsson, N., et al., 2020. The Internet of Things enhancing animal welfare and farm operational efficiency. Journal of Dairy Research 87 (S1), 20–27.

Mirsalami, S.M., Mirsalami, M., 2025. Advances in genetically engineered microorganisms: Transforming food production through precision fermentation and synthetic biology. Future Foods 11, 100601.

Mohammadi, S., Gowen, A., Luo, J., O'Donnell, C., 2024. Prediction of milk composition using multivariate chemometric modelling of infrared, Raman, and fluorescence spectroscopic data: A review. Food Control 165, 110658.

Moran, F., Sullivan, C., Keener, K., Cullen, P., 2017. Facilitating smart HACCP strategies with process analytical technology. Current Opinion in Food Science 17, 94–99. Available at https://doi.org/10.1016/j.cofs.2017.11.003 ISSN 2214-7993.

Negri, E., Fumagalli, L., Macchi, M., 2017. A review of the roles of digital twin in CPS-based production systems. Procedia Manufacturing 11, 939–948.

Panda, B.K., Panigrahi, S.S., Mishra, G., Kumar, V., 2023. Robotics for general material handling machines in food plants. In: Jafari, S.M., Malekjani, N. (Eds.), Transporting Operations of Food Materials Within Food Factories. Elsevier, pp. 341–372. Available at https://doi.org/10.1016/b978-0-12-818585-8.00005-2.

van Nuenen, A., Diaz, V.F., Diaz-Olivares, J.A., Saeys, W., Aernouts, B., 2024. On-farm NIR sensor for milk analysis: Exploitation of bias monitoring and bias correction. Spectrochimica Acta, Part A: Molecular and Biomolecular Spectroscopy 320, 124544.

Pu, Y., Pérez-Marín, D., O'Shea, N., Garrido-Varo, A., 2021. Recent advances in portable and handheld NIR spectrometers and applications in milk, cheese and dairy powders. Foods 10 (10), 2377.

Pu, Y.Y., O'Donnell, C., Tobin, J.T., O'Shea, N., 2020. Review of near-infrared spectroscopy as a process analytical technology for real-time product monitoring in dairy processing. International Dairy Journal 103, 104623.

Purlis, E., 2024. Digital twin methodology in food processing: Basic concepts and applications. Current Nutrition Reports 13 (4), 914–920.

Rao, S., Neethirajan, S., 2025. Computational architectures for precision dairy nutrition digital twins: A technical review and implementation framework. Sensors 25 (16), 4899.

Rathod, P., Dixit, S., 2020. Precision dairy farming: Opportunities and challenges for India. Indian Journal of Animal Sciences 90 (8), 1083–1094.

Ronchetti, F., Springer, L., & Purnhagen, K., 2024. The regulatory Landscape in the EU for Dairy products derived from precision fermentation. SpringerBriefs in Law.

Rose, K., Eldridge, S., Chapin, L., 2015. The internet of things: An overview. The Internet Society (ISOC) 80 (15), 1–53.

Rout, S., Srivastav, P.P., 2025. Recent trends in the production of proteins by precision fermentation for improving the quality and attributes of food: Role of genetic engineering towards next generation of food production. The Microbe 8, 100551.

Sandey, K.K., Qureshi, M.A., Meshram, B.D., Agrawal, A.K., Uprit, S., 2017. Robotics—An emerging technology in dairy industry. International Journal of Engineering Trends and Technology 43 (1), 58–62.

Sauer, P.C., Orzes, G., Davi, L., 2021. Toward SME 4.0: The Impact of Industry 4.0 technologies on SMEs' business models. Implementing Industry 4.0 in SMEs. Palgrave Macmillan, Cham, pp. 293–343.

Serrano-Torres, G.J., López-Naranjo, A.L., Larrea-Cuadrado, P.L., Mazón-Fierro, G., 2025. Transformation of the dairy supply chain through artificial intelligence: A systematic review. Sustainability 17 (3), 982.

Sibono, L., Tronci, S., Hedegaard, M.A.B., Errico, M., Grosso, M., 2025. PAT-driven dairy processing: A rheo-Raman-based kinetic model for in-line prediction of milk coagulation dynamics. Analytical and Bioanalytical Chemistry 417 (25), 5691–5702.

Siddiqui, M.N., Malik, J.A., Rehman, M.A.U., Khan, M.Y., Shahid, S., Saleem, M., 2024. Blockchain enhanced IoT integration for transparency and efficiency in smart agriculture and food supply chain Systems. Journal of Computing & Biomedical Informatics. Available at https://www.jcbi.org/index.php/Main/article/view/486.

Singh, L., Sharanagat, V.S., 2024. Application of biosensors against food-borne pathogens. Nutrition & Food Science 54 (1), 207–237.

Srivastava, A., Atri, C., 2025. The role of AI in microbial fermentation: Transforming industrial applications. Methods in Microbiology Vol. 56. Academic Press, pp. 267–285.

Stephens, N., Sexton, A.E., Driessen, C., 2019. Making sense of making meat: Key moments in the first 20 years of tissue engineering muscle to make food. Frontiers in Sustainable Food Systems 3, 45.

Strani, L. (2021). Process analytical technology approaches for dairy industry (Doctoral dissertation, Università degli Studi di Milano). AIR—Archivio istituzionale della ricerca, Università degli Studi di Milano.

Szpicer, A., Bińkowska, W., Stelmasiak, A., Zalewska, M., Wojtasik-Kalinowska, I., Piwowarski, K., et al., 2025. Computational fluid dynamics simulation of thermal processes in food technology and their applications in the food industry. Applied Sciences 15 (1), 424.

Tao, F., Zhang, H., Liu, A., Nee, A.Y., 2018. Digital twin in industry: State-of-the-art. IEEE Transactions on Industrial Informatics 15 (4), 2405–2415.

Tarapata, J., Murphy, T.R., Finnegan, E.W., O'Callaghan, T.F., O'Mahony, J.A., 2025. Approaches for measuring and predicting fouling during thermal processing of dairy solutions. Comprehensive Reviews in Food Science and Food Safety 24 (4), e70209.

Tarapoulouzi, M., Jia, W., Koidis, A., 2026. New developments and future challenges of non-destructive near-infrared spectroscopy sensors in the cheese industry. Sensors 26 (2), 556.

Thostrup, T.M., Pedersen, K.E., Tronci, S., Errico, M., 2022. Dynamic simulator and model predictive control of a milk pasteurizer. IFAC-PapersOnLine 55 (7), 538–543.

Treiblmaier, H., 2018. The impact of the blockchain on the supply chain: A theory-based research framework and a call for action. Supply Chain Management: An International Journal 23 (6), 545–559.

Tubb, C., Seba, T., 2021. Rethinking food and agriculture 2020-2030: The second domestication of plants and animals, the disruption of the cow, and the collapse of industrial livestock farming. Industrial Biotechnology 17 (2), 57–72 A.

Vargas-Bello-Pérez, E., Sandoval, O.R.E., Rodríguez-Piñeros, S., Ghavipanje, N., 2025. Blockchain and its future in the supply and value chains of the dairy industry. Journal of Dairy Research 92 (2), 127–134.

Verboven, P., Defraeye, T., Datta, A.K., Nicolai, B., 2020. Digital twins of food process operations: The next step for food process models? Current Opinion in Food Science 35, 79–87.

Verma, K., Duhan, P., Pal, D., Verma, P., Bansal, P., 2025. Precision fermentation for the next generation of food ingredients: Opportunities and challenges. Future Foods 100750.

Wolfert, S., Ge, L., Verdouw, C., Bogaardt, M.J., 2017. Big data in smart farming—A review. Agricultural Systems 153, 69–80.

Wu, Z., Christofides, P., Wu, W., Wang, Y., Abdullah, F., Alnajdi, A., et al., 2025. A tutorial review of machine learning-based model predictive control methods. Reviews in Chemical Engineering 41 (4), 359–400.

Yang, L., Sun, X., Chen, J., Zhang, J., Li, X., Qu, S., et al., 2024. Simultaneous determination of somatic cell count and total plate count in raw milk based on ATP bioluminescence assay. Analytica Chimica Acta 1331, 343338.

Zhang, Z.Y., Su, J.S., Xiong, H.M., 2025. Technology for the quantitative identification of dairy products based on Raman spectroscopy, chemometrics, and machine learning. Molecules 30 (2), 239.

Chapter 11

The differentiated socioeconomic impacts of Dairy 4.0 in biologically variable production regimes

Yassine Louzzani[1] **and Alexandre Cabagnols**[2]
[1]*Department of Sustainable Food Systems, Management and Innovation, VetAgro Sup, CleRMa (Clermont Recherche Management—Research in Sustainable Management), Lempdes, France,* [2]*Department of Humanities, INP Clermont-Auvergne, Polytech Clermont-Ferrand, CleRMa (Clermont Recherche Management—Research in Sustainable Management), Clermont-Ferrand, France*

11.1 Introduction

The Food and Agriculture Organization of the United Nations (FAO) defines digital transformation in agri-food systems as the integration of data-driven technologies to enhance productivity, sustainability, and resilience across agricultural value chains (Santos Valle & Kienzle, 2020). Within this framework, the concept of Dairy 4.0 has emerged to describe the application of Industry 4.0 technologies, such as smart sensors, cyber-physical systems, artificial intelligence, and advanced analytics, to dairy production and processing systems. The primary objective of Dairy 4.0 is generally framed in terms of performance optimization: improving production efficiency, strengthening quality control, enhancing traceability, and supporting real-time decision-making (Hassoun et al., 2023; Gehlot et al., 2022).

As dairy supply chains become increasingly interconnected and data-intensive, digital technologies are often presented as necessary tools to manage complexity and ensure competitiveness. From this perspective, digitalization appears as a strategic requirement for maintaining operational efficiency, ensuring compliance with safety standards, and reinforcing consumer trust. In line with broader Industry 4.0 narratives, the transformation is frequently interpreted as a progressive modernization pathway driven by technological upgrading.

Dairy 4.0. DOI: https://doi.org/10.1016/B978-0-443-44730-3.00013-4

However, such interpretations tend to assume that dairy production systems constitute stabilizable environments in which variability can be progressively reduced through improved monitoring and automation. Although financial constraints, skills gaps, and organizational barriers to adoption are widely recognized (Malik et al., 2024; Wojcicki et al., 2025), socioeconomic impacts are often assumed to increase with the degreeof technological integration. However, this assumption may overlook key structural characteristics of dairy production systems, particularly in segments where biological processes constrain the extent to which variability can be reduced.

In dairy activities structured by complex biological dynamics, variability is not merely a temporary inefficiency but an intrinsic component of production. Raw milk is a living matter, microbial ecosystems evolve nonlinearly, and transformation processes such as coagulation and ripening, in cheese technology, involve irreversible biochemical sequences. These characteristics limit the possibility of full standardization and generate asymmetric risk structures, where rare deviations may have disproportionate economic or sanitary consequences. In such contexts, production rationality extends beyond average optimization and is fundamentally oriented toward robustness (i.e., the capacity to absorb variability while preventing critical failures).

The emergence of Dairy 4.0 technologies therefore raises a broader analytical question: do digital tools primarily function as optimization devices, or do they become components of robust architectures designed to manage irreducible biological variability? Addressing this issue requires moving beyond diffusion-centered interpretations of digital transformation and examining the production regimes into which technologies are embedded.

This chapter introduces the concept of a biologically variable production regime (BVPR) to provide such an analytical framework. A BVPR is defined as a production configuration characterized by (i) constitutive biological variability, (ii) asymmetric loss structures associated with rare but high-impact deviations, and (iii) a structural requirement for robustness that organizes the interaction between digital instrumentation and human expertise. Within this framework, digital technologies are not neutral performance enhancers but elements whose socioeconomic effects depend on their articulation with regime-specific constraints.

Drawing on an in-depth case study of Protected Designation of Origin (PDO) cheese dairies in Auvergne, the chapter analyses how BVPR properties shape the trajectories of Dairy 4.0 adoption and generate differentiated socioeconomic impacts at organizational, sectoral, and territorial levels. It proposes an integrated model linking biological variability, robustness-oriented decision rationality, and path-dependent investment dynamics. By doing so, the chapter contributes to the Dairy 4.0 literature through a shift in analytical focus from technology-driven diffusion to regime-dependent transformation dynamics. It demonstrates that digital transformation in dairy systems is not technologically deterministic but contingent upon the structural properties of the production regimes in which it unfolds.

The chapter first discusses the limitations of the implicit assumptions of uniform diffusion of Dairy 4.0 and proposes an integrated analytical framework. It then formalizes the concept of BVPR and examines how its properties are translated into organizational practices through path dependence and decision-making rationality. The empirical study of PDO cheese dairies in Auvergne is presented to illustrate these mechanisms. Finally, the model is consolidated, and its implications are discussed, with particular attention to economic evaluation and public policy.

11.2 Rethinking Dairy 4.0 beyond the uniform diffusion

11.2.1 The promises of Dairy 4.0 and their implicit assumptions

Recent literature on Dairy 4.0 largely follows on from work on Industry 4.0, emphasizing the integration of IoT (Internet of Things) sensors, cyber-physical systems, advanced analytics, and artificial intelligence into dairy processing chains (Hassoun et al., 2023; Gehlot et al., 2022). This highlights the expected gains in terms of yield optimization, enhanced traceability, reduced losses, and improved quality.

From this perspective, Dairy 4.0 transformation is mainly seen as a lever for increasing average performance (i.e., better resource allocation, reduction of uncontrolled variability, and continuous improvement of productivity indicators). Even when the authors acknowledge the existence of barriers to adoption—financial constraints, skills shortages, or organizational fragmentation (Malik et al., 2024; Wojcicki et al., 2025)—the implicit trajectory remains one of gradual diffusion toward a more integrated, automated, and optimized system that exploits the full functional potential of these technologies.

However, this approach relies on an assumption that is rarely made explicit: that the production regime is sufficiently stable for algorithmic optimization to be a rational objective from the point of view of economic actors. In our view, this assumption warrants discussion.

11.2.2 Limitations of a uniform diffusion interpretation: the question of the production regime

Work in bioprocessing and biological systems engineering has demonstrated that, in certain production environments, heterogeneity is not a transient imperfection to be corrected, but a constitutive property of the system (Mu & Zhang, 2023). Nonlinear interactions between biological parameters, sensitivity to micro-environmental variations, and the possibility of rare but high-impact deviations structurally limit the full stabilization of the process.

In these configurations, engineering rationality cannot be thought solely in terms of average optimization; it must incorporate a logic of robustness, understood as the system's ability to maintain functional viability in the face of inevitable disturbances (Becker et al., 2023). The distinction between optimization and robustness then becomes analytical: the former seeks to maximize performance under the assumption of stability, whereas the latter aims to preserve system viability under structural uncertainty.

This perspective invites a shift in analytical focus: rather than questioning the speed of adoption of Dairy 4.0, attention should be directed toward the structural properties of the productive regime within which digital adoption takes place.

11.2.3 Toward an integrated analytical framework: BVPR, path dependence, and decision-making rationality

This chapter proposes to analyze the Dairy 4.0 transformation in specialized dairy sectors through three interdependent dimensions: (1) the properties of the biologically variable production regimes (BVPR); (2) the decision-making rationality of actors under constraints of robustness; and (3) cumulative path dependence at the technical, organizational, and cognitive levels.

The concept of BVPR refers to production regimes characterized by *(i)* constitutive and multilevel biological variability, *(ii)* asymmetric loss structures dominated by low probability but high impact events, and *(iii)* a persistent centrality of human judgment in the regulation of nonstandardized situations.

In such systems, the adoption of Dairy 4.0 cannot be conceived as a simple trajectory of uniform diffusion. Fig. 11.1 formalizes this mechanism by showing how Dairy 4.0 adoption is embedded in inherited architectures shaped by successive adjustments to biological variability.

11.2.4 From decision to socioeconomic impacts: a structured multilevel mechanism

The proposed framework implies that the socioeconomic impacts of Dairy 4.0 cannot be analyzed as the direct effects of a technology applied to a supposedly neutral production regime. They result from a sequence structured by three levels of analysis: properties of the production regime, inherited trajectories, and decision-making rationality under robustness constraints.

At the micro-organizational level, investment decisions are primarily driven by the mitigation of identifiable vulnerabilities rather than by the identification of a theoretical optimum. This initial filtering effect shapes the nature of the technologies effectively implemented. Some process sequences are readily automated insofar they are unlikely to compromise system robustness, whereas more risk-sensitive stages remain under human supervision, resulting in persistent forms of hybridization.

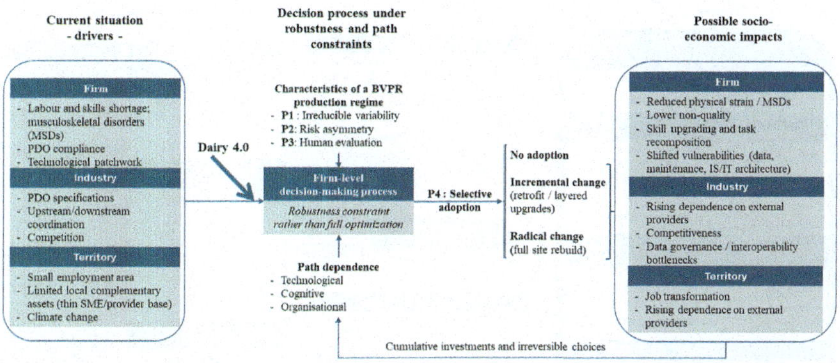

FIGURE 11.1 Decision process under robustness constraint (BVPR). This figure represents a causal chain: company–sector–territory constraints feed into a decision-making process guided by a robustness constraint (preventing critical losses) rather than by average optimization. This process is embedded in a path dependence: inherited choices and infrastructures (equipment, information systems, routines, skills) limit the range of options and make certain trajectories costly to revise. The result is selective adoption of Dairy 4.0, which has differentiated impacts on the company, the sector (interoperability/supplier dependence) and the territories (job restructuring, dependencies).

At the sectoral level, this selective adoption unfolds along differentiated trajectories depending on the company's absorption capacities. Preexisting stratified architectures and firm-specific assets reinforce cumulative effects, which are likely to accentuate the gaps between actors capable of integrating complex systems and those constrained to incremental or partial adjustments (Wojcicki et al., 2025).

At the territorial level, Dairy 4.0 transformation interacts with the availability of complementary assets—skills, training programmes, service providers networks—which determine local absorption capacity (Sharafizad et al., 2022). Regional ecosystems thus mediate the diffusion and effective use of digital technologies. The observed impacts are therefore neither uniform nor mechanical; they depend on institutional and organizational embeddedness of firms within regional ecosystems. Fig. 11.2 summarizes the interactions between the productive context, adoption decisions, and potential socio-economic consequences, explicitly integrating micro, sectoral, and territorial dimensions into a coherent analytical mechanism.

11.3 Biologically variable production regimes

11.3.1 Constitutive variability and partial controllability

BVPR are distinguished by a structuring property: variability is not a residual imperfection of the system but a constitutive characteristic of the production

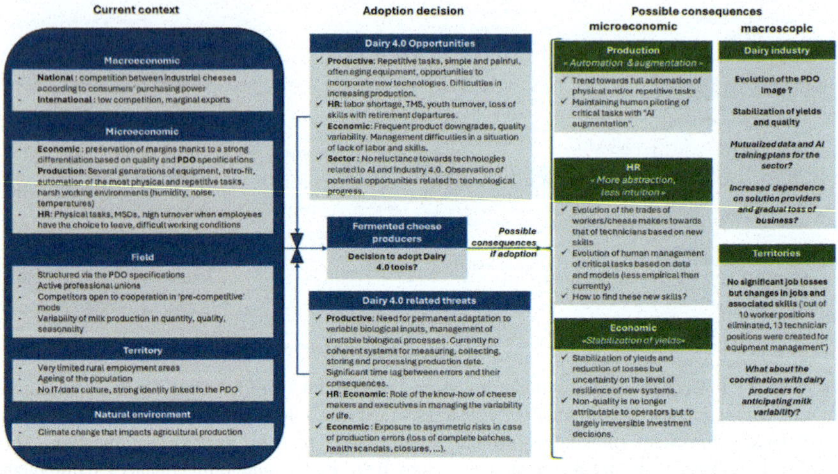

FIGURE 11.2 Multilevel analytical framework for the adoption of Dairy 4.0 and its socioeconomic impacts. This figure summarizes a multilevel chain: context (macro, micro, PDO sector, territory, environment) → trade-off between opportunities and threats of Dairy 4.0 → adoption decision (often selective) → socioeconomic impacts. These impacts are broken down into three levels: company (automation/augmentation, reskilling, stabilization of yields/quality), sector (pooling vs supplier dependence), and territories (reskilling and coordination needs). This multilevel articulation justifies an empirical analysis that pays close attention to the interactions between production constraints, cumulative trajectories, and decision-making trade-offs. It is precisely this approach that we are using in the illustrative example of PDO cheese dairies in Auvergne.

process. Unlike industrial systems based on standardized inputs and stabilizable technical sequences, BVPRs rely on living substrates whose properties evolve under the influence of endogenous dynamics and changing environmental conditions.

Recent work in bioprocessing highlights the existence of multilevel heterogeneity in biological production regimes: cellular variability, metabolic fluctuations, genetic instability and sensitivity to micro-variations in the environment (Mu & Zhang, 2023; Eigenfeld & Schwaminger, 2025). This heterogeneity cannot be eliminated by increased instrumentation or more refined modeling; it is inherent to the nonlinear interactions specific to living systems.

In such configurations, the process is only partially controllable. Engineering interventions can attenuate some sources of variance, but it cannot ensure complete and definitive stabilization. Becker et al. (2023) argue that, in bioprocesses, robustness should be understood as the ability to maintain operation within acceptable performance ranges despite inevitable disturbances, rather than as the eradication of variability itself.

The distinction between controllability and stabilization is therefore central. A BVPR can be monitored, adjusted, and supervised; it cannot be entirely determined. This structural limitation changes the very meaning of optimization. In highly standardized production regimes, optimization may approximate an asymptotic reduction of deviations from a predefined target. In a BVPR, by contrast, the objective shifts from absolute variance minimization to the dynamic regulation of unavoidable fluctuations.

This constitutive variability has several analytical consequences. First, it increases informational management complexity: not all relevant variables are observable, measurable or interpretable in real-time. Second, it increases the sensitivity to cross-interactions between technical parameters and biological dynamics. Third, it makes the emergence of unexpected drift structurally plausible, even in the presence of formalized control architectures.

Accordingly, Dairy 4.0 within a BVPR cannot be conceptualized as a linear progression toward greater instrumental precision. It unfolds in an environment where uncertainty does not vanish with the accumulation of data; it is reconfigured. While data collection and advanced analytic improve diagnostic and predictive capabilities, they do not convert a biologically variable regime into a fully stabilized one.

Recognition of this partial controllability is the first conceptual building block of the BVPR. It implies that the numerical trajectories observed in these regimes cannot be evaluated solely in terms of progress toward full control. Biological variability sets structural limits on standardization and opens a sustainable space for adaptive adjustment.

11.3.2 Asymmetry of losses and severity of deviations

Beyond their inherent variability, BVPRs are characterized by a distinct risk architecture: losses are distributed asymmetrically. While the ordinary functioning of the system may appear relatively stable, certain rare disturbances can generate disproportionate technical, sanitary, or economic consequences. This asymmetric risk structure fundamentally reshapes productive rationality.

In complex biological systems, nonlinear dynamics and cross-interactions between parameters can produce threshold effects, whereby an apparently marginal variation triggers a qualitative shift in the process. Becker et al. (2023) emphasize that robustness in bioprocesses involves not only absorbing continuous fluctuations but also preventing or containing critical deviations capable of compromising the system viability. The analytical focus therefore extends beyond average variance to the management of extreme events.

This asymmetry distinguishes BVPRs from industrial systems in which deviations are typically gradual and broadly proportional to their causes. In a BVPR, certain failures may entail costs far disproportionate to both the likelihood and the magnitude of the initial disturbance, such as the loss of an entire production batch, irreversible product degradation, breaches of sanitary compliance, or damage to collective reputation. Risk exposure is thus structurally nonlinear rather than incrementally distributed.

Research on genetic stability in biological systems illustrates this dynamic. Even under advanced control architectures, escape or drift phenomena may occur infrequently yet produce substantial consequences (Clavier et al., 2024). The persistence of such rare but critical events demonstrates the effective control cannot rely exclusively on reducing routine variability; it must incorporate the structural possibility of exceptional deviations.

In this context, productive performance cannot be assessed exclusively through average efficiency indicators. The prevention of critical failures becomes a primary objective, reinforcing the centrality of organizational vigilance, anticipatory monitoring, and responsive intervention capacities. Optimization is thus subordinated to a broader logic of operational robustness.

The asymmetry of losses constitutes the second conceptual pillar of BVPR framework. Combined with constitutive biological variability, it explains why digitalization in these regimes is oriented less toward extensive automation than toward the securing of critical control points. The underlying risk structure consequently shapes investment appraisal and reveals a decision-making rationality centered on robustness, a dynamic examined in the following section.

11.3.3 Persistent hybridization and the centrality of human judgment

The combination of constitutive variability and loss asymmetry leads to a third structuring property of BVPRs: the persistent centrality of human judgment in the regulation of the production process. This centrality is neither a symbolic attachment to traditional practices nor a simple technological lag; it stems from the structural limits of controllability in biologically variable systems and the asymmetry distribution of risks.

In complex bioprocesses, which typically resemble rugged adaptation landscapes (NK models), control does not rely exclusively on the mechanical application of fixed instructions (univariate optimization). Instead, it relies on the ability to navigate cautiously within acceptable multivariate ranges, where parameters interact dynamically (Kunzelmann et al., 2024). Defining these ranges does not eliminate uncertainty; it provides a structured framework for its expression. Interpreting weak signals, identifying significant deviations, and adjusting to evolving situations therefore requires situational competence.

Advances in Process Analytics 4.0 and real-time instrumentation demonstrate that the integration of sensors, predictive models and advanced analytics significantly improves diagnostic and anticipatory capabilities (Wasalathanthri et al., 2021). However, these technological tools do not transform biological systems into perfectly deterministic. They increase informational visibility, but the final decisions remain subject to human interpretation and arbitration.

Within a BVPR, the hybridization between technical instrumentation and human expertise is therefore a structural and persistent configuration. Automation can stabilize routine or repetitive sequences, but critical milestones—where biological and technical interactions can produce irreversible drifts—require adaptive capacity that algorithms cannot fully formalize. Human presence functions as an adaptive adjustment mechanism in the face of partially observable uncertainties.

This persistent hybridization has several analytical implications. First, it challenges the model relevance of full automation as the universal horizon of Dairy 4.0. Second, it redefines the nature of required skills: value lies not only in the execution of technical tasks, but in the ability to interpret heterogeneous information and intervene contextually. Third, it reinforces the organizational dimension of robustness: system stability depends as much on the articulation between humans and machines as on the intrinsic performance of technical devices.

Consequently, the centrality of human judgment complements the two previous properties of BVPR. Constitutive variability makes full stabilization impossible; the asymmetry of losses makes certain deviations critical; persistent hybridization ensures the ability to adjust in the face of these limitations. Together, these three dimensions define a productive regime in which Dairy 4.0 cannot be understood as a simple substitution of human labor, but as a structured reconfiguration of the relationship between technology and expertise.

11.3.4 Summary definition and analytical scope of the BVPR concept

The preceding analysis stabilizes and formalizes the concept of BVPR as a distinct analytical category. A BVPR is not defined by a specific sector or a level of technological sophistication, but by the combination of three interdependent properties: (i) constitutive variability of the substrate and processes, resulting in partial controllability of the outcome; (ii) an asymmetry of losses, in which rare events can have disproportionate consequences; and (iii) a persistent hybridization between technical instrumentation and human judgment, necessary to keep the system within viable ranges.

These properties are not additive; they reinforce each other. Constitutive variability increases the probability of unforeseen deviations; loss asymmetry makes certain deviations critical; hybridization ensures the capacity for adaptive adjustment in the face of these structural limitations. Together, they configure a productive regime in which full optimization is structurally limited, and robustness becomes an organizing principle.

This formalization makes it possible to distinguish BVPRs from highly standardized production regimes. In the latter, the gradual stabilization of inputs and technical sequences makes it plausible to pursue extensive automation aimed at maximizing average performance. Conversely, in a BVPR, the very nature of the processed substrate introduces an intrinsic limit to variance reduction, which alters the hierarchy of production objectives.

The concept of BVPR thus has a scope that extends beyond cheese making. It can be applied not only to activities based on complex biological processes—industrial fermentation (Mu & Zhang, 2023), pharmaceutical bioproduction (Kunzelmann et al., 2024; Gasset et al., 2024), certain specialized agricultural production—but also to any process involving non-linear interactions and dynamic equilibria (Becker et al., 2023; Levinthal, 1997) accompanied by an asymmetric risk distribution (Clavier et al., 2024). It provides a framework for analyzing technological adoption trajectories not in terms of delay or advance, but in terms of the productive constraints specific to each configuration.

By defining BVPR as a regime characterized by partial controllability, asymmetry of losses and structural hybridization, we now have a conceptual tool for analyzing Dairy 4.0 transformation in relation to the nature of the productive system. The following sections examine how these intrinsic properties translate organizationally through cumulative trajectories and decision-making trade-offs under robustness constraints.

11.4 Path dependence and decision-making rationality

11.4.1 Cumulative trajectories and stratified architectures

The notion of path dependence refers to processes whereby initial choices exert lasting impact on subsequent trajectories, due to self-reinforcing mechanisms and sunk costs (David, 1985; Arthur, 1989). Subsequent research has clarified that organizational lock-in results from cumulative sequences in which routines, specific investments and learning effects gradually stabilize certain configurations at the expense of alternatives (Vergne & Durand, 2010; Sydow et al., 2009). However, in the context of a BVPR, path dependence cannot be reduced to organizational and technological inertia alone. It must be understood as the cumulative materialization of adjustments designed to manage inherent variability and asymmetric risk structures. In such regimes, each technical or organizational innovation aims primarily to stabilize a fragile equilibrium.

The measures introduced—whether equipment, procedures, or information systems—are calibrated to absorb specific forms of variability and prevent critical deviations. They are therefore highly contextualized and integrated into an already structured configuration. This gradual integration produces stratified architectures. New solutions rarely displace existing ones entirely; instead, they overlap, complement, or selectively reconfigure them. The productive trajectory becomes cumulative: past choices constrain the scope for future change, not only because of sunk costs, but also because they actively contribute to maintaining partial controllability.

Path dependence in a BVPR is therefore co-produced by the search for robustness. Accumulated investments are not merely rigidities; they constitute stabilizing supports of present viability. Radical bifurcations become risky insofar as they may destabilize equilibria built to absorb persistent biological variability.

11.4.2 Specific assets, lock-in, and bifurcation limits

The cumulative trajectory described above materializes in specific assets that reinforce lock-in effects. In the path dependence literature, lock-in refers to the durable stabilization of a technological or organizational configuration, made difficult to reverse by the accumulation of sunk costs, specialized learning and systemic interdependencies (Arthur, 1989; Vergne & Durand, 2010). In a BVPR, however, lock-in assumes a distinctive character. Specific assets are not simply limited to standardized material investments; they are adjusted to a partially stabilized biological process and to a site specific environmental configuration. Equipment, settings, diagnostic routines, and embedded skills are calibrated to absorb variability and secure critical milestones. Their value is therefore highly contextual and relational.

This contextual specificity increases bifurcation costs. Replacing a technological component does not simply mean substituting one technology for another; it involves reconfiguring an interdependent set of technical and organizational balances underpinning system robustness. The costs involved are not only financial, but also cognitive and operational. The transition period associated to a radical redesign heightens exposure to systemic risk, a particularly sensitive issue in regimes characterized by loss asymmetry.

Lock-in in a BVPR is thus ambivalent. On the one hand, it constrains rapid adoption of radical innovations. On the other hand, it acts as a protective mechanism: accumulated assets embody proven solutions for absorbing variability and preventing critical drifts. Apparent stability is therefore not only inertial; it is functional. This ambivalence distinguishes BVPRs lock-in from industrial configurations where path dependence is mainly interpreted as inefficiency or rigidity. In BVPRs, architectural continuity contributes directly to robustness. Technological bifurcation becomes a delicate trade-off between potential optimization and the risk of destabilizing a system that is already partially under control.

In this sense, path dependence in a BVPR cannot be interpreted solely as a historical constraint; it is co-produced by the need to preserve biologically sensitive balances. This configuration paves the way for the emergence of a specific decision-making rationality, in which investments are prioritized according to their contribution to robustness rather than to the abstract maximization of performance.

11.4.3 Decision-making rationality under robustness constraints

In a BVPR, the structural properties of the system—constitutive variability, asymmetry of losses, and persistent hybridization—frame the boundaries of organizational decision-making. Technological and organizational choices are not made in a context of abstract optimization, but within the constraints of preserving a partially stabilized and biologically sensitive process. The decision-making rationality observed in these regimes can be described as robustness-oriented rationality. Its primary goal is not the maximization of average performance indicators, but the reduction of vulnerabilities that could lead to critical failures. This orientation is consistent with the concept of robustness developed in bioprocess engineering, where the central challenge is to maintain the system within acceptable operating ranges despite inevitable disturbances (Becker et al., 2023). This rationality changes the hierarchy of investments. Projects are evaluated not only in terms of potential productivity gains, but also in terms of their contribution to securing sensitive milestones, limiting extreme losses and ensuring business continuity. In such contexts, return on investment (ROI) can be difficult to isolate using traditional criteria, as a significant part of the value lies in avoiding rare events and stabilizing the organization.

Furthermore, the previously described path dependence constrains radical transformation. Decisions are made in an environment of specific assets and layered architectures, where any major transformation carries the risk of temporarily destabilizing the system. However, path dependence does not only refer to a constraint inherited from the past; it can also be understood as a rational strategy in a context of high uncertainty. In systems characterized by strong technical and biological interdependencies, global transformations simultaneously alter multiple variables whose interactions remain only partially predictable. Decision-making rationality therefore incorporates a prudential dimension: technological adoption tends to be gradual, selective, and hierarchically sequenced to minimize systemic exposure.

The result is a configuration in which digital transformation does not follow a uniform or linear trajectory toward increasing automation. Instead, it is part of a cumulative dynamic, driven by the search for robust equilibria rather than theoretical optima. Organizational choices reflect a continuous trade-off between optimization potential and the preservation of stability. This rationality, constrained by the need for robustness, does not only produce technical effects; it also shapes the socioeconomic trajectories observable at different levels of analysis – micro-organizational, sectoral and territorial. From this perspective, the empirical study of PDO cheese dairies in Auvergne provides a concrete illustration of how these mechanisms operate within a BVPR.

11.5 Case study: Auvergne protected designation of origin cheese dairies as a biologically variable production regime

The analysis of the socioeconomic impacts of Dairy 4.0 conducted in this chapter falls within the framework of BVPR. In such regimes, complete stabilization of the process is structurally constrained, not by a lack of information, but by the intrinsic nonlinearities and interdependencies of biological systems (Mu & Zhang, 2023). Therefore, the engineering challenge is not only to maximize average performance, but also to build robustness against rare yet potentially critical deviations (Becker et al., 2023).

The choice of Auvergne PDO cheese dairies is relevant for two reasons. On the one hand, the sector is a regional socioeconomic pillar. The five main PDOs (Saint-Nectaire, Cantal, Salers, Bleu d'Auvergne, Fourme d'Ambert) represent more than 60,000 tons of annual production, involve several thousand dairy farms and constitute a dense network of artisanal workshops, cooperatives and private companies. In certain areas of Puy-de-Dôme and Cantal, cheese processing is one of the main industrial employers. As a result, Dairy 4.0 adoption trajectories have tangible territorial implications. On the other hand, this setting offers a particularly suitable empirical laboratory to observe the three theoretical dimensions previously formalized: *(i)* the properties of the production regime, *(ii)* the role of stratified (layered) technical architectures , and *(iii)* decision-making under constraints of robustness. This articulation allows the analysis of Dairy 4.0 not as a linear diffusion of technologies, but as a transformation embedded in specific productive and institutional constraints. The methodology adopted for this empirical setting is presented in Table 11.1.

11.5.1 Structural properties of the protected designation of origin production regime: empirical confirmation of the biologically variable production regime

This section empirically examines the three constituent properties of BVPR defined above: constitutive variability, losses asymmetry, and structural hybridization.

11.5.1.1 Irreducible biological variability and limits to stabilization

The first structural property relates to the constitutive variability of the living substrate. Milk is not a homogeneous input; it evolves under the combined influence of the seasons, climatic conditions, farming practices and microbial dynamics.

E4 describes this instability and its operational implications very directly: "Milk is a living material that changes throughout the year. we are constantly

TABLE 11.1 Data and methods.

Dimension	Description	Analytical justification
Field of study	Auvergne PDO cheese dairies	Relevant case of BVPR with strong regional socioeconomic embeddedness
Data collection period	January 2025–February 2026	Allow the observation of relatively stabilized Dairy 4.0 adoption trajectories
Number of interviews	5 semi-structured interviews (with key informants)	In-depth qualitative approach focused on identifying mechanisms
Interview profiles	E1–E2: ExecutivesE3: Industrial site managerE4: Industrial/group managementExpert: Sector specialist	Ensures multilevel coverage (workshop, site, group,)
Sampling strategy	Purposive (theoretical) sampling	Organizational diversity (cooperatives, regional groups, national groups, sector-level actor) to identify recurrent mechanisms rather than exceptional cases
Coding—Stage 1	Deductive coding	Guided by the analytical sequence: BVPR → robustness constraint → selective adoption → socioeconomic impacts
Coding—Stage 2	Inductive coding	Identification of recurring subthemes: interoperability, maintenance, hybrid skills, investment trade-offs, supplier dependency
Validation strategy	Cross-profile triangulation	Consistency check of interpretations across decision-making levels
Status of figures	Synthetic analytical models	Aim at analytical generalization (transferable mechanisms), not statistical generalization
Limitations	Limited sample size	Focus on plausibility and inter-actor consistency rather than exhaustiveness

chasing after technological modifications to keep up with the variations in milk, because it no longer meets our technical parameters as we had defined them."

Variability is not only a question of average levels (fat, protein), but also of parameters that are sometimes unobservable and difficult to anticipate. E2 emphasises this limitation: "We have a process that still has variables that are not measurable and that we cannot really anticipate."

At the same time, the raw material is not treated as a standardized input, which strongly distinguishes PDO production from other segments of the dairy industry. E1 states: "We are not dealing with products that are standardised in terms of fat and protein content, as other commercial products may be."

Variability also affects the production environment, particularly through delicate microbial balances: As Expert explains: "We are not in sterile chambers, we are in living, variable environments. these are delicate microbial balances. that we do not always control."

These elements converge toward a form of systemic variability affecting inputs, processes, and environments, that cannot be eliminated. It must be managed and absorbed. In this sense, heterogeneity is not a temporary imperfection but a structural condition of biological production regimes (Mu & Zhang, 2023), requiring management oriented toward robustness rather than full standardization (Becker et al., 2023).

11.5.1.2 Risk asymmetry and milestone criticality

The second property of BVPR lies in its asymmetric risk structure. Most production days are "similar"; yet rare deviations may generate disproportionate consequences.

E1 describes a particularly revealing situation: deviations can occur despite compliant intermediate controls:

"We have products that deviate even though all our control points during the process are good, and we can't find an explanation. we realise that it's really multifactorial."

This multifactorial dimension explains why certain deviations escape strictly instrumental control logic. Health risks are another aspect of this asymmetry. E3 explicitly mentions the prospect of a rare but major event: "With the fear [.] of pathogenic bacterial contamination and potentially a crisis involving huge recalls to manage."

Finally, some of the risks are expressed at the product level itself, through micro-imbalances that can produce irreversible defects. E4 gives a concrete example of ripening:

"If my cheeses have a slight tendency to release a little whey during ripening [.] on the side that is on the board, the whey will stagnate and damage the rind. and then my rind is ruined."

Under these conditions, performance is not primarily about marginal maximization of yields; it depends on the ability to avoid deviations outside acceptable ranges. This logic is consistent with a conception of robustness as the ability to maintain viability in the face of rare and critical disturbances (Becker et al., 2023).

Dairy 4.0 technologies are therefore mobilized as instruments for securing milestones and reducing vulnerabilities, rather than as tools for average productivity optimization. Robustness, maintaining viability in the face of rare and critical disturbances, becomes the organizing principle structuring digital investment.

11.5.1.3 Centrality of human judgment and tacit knowledge as a mechanism of robustness

The third structuring property of BVPRs is the centrality of human judgment. This is not a symbolic attachment to "know–how," but a functional necessity. Expertise becomes decisive when variability and risks exceed what formalized routines and instruments can guarantee.

E3 clearly expresses this role in situations of deviation:

"On the day we stray too far from the specifications, the only person who can ensure that we don't throw everything away is the cheesemaker."

Expertise is built through confrontation with events and is difficult to transfer solely through written procedures. E2 highlights this experiential dimension:

"As long as they haven't had the problem, even if we tell them about it verbally, it's not registered. They have to face the problem to know how to react."

This centrality creates organisational vulnerability, as key knowledge may be concentrated in specific individuals. E1 notes:

"We often realise when people retire that they held a lot of the keys to our technical processes. And that's dangerous for our companies."

The expert further indicates that this human role is institutionally maintained: "We have reinforced and maintained the role of humans as controllers of certain stages of manufacturing."

In this context, Dairy 4.0 does not lead to the marginalization of human expertise. Instead, digital instrumentation and tacit knowledge form a persistent hybrid architecture. Automation supports monitoring and traceability, but decision-making authority remains anchored in situated expertise at critical milestones. The PDO production regime thus confirms the BVPR configuration: a structurally hybrid system in which humans remain a core component of control, consistent with the idea that the management of biological processes falls within acceptable multivariate ranges rather than the mechanical execution of instructions (Kunzelmann et al., 2024).

11.5.2 Cumulative trajectories and path dependence in protected designation of origin cheese dairies

The Dairy 4.0 transition observed in PDO cheese dairies cannot be understood as the adoption of digital technologies ex nihilo. It is part of a long industrial trajectory, where technical and informational architecture has accumulated in successive layers. Understanding Dairy 4.0 therefore requires situating digital technologies within these inherited configurations.

Workshops are rarely homogeneous. They combine recent equipment with older devices that remain functional but are difficult to integrate into modern digital architectures. E2 refers to the existence of "30- or 40-year-old controllers" that "cannot export data." This persistence of ageing is not merely a sign of delay; it creates a structural integration constraint. Digitization involves making heterogeneous technical layers interoperable. Information systems reflect similar stratification. E1 summarizes: "We're navigating through lots of databases. it's a mess." Digital transformation thus becomes as much an issue of data architecture and governance as of physical equipment.

Large-scale projects involve workshop shutdowns, revalidations and involve managing disruption risks at sensitive milestones. E4 highlights the continuity constraint: "Large projects require the factory to shut down. you have to plan one or two years in advance."

Although a complete rebuild may appear technically coherent, it is often economically and operationally unrealistic. As E2 notes: "It would be much simpler to start from scratch everywhere. but the entry ticket is enormous."

Dairy 4.0 in this configuration takes the form of gradual integration of new technology building blocks into the existing structure. It is cumulative rather than disruptive.

Path dependency is reinforced by the specificity of assets and skills. Lines and equipment are often site-specific and product-specific. E2 emphasizes: "We don't buy many machines from catalogues. Factories are really giant prototypes of specialised machinery." Such specificity increases switching costs and deepens dependence on integrators. Maintenance becomes strategic. It explains why digital adaptation often takes the form of gradual organizational learning, typical of the gradual transformations of agrifood SMEs (Cannas, 2023).

Dairy 4.0 must therefore be inserted into a legacy system without increasing process vulnerability. Digital transformation materializes as layered adaptation rather than systemic rupture.

11.5.3 Decision-making process: securing, prioritizing, recomposing

The properties of BVPR (5.1) and path dependence (5.2) do not imply a given level of digitalization; they provide a framework for a specific decision-making rationality. The technological choices observed are prioritized according to their contribution to robustness: reducing vulnerabilities, securing critical milestones, sustainability of work, and business continuity.

11.5.3.1 Securing vulnerabilities and selective hybridization

Investments are not triggered by the search for marginal productivity gains, but by the identification of specific points of fragility for which solutions do not increase systemic risks in the event of failure. E1 explains:

"We are in the process of mechanising, installing sensors and using automation much more to reduce risks."

The aim is to avoid simple but costly mistakes:

"The operator might put a Cantal label on a Saint-Nectaire label. These are things that cost us a lot of money."

Mechanization also aims to reduce drudgery and stabilize the workforce. As the expert points out:

"Mechanisation was one of the ideas for removing this constraint."

However, this security does not lead to complete automation. A distinction is made between tasks that can be stabilized and decision-making functions linked to the variability of living organisms. E3 formulates this boundary:

"Tasks that really have no added value, these are not a problem to automate. On the other hand, real tasks with real technological decision-making power cannot be automated."

E2 confirms that hybridization provides a stable balance:

"We cannot be completely automatic. Each time, we use semi-automation so that we can still adjust to the product."

Dairy 4.0 thus appears to be selective: it enhances the reliability of peripheral routines while maintaining human decision-making space on critical milestones.

11.5.3.2 Budgetary constraints and organizational restructuring

Digital investment competes with other essential expenditures. E1 mentions a change in the information system representing "a budget of €500,000." The Expert reminds that certain infrastructures are essential: "If my wastewater treatment plant doesn't work, I can't manufacture anything."

Digital decision is therefore relative, embedded in a hierarchy of critical investments. Organizational effects are qualitative rather than purely quantitative. E2 states: "I eliminated 10 operator positions, but I created 12 electromechanical engineer positions."

E4 describes a transformation of work constraints: "Before, they exerted themselves physically, now they are more mentally fatigued."

Dairy 4.0 does not eliminate expertise; it transforms its nature and redistributes vulnerabilities toward increased dependence on scarce intermediate technical skills. Robustness now depends as much on maintaining complex technical architectures as on mastering biological processes.

The decision-making trade-offs analyzed above go beyond the strictly technical framework of the workshop. They shape the conditions under which digital transformation is spreading and, in doing so, pave the way for organizational and territorial effects that now need to be examined.

11.5.4 Sectoral and territorial socioeconomic impacts of Dairy 4.0 under robustness constraints

The socioeconomic impacts observed in Auvergne PDO cheese dairies present a specific configuration of Dairy 4.0, filtered through biological variability, path dependence, and robustness-oriented decision-making.

In this sector, automation remains selective. Biologically sensitive milestones remain under human control, while stabilizable sequences are instrumented and fully or partially automated. As a result, Dairy 4.0 does not eliminate cheesemaking expertise; it reconfigures it toward integration, supervision, and maintenance.

At the sectoral level, differentiated trajectories emerge. Companies with sufficient engineering capabilities and financial resources can absorb the costs of achieving interoperability across heterogeneous technical and informational systems and strengthen organizational robustness. Others remain in partial adjustment, accentuating intra-sector asymmetries. Field observations thus corroborate research showing that the benefits and barriers of Industry 4.0 in the dairy sector are highly dependent on companies' organizational and financial capabilities (Wojcicki et al., 2025).

At the regional level, digital transformation interacts with local training systems and recruitment dynamics. The demand for hybrid profiles, combining biological knowledge, digital literacy, and technical maintenance skills, is increasing. Several stakeholders report declining enrollment in traditional cheesemaking schools. Public initiatives aimed at developing new specialised training programmes can be interpreted as institutional responses to this emerging skills gap.

Dairy 4.0 in this context therefore contributes to a selective modernization of the dairy sector: technologically advanced yet structurally hybrid; digitally instrumented yet still dependent on human expertise and territorial ecosystems. Territorial resilience thus appears to be conditioned by the local ecosystem's ability to adjust its training apparatus and maintain a pool of intermediate skills. Analyses highlighting the role of local embeddedness in the performance of regional SMEs find concrete expression here: the adoption of Dairy 4.0 depends not only on internal choices made by companies, but also on the density and adaptability of territorial institutions (Sharafizad et al., 2022).

11.6 Consolidating the model: differentiated socioeconomic impacts of Dairy 4.0 in BVPRs

11.6.1 The biologically variable production regime as a structuring productive configuration

The study of Auvergne PDO cheese dairies makes it possible to go beyond a sectoral reading of *Dairy 4.0* to formalize a specific productive configuration: the BVPR. The case is not a regional exception or a simple agrifood example; it provides empirical evidence of a regime in which biological variability, asymmetry of losses, and the centrality of human judgment structurally shape organizational and technological choices, thereby generating differentiated socioeconomic impacts.

Firstly, the variability observed is neither residual nor cyclical. It is not due to a lack of instrumentation or deficiencies in standardization. It is an integral part of the production regime itself: fluctuating milk composition, microbiological dynamics, and nonlinear interactions between technical parameters. Observations from Auvergne confirm analyses in bioprocesses showing that biological heterogeneity is a constitutive property to be governed rather than an anomaly to eliminate (Mu & Zhang, 2023). The concept of BVPR integrates this constraint into organizational analysis, consistent with robustness as an engineering principle (Becker et al., 2023).

Secondly, the regime, the regime exhibitss an asymmetric distribution of losses. Most production sequences proceed smoothly; however, rare deviations may generate disproportionately high sanitary, reputational, or economic risks. This risk structure alters production rationality: the primary objective is not marginal yield gains but the prevention of critical failures, making robustness the organizing principle rather than optimization.

Thirdly, the persistent centrality of human judgment confirms that *Dairy 4.0*, in a BVPR, cannot entirely replace local expertise. Human intervention is an essential adjustment mechanism when instrumented parameters cannot guarantee system stability. Control relies on a peristent hybridization of technical instrumentation and tacit knowledge.

The Auvergne case shows that these dimensions are not independent. Biological variability generates risk asymmetry which justifies maintaining human decision-making space. Together, these factors constrain *Dairy 4.0* adoption and shape its socioeconomic effects. The BVPR therefore appears as an integrated explanatory configuration, rather than a simple sectoral description.

This formalization allows us to reformulate the question of Dairy 4.0. In a BVPR, digitalization cannot be conceived as a linear automation process aimed at complete system optimization. It is structurally oriented toward robustness, implying selective adoption, hierarchical deployment, and alignment with the constraints of living processes.

Consequently, the BVPR becomes an explanatory variable that helps explain why PDO cheese fermentation and bioproduction follow digital trajectories that differ from those observed in highly standardized production regimes. It also clarifies why the socioeconomic impacts of Dairy 4.0 are differentiated: value creation is less associated with full automation or average optimization than with robustness, loss avoidance, the securing of critical milestones, and the preservation of human-machine complementarity.

11.6.2 Path dependence as the organizational materialization of the biologically variable production regime

Analysis of the Auvergne case shows that path dependence is not simply a historical legacy or generic organizational inertia. It appears to be the result of gradual adjustments required to manage biological variability. In other words, the path dependence observed is not independent of the production regime: it is partly generated by it.

Stratified industrial architectures—the coexistence of old and new machines, successive layers of information systems, and site-specific equipment—result from incremental adaptations oriented toward continuity and risk control rather than systemic redesign. This cumulative dynamic is consistent with analyses showing that agrifood SMEs engaged in digital transformation favor the gradual integration of technological building blocks into existing architectures, rather than radical breaks (Cannas, 2023).

In a BVPR, however, this path dependency is reinforced by the need to absorb inherent variability and preserve fragile production balances. Equipment and organizational systems are not interchangeable: they are adjusted to local parameters (raw materials, workshop configuration, critical sequences) and thus become specific assets. This specificity increases switching costs and limits the feasibility of complete reconstructions, even when these would be technically optimal.

Path dependence thus appears as a structural constraint produced by the interaction between biological variability and robustness requirements. Past choices not only determine current margins for change; they also crystallize previous compromises between production continuity, risk control, and investment capacity. *Dairy 4.0* must fit into this cumulative architecture, which explains its selective and often partial character and contributes to differentiated adoption trajectories across firms.

This interpretation allows us to reformulate path dependence in BVPRs: it does not refer solely to organizational rigidities, but to trajectories shaped by the need to stabilize an intrinsically unstable system. Thus, digital transformation does not break with the past; it extends and reconfigures already established robustness strategies. Consequently, the socioeconomic impacts of Dairy 4.0 are differentiated by firms' inherited architectures, specific assets, integration capabilities, and capacity to absorb the operational risks associated with technological change.

11.6.3 Decision-making rationality as a mediation mechanism geared toward robustness and differentiated socioeconomic effects

The interaction between BVPR and cumulative path dependence does not mechanically determine a specific form of *Dairy 4.0*. It defines a constrained decision space within which firms arbitrate. In the Auvergne PDO cheese dairies, this arbitration reflects a rationality oriented toward robustness, rather than abstract optimization.

In a BVPR, performance depends not only on marginal yield improvements but on the ability to avoid critical tipping points. Investment decisions are therefore triggered less by the search for theoretical optima than by the identification of concrete vulnerabilities: quality instability, costly errors, physical strain, occupational health risks, regulatory exposure, or information gaps.

Decision-making rationality thus mediates between biological constraints and inherited architectures. Rather than pursuing radical redesign, firms introduce targeted digital solutions aimed at reducing risks without destabilizing the accumulated productive balances. This logic explains the selective form of Dairy 4.0 adoption observed: automation of stabilizable sequences, combined with the preservation of human oversight at critical milestones.

Such an orientation is consistent with analyses in bioprocess engineering, which emphasize maintaining complex control systems within acceptable multivariate ranges rather than imposing uniform control schemes (Kunzelmann et al., 2024). At the organizational level, this approach translates into adaptive and hybrid arrangements.

Decision-making rationality also plays a distributive role. Under budgetary constraints, digital investments compete with other critical expenditures (regulatory compliance, environmental infrastructure, heavy maintenance). Dairy 4.0 adoption therefore remains embedded in a broader matrix of interdependent risks rather than constituting an autonomous strategic priority. This prioritization contributes to differentiated socioeconomic impacts, since firms do not invest in the same technologies, at the same pace, or for the same vulnerabilities.

In summary, decision-making rationality geared toward robustness is the central mechanism linking BVPR and path dependence. It transforms biological and historical constraints into concrete organizational choices and explains why Dairy 4.0, in this type of production regime, takes the form of a cumulative, hierarchical, and partially hybrid transformation. It also explains why its socioeconomic effects vary across firms, work organizations, and production sites, depending on the vulnerabilities that digitalization is designed to reduce.

11.6.4 Summary formalization of the model and general scope and differentiated socioeconomic impacts

The Auvergne PDO cheese case allows us to articulate a coherent explanatory framework linking three interdependent dimensions:

(1) the structural properties of biologically variable production regimes (BVPR)
(2) cumulative path dependence,
(3) robustness-oriented decision-making rationality.

In such regimes, the inherent variability of the living substrate, the asymmetry of losses, and the enduring role of human judgment delimit the scope of technological stabilization. These constraints materialize organizationally through incremental and stratified industrial architectures. Within this framework, firms adopt digital technologies selectively, prioritizing the reduction of identified vulnerabilities over abstract optimization targets.

Dairy 4.0 thus appears, in a BVPR, not as a trajectory toward full automation but as a cumulative process of robustness building. Digital tools reinforce monitoring, traceability, and control capacities while preserving hybrid configurations in which human expertise remains central at critical milestones. This mechanism also explains why the socioeconomic impacts of Dairy 4.0 are differentiated rather than uniform.

This interpretation moves beyond diffusionist readings of digital transformation. It suggests that the form, pace and socioeconomic effects of Dairy 4.0 depend fundamentally on the structural characteristics of production regimes. In biologically variable contexts, digital investments frequently generate value through the prevention of rare but high-impact failures rather than through marginal productivity gains. As highlighted in recent analyses of Industry 4.0 in the dairy sector (Wojcicki et al., 2025), the economic benefits of digitalization are often indirect, diffuse, and strongly conditioned by organizational absorption capacities.

Economic performance in a BVPR must therefore be understood in relation to system stability and continuity. Return on investment is not limited to efficiency gains but includes the avoidance of potential losses, the safeguarding of regulatory compliance, and the stabilization of production teams and organizational routines. At the same time, these effects vary across firms according to their inherited architectures, financial constraints, technical capabilities, and ability to integrate digital tools without destabilizing existing robustness mechanisms.

More broadly, the model proposed here invites comparative exploration. Other sectors relying on complex biological or process variability—such as fermentation industries, biotechnologies, or specialized agricultural production—may display similar robustness-oriented digital trajectories and similarly differentiated socioeconomic impacts.

Taken together, these findings reinforce a central argument of this chapter: Dairy 4.0 cannot be analyzed independently of the biological, organizational, and historical configurations within which it unfolds. Digital transformation in the dairy sector is not uniform; it is structured by the productive regimes that condition its possibilities and by the firm-level, sectoral, and territorial capacities that shape its socioeconomic consequences.

11.7 Conclusion

This chapter has examined the differentiated socioeconomic impacts of Dairy 4.0 by moving beyond interpretations that assume a uniform and linear diffusion of digital technologies. Rather than treating digitalization as a universal lever for optimization and automation, the analysis has highlighted that its effects are structured by the intrinsic properties of the processing regime in which it operates. Through the case of PDO cheese dairies in Auvergne, we have demonstrated that digital trajectories and their socio-economic consequences are deeply embedded in a biologically variable production regime (BVPR).

The chapter introduced and developed this BVPR concept, characterized by three interrelated features: *(i)* inherent variability of substrates and processes, *(ii)* asymmetry in losses concentrated in rare but critical deviations, and *(iii)* persistent hybridization of technical instrumentation and human judgment. In such regimes, complete stabilization of the process is structurally limited. As a result, digital transformation is not primarily oriented toward maximizing average performance, but toward maintaining robustness in the face of variability and risk.

From this perspective, Dairy 4.0 adoption appears as a selective and cumulative process. Investment decisions are shaped by the need to secure critical milestones rather than to automate uniformly. Stabilized or repetitive sequences may lend themselves to substitution technologies, whereas biologically sensitive stages remain structured around augmentation strategies that preserve the central role of human expertise. Automation therefore transforms work more than it replaces it: it reduces certain forms of physical strain while increasing the technicality of maintenance, coordination, and system supervision.

At the sectoral level, the combination of path dependency, specific assets, and financial capacity leads to differentiated trajectories. Some companies are able to integrate digital tools and strengthen their organizational robustness, while others remain engaged in partial adjustments, accentuating intra-sector asymmetries. At the regional level, digital transformation depends on the availability of complementary assets—training programmes, networks of service providers, and a pool of hybrid skills. It thus appears not only as a technical issue but also as a challenge in terms of reproducing and renewing local skills and sustaining territorial resilience.

The analysis also invites a more nuanced approach to the economic evaluation of Dairy 4.0 in biologically variable contexts (i.e., BVPRs). Much of the value created lies in the avoidance of extreme losses, the securing of critical control points, and the stabilization of production systems, dimensions that are not always easily captured by conventional productivity indicators. Supporting digital transformation in biologically variable sectors may therefore require attention not only to equipment, but also to skills (linked to training), organizational integration, and territorial coordination.

More broadly, while the empirical focus here has been on PDO cheese production, the mechanisms highlighted are relevant to other production systems characterized by irreducible variability and asymmetric risks. In these environments, digital technologies do not eliminate uncertainty; they reshape how it is managed. Human judgment remains central at critical stages, even as its forms and supports evolve.

This research nevertheless has limitations. Based on a specific sectoral and territorial case study, it calls for comparative work to examine the extent to which the concept of BVPR, as a particular productive configuration, can shed light on other sectors characterized by irreducible variability and asymmetric risks. Future research could further explore the empirical measurement of the indirect effects of robustness on economic performance and the conditions under which Dairy 4.0 generates differentiated firm-level, sectoral, and territorial outcomes.

Ultimately, the socioeconomic impacts of Dairy 4.0 in biologically variable regimes should not be understood as a uniform technological disruption. They represent a gradual and cumulative recomposition of productive, organizational, and territorial equilibria shaped as much by the nature of the living systems as by the capabilities of digital technologies and by the capacities of firms and territories to absorb them.

AI disclosure

During the preparation of this work, the authors used ChatGPT to assist in the reformulation of certain sentences. All outputs were carefully reviewed and edited by the authors, who take full responsibility for the content of this publication.

References

Arthur, W.B., 1989. Competing technologies, increasing returns, and lock-in by historical events. The Economic Journal 99 (394), 116–131. Available at https://doi.org/10.2307/2234208.

Becker, L., Sturm, J., Eiden, F., Holtmann, D., 2023. Analyzing and understanding the robustness of bioprocesses. Trends in Biotechnology 41 (8), 1013–1026. Available at https://doi.org/10.1016/j.tibtech.2023.03.002.

Cannas, R., 2023. Digital transformation and dynamic capabilities in agri-food SMEs. Technological Forecasting and Social Change 189, 122358. Available at https://doi.org/10.1016/j.techfore.2023.122358.

Clavier, T., Pinel, C., de Jong, H., Geiselmann, J., 2024. Improving the genetic stability of bacterial growth control for long-term bioproduction. Biotechnology and Bioengineering 121 (9), 2808–2819. Available at https://doi.org/10.1002/bit.28756.

David, P.A., 1985. Clio and the economics of QWERTY. The American Economic Review 75 (2), 332–337.

Eigenfeld, M., Schwaminger, S.P., 2025. Cellular variability as a driver for bioprocess innovation and optimization. Biotechnology Advances 79, 108528. Available at https://doi.org/10.1016/j.biotechadv.2025.108528.

Gasset, A., Van Wijngaarden, J., Mirabent, F., Sales-Vallverdu, A., Garcia-Ortega, X., Montesinos-Segui, J.L., et al., 2024. Continuous process verification 4.0 application in upstream: adaptiveness implementation managed by AI in the hypoxic bioprocess of the Pichia pastoris cell factory. Frontiers in Bioengineering and Biotechnology 12. Available at https://doi.org/10.3389/fbioe.2024.1439638.

Gehlot, A., et al., 2022. Industry 4.0 technologies in food and dairy processing: A review. Sustainable Computing: Informatics and Systems 35, 100693. Available at https://doi.org/10.1016/j.suscom.2022.100693.

Hassoun, A., Garcia-Garcia, G., Trollman, H., Jagtap, S., Parra-Lopez, C., Cropotova, J., et al., 2023. Birth of Dairy 4.0: Opportunities and challenges in adoption of fourth industrial revolution technologies in the production of milk and its derivatives. Current Research in Food Science 7, 100535. Available at https://doi.org/10.1016/j.crfs.2023.100535.

Kunzelmann, M., Thoma, J., Laibacher, S., Studts, J.M., Presser, B., Spitz, J., 2024. An in-silico approach towards multivariate acceptable ranges in biopharmaceutical manufacturing. AAPS Open 10 (1). Available at https://doi.org/10.1186/s41120-024-00095-y.

Levinthal, D.A., 1997. Adaptation on rugged landscapes. Management Science 43 (7), 934–950. Available at https://doi.org/10.1287/mnsc.43.7.934.

Malik, M., Gahlawat, V.K., Mor, R.S., Hosseinian-Far, A., 2024. Towards white revolution 2.0: challenges and opportunities for the industry 4.0 technologies in Indian dairy industry. Oper Manag Res. https://doi.org/10.1007/s12063-024-00482-4.

Mu, X., Zhang, F., 2023. Diverse mechanisms of bioproduction heterogeneity in fermentation and their control strategies. Journal of Industrial Microbiology and Biotechnology 50 (1). Available at https://doi.org/10.1093/jimb/kuad033.

Santos Valle, S., Kienzle, J., 2020. Agriculture 4.0 – Agricultural robotics and automated equipment for sustainable crop production. Integrated Crop Management, 24, Rome, FAO.

Sharafizad, J., Redmond, J., Parker, C., 2022. The influence of local embeddedness on the economic, social, and environmental sustainability practices of regional small firms. Entrepreneurship & Regional Development 34 (5-6), 1234–1256. Available at https://doi.org/10.1080/08985626.2021.2024889.

Sydow, J., Schreyögg, G., Koch, J., 2009. Organisational path dependence: Opening the black box. Academy of Management Review 34 (4), 689–709. Available at https://doi.org/10.5465/amr.34.4.zok689.

Vergne, J.-P., Durand, R., 2010. The missing link between lock-in and path dependence. Journal of Management Studies 47 (4), 736–759. Available at https://doi.org/10.1111/j.1467-6486.2009.00886.x.

Wasalathanthri, D.P., Shah, R., Ding, J., Leone, A., Li, Z.J., 2021. Process analytics 4.0: A paradigm shift in rapid analytics for biologics development. Biotechnology Progress 37 (4). Available at https://doi.org/10.1002/btpr.3177.

Wójcicki, K., Młody, M., Sajdak, M., 2025. Opportunities and barriers in the digital transformation of dairy industry and its impact on sustainable production management. EiS 94, 1081. https://doi.org/10.34659/eis.2025.94.3.1081.

Chapter 12

Barriers to Industry 4.0 adoption in the dairy sector: current opportunities and future perspectives

Abdo Hassoun[1,2], Ahmed Hamad[3], and Yostina Atef Zaky[3]
[1]*Sustainable AgriFoodtech Innovation & Research (SAFIR), Arras, France,* [2]*Faculty of Agricultural Engineering, University of Aleppo, Aleppo, Syria,* [3]*Food Hygiene and Control Department, Faculty of Veterinary Medicine, Benha University, Benha, Egypt*

12.1 Introduction

Industry 4.0, also known as the Fourth Industrial Revolution, represents the integration of digital, physical, and biological systems through the convergence of disruptive technologies such as the Internet of Things (IoT), smart sensors, machine learning and artificial intelligence (AI), cloud computing, big data analytics, advanced robotics, augmented reality, 3D printing, and smart adaptive communication (Hassoun et al., 2023a; Dutta et al., 2026; Jabbar et al., 2023). A central element of this paradigm is the "smart factory," in which cyber-physical systems and IoT interconnect processes and operations to enable modular, flexible, and customized production (Hamad and Awad, 2025; Ryalat et al., 2023). Through this extensive digital transformation, Industry 4.0 supports intelligent manufacturing characterized by automation, real-time monitoring, data-driven decision-making, and sustainable industrial practices (Rejeb et al., 2025; Lampropoulos et al., 2019). This is supported by technologies such as digital twins, which are virtual replicas of physical systems that allow real-time simulation and optimization of manufacturing processes (Fantozzi et al., 2025), and Industrial IoT (IIoT), which enables real-time data acquisition and analytics from sensors distributed in smart factories (Kamath et al., 2020). The implementation of Industry 4.0 involves transitioning to intelligent automation, including self-automation, self-configuration, self-diagnosis, and intelligent decision-making capabilities across manufacturing and supply chains (Karabegović et al., 2019). It brings benefits such as

Dairy 4.0. DOI: https://doi.org/10.1016/B978-0-443-44730-3.00011-0

improved operational efficiency, production flexibility, and competitiveness while aiming to meet sustainability goals aligned with the UN Sustainability 2030 Agenda (Jamwal et al., 2021).

Nevertheless, Industry 4.0 also presents challenges, including changes in business paradigms, legal and security issues, resource planning, and standardization needs (Karabegović et al., 2019). Its success depends on collaborative efforts across all participants in the production chain, emphasizing a systemic and integrative approach to achieving operational excellence in this new digital era (Luz Tortorella et al., 2021). The adoption of Industry 4.0 in the dairy sector faces multifaceted barriers, including knowledge gaps, high costs, skill shortages, infrastructure limitations, and policy inadequacies, particularly in regions such as India and Africa (Annamalai et al., 2024; Vuvor et al., 2025). These challenges are interconnected with foundational issues such as unfamiliarity driving higher-level obstacles (Annamalai et al., 2024).

Despite these structural and institutional constraints, the rapid evolution of digital technologies is opening new pathways for transformation within the dairy sector. Current opportunities lie in digital tools such as IoT, AI, mobile technologies, and precision livestock farming (PLF) to enhance efficiency, traceability, and value chains (Morrone et al., 2022; Vuvor et al., 2025; Göller et al., 2021; Akhigbe et al., 2021). Prospects emphasize comprehensive strategies involving training, policy support, multi-stakeholder collaboration, and Responsible Research and Innovation (RRI) to achieve sustainable, resilient dairy systems (Annamalai et al., 2024; Vuvor et al., 2025).

Therefore, this chapter aims to critically examine the conceptual foundations, enabling technologies, practical applications, and implementation challenges of Industry 4.0 within the dairy sector. It seeks to analyze how digital transformation, through IoT, AI, smart sensors, digital twins, and precision livestock farming, can enhance productivity, traceability, animal health management, product quality, and sustainability across the dairy value chain. Furthermore, the chapter intends to identify the technical, economic, and institutional barriers limiting adoption, particularly in emerging and resource-constrained regions, and to propose integrative strategies grounded in capacity building, policy alignment, and multi-stakeholder collaboration. By synthesizing current evidence and practical insights, this chapter provides a comprehensive framework to guide researchers, industry stakeholders, and policymakers toward building resilient, data-driven, and sustainable dairy systems in the era of Industry 4.0.

12.2 Major challenges in implementing Industry 4.0 in the dairy sector

In essence, the transition to Industry 4.0 is challenged by technological integration demands, financial and administrative hurdles, legal and security concerns, standardization issues, workforce and process transformations, and the need for updated product-design approaches. Overcoming these challenges requires systemic change, collaboration among all participants, and strategic planning to achieve sustainable Industry 4.0 adoption (Bousdekis and Mentzas, 2021; Rosário and Dias, 2023; Serumaga-Zake and van der Poll, 2021; Jankowska et al., 2022; Karabegović et al., 2019).

12.2.1 Technical and infrastructure barriers

The adoption of Industry 4.0 in the dairy sector, often termed "Dairy 4.0," faces several technical and infrastructure barriers that impede its full-scale implementation. Key technical challenges include the difficult adaptation and integration of advanced technologies, such as robotics, 3D printing, AI, IoT, big data, and blockchain, into dairy production systems (Dutta et al., 2026). These technologies require sophisticated technical capabilities that may be lacking in current dairy production environments (Fig. 12.1).

In terms of infrastructure, one major barrier is insufficient and uneven digital infrastructure, particularly inadequate Internet access, which is critical for enabling smart farming technologies and real-time data connectivity necessary for Industry 4.0 applications in the agriculture and dairy sectors (Fragomeli et al., 2024). Without reliable connectivity, the deployment and efficient functioning of IoT devices, data analytics, and digital monitoring systems are all compromised.

Additionally, the dairy sector faces challenges in integrating disparate legacy systems with new digital platforms, which require considerable investment and the overhaul of existing infrastructures, often lacking in smaller or less technologically advanced farms (Hassoun et al., 2023b). This complexity underscores the need for infrastructure that supports interoperability and seamless digital communication across the entire production-to-distribution

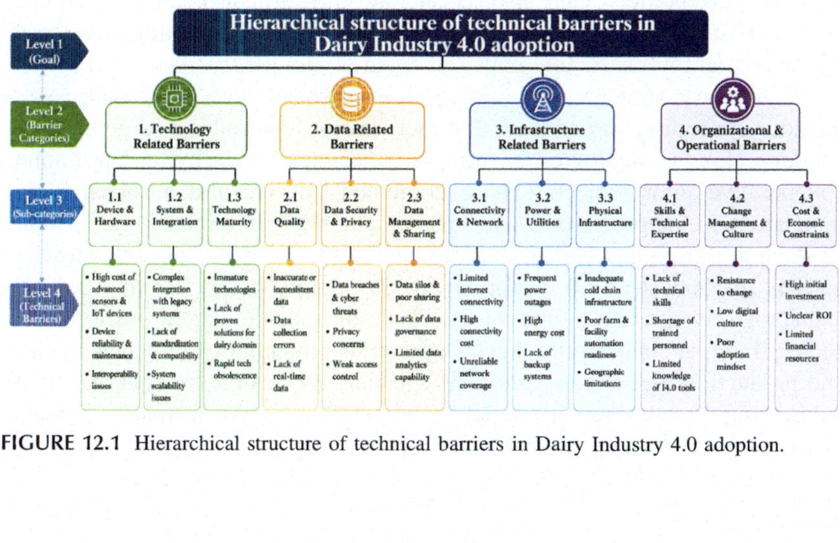

FIGURE 12.1 Hierarchical structure of technical barriers in Dairy Industry 4.0 adoption.

chain. Furthermore, the technical skill gaps among workers and managers hinder the effective use of cutting-edge Industry 4.0 technologies in the dairy sector, while logistical limitations in production infrastructure constrain adoption (Ingaldi and Ulewicz, 2025). These human and physical resource constraints tend to slow the progress of digital transformation and reduce the potential benefits of Industry 4.0.

To improve the digital infrastructure in the dairy sector for Industry 4.0 adoption, several strategies can be implemented, focusing on enhancing connectivity, skills, and technology integration. First, expanding high-speed Internet coverage is critical, as widespread and reliable connectivity supports the seamless interconnection of smart devices and systems that underpin Industry 4.0 applications in the dairy sector (Aleca and Mihai, 2025). Investment in broadband infrastructure, especially in rural dairy farming areas, is foundational for enabling IoT devices, real-time data transmission, and cloud computing. Second, fostering digital skill development among the dairy workforce is essential. Training programs aimed at improving employees' and managers' proficiency with Industry 4.0 technologies and digital tools empower better utilization of smart farming systems and data analytics, ultimately boosting productivity (Aleca and Mihai, 2025). Third, leveraging cloud solutions can catalyze process efficiency by providing scalable and flexible platforms for data storage, processing, and analytics. Cloud adoption removes barriers related to local server maintenance and supports the integration of complex data from IoT sensors and AI-driven systems in dairy operations (Aleca and Mihai, 2025).

Additionally, developing interoperable and standardized frameworks to ensure compatibility between legacy dairy production systems and new digital technologies facilitates smoother technological transitions and the full realization of automation benefits (Dutta et al., 2026). Pilot projects and collaborative initiatives involving technology providers, government bodies, and dairy industry stakeholders can also accelerate infrastructure enhancement by sharing costs, knowledge, and best practices, thus overcoming investment limitations and infrastructure gaps (Hassoun et al., 2023b).

In summary, strategies to improve digital infrastructure in the dairy sector include expanding broadband Internet access, enhancing digital skills, adopting cloud technologies, ensuring system interoperability, and promoting collaborative pilot projects. These combined efforts build a robust digital foundation to support the transformative potential of Industry 4.0 in dairy production (Aleca and Mihai, 2025). Overall, the technical and infrastructure barriers to Industry 4.0 adoption in the dairy sector include complex technology integration requirements, lack of reliable digital infrastructure such as Internet connectivity, legacy system incompatibility, and gaps in skills and production logistics (Ingaldi and Ulewicz, 2025; Fragomeli et al., 2024). Addressing these barriers is crucial for harnessing the transformative potential of Industry 4.0 in dairy production and processing.

12.2.2 Economic and financial barriers

The adoption of Industry 4.0 technologies in the dairy sector faces significant economic and financial barriers that hinder their full-scale implementation. A primary economic challenge is the high initial capital investment required to acquire advanced technologies such as robotics, IoT devices, AI systems, and blockchain, which are central to "Dairy 4.0f" operations (Hamad and Edris, 2026). Many dairy producers, particularly SMEs, struggle to afford these upfront costs, limiting their broad adoption.

Furthermore, uncertainties regarding economic benefits and return on investment act as obstacles. Businesses often find it difficult to quantify the short- and long-term financial gains from digitization, automation, and data integration, which reduces their willingness to commit substantial resources (Sayem et al., 2022). Operational and recurring costs related to maintaining, upgrading, and integrating Industry 4.0 digital infrastructure, including software licensing, cybersecurity, data management, system integration, and workforce training, place significant financial pressure on dairy enterprises, particularly in the absence of targeted subsidies, low-interest loans, or grant mechanisms to support technology adoption (Khurshid et al., 2024). This financial strain is further intensified by market volatility, fluctuating input costs, and competing investment priorities, all of which discourage producers from committing capital to digital transformation initiatives.

Market dynamics profoundly influence dairy producers' financial decision-making regarding technology investments. One key factor is the growing consumer demand for higher-quality, sustainably produced dairy products, which pressures producers to adopt advanced technologies such as IoT, sensors, and AI to improve productivity, food safety, and traceability (Kazi, 2024). This demand-driven market environment encourages investment in technology to remain competitive in the market. Market access, particularly to formal and high-value markets, also plays a role. Producers engaged in formalized milk markets tend to have more incentives and financial means to invest in technology because of better price premiums and contract stability. Conversely, those reliant on informal markets may face economic constraints that limit their investment options (Kanire et al., 2024).

Additionally, dairy producers face volatility in milk prices and market uncertainties that affect their financial capacity and willingness to invest in their farms. Price fluctuations reduce financial predictability, making producers cautious about committing substantial capital to new technologies (Douphrate et al., 2013). Producers may delay or scale back technology investments to conserve resources when market conditions are unstable or prices are low. Finally, the availability of digital financial inclusion and supportive financial environments can moderate market effects by increasing farmers' access to credit, subsidies, and technical training, thereby enhancing their willingness and ability to invest in Industry 4.0 technologies (Zhou et al., 2022). Thus, policy initiatives fostering financial access and stable market conditions can positively impact technology adoption decisions.

In summary, economic and financial barriers in the dairy sector revolve around the high costs of advanced technology acquisition and maintenance, uncertainty over investment gains, limited financial aid, and challenging market conditions. Addressing these economic issues is crucial for enabling the dairy industry's successful digital transformation under Industry 4.0 (Khurshid et al., 2024; Sayem et al., 2022). In addition, market dynamics, such as consumer demand, price volatility, market access, production scale, and financial inclusion, critically shape dairy producers' financial decisions regarding technology investments by influencing perceived risks, potential returns, and available resources (Douphrate et al., 2013; Kazi, 2024; Eastwood et al., 2015; Kanire et al., 2024; Zhou et al., 2022).

12.2.3 Human and knowledge barriers

Human and knowledge barriers significantly impact the adoption of Industry 4.0 technologies in the dairy industry. A primary obstacle is the lack of adequate knowledge and understanding of Industry 4.0 concepts and technologies among dairy farmers and workers, which hinders their ability to embrace and effectively use advanced tools (Shabur et al., 2024).

Furthermore, resistance to change and attitudinal barriers among human resources contribute to slow adoption rates. Employees may fear job displacement or face trust and privacy concerns related to automation and digital technology (Leesakul et al., 2022). Acceptance issues, low motivation, and uncertainty about the impact of new technologies on daily routines also restrict the willingness to adopt Industry 4.0 innovations. A shortage of qualified workforce and expertise in digital technologies is another critical human barrier. The dairy sector often struggles to recruit or train personnel with the skills necessary to implement, manage, and optimize Industry 4.0 systems effectively, especially in smaller or rural operations (Sayem et al., 2022). Without proper training programs and continuous upskilling initiatives, employees' competency levels remain insufficient.

In addition, inadequate extension services and a lack of tailored training opportunities limit farmers' access to the practical knowledge and technical assistance necessary for the adoption of Industry 4.0 (Quddus, 2013). Many farmers expressed the need for more reliable, ongoing support and education to better understand improved technologies in animal health, feeding, breeding, and farm management. Effective training programs to address the knowledge gap in the dairy sector for Industry 4.0 adoption should combine technical skill development with adaptive learning approaches tailored to the agricultural context of the region.

Six key training components can include the following:

i. Technology-specific training: Programs focused on the core enabling technologies of Dairy 4.0, such as robotics, AI, IoT, big data, and blockchain, are essential. This includes hands-on training in the operation, troubleshooting, and maintenance of these digital tools to enhance farm automation and data-driven decision-making (Hamad and Awad, 2025).
ii. Digital literacy and competency development: Building foundational digital skills among dairy farmers and workers is critical. Training should cover digital device handling, data analytics basics, cybersecurity awareness, and cloud technology utilization to empower users for digital transformation (Hamad and Tayel, 2026).
iii. Extended technical vocational education and training (TVET): Incorporating Industry 4.0 competencies into vocational and technical education curricula prepares a future-ready workforce. This involves updating training materials and methodologies in agricultural colleges and extension services to reflect the emerging digital demands (George et al., 2024).
iv. Blended learning and challenge-based approaches: Utilizing mixed methodologies, such as virtual labs, physical practice sessions, and challenge-based learning, can improve engagement and knowledge retention. These approaches link theoretical concepts to practical farm scenarios to enhance technology adoption readiness (Cantú-Ortiz et al., 2020).
v. Leadership and change management training: Equipping farm managers and industry leaders with transformational leadership skills facilitates organizational acceptance and employee motivation toward adopting Industry 4.0 technologies, addressing human factors, and overcoming resistance barriers (Mayer, 2024).
vi. Continuous upskilling and knowledge management: Programs that emphasize lifelong learning, iterative knowledge development, and skill evolution ensure that the workforce adapts continuously to technological advances and changing industry requirements (Anshari and Hamdan, 2022).

By integrating these strategies, training initiatives can effectively bridge the knowledge gap, empower the dairy workforce, and accelerate Industry 4.0 technology adoption, ultimately leading to increased productivity, sustainability, and competitiveness in the dairy sector (Cantú-Ortiz et al., 2020; George et al., 2024). Overall, these human and knowledge barriers underscore the necessity of comprehensive strategies that include awareness-raising, dedicated training programs, change management, and expanded technical support services. Such interventions can foster digital literacy, build trust in innovative technologies, and enhance workforce readiness, thereby enabling smoother and more effective Industry 4.0 adoption in the dairy sector (Leesakul et al., 2022; Quddus, 2013; Shabur et al., 2024).

12.2.4 Regulatory, policy, and ethical barriers

Regulatory, policy, and ethical barriers pose significant challenges to the adoption of Industry 4.0 technologies in the dairy sector. At the regulatory level, there is often a lack of clear or harmonized frameworks specifically addressing the deployment of advanced digital technologies in agriculture, including the dairy sector. This results in uncertainties regarding compliance, data governance, and standards, which complicate technology integration and can discourage investment (Baldin et al., 2021; Batra et al., 2024). Issues related to data collection, integration, and sharing raise concerns over privacy, ownership, and security, necessitating robust regulatory mechanisms that are often not yet fully developed in this sector (Baldin et al., 2021).

Policy constraints include insufficient incentives and support programs that are vital for offsetting the high costs and technical complexities of Industry 4.0 technology. The absence of streamlined funding, training initiatives, and innovation-friendly policies limits the ability of dairy farmers, especially SMEs, to access, adopt, and sustain digital transformation technologies (Alkhodair and Alkhudhayr, 2025; Ortega-Gras et al., 2025). Moreover, the slow pace of policy adaptation relative to rapid technological advancements hampers effective governance and sectoral development in the country.

Ethical barriers manifest as concerns about job displacement due to automation, impacts on rural communities and labor practices, and trust in AI and automated decision systems. There is a need for ethical guidelines and frameworks that govern the responsible use of AI, robotics, and data analytics to ensure the fairness, transparency, and social acceptance of Industry 4.0 implementations in dairy farming (Sumarlin and Kusumajaya, 2024; Shabur et al., 2025).

12.3 Current opportunities and innovation pathways

The current opportunities in Dairy Industry 4.0 stem from the integration of advanced digital technologies that promise to revolutionize dairy production, processing, and sustainability in the dairy industry. Central enabling technologies include robotics, AI, the IoT, big data analytics, and blockchain, which are progressively adopted throughout the dairy supply chain from farm to table (Hamad and Tayel, 2026) (Fig. 12.2).

These technologies offer the opportunity to optimize production efficiency through real-time monitoring and automation of complex processes, such as herd management, milking, and feed optimization, resulting in improved animal welfare, reduced operational costs, and productivity gains (Rejeb et al., 2025). IoT devices and sensors provide continuous data streams, enabling precise decision-making, predictive maintenance, and proactive health management.

Big data and AI enable sophisticated analyses of production data to improve quality control, yield prediction, and supply chain optimization.

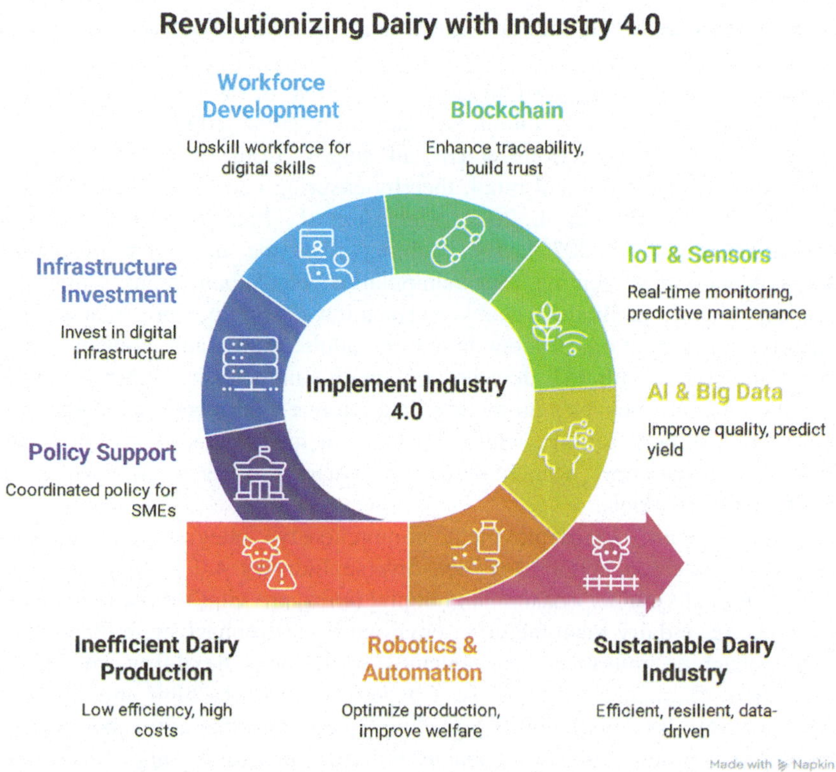

FIGURE 12.2 Ecosystem of Industry 4.0 technologies showing integration points across dairy value chains from farm-level mobile/IoT tools to processing AI/robotics, converging on traceability and efficiency.

Blockchain offers enhanced traceability and transparency, which are important for food safety compliance and building consumer trust in dairy product authenticity (Javaid et al., 2021).

There is also a significant opportunity to enhance environmental sustainability through the application of industry 4.0. Technologies can facilitate energy-efficient operations, waste reduction, and promote circular economy principles in dairy manufacturing, thereby reducing the ecological footprint and aligning with global sustainability goals (Javaid et al., 2022; Muzammal et al., 2024). Furthermore, Industry 4.0 enables the creation of "smart factories" and digitally connected ecosystems that facilitate adaptive production systems capable of scaling and customizing to meet varying market demands efficiently (Rejeb et al., 2025). With these advances, the dairy sector can expect improved competitiveness in the marketplace, greater operational resilience, and the ability to respond dynamically to consumer preferences and regulatory requirements, marking a transformative shift toward a more sustainable, data-driven, and automated dairy industry (Jamwal et al., 2021; Singh et al., 2025).

Blockchain technology strengthens traceability in the dairy supply chain through its decentralized, immutable, and transparent ledger system, which securely records transactions across all stages, from farm production to processing, distribution, and retail, thereby ensuring end-to-end visibility and data integrity (Khanna et al., 2022; Vignesh et al., 2025; Longo et al., 2019; Baranidharan et al., 2025). By enabling real-time data sharing among stakeholders and preventing data manipulation, blockchain reduces information asymmetry, facilitates reliable verification of product provenance, and supports the detection of contamination, adulteration, and counterfeiting. When integrated with IoT devices and smart contracts, it further enhances operational efficiency by automating data capture and triggering predefined quality or payment actions, while consumer-facing tools such as QR codes promote transparency, accountability, and market differentiation of high-quality dairy products.

Beyond technology deployment, comprehensive integration across processes and interconnections between diverse Industry 4.0 systems, such as cyber-physical systems, sensors, and digital twins, are vital for realizing smart and responsive dairy manufacturing environments. Emphasizing collaborative human–machine interfaces can enhance operational flexibility and foster innovation (Turner et al., 2020). Challenges such as high initial costs, cybersecurity risks, and regulatory compliance must be mitigated through coordinated policy support, capacity-building programs, and investment incentives that aim to lower barriers, especially for SMEs in the dairy sector (Sayem et al., 2022).

Looking ahead, industry stakeholders should adopt a human-centric approach by aligning Industry 4.0 initiatives with sustainability goals, ethical AI use, and workforce adaptability, thereby transitioning to the emerging Industry 5.0 paradigm. This approach prioritizes mass customization, environmental responsibility, and resilient supply chains, while leveraging human creativity alongside intelligent technologies (Shabur et al., 2025).

12.4 Strategic directions for Dairy Industry 4.0

The prospects for Dairy Industry 4.0 are strongly dependent on addressing existing barriers through targeted strategies that encompass holistic approaches, capacity building, sustainability and resilience, and regional focus (Fig. 12.3).

Holistic approaches emphasize the integration of technological innovations with environmental, social, and economic considerations to create sustainable production systems. Cleaner production principles and collaborative efforts between social stakeholders can help design tailored Industry 4.0 implementations that avoid one-size-fits-all solutions and promote social harmony while

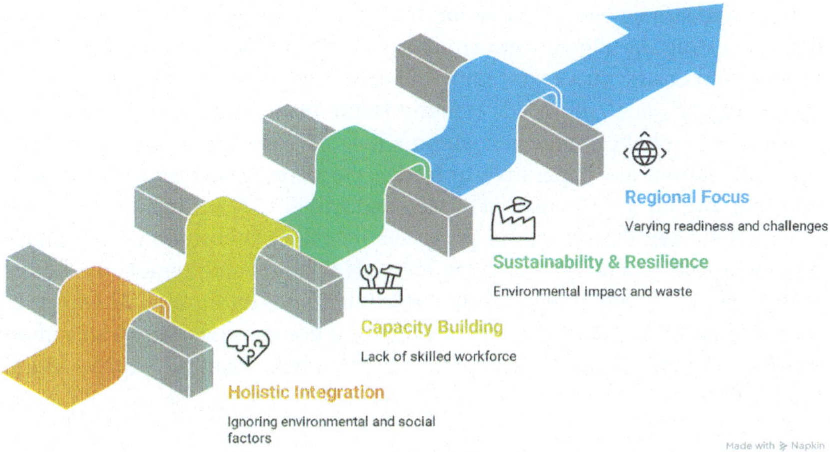

FIGURE 12.3 Overcoming dairy industry 4.0 barriers through phased strategies, supported by key enablers like training and collaboration, leading to sustainable systems.

advancing digital transformation (Chovancová et al., 2023; Satyro et al., 2021). Such approaches align industry efforts with the Sustainable Development Goals (SDGs), particularly emphasizing infrastructure development, quality education, innovation, and responsible consumption, which are crucial for dairy sustainability (Franco and Tracey, 2019). Capacity building involves empowering the workforce and management to adopt and utilize Industry 4.0 technologies effectively. Developing skills through training, knowledge management, and organizational learning promotes successful adoption and competitiveness within dairy SMEs, ensuring that human capital meets evolving technological demands (Cheng et al., 2024; Krisciukaitiene and Bathaei, 2025).

Sustainability and resilience are essential pillars where Industry 4.0 can help improve milk quality, reduce waste, and lower greenhouse gas emissions through precision farming, automation, AI, and big data analytics. Sustainable dairy production requires integrated strategies focusing on environmental footprints, circular economy practices such as whey valorization, and climate-resilient systems to foster long-term industry viability (Neethirajan, 2024; Priyashantha, 2025; Soumati et al., 2023). Regional focus is critical, as sustainability challenges and technology readiness vary geographically. Local governments and stakeholders must develop policies and resource allocation frameworks adapted to regional contexts to enable the adoption and sustainability of Industry 4.0. Coordinated multilevel governance and inclusive strategies strengthen cross-sector cooperation and address fragmentation in public institutional structures, enhancing implementation outcomes for the dairy sector in diverse regions (Bianchi and Richiedei, 2023; Krisciukaitiene and Bathaei, 2025).

Effective strategies for advancing Industry 4.0 adoption in the dairy sector entail investing in digital infrastructure and cultivating a digitally skilled workforce to ensure successful implementation. Upskilling and requalification of existing personnel, alongside training future professionals in data analytics, collaborative robotics, machine automation, cybersecurity, and human–machine interaction, are imperative to bridge the current skill gaps and promote successful digital transformation (Aleca and Mihai, 2025; Leitao et al., 2020).

In conclusion, the successful prospects for Dairy Industry 4.0 necessitate a comprehensive, capacity-strengthening, sustainable, and regionally tailored strategic approach that integrates digital technologies harmoniously with social and environmental goals to build resilient and competitive dairy value chains globally (Bianchi and Richiedei, 2023; Priyashantha, 2025; Satyro et al., 2021).

12.5 Mitigation strategies

SMEs can effectively overcome challenges in adopting Industry 4.0 technologies through a multifaceted approach that addresses key barriers, such as limited resources, skills gaps, and technological complexity.

First, addressing the critical lack of resources, both financial and infrastructural, can be achieved by leveraging government incentives, subsidies, or partnerships that offer financial support and technical guidance. Public policies that provide capital allocation and financial incentives specifically targeting SMEs can encourage digital transformation efforts (Kumar et al., 2022). Second, overcoming workforce competence and expertise gaps are essential. SMEs should invest in continuous employee training, upskilling, and knowledge management to build digital capabilities related to Industry 4.0 technologies, such as AI, IoT, robotics, and data analytics. Organizational learning mechanisms promote effective technology adoption and enhance competitiveness (Cheng et al., 2024; Kumar et al., 2022).

Third, SMEs can follow tailored digital maturity frameworks and gradual implementation strategies designed specifically for their size and context. Starting with basic digital tools and progressively integrating more advanced Industry 4.0 technologies minimizes disruption and allows for learning and adaptation over time (Amaral and Peças, 2021). Fourth, mitigating cybersecurity risks through dedicated assessment and management models protects SMEs against threats emerging from heightened connectivity and data digitalization, fostering trust and operational continuity (Emer et al., 2021). Finally, collaboration and knowledge sharing among SMEs, technology providers, and policymakers can foster an ecosystem that supports innovation adoption with shared expertise and reduces complexity and cost barriers (Ghobakhloo et al., 2022).

In summary, effectively overcoming Industry 4.0 adoption challenges by SMEs requires strategic access to financial resources, focused workforce upskilling, incremental technology integration, cybersecurity measures, and supportive collaborative frameworks to enable sustainable digital transformation and competitive market positioning (Amaral and Peças, 2021; Cheng et al., 2024; Ghobakhloo et al., 2022; Emer et al., 2021; Kumar et al., 2022).

12.6 Conclusion

This chapter examines the persistent constraints limiting the implementation of "Dairy 4.0," the integration of Industry 4.0 technologies across the dairy value chain, particularly in developing contexts such as India and Africa, while outlining strategic pathways for sustainable adoption. The identified barriers are multidimensional and interrelated, encompassing technological, economic, human–capital, and regulatory domains. Foundational knowledge gaps, inadequate digital infrastructure, especially limited broadband connectivity, weak interoperability, and shortages of skilled personnel constrain effective technological integration. These challenges are further intensified by high capital and operational costs, uncertain returns on investment, market volatility, small-scale production structures, resistance to organizational change, and insufficiently harmonized data governance and regulatory frameworks. Despite these constraints, significant opportunities exist in precision livestock farming, AI-enabled decision support systems, predictive maintenance, digital twins, and blockchain-based traceability to enhance productivity, sustainability, and consumer trust. Accordingly, the chapter emphasizes the necessity of coordinated, context-specific strategies integrating infrastructure investment, interoperable standards, targeted financing mechanisms, capacity development, cybersecurity governance, and enabling policy frameworks to support resilient, sustainable, and ethically guided digital transformation in the dairy sector.

AI disclosure

During the preparation of this work, the authors used ChatGPT to assist in the reformulation of certain sentences. All outputs were carefully reviewed and edited by the authors, who take full responsibility for the content of this publication.

References

Aleca, O.E., Mihai, F., 2025. The role of digital infrastructure and skills in enhancing labor productivity: Insights from Industry 4.0 in the European Union. Systems 13 (2). Available at https://doi.org/10.3390/systems13020113.

Akhigbe, B.I., Munir, K., Akinade, O., Akanbi, L., Oyedele, L.O., 2021. IoT technologies for livestock management: A review of present status, opportunities, and future trends. Big Data and Cognitive Computing 5, 10. Available at https://doi.org/10.3390/bdcc5010010.

Alkhodair, M., Alkhudhayr, H., 2025. Harnessing Industry 4.0 for SMEs: Advancing smart manufacturing and logistics for sustainable supply chains. Sustainability 17 (3), 813. Available at https://doi.org/10.3390/su17030813.

Amaral, A., Peças, P., 2021. A framework for assessing manufacturing SMEs Industry 4.0 maturity. Applied Sciences 11 (13), 6127. Available at https://doi.org/10.3390/app11136127.

Annamalai, S., Ganeshkumar, C., David, A., 2024. Overcoming barriers to Industry 4.0 adoption in the Indian dairy sector. Advances in marketing, customer relationship management, and e-services book series. Available at https://doi.org/10.4018/979-8-3693-6715-5.ch011.

Anshari, M., Hamdan, M., 2022. Understanding knowledge management and upskilling in Fourth Industrial Revolution: Transformational shift and SECI model. Darussalam VINE Journal of Information and Knowledge Management Systems 52 (3), 373–393. Available at https://doi.org/10.1108/VJIKMS-09–2021-0203.

Baldin, M., Breunig, T., Cue, R., De Vries, A., Doornink, M., Drevenak, J., et al., 2021. Integrated decision support systems (IDSS) for dairy farming: A discussion on how to improve their sustained adoption. Animals 11 (7). Available at https://doi.org/10.3390/ani11072025.

Baranidharan, K., Mahalakshmi, R., Anagha, S., 2025. Blockchain technology for enhancing supply chain transparency: Opportunities and challenges. International Journal of Advanced Research in Science, Communication and Technology 5, 345–353. Available at https://doi.org/10.48175/ijarsct-23039.

Batra, I., Sharma, C., Malik, A., Sharma, S., Kaswan, M.S., Garza-Reyes, J.A., 2024. Industrial revolution and smart farming: a critical analysis of research components in Industry 4.0. The TQM Journal 37 (6), 1497–1525. Available at https://doi.org/10.1108/tqm-10-2023-0317.

Bianchi, S., Richiedei, A., 2023. Territorial governance for sustainable development: A multi-level governance analysis in the Italian context. Sustainability 15 (3), 2526. Available at https://doi.org/10.3390/su15032526.

Bousdekis, A., Mentzas, G., 2021. Enterprise integration and interoperability for big data-driven processes in the frame of Industry 4.0. Frontiers in Big Data 4. Available at https://doi.org/10.3389/fdata.2021.644651.

Cantú-Ortiz, F.J., Galeano Sánchez, N., Garrido, L., Terashima-Marin, H., Brena, R.F., 2020. An artificial intelligence educational strategy for the digital transformation. International Journal on Interactive Design and Manufacturing 14 (4), 1195–1209. Available at https://doi.org/10.1007/s12008-020-00702-8.

Cheng, C.H., Li, M.H., Tang, B.J., Cheng, Y.R., 2024. The impact of knowledge management and organizational learning promotion in small and medium enterprises on the implementation of industry 4.0 and competitiveness. Administrative Sciences 14 (8). Available at https://doi.org/10.3390/admsci14080161.

Chovancová, J., Majerník, M., Drábik, P., Štofková, Z., 2023. Environmental technological innovations and the sustainability of their development. Ecological Engineering & Environmental Technology 24 (4), 245–252. Available at https://doi.org/10.12912/27197050/162708.

Douphrate, D.I., Hagevoort, G.R., Nonnenmann, M.W., Lunner Kolstrup, C., Reynolds, S.J., Jakob, M., et al., 2013. The dairy industry: A brief description of production practices, trends, and farm characteristics around the world. Journal of Agromedicine 18 (3), 187–197. Available at https://doi.org/10.1080/1059924X.2013.796901.

Dutta, P., Singh, R., Haghi, A.K., Hamad, A., 2026. Harnessing AI to reshape the future of agriculture, In: Pushan. Dutta, Rashmi. Singh, A.K. Haghi, Ahmed. Hamad. 1st edition. Springer Cham, India. Available at https://doi.org/10.1007/978-3-032–12118-9.

Eastwood, C.R., Jago, J.G., Edwards, J.P., Burke, J.K., 2015. Getting the most out of advanced farm management technologies: Roles of technology suppliers and dairy industry organisations in supporting precision dairy farmers. Animal Production Science 56 (10), 1752–1760. Available at https://doi.org/10.1071/an141015.

Hamad, A., Edris, S., 2026. Robotic systems and automation for quality inspection in food production. In Artificial Intelligence in Food Science. Academic Press, pp. 523–536.

Emer, A., Unterhofer, M., Rauch, E., 2021. A cybersecurity assessment model for small and medium-sized enterprises. IEEE Engineering Management Review 49 (2), 98–109. Available at https://doi.org/10.1109/emr.2021.3078077.

Fantozzi, I.C., Santolamazza, A., Loy, G., Schiraldi, M.M., 2025. Digital twins: Strategic guide to utilize digital twins to improve operational efficiency in Industry 4.0. Multidisciplinary Digital Future Internet 17 (1). Available at https://doi.org/10.3390/fi17010041.

Fragomeli, R., Annunziata, A., Punzo, G., 2024. Promoting the transition towards Agriculture 4.0: A systematic literature review on drivers and barriers. Sustainability 16 (6), 2425. Available at https://doi.org/10.3390/su16062425.

Franco, I.B., Tracey, J., 2019. Community capacity-building for sustainable development. International Journal of Sustainability in Higher Education 20 (4), 691–725. Available at https://doi.org/10.1108/ijshe-02–2019-0052.

George, W.K., Silas, E.I., Pandey, D., & Pandey, B.K. (2024). Utilization of Industry 4.0 technologies in Nigerian technical and vocational education. In *Examining the rapid advance of digital technology in Africa* (pp. 272–295). IGI Global. Available at https://doi.org/10.4018/978-1-6684–9962-7.ch014.

Ghobakhloo, M., Iranmanesh, M., Vilkas, M., Grybauskas, A., Amran, A., 2022. Drivers and barriers of Industry 4.0 technology adoption among manufacturing SMEs: A systematic review and transformation roadmap. Journal of Manufacturing Technology Management 33 (6), 1029–1058. Available at https://doi.org/10.1108/jmtm-12–2021-0505.

Göller, M., Caruso, C., Harteis, C., 2021. Digitalisation in agriculture: Knowledge and learning requirements of German dairy farmers. International Journal for Research in Vocational Education and Training 8, 208–223. Available at https://doi.org/10.13152/ijrvet.8.2.4.

Hamad, A.M.A., Awad, D.A.B., 2025. Integration of IoT in dairy facilities. Next-generation technologies in dairy processing and production. IGI Global, Egypt, pp. 139–170. Available at https://doi.org/10.4018/979-8-3693-9949-1.ch005.

Hamad, A., Tayel, A., 2026. Food 2050 concept: the trends that shape the future of our food. Journal of Future Foods 6 (6), 1053–1066. Available at https://doi.org/10.1016/j.jfutfo.2025.03.003.

Hassoun, A., Aït-Kaddour, A., Abu-Mahfouz, A.M., Rathod, N.B., Bader, F., Barba, F.J., et al., 2023a. The fourth industrial revolution in the food industry—Part I: Industry 4.0 technologies. Critical Reviews in Food Science and Nutrition 63 (23), 6547–6563. Available at https://doi.org/10.1080/10408398.2022.2034735.

Hassoun, A., et al., 2023b. Birth of dairy 4.0: Opportunities and challenges in adoption of fourth industrial revolution technologies in the production of milk and its derivatives. Current Research in Food Science. Available at https://doi.org/10.1016/j.crfs.2023.100535.

Ingaldi, M., Ulewicz, R., 2025. Sustainable development and technological advancements in Industry 4.0: Overcoming barriers in SME sector integration. Management Systems in Production Engineering 33 (1), 144–162. Available at https://doi.org/10.2478/mspe-2025-0015.

Jabbar, A., Jamshed, M.A., Abbasi, Q., Imran, M.A., Ur-Rehman, M., 2023. Leveraging the role of dynamic reconfigurable antennas in viewpoint of Industry 4.0 and beyond. Research; a Journal of Science and its Applications 6. Available at https://doi.org/10.34133/research.0110.

Jamwal, A., Agrawal, R., Sharma, M., Giallanza, A., 2021. Industry 4.0 technologies for manufacturing sustainability: A systematic review and future research directions. Applied Sciences 1 (12), 5725. Available at https://doi.org/10.3390/app11125725.

Jankowska, B., Mińska-Struzik, E., Bartosik-Purgat, M., Götz, M., Olejnik, I., 2022. Industry 4.0 technologies adoption: Barriers and their impact on Polish companies' innovation performance. European Planning Studies, ahead-of-print(ahead-of-print) 1029–1049. Available at https://doi.org/10.1080/09654313.2022.2068347.

Javaid, M., Haleem, A., Singh, R.P., Suman, R., 2021. Significant applications of big data in Industry 4.0. Journal of Industrial Integration and Management 6 (4), 429–447. Available at https://doi.org/10.1142/S2424862221500135.

Javaid, M., Haleem, A., Singh, R.P., Suman, R., Gonzalez, E.S., 2022. Understanding the adoption of Industry 4.0 technologies in improving environmental sustainability. Sustainable Operations and Computers 3, 203–217. Available at https://doi.org/10.1016/j.susoc.2022.01.008.

Kamath, V., Morgan, J., & Ali, M.I. (2020, June 1). *Industrial IoT and digital twins for a smart factory : An open source toolkit for application design and benchmarking*. In *2020 Global Internet of Things Summit (GIoTS)*. IEEE. https://doi.org/10.1109/giots49054.2020.9119497.

Kanire, E., Msuya, E., Alphonce, R., 2024. Drivers of dairy farmers' engagement in informal milk markets: Policy implications for developing countries. Journal of Agriculture and Food Research 16, 101128. Available at https://doi.org/10.1016/j.jafr.2024.101128.

Karabegović, I., Karabegović, E., Mahmić, M., Husak, E., 2019. Implementation of Industry 4.0 and Industrial robots in the manufacturing processes. Springer, pp. 3–14. Available at https://doi.org/10.1007/978-3-030-18072-0_1.

Kazi, K.S.L. (2024). *IoT technologies for the intelligent dairy industry* (pp. 321–350). IGI Global. Available at https://doi.org/10.4018/979-8-3693-5498-8.ch012.

Khanna, A., Jain, S., Burgio, A., Bolshev, V., Panchenko, V., 2022. Blockchain-enabled supply chain platform for Indian dairy industry: Safety and traceability. Foods 11 (17), 2716. Available at https://doi.org/10.3390/foods11172716.

Khurshid, A., Khan, K., Khan, S., Cifuentes-Faura, J., 2024. Tackling sustainable development challenges: A fuzzy set examination of textile industry 4.0 and green supply chain barriers. Sustainable Development 32 (6), 6819–6835. Available at https://doi.org/10.1002/sd.3061.

Krisciukaitiene, I., Bathaei, A., 2025. Developing a tailored sustainable supply chain framework for dairy SMEs in Lithuania using Best-Worst Method. Transformations and Sustainability 1 (1), 43–62. Available at https://doi.org/10.63775/sv982x31.

Kumar, S., Raut, R.D., Aktas, E., Narkhede, B.E., Gedam, V.V., 2022. Barriers to adoption of industry 4.0 and sustainability: A case study with SMEs. International Journal of Computer Integrated Manufacturing 36 (5), 657–677. Available at https://doi.org/10.1080/0951192x.2022.2128217.

Lampropoulos, G., Siakas, K., Anastasiadis, T., 2019. Internet of Things in the context of Industry 4.0: An overview. International Journal of Entrepreneurial Knowledge 7 (1). Available at https://doi.org/10.37335/ijek.v7i1.84.

Leesakul, N., Oostveen, A.-M., Eimontaite, I., Wilson, M.L., Hyde, R., 2022. Workplace 4.0: Exploring the implications of technology adoption in digital manufacturing on a sustainable workforce. Sustainability 14 (6), 3311. Available at https://doi.org/10.3390/su14063311.

Leitao, P., Geraldes, C.A.S., Fernandes, F.P., & Badikyan, H. (2020). Analysis of the workforce skills for the factories of the future. In *Proceedings of the 2020 IEEE Conference on Industrial Cyberphysical Systems, ICPS 2020* (pp. 353–358). Institute of Electrical and Electronics Engineers Inc. Portugal. https://doi.org/10.1109/ICPS48405.2020.9274757.

Longo, F., Nicoletti, L., Padovano, A., 2019. Estimating the impact of blockchain adoption in the food processing industry and supply chain. International Journal of Food Engineering 16 (5–6). Available at https://doi.org/10.1515/ijfe-2019-0109.

Luz Tortorella, G., Cauchick-Miguel, P.A., Li, W., Staines, J., Mcfarlane, D., 2021. What does operational excellence mean in the Fourth Industrial Revolution era? International Journal of Production Research 60 (9), 2901–2917. Available at https://doi.org/10.1080/00207543.2021.1905903.

Mayer, C.H., 2024. Leaders' views on leadership and skills development in the Fourth Industrial Revolution. SA Journal of Industrial Psychology 50. Available at https://doi.org/10.4102/sajip.v50i0.2172.

Morrone, S., Dimauro, C., Gambella, F., Cappai, M.G., 2022. Industry 4.0 and Precision Livestock Farming (PLF): An up to date overview across animal productions. Sensors 22 (12), 4319. Available at https://doi.org/10.3390/s22124319.

Muzammal, S.M., Tayyab, M., Zahra, F., Jhanjhi, N.Z., & Ashraf, H.,2024. Smart factories greener future (pp. 70–92). IGI Global. https://doi.org/10.4018/979-8-3693–5375-2.ch004.

Neethirajan, S., 2024. Net zero dairy farming—Advancing climate goals with big data and artificial intelligence. Climate 12 (2), 15. Available at https://doi.org/10.3390/cli12020015.

Ortega-Gras, J.-J., Bueno-Delgado, M.-V., Puche-Forte, J.-F., Garrido-Lova, J., Martínez-Fernández, R., 2025. Exploring Industry 4.0 technologies implementation to enhance circularity in Spanish manufacturing enterprises. Sustainability 17 (17), 7648. Available at https://doi.org/10.3390/su17177648.

Priyashantha, H., 2025. World dairy system sustainability: A milk quality perspective. Frontiers in Sustainable Resource Management 4. Available at https://doi.org/10.3389/fsrma.2025.1572962.

Quddus, M.A., 2013. Adoption of dairy farming technologies by small farm holders: Practices and constraints. Bangladesh Journal of Animal Science 41 (2), 124–135. Available at https://doi.org/10.3329/bjas.v41i2.14132.

Rejeb, A., Rejeb, K., Süle, E., Hassoun, A., Keogh, J.G., 2025. Knowledge flows in industry 4.0 research: A longitudinal and dynamic analysis. Journal of Data, Information and Management 7 (2), 123–145. Available at https://doi.org/10.1007/s42488-025-00146-3.

Rosário, A.T., Dias, J.C., 2023. How Industry 4.0 and sensors can leverage product design: Opportunities and challenges. Sensors 23 (3). Available at https://doi.org/10.3390/s23031165.

Ryalat, M., Elmoaqet, H., Alfaouri, M., 2023. Design of a smart factory based on cyber-physical systems and Internet of Things towards Industry 4.0. Applied Sciences 13 (4), 2156. Available at https://doi.org/10.3390/app13042156.

Satyro, W.C., Contador, J.C., Contador, J.L., Fragomeni, M.A., Monken, S.Fd.P., Ribeiro, A.F., et al., 2021. Implementing industry 4.0 through cleaner production and social stakeholders: Holistic and sustainable model. Sustainability (Switzerland) 13 (22). Available at https://doi.org/10.3390/su132212479.

Sayem, A., Biswas, P.K., Khan, M.M.A., Romoli, L., Dalle Mura, M., 2022. Critical barriers to Industry 4.0 adoption in manufacturing organizations and their mitigation strategies. Journal of Manufacturing and Materials Processing 6 (6). Available at https://doi.org/10.3390/jmmp6060136.

Serumaga-Zake, J.M., van der Poll, J.A., 2021. Addressing the impact of Fourth Industrial Revolution on South African manufacturing small and medium enterprises (SMEs). Sustainability 13 (21), 11703. Available at https://doi.org/10.3390/su132111703.

Shabur, M.A., Ali, M.F., Alam, M.M., 2024. Analysis of the barriers and possible approaches for adopting Industry 4.0 in the fertilizer sector of Bangladesh. Discover Applied Sciences 6 (7). Available at https://doi.org/10.1007/s42452-024-06074-y.

Shabur, M.A., Shahriar, A., Ara, M.A., 2025. From automation to collaboration: Exploring the impact of industry 5.0 on sustainable manufacturing. Discover Sustainability 6 (1). Available at https://doi.org/10.1007/s43621-025-01201-0.

Singh, R., Joshi, A., Dissanayake, H., Iddagoda, A., Khan, S., Félix, M.J., et al., 2025. Integrating Industry 4.0, circular economy, and green HRM: A framework for sustainable transformation. Sustainability 17 (7), 3082. Available at https://doi.org/10.3390/su17073082.

Soumati, B., Atmani, M., Benabderrahmane, A., Benjelloun, M., 2023. Whey valorization—Innovative strategies for sustainable development and value-added product creation. Journal of Ecological Engineering 24 (10), 86–104. Available at https://doi.org/10.12911/22998993/169505.

Sumarlin, T., Kusumajaya, R.A., 2024. AI challenges and strategies for business process optimization in Industry 4.0: Systematic literature review. Journal of Management and Informatics 3 (2), 195–211. Available at https://doi.org/10.51903/jmi.v3i2.25.

Turner, C.J., Oyekan, J., Stergioulas, L., Griffin, D., 2020. Utilizing Industry 4.0 on the construction site: Challenges and opportunities. IEEE Transactions on Industrial Informatics 17 (2), 746–756. Available at https://doi.org/10.1109/tii.2020.3002197.

Vignesh, B., Chandrakumar, M., Divya, K., Prahadeeswaran, M., Vanitha, G., 2025. Blockchain technology in agriculture: Ensuring transparency and traceability in the food supply chain. Plant Science Today 12 (sp1). Available at https://doi.org/10.14719/pst.5970.

Vuvor, C.M., Epifani, S., Pulina, G., 2025. Exploring opportunities in Africa's dairy production: Digital transition as a catalyst for marketing innovation and value chain synergies—A systematic review laying foundations for future research. Frontiers in Animal Science. Available at https://doi.org/10.3389/fanim.2025.1632532.

Zhou, Z., Zhang, Y., Yan, Z., 2022. Will digital financial inclusion increase Chinese farmers' willingness to adopt agricultural technology? Agriculture 12 (10), 1514. Available at https://doi.org/10.3390/agriculture12101514.

Chapter 13

Transition from Dairy 4.0 to Dairy 5.0

Abdo Hassoun[1,2]
[1]*Faculty of Agricultural Engineering, University of Aleppo, Aleppo, Syria,* [2]*Sustainable AgriFoodtech Innovation & Research (SAFIR), Arras, France*

13.1 Introduction

The global food sector is currently facing an unprecedented convergence of challenges that threaten its long-term sustainability, resilience, and ethical integrity. Climate change, resource depletion, biodiversity loss, water scarcity, rising energy costs, labor shortages, and increasing consumer demands for transparency and ethical production are collectively reshaping how food systems must operate. In parallel, food supply chains have become more vulnerable to shocks, as evidenced by recent pandemics, geopolitical conflicts, and market volatility (Iqbal et al., 2025; Mkumbukiy et al., 2025; Priyashantha, 2025). Within this broader context, the dairy sector occupies a particularly complex position. Dairy systems are resource-intensive, highly regulated, and biologically sensitive, while simultaneously being expected to deliver safe, affordable, nutritious, and increasingly sustainable food products. These pressures have accelerated the adoption of digital technologies and related innovations, paving the way for the emergence of Dairy 4.0 (Alrhmoun et al., 2025; Bhat and Infascelli, 2025; Gehlot et al., 2022; Malik et al., 2025; Sangode, 2025).

Dairy 4.0 can be understood as the sector-specific application of Industry 4.0 principles across the dairy value chain, encompassing primary production, milk collection, processing, quality control, logistics, and market interaction. Rooted in the broader Industry 4.0 paradigm, Dairy 4.0 emphasizes automation, digitalization, and connectivity through the integration of the Internet of Things (IoT), artificial intelligence (AI), big data analytics, robotics, blockchain, and digital twins, among others. Similar to Food Industry 4.0, the core objective of Dairy 4.0 has been to enhance operational efficiency, productivity, and consistency by enabling real-time data acquisition, automated decision-

Dairy 4.0. DOI: https://doi.org/10.1016/B978-0-443-44730-3.00004-3

making, and system-level optimization (Hassoun et al., 2023; Hassoun et al., 2024a; Hostens et al., 2025; Peres et al., 2025; Sangode, 2025).

Over the past decade, the implementation of Dairy 4.0 technologies has delivered substantial achievements. At the farm level, precision dairy farming systems have enabled continuous monitoring of animal health, behavior, feeding, fertility, and milk yield through sensors, wearable devices, and automated milking systems. These tools have contributed to improved herd management, early disease detection, reduced labor dependency, and more consistent production performance (Brito et al., 2025; Dell'Unto et al., 2025; Michelena et al., 2025). At the processing and manufacturing stages, automation, smart sensors, and AI-driven quality monitoring have enhanced process control, reduced variability, and improved dairy safety outcomes. In addition, blockchain-based and data-enabled traceability systems have strengthened transparency across dairy supply chains, allowing stakeholders and consumers to access reliable information on origin, processing conditions, and quality attributes (Khanashyam et al., 2025; Malik et al., 2026; Mhapsekar et al., 2025).

Despite these advances, the Dairy 4.0 paradigm exhibits structural limitations that increasingly restrict its ability to address the systemic challenges facing the dairy sector. Dairy 4.0 remains, in some cases, technology-centric, prioritizing efficiency, productivity, and economic performance, often at the expense of broader environmental and social considerations. Although digital tools can reduce waste and optimize resource use, they do not inherently ensure regenerative practices or equitable outcomes, and may introduce new risks related to energy-intensive infrastructures, electronic waste, and technological dependency (Cantarelli Da Silva et al., 2025; Hassoun et al., 2023; Hassoun et al., 2024a; Neves et al., 2025; Singh et al., 2025). In parallel, the human dimension of dairy systems is frequently underrepresented. While automation can reduce physical labor, it may also lead to skill mismatches, workforce displacement, and diminished farmer autonomy when decision-making becomes overly algorithm-driven. Animal welfare, despite improved monitoring, is not always embedded as a core ethical objective within optimization-focused frameworks (Bazargani and Deemyad, 2024; Neculai-Văleanu et al., 2025; Závodná et al., 2025). Moreover, Dairy 4.0 could show limited capacity to foster systemic resilience. Real-time monitoring and predictive analytics enhance short-term responsiveness but do not necessarily strengthen long-term adaptability to climate variability, supply chain disruptions, or market instability, often resulting in fragmented efficiency gains rather than integrated resilience strategies (Mkumbukiy et al., 2025; Neethirajan, 2023; Serrano-Torres et al., 2025).

In response to these limitations, the concept of Dairy 5.0 is emerging as an evolutionary extension of Dairy 4.0 rather than a disruptive replacement. Inspired by the principles of Industry 5.0, Dairy 5.0 shifts the focus from purely technology-driven optimization toward a value-driven, human-centric, sustainable, and resilient dairy system. This transition emphasizes the collaborative interaction between humans and intelligent machines, the integration of ethical and environmental considerations into technological design, and the alignment of dairy innovation with long-term societal goals (Bissadu et al., 2025; Haloui et al., 2024; Hassoun et al., 2024a; Malik et al., 2025).

The rationale for transitioning toward Dairy 5.0 lies in recognizing that digital transformation alone is insufficient to address the complex, interconnected challenges facing the dairy sector. Dairy 5.0 seeks to build upon the digital foundations of Dairy 4.0 while embedding human well-being, environmental stewardship, and adaptive capacity at the core of system design. Rather than maximizing output or efficiency in isolation, Dairy 5.0 will aim to optimize value creation across economic, social, and ecological dimensions. As such, the transition from Dairy 4.0 to Dairy 5.0 represents a necessary evolutionary step toward more inclusive, resilient, and future-proof dairy systems aligned with global sustainability objectives and societal expectations. This chapter first revisits the conceptual foundations of Dairy 4.0, then introduces Dairy 5.0 as a human-centric, sustainable, and resilient paradigm, before reviewing its key enabling technologies and concluding with a synthesis of the main insights.

13.2 Conceptual foundations of Dairy 4.0

The concept of Dairy 4.0 originates from the broader paradigm of the fourth industrial revolution (Industry 4.0) and represents the systematic digital transformation of dairy systems across the entire value chain. Rather than being limited to isolated technological upgrades, Dairy 4.0 reflects a structural reconfiguration of how data, processes, and decisions are generated, shared, and operationalized within dairy production, processing, logistics, and market interfaces (Hassoun et al., 2023; Malik et al., 2024; Sangode, 2025). At its core, Dairy 4.0 applies Industry 4.0 principles to dairy systems that are biologically complex, highly regulated, and sensitive to safety risks. In this context, natural variability in raw milk, animal physiology, and microbial activity must be effectively managed while consistently meeting strict quality and food safety requirements (Brito et al., 2025; Murphy et al., 2016; Serrano-Torres et al., 2025).

The enabling technologies of Dairy 4.0 can be conceptually organized into three tightly interconnected domains: digital sciences, physical and engineering sciences, and biological sciences. Digital sciences form the intelligence layer of Dairy 4.0 and are responsible for transforming raw data into actionable knowledge and decision support. Technologies such as AI, big data analytics, blockchain, digital twins, and extended reality operate primarily in the computational and informational space. AI and big data analytics enable learning, prediction, and optimization by extracting patterns from large and heterogeneous datasets generated across farms, processing plants, and dairy supply chains (Hostens et al., 2025; Oztuna Taner and Çolak, 2024; Palma et al., 2025; Serrano-Torres et al., 2025). Blockchain provides a secure and decentralized digital infrastructure for trusted data exchange, traceability, and transparency among multiple stakeholders in the dairy sector (Khanna et al., 2022; Malik et al., 2026; Sharma et al., 2025). Digital twins create virtual representations of dairy systems that allow real time monitoring, simulation, and scenario analysis, while extended reality technologies support visualization, training, and remote assistance (Anastasiou et al., 2023; Rao and Neethirajan, 2025; Woo et al., 2025). Collectively, these digital technologies enable a shift from reactive and experience-based management toward predictive, data-driven, and system-level decision making, which is a defining characteristic of Dairy 4.0

Physical and engineering sciences constitute the operational interface between digital intelligence and real-world dairy systems. This category includes IoT infrastructures, smart sensors, robotics, automation, and additive manufacturing technologies. Smart sensors and connected IoT devices are embedded in animals, equipment, processing lines, and environments to continuously measure physical, chemical, and biological parameters and transmit data to digital platforms (Mhapsekar et al., 2025; Tangorra et al., 2024). Robotics and automated control systems translate digital decisions into precise physical actions, enabling automated milking, feeding, processing, handling, and packaging with high levels of accuracy and repeatability (Johansen et al., 2025; Woo et al., 2025). Additive manufacturing, while driven by digital design models, belongs primarily to the physical domain because it materializes digital instructions into tangible structures, customized products, or functional food matrices (Kanwal et al., 2025; Uribe-Alvarez et al., 2025). These physical technologies are essential for closing the loop between sensing, analysis, and action, allowing Dairy 4.0 systems to operate as cyber-physical ecosystems rather than disconnected digital tools. Their effectiveness, however, depends on robust integration with digital analytics and a deep understanding of the biological processes they interact with.

Biological sciences represent the foundational domain upon which both digital and physical Dairy 4.0 technologies ultimately depend, and this category is increasingly expanding through advances in biotechnology, breeding sciences, and nanotechnology. Dairy systems are inherently biological, involving living animals, microbial ecosystems, and complex biochemical transformations. A sound understanding of animal physiology, behavior, genetics, and welfare is essential for designing meaningful AI models, interpreting sensor-derived data, and deploying automation in a responsible and effective manner (Pacheco et al., 2025; Priyashantha, 2025). In this context, modern breeding and genetic improvement programs play a central role, as genomic selection, precision breeding, and gene editing technologies enable the targeted enhancement of traits related to productivity, health, fertility, disease resistance, robustness, and environmental efficiency. These biological insights are increasingly integrated with digital tools, allowing real-time phenotypic data from sensors to be linked with genomic information to support data-driven breeding decisions and long-term herd optimization (Billah et al., 2025; Brito et al., 2025).

Microbiology and dairy biochemistry further underpin quality and safety management, particularly in fermentation control, spoilage prevention, and shelf-life prediction. Recent advances in biotechnology, including precision fermentation and the development of functional microbial cultures, strengthen this biological foundation by enabling greater control over product characteristics and process consistency (Aprilia Patindra Purba et al., 2025; Knychala et al., 2024; Ray et al., 2026). In parallel, nanotechnologies are playing a growing role by supporting advanced biosensing, smart and active packaging, targeted delivery of nutrients or antimicrobials, and improved control of material and functional properties at the nanoscale (Abedi-Firoozjah et al., 2024; Donkor et al., 2025; Pan et al., 2024). Within the Dairy 4.0 paradigm, biological sciences are not standalone technologies but embedded knowledge systems that shape how digital models are constructed and how physical interventions are applied. As such, the biological domain provides the essential scientific context that ensures Dairy 4.0 technologies remain biologically meaningful, ethically grounded, and aligned with animal welfare, food safety, sustainability, and long-term genetic improvement objectives.

From a conceptual standpoint, Dairy 4.0 has primarily been driven by objectives related to efficiency, productivity, quality consistency, and cost reduction. These objectives align closely with the original Industry 4.0 focus on operational excellence and competitiveness. In the dairy sector, such goals are particularly salient given narrow profit margins, high energy and resource intensity, and strong regulatory pressures. As a result, Dairy 4.0 has delivered tangible improvements in herd management, process control, traceability, and quality assurance across many contexts (Gehlot et al., 2022; Hassoun et al., 2023; Michelena et al., 2025).

However, the conceptual foundations of Dairy 4.0 also reveal inherent boundaries. The paradigm largely assumes that improved data, automation, and optimization will naturally translate into broader sustainability and resilience outcomes. While this assumption is supported in certain operational contexts, a growing body of research suggests that it does not fully address systemic challenges related to environmental regeneration, social sustainability, ethical considerations, and long-term adaptability. Human roles within Dairy 4.0 systems are often framed in terms of supervision and exception handling, rather than active collaboration and empowerment. Similarly, animal welfare and environmental stewardship are frequently treated as constraints to be optimized rather than core values shaping system design. Understanding these conceptual foundations is essential for contextualizing the transition toward Dairy 5.0. Dairy 4.0 provides the indispensable technological and organizational backbone upon which more human-centric, sustainable, and resilient dairy systems can be built. At the same time, its technology-centric orientation highlights the need for an expanded paradigm that rebalances efficiency with values, automation with collaboration, and short-term optimization with long-term system viability.

13.3 Dairy 5.0 as a human-centric, sustainable, and resilient paradigm

Dairy 5.0 can be understood as the dairy sector expression of Industry 5.0, building on the digital, physical, and biological foundations of Dairy 4.0 while redefining system objectives beyond efficiency toward shared value creation. In this paradigm, advanced technologies are not deployed primarily to automate decisions, but to support a human-led, ethically grounded, and future-ready dairy system structured around three interdependent pillars: human-centricity, sustainability, and resilience (Fig. 13.1).

13.3.1 Human-centricity in Dairy 5.0: empowering farmers, workers, and stakeholders

Human-centricity constitutes a core pillar of Dairy 5.0 and represents a fundamental departure from the predominantly automation-centric logic of Dairy 4.0. Rather than positioning humans as passive supervisors of automated systems, Dairy 5.0 reframes farmers, workers, veterinarians, and supply-chain actors as active cocreators of value within digitally augmented dairy ecosystems (Bonamigo et al., 2025; Ivanov, 2023; Malik et al., 2025; Sheikh et al., 2024). In practical terms, this involves designing digital tools that enhance human decision-making, situational awareness, and autonomy rather than substituting experiential knowledge with opaque algorithmic outputs. Advanced analytics, generative AI, and digital twins are thus envisioned as decision-support companions that contextualize data, explain system behavior, and support learning, rather than imposing prescriptive control. This approach resonates with Industry 5.0's emphasis on meaningful human–machine collaboration and the transition from Operator 4.0 to Operator 5.0 concepts observed across industrial sectors (Gladysz et al., 2023; Jaime et al., 2026; Piccarozzi et al., 2024).

In dairy systems, human-centricity also extends to ethical responsibility toward animals and rural communities. Farmer well-being, occupational health, skills development, and social inclusion become explicit design objectives rather than secondary outcomes. Empirical evidence from participatory welfare research further indicates that farmers strongly associate animal welfare with providing a "good life" for dairy cattle, valuing not only productivity but also animals' quality of life and their own sense of professional pride and purpose. Within this context, smart farming

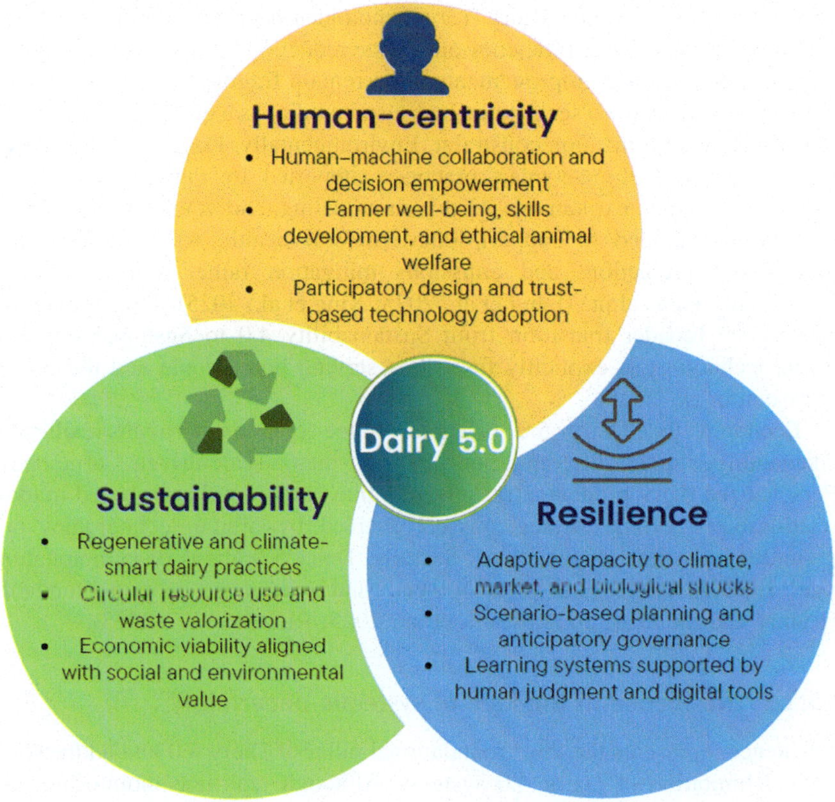

FIGURE 13.1 Conceptual framework of Dairy 5.0 based on human-centricity, sustainability, and resilience.

technologies, when designed and implemented in a supportive and participatory manner, have the potential to enhance animal welfare by enabling earlier health detection, improved monitoring of behavior and comfort, and more informed, timely management decisions (Dawkins, 2025; Doidge et al., 2023; Stokes et al., 2022). Studies consistently indicate that farmer adoption of positive animal welfare practices is highest when these practices are codeveloped with scientists and respect intrinsic animal welfare values, whereas overly technology-driven approaches that marginalize farmer engagement tend to erode trust, reduce uptake, and weaken long-term performance, particularly in small and medium dairy enterprises.

13.3.2 Sustainability as a multidimensional value proposition

Sustainability in Dairy 5.0 is conceptualized as a multidimensional and integrated value proposition encompassing environmental, economic, and social dimensions. While Dairy 4.0 technologies have contributed to incremental gains in resource efficiency and waste reduction (Hassoun et al., 2024b; Sangode, 2025), these improvements often remain fragmented and operationally bounded. Dairy 5.0 seeks to embed sustainability as a guiding system logic rather than an optimization constraint. Environmentally, Dairy 5.0 emphasizes regenerative and climate-smart practices supported by digital intelligence. Advanced sensing, explainable AI-driven modeling, and whole-system digital twins are leveraged to support long-term soil health, water stewardship, biodiversity protection, and emissions mitigation rather than short-term productivity gains alone (Feng et al., 2025; Xing et al., 2025). This orientation reflects the broader transition from Sustainability 4.0 to Sustainability 5.0, where technology is explicitly framed as serving both people and the planet (Hassoun et al., 2025).

Economically, Dairy 5.0 promotes value creation through resilience, adaptability, and differentiation rather than scale-driven efficiency. Personalized dairy products, transparent value chains, and trust-based market relationships become strategic assets enabled by digital technologies but grounded in societal expectations. Socially, Dairy 5.0 aligns dairy innovation with food security, rural development, and ethical production standards (Kumar and Shankar, 2024; Malik et al., 2025; Richter et al., 2025).

13.3.3 Resilience in the face of systemic uncertainty

Resilience represents the third foundational pillar of Dairy 5.0 and addresses a critical limitation of Dairy 4.0 systems. Although real-time monitoring and predictive analytics enhance short-term responsiveness, they do not necessarily confer long-term adaptive capacity to cope with climate variability, market shocks, geopolitical disruptions, or biological crises (Espinoza-Sandoval et al., 2024; Li et al., 2024; van Erp et al., 2024).

Dairy 5.0 conceptualizes resilience as an emergent system property arising from diversity, redundancy, learning, and human–machine collaboration. Digital twins, scenario simulation, and system-level modeling enable anticipatory governance by exploring alternative futures and stress-testing dairy systems under uncertain conditions. Importantly, resilience is not achieved through technological sophistication alone, but through institutional flexibility, human judgment, and participatory decision-making frameworks, as emphasized in Industry 5.0 resilience literature (Khan et al., 2026; Neethirajan, 2025; Nsabiyeze et al., 2025).

13.4 Enabling technologies for Dairy 5.0

The transition from Dairy 4.0 to Dairy 5.0 is not defined by a completely new technological stack, but by a reconfiguration of existing and emerging technologies toward the three Industry 5.0 pillars, human-centricity, sustainability, and resilience. In Dairy 4.0, technologies are mainly deployed to automate operations, stabilize quality, and optimize productivity. In Dairy 5.0, the same technological foundations are extended and complemented to support human decision empowerment, ethical and welfare aligned outcomes, and whole system adaptive capacity across farms, processing plants, and supply chains (Fig. 13.2).

13.4.1 Generative and agentic artificial intelligence

Like in Food Industry 5.0, in Dairy 5.0, AI evolves toward generative and agentic forms that can support knowledge co creation and explainable decision support. Generative AI can synthesize heterogeneous dairy data streams, including sensor data, lab analytics, veterinary records, feed information, and market signals, then translate them into human understandable guidance, options, and trade off narratives. Agentic AI or autonomous AI agents are systems that sense and interpret their environment and take actions independently, with autonomy defined by their capacity to decide and act without direct external control (Gonzalez et al., 2026; Hostens et al., 2025; Liu et al., 2025; Piccialli et al., 2025; Rojek et al., 2025; Sharifmousavi et al., 2024). In addition, Dairy 5.0 increasingly calls for regenerative intelligence, meaning that AI is not only optimizing performance but also supporting regeneration targets, including emissions reduction, circularity, and welfare improvement through continuous learning and multi objective decision support. The Food Industry 5.0 enabling technology synthesis explicitly highlights regenerative

FIGURE 13.2 Enabling technologies for Dairy 5.0: toward human-centric, sustainable, and resilient dairy systems.

and conversational AI as a key differentiator for the fifth paradigm (Hassoun, 2024; Hassoun et al., 2024a), and dairy oriented Industry 5.0 discussions also place generative AI among the defining technologies of the transition. Large language models (LLM) extend these capabilities by enabling the integration of heterogeneous and multimodal dairy data, including text, sensor outputs, images, and tabular records, into unified analytical and decision support frameworks (Ferreira and Dórea, 2025). In Dairy 5.0 contexts, LLM complement computer vision systems by translating complex phenotypic, health, and management information into interpretable insights that support individual animal monitoring, farm management, and knowledge-driven phenotyping.

13.4.2 The Internet of Everything

Dairy 4.0 is enabled by IoT infrastructures that connect sensors, milking systems, feeding stations, barn environment monitoring, processing line sensors, and logistics trackers. Dairy 5.0 extends this model toward the Internet of Everything (IoE), where not only devices but also people, processes, data semantics, and organizations are interconnected in a more meaningful, interoperable way (Hassoun et al., 2024a; Suresh Kumar et al., 2021). IoE implies richer context awareness, better integration across the dairy value chain, and stronger participation of human stakeholders in the data ecosystem. In practical terms, IoE supports cross boundary integration, for example linking animal level phenotypes with genomic and breeding databases, connecting processing quality analytics with supplier practices, and enabling multi stakeholder traceability and risk governance from farm to consumer (Malik et al., 2025; Nozari et al., 2021; Singh et al., 2025).

13.4.3 Collaborative robotics (cobots)

Robotics and automation in Dairy 4.0 have focused on automated milking, feeding, manure handling, and repetitive tasks in processing and packaging. These systems improve labor efficiency and reduce variability, but they can also reinforce a replacement narrative and create barriers for smaller farms due to investment, maintenance skills, and interoperability constraints (Konstantinidis et al., 2023b; Lavrijsen-Kromwijk et al., 2024; Lee et al., 2024; Nettle and Ingram, 2025; Shweta et al., 2025). Dairy 5.0 reframes robotics toward collaborative robotics, where cobots are designed to work safely with humans, complementing skills rather than substituting labor. In dairy farms, cobots can support ergonomically demanding tasks, assist in animal handling under welfare oriented protocols, and reduce injury risks, while keeping farmers in control of ethical and contextual decisions. In processing, cobots can assist with flexible operations, sanitation sensitive tasks, and rapid reconfiguration for product personalization (Nikolova-Alexieva et al., 2025; Wu et al., 2025). The Industry 5.0 research landscape emphasizes human robot collaboration as a central theme, aligning with the broader shift to human-centric production systems. The Food Industry 5.0 enabling technology mapping explicitly identifies cobots and advanced robotics as core enablers, and this logic transfers directly into Dairy 5.0 (Dhanda et al., 2025; Malik et al., 2025).

13.4.4 Advanced additive manufacturing

Within the Dairy 4.0 paradigm, additive manufacturing has primarily been explored through 3D food printing as a flexible production approach enabling the shaping, structuring, and compositional customization of dairy-based products. Early applications demonstrated the potential of extrusion-based 3D printing to valorize dairy ingredients, reduce food waste, and support personalized nutrition, positioning 3D printing as an innovative, efficient, and sustainability-oriented tool within digitally enabled dairy systems (Hassoun et al., 2024b; Sangode, 2025).

In Dairy 5.0, as in Food Industry 5.0 more broadly, additive manufacturing advances from 3D food printing to 4D printing, where the fourth dimension is time and printed edible structures are designed to respond to stimuli such as moisture, temperature, pH, or microwave heating. This enables dynamic functionality, including programmable texture transitions, on demand release profiles for flavors, and smart, personalized structures tailored to specialized nutrition and individual consumer needs (Forsan, 2025; Kok Wah, 2025; Kokane et al., 2024). In parallel, the strategic shift is not only personalization for market value but also sustainability and resilience, including localized manufacturing, reduction of waste, and novel valorization pathways for dairy side streams. Food Industry 5.0 frameworks explicitly highlight 4D printing and other advanced additive manufacturing as part of the next enabling technology portfolio, and this provides a strong conceptual basis for positioning advanced additive manufacturing as a Dairy 5.0 enabler (Arshad et al., 2025; Dong et al., 2025; Sala and Richert, 2025; Singh et al., 2025).

13.4.5 Human-centric digital twins

Digital twins are data-driven virtual representations of physical assets, processes, or systems that synchronize with their real-world counterparts through sensor data, computational models, and feedback loops. By integrating real-time monitoring with simulation and predictive analytics, digital twins support understanding, diagnosis, and informed decision-making across system life cycles. Within the Industry 4.0 paradigm, digital twins have been primarily deployed as asset- or process-centric tools for condition monitoring, fault diagnosis, predictive maintenance, and operational optimization. In food and dairy manufacturing, these applications have largely focused on enhancing operational stability, quality consistency, and efficiency through tight coupling between physical systems and their digital replicas (Adeleke et al., 2026; Feng et al., 2025; Rao and Neethirajan, 2025). For instance, Dairy 4.0 architectures integrating digital twins and machine vision have demonstrated the ability to support zero-defect manufacturing by linking physical inspection systems with predictive digital layers across production and logistics (Konstantinidis et al., 2023a).

As dairy systems transition toward Dairy 5.0 in alignment with Industry 5.0 principles, the role of digital twins expands beyond localized optimization to encompass human-centric cognition, system integration, and governance-oriented decision support. In this paradigm, digital twins are reframed as system-level and knowledge-oriented instruments that explicitly embed human judgment, sustainability objectives, and resilience considerations into their design (Barata and Kayser, 2024; Hassoun et al., 2024a). Rather than serving solely as high-fidelity mirrors of physical operations, human-centric digital twins increasingly function as exploratory and anticipatory environments, enabling stakeholders to evaluate alternative scenarios, trade-offs, and adaptive strategies under uncertainty. This evolution positions digital twins as enablers of participatory decision-making, long-term system learning, and shared responsibility, aligning technological intelligence with human values and societal expectations (Fonseca i Casas and Pi i Palomes, 2026; Ghobakhloo et al., 2025).

In the dairy sector, this shift is already evident in precision nutrition and processing applications. Digital twin frameworks integrating real-time barn sensor data with biophysical models through hybrid edge–cloud architectures support dynamic modeling of feeding and rumination at both individual animal and herd levels, facilitating human-informed nutritional assessment and management. Reported recognition accuracies exceeding 90% for feeding and rumination behaviors highlight their potential to strengthen welfare-oriented, climate-smart, and resilient dairy management practices (Rao and Neethirajan, 2025). At the processing level, digital twins combined with reinforcement learning have been applied to milk pasteurization systems to support fault diagnosis, data reconstruction, and adaptive control, enhancing process reliability and product quality under real industrial conditions (Kadri and Abdelhadi, 2024).

13.4.6 Next-generation blockchain systems

Blockchain can be defined as a decentralized digital ledger technology that records transactions in a secure, transparent, and tamper-resistant manner through cryptographic mechanisms and consensus protocols. By enabling shared and immutable records across multiple stakeholders without reliance on a central authority, blockchain has emerged as a key trust technology within Industry 4.0, particularly for food supply-chain traceability, certification, and data integrity. In this context, blockchain adoption has primarily focused on provenance verification, fraud prevention, and regulatory compliance, with immutability and transparency serving as its core value propositions (Keramati et al., 2025; Pavithra et al., 2025; Ruan et al., 2026; Sharma et al., 2025). In the dairy sector, blockchain applications have already demonstrated value across multiple stages of the supply and value chains. Dairy 4.0 studies highlight the use of blockchain-enabled platforms to support real-time monitoring, quality assurance, and animal welfare information exchange by integrating data from sensors, farm operations, and logistics into transparent traceability systems (Gehlot et al., 2022; Hawashin et al., 2025; Khanna et al., 2022; Vargas-Bello-Pérez et al., 2025).

Review studies further show how blockchain can reduce information asymmetry, strengthen consumer trust, and enhance food safety by enabling secure backtracking of dairy products from farm to market (Rajput et al., 2025; Vasileiou et al., 2025). More recent work in the dairy sector investigates hybrid blockchain and AI-enabled traceability frameworks, demonstrating how the integration of blockchain with advanced data analytics can enable automated data collection, real-time monitoring, improved transparency, and more effective collaboration and compliance across dairy supply-chain operations (Malik et al., 2026). These developments underscore the relevance of blockchain as a key enabling technology for Dairy 5.0, where trust, regulatory alignment, and cooperative value creation become as important as efficiency and automation. The fundamental evolution of blockchain from Dairy 4.0 to Dairy 5.0 lies in its shift from a purely technical tool for efficiency and compliance toward a human-centric, sustainability-aligned infrastructure that supports collaboration, ethical data governance, and long-term system resilience.

13.4.7 Immersive and experiential technologies

Within the Dairy 4.0 paradigm, immersive technologies such as augmented reality (AR), virtual reality (VR), and mixed reality (MR) have primarily been applied as supportive tools for visualization, training, and technical assistance. Typical applications include operator training for milking and processing equipment, visualization of production lines, and remote maintenance support, enabling more efficient knowledge transfer and reduced downtime (Anastasiou et al., 2023; Anderson et al., 2025; Nguyen et al., 2024).

As dairy systems evolve toward Dairy 5.0, immersive technologies move beyond training and visualization toward human-centric empowerment and experiential decision support. In alignment with Industry 5.0 principles, AR, VR, MR, and metaverse-like environments are designed to enhance human understanding, situational awareness, ethical engagement, and collective sense-making rather than merely transmitting technical instructions (Chai et al., 2022; Mourtzis, 2023; Zahid et al., 2025). In dairy farms and processing facilities, these technologies can guide welfare-oriented protocols, support participatory learning, facilitate real-time collaboration among stakeholders, and enable scenario-based exploration of management strategies under changing climatic, operational, or market conditions (Hostens et al., 2025; Kaushik and Rajwanshi, 2025). By strengthening learning, trust, and adaptive human capabilities across generations, immersive technologies complement digital intelligence and reinforce the human-centric and resilience-oriented objectives of Dairy 5.0.

13.4.8 Advanced connectivity and resilient communication infrastructures

In Dairy 4.0, connectivity technologies have served as a foundational enabler by supporting data transmission between sensors, machines, and digital platforms. Wireless networks, cloud infrastructures, and early 5 G deployments enabled real-time monitoring, automated control, and centralized analytics across farms, processing plants, and logistics systems, with connectivity primarily treated as a technical backbone for data availability, responsiveness, and operational efficiency (Naqvi et al., 2022; Shu et al., 2022; Zhang et al., 2021).

In Dairy 5.0, connectivity evolves from a passive data transport layer into a strategic enabler of human-centric, resilient, and distributed dairy systems. Ultra-reliable and low-latency communication supported by advanced 5 G and emerging beyond-5G architectures enables real-time human–machine collaboration, edge intelligence, and synchronized digital twin operations closer to farms and processing units (Hassoun et al., 2024a; Rehman et al., 2025). This shift strengthens resilience through decentralization and rapid system reconfiguration while also supporting continuous stakeholder interaction, remote advisory services, and more inclusive participation of small and medium dairy enterprises.

13.5 Conclusion and outlook

Dairy 5.0 represents an evolutionary maturation of Dairy 4.0 rather than a disruptive technological break. Building on the digital, physical, and biological foundations established under Dairy 4.0, the fifth paradigm reframes digital transformation around human-centricity, sustainability, and resilience. While Dairy 4.0 has delivered substantial gains in efficiency, automation, traceability, and quality control, its technology-centric orientation could be insufficient to fully address systemic challenges related to environmental regeneration, social sustainability, ethical governance, and long-term adaptability. Dairy 5.0 responds to these limitations by redefining the purpose of technology as an enabler of shared value creation rather than narrow operational optimization.

From a strategic perspective, Dairy 5.0 emphasizes meaningful human–machine collaboration, where advanced technologies such as artificial intelligence, digital twins, blockchain, collaborative robotics, and immersive tools support human judgment, transparency, and learning. Farmers, workers, and other stakeholders are repositioned as active participants in digitally augmented dairy ecosystems, with animal welfare, farmer well-being, and trust embedded as explicit system objectives. At the same time, sustainability in Dairy 5.0 extends beyond incremental efficiency improvements toward regenerative, climate-adaptive, and circular practices that align dairy innovation with broader food system transformation and sustainable development goals.

Looking forward, the successful implementation of Dairy 5.0 will depend on coordinated efforts across academia, industry, and policymaking. Research must advance toward robust metrics for human-centric performance, resilience, and sustainability, alongside governance models that address data ownership, ethics, and interoperability. Industry actors will need to integrate technologies into coherent sociotechnical architectures rather than isolated solutions, ensuring accessibility for small and medium dairy enterprises and diverse regional contexts. Policymakers and educational institutions play a critical role in enabling this transition through supportive regulation, skills development, and innovation ecosystems that foster responsible adoption. By aligning digital innovation with human values, ethical responsibility, and long-term system resilience, Dairy 5.0 offers a promising pathway toward more inclusive, sustainable, and future-proof dairy systems within an increasingly uncertain global food landscape.

AI disclosure

During the preparation of this work, the authors used ChatGPT to assist in the reformulation of certain sentences. All outputs were carefully reviewed and edited by the authors, who take full responsibility for the content of this publication.

References

Abedi-Firoozjah, R., Alizadeh-Sani, M., Zare, L., Rostami, O., Azimi Salim, S., Assadpour, E., et al., 2024. State-of-the-art nanosensors and kits for the detection of antibiotic residues in milk and dairy products. Advances in Colloid and Interface Science 328 (April), 103164. Available at https://doi.org/10.1016/j.cis.2024.103164.

Adeleke, I., Nwulu, N., Adebo, O.A., 2026. A digital twin approach for real-time monitoring of amasi acidity using non-invasive computer vision, IoT, and machine learning. Food Control 182, 111809. Available at https://doi.org/10.1016/J.FOODCONT.2025.111809.

Alrhmoun, M., Gauly, M., Zanon, T., 2025. Transitioning toward sustainable dairy systems in Europe: A systematic literature review. Journal of Dairy Science 108 (11), 12373–12390. Available at https://doi.org/10.3168/jds.2025-27018.

Anastasiou, E., Balafoutis, A.T., Fountas, S., 2023. Applications of extended reality (XR) in agriculture, livestock farming, and aquaculture: A review. Smart Agricultural Technology 3 (July 2022), 100105. Available at https://doi.org/10.1016/j.atech.2022.100105.

Anderson, N.C., Underwood, L., Byrd, C.J., 2025. Student reported learning of swine and dairy welfare concepts following a virtual reality livestock farm experience. Journal of Applied Animal Welfare Science 00 (00), 1–13. Available at https://doi.org/10.1080/10888705.2025.2559700.

Aprilia Patindra Purba, R., Sangsawad, P., Rayudika Aprilia Patindra Purba, O., 2025. Biotechnology approaches to dairy alternatives through precision fermentation and cellular agriculture. Food Science of Animal Resources 45 (5), 1221–1250. Available at https://doi.org/10.5851/KOSFA.2025.E33.

Arshad, R., Saqib, A., Sharif, H.R., Liaqat, A., Xu, B., 2025. Recent advances in 3D food printing: Therapeutic implications, opportunities, potential applications, and challenges in the food industry. Food Research International 203 (January), 115791. Available at https://doi.org/10.1016/j.foodres.2025.115791.

Barata, J., Kayser, I., 2024. How will the digital twin shape the future of industry 5.0? Technovation 134, 103025. Available at https://doi.org/10.1016/J.TECHNOVATION.2024.103025.

Bazargani, K., Deemyad, T., 2024. Automation's impact on agriculture: Opportunities, challenges, and economic effects. Robotics 13 (2), 33. Available at https://doi.org/10.3390/ROBOTICS13020033.

Bhat, R., Infascelli, F., 2025. The path to sustainable dairy industry: Addressing challenges and embracing opportunities. Sustainability 17 (9), 3766. Available at https://doi.org/10.3390/SU17093766.

Billah, M., Bermann, M., Hollifield, M.K., Tsuruta, S., Chen, C.Y., Psota, E., et al., 2025. Review: Genomic selection in the era of phenotyping based on digital images. Animal 19, 101486. Available at https://doi.org/10.1016/J.ANIMAL.2025.101486.

Bissadu, K.D., Sonko, S., Hossain, G., 2025. Society 5.0 enabled agriculture: Drivers, enabling technologies, architectures, opportunities, and challenges. Information Processing in Agriculture 12 (1), 112–124. Available at https://doi.org/10.1016/J.INPA.2024.04.003.

Bonamigo, A., da Silva Pereira, E., de Oliveira, J.M., Antony, J., Garza-Reyes, J.A., 2025. Lean Service 5.0 in dairy production: An empirical evaluation from the perspective of the sociotechnical and circular economy approach. Business Strategy and the Environment 34 (7), 8134–8151. Available at https://doi.org/10.1002/BSE.70015.

Brito, L.F., Heringstad, B., Klaas, I.C., Schodl, K., Cabrera, V.E., Stygar, A., et al., 2025. Invited review: Using data from sensors and other precision farming technologies to enhance the sustainability of dairy cattle breeding programs. Journal of Dairy Science 108 (10), 10447–10474. Available at https://doi.org/10.3168/jds.2025-26554.

Cantarelli Da Silva, R., Lima, L.B., Batistela, E., Santos, D., De Cássia Akutsu, R., 2025. The use of Industry 4.0 and 5.0 technologies in the transformation of food services: An integrative review. Foods 14 (24), 4320. Available at https://doi.org/10.3390/FOODS14244320.

Chai, J.J.K., O'Sullivan, C., Gowen, A.A., Rooney, B., Xu, J.-L., 2022. Augmented/mixed reality technologies for food: A review. Trends in Food Science & Technology 124, 182–194. Available at https://doi.org/10.1016/J.TIFS.2022.04.021.

Dawkins, M.S., 2025. Smart farming and Artificial Intelligence (AI): How can we ensure that animal welfare is a priority? Applied Animal Behaviour Science 283, 106519. Available at https://doi.org/10.1016/J.APPLANIM.2025.106519.

Dell'Unto, D., Selvaggi, R., Pappalardo, G., Cortignani, R., 2025. Adoption of precision livestock farming devices in the dairy cattle sector: An assessment based on agroeconomic modelling. Science of The Total Environment 1002, 180555. Available at https://doi.org/10.1016/J.SCITOTENV.2025.180555.

Dhanda, M., Rogers, B.A., Hall, S., Dekoninck, E., Dhokia, V., 2025. Reviewing human-robot collaboration in manufacturing: Opportunities and challenges in the context of industry 5.0. Robotics and Computer-Integrated Manufacturing 93 (October 2024), 102937. Available at https://doi.org/10.1016/j.rcim.2024.102937.

Doidge, C., Frössling, J., Dórea, F.C., Ordell, A., Vidal, G., Kaler, J., 2023. Social and ethical implications of data and technology use on farms: A qualitative study of Swedish dairy and pig farmers. Frontiers in Veterinary Science 10, 1171107. Available at https://doi.org/10.3389/fvets.2023.1171107.

Dong, Y., Dai, J., Han, Jingjing, Zhang, Y., Xu, Z., et al., 2025. Dysphagia foods: Current research, technological innovations, and future directions. Trends in Food Science and Technology 165 (July), 105330. Available at https://doi.org/10.1016/j.tifs.2025.105330.

Donkor, F., Okafor, M.N., Enyejo, J.O., 2025. Investigating nanotechnology-based smart packaging for extending dairy product shelf life and improving food quality assurance. International Journal of Healthcare Sciences 13 (2), 17–34. Available at https://doi.org/10.5281/zenodo.17381311.

van Erp, T., Carvalho, N.G.P., Gerolamo, M.C., Gonçalves, R., Rytter, N.G.M., Gladysz, B., 2024. Industry 5.0: A new strategy framework for sustainability management and beyond. Journal of Cleaner Production 461, 142271. Available at https://doi.org/10.1016/J.JCLEPRO.2024.142271.

Espinoza-Sandoval, O.R., Angeles-Hernandez, J.C., Gonzalez-Ronquillo, M., Ghavipanje, N., Zhang, N., Bayat, A.R., et al., 2024. Dairy farming in the era of artificial intelligence: Trend or a real game changer? Journal of Dairy Research 91 (2), 139–145. Available at https://doi.org/10.1017/S0022029924000426.

Feng, C., Reed, K.F., Giordano, J.O., You, F., 2025. Precision decarbonization for clean dairy farming with digital twin and edge intelligence. Nexus 2 (4), 100105. Available at https://doi.org/10.1016/j.ynexs.2025.100105.

Ferreira, R.E.P., Dórea, J.R.R., 2025. International symposium on ruminant physiology: Leveraging computer vision, large language models, and multimodal machine learning for optimal decision making in dairy farming. Journal of Dairy Science 108 (7), 7493–7510. Available at https://doi.org/10.3168/jds.2024-25650.

Fonseca i Casas, P., Pi i Palomes, X., 2026. Building Society 5.0: A foundation for decision-making based on open models and digital twins. Advanced Engineering Informatics 69, 103970. Available at https://doi.org/10.1016/J.AEI.2025.103970.

Forsan, H.F., 2025. Utilizing the potential of technologies from the Fourth Industrial Revolution to decrease waste and increase the value of byproducts in the dairy industry. In: Aguilar-Zárate, R., Haghi, P., Gómez-García, A.K. (Eds.), Reducing Food Loss and Waste. Springer, pp. 85–101. Available at https://doi.org/10.1007/978-3-031-91693-9_4.

Gehlot, A., Malik, P.K., Singh, R., Akram, S.V., Alsuwian, T., 2022. Dairy 4.0: Intelligent communication ecosystem for the cattle animal welfare with blockchain and IoT enabled technologies. Applied Sciences 12 (14), 7316. Available at https://doi.org/10.3390/APP12147316.

Ghobakhloo, M., Fathi, M., Okwir, S., Al-Emran, M., Ivanov, D., 2025. Adaptive social manufacturing: A human-centric, resilient, and sustainable framework for advancing

Industry 5.0. International Journal of Production Research 7543, 0–34. Available at https://doi.org/10.1080/00207543.2025.2559137.

Gladysz, B., Tran, T.A., Romero, D., van Erp, T., Abonyi, J., Ruppert, T., 2023. Current development on the Operator 4.0 and transition towards the Operator 5.0: A systematic literature review in light of Industry 5.0. Journal of Manufacturing Systems 70, 160–185. Available at https://doi.org/10.1016/J.JMSY.2023.07.008.

Gonzalez, G.R., Habel, J., Hunter, G.K., 2026. AI agents, agentic AI, and the future of sales. Journal of Business Research 202 (October 2025), 115799. Available at https://doi.org/10.1016/j.jbusres.2025.115799.

Haloui, D., Oufaska, K., Oudani, M., Yassini, K.E., 2024. Bridging Industry 5.0 and Agriculture 5.0: Historical perspectives, opportunities, and future perspectives. Sustainability 16 (9), 3507. Available at https://doi.org/10.3390/SU16093507.

Hassoun, A., 2024. From Food Industry 4.0 toward Food Industry 5.0: Human-centric approach for enhanced sustainability and resilience in the food and agriculture sector. *Food Industry 4.0: Emerging Trends and Technologies in Sustainable Food Production and Consumption*. Elsevier. Available at https://doi.org/10.1016/B978-0-443-15516-1.00014-1.

Hassoun, A., Garcia-garcia, G., Trollman, H., Jagtap, S., Cropotova, J., Bhat, Z., et al., 2023. Birth of dairy 4.0: Opportunities and challenges in adoption of fourth industrial revolution technologies in the production of milk and its derivatives. Current Research in Food Science 7 (May), 100535. Available at https://doi.org/10.1016/j.crfs.2023.100535.

Hassoun, A., Jagtap, S., Trollman, H., Guillermo, G.-G., Prateek, S., Duong, L.N.K., et al., 2024a. From Food Industry 4.0 to Food Industry 5.0: Identifying technological enablers and potential future applications in the food sector. Comprehensive Reviews in Food Science and Food Safety 23, e370040. Available at https://doi.org/10.1111/1541-4337.70040.

Hassoun, A., Tarchi, I., Aït-kaddour, A., 2024b. Leveraging the potential of fourth industrial revolution technologies to reduce and valorize waste and by- products in the dairy sector. Current Opinion in Green and Sustainable Chemistry 47, 100927. Available at https://doi.org/10.1016/j.cogsc.2024.100927.

Hassoun, A., López, C.P., Kaddour, A.A., Abdallah, S.B., 2025. Transition from Food Sustainability 4.0 to Food Sustainability 5.0: A current trend or a future vision? In: Hassoun, A. (Ed.), Food Industry 4.0 and Food Sustainability. Springer, pp. 269–284. Available at https://doi.org/10.1007/978-3-032-06527-8.

Hawashin, D., Salah, K., Jayaraman, R., Ahmad, R.W., Kanhere, S., Damiani, E., 2025. Enhancing sustainability in dairy industry: Blockchain-based waste reduction. Sustainable Futures 9, 100529. Available at https://doi.org/10.1016/J.SFTR.2025.100529.

Hostens, M., Franceschini, S., van Leerdam, M., Yang, H., Pokharel, S., Liu, E., et al., 2025. The future of big data and artificial intelligence on dairy farms: A proposed dairy data ecosystem. JDS Communications 6, S9–S14. Available at https://doi.org/10.3168/jdsc.2025-0843.

Iqbal, B., Alabbosh, K.F., Jalal, A., Suboktagin, S., Elboughdiri, N., 2025. Sustainable food systems transformation in the face of climate change: Strategies, challenges, and policy implications. Food Science and Biotechnology 34 (4), 871–883. Available at https://doi.org/10.1007/s10068-024-01712-y.

Ivanov, D., 2023. The Industry 5.0 framework: Viability-based integration of the resilience, sustainability, and human-centricity perspectives. International Journal of Production Research 61 (5), 1683–1695. Available at https://doi.org/10.1080/00207543.2022.2118892.

Jaime, A., Osorio-sanabria, M.A., Yohana, D., Torres, B., 2026. Similarities and differences between Industry 4.0 and Industry 5.0: Towards a transitioning model. Journal of Innovation

& Knowledge 13 (December 2025), 100921. Available at https://doi.org/10.1016/j.jik.2025.100921.

Johansen, F.P., Donnelly, P., Arnott, G., Buijs, S., 2025. Use of a priority lane to increase voluntary visits to a milking robot in dairy cows. Journal of Dairy Science 108 (7), 7387–7400. Available at https://doi.org/10.3168/jds.2024-26081.

Kadri, O., Abdelhadi, A., 2024. Optimizing milk pasteurization diagnosis through deep Q-networks and digital twin technology. International Journal of Web Services Research 21 (1), 1–22. Available at https://doi.org/10.4018/IJWSR.366586.

Kanwal, N., Zhang, M., Law, C.L., khan, S., Huang, J., 2025. Integrating cellular agriculture with 3D food printing: A pathway to customized, sustainable animal-free dairy foods. Food Bioscience 69 (December 2024), 107017. Available at https://doi.org/10.1016/j.fbio.2025.107017.

Kaushik, H., Rajwanshi, R., 2025. Exploring technology adoption in dairy farms: Interactive management perspective. Journal of Science and Technology Policy Management. Available at https://doi.org/10.1108/JSTPM-06-2024-0229.

Keramati, A., Siau, B., Bellitto, T., Heydari, J., Panchal, T., 2025. Blockchains effects on responsiveness to recalls in the food and beverage industry. Journal of Economy and Technology 3, 283–298. Available at https://doi.org/10.1016/J.JECT.2025.05.001.

Khan, M.U., Zhen, L., Tian, J., Farid, A., Chen, Z., Liu, H., 2026. From detection to prevention: A cyber-secure AI ecosystem for mitigating milk adulteration and dairy fraud. Trends in Food Science and Technology 168 (October 2025), 105503. Available at https://doi.org/10.1016/j.tifs.2025.105503.

Khanashyam, A.C., Jagtap, S., Agrawal, T.K., Thorakkattu, P., Malav, O.P., Trollman, H., et al., 2025. Applications of artificial intelligence in the dairy Industry: From farm to product development. Computers and Electronics in Agriculture 238 (April), 110879. Available at https://doi.org/10.1016/j.compag.2025.110879.

Khanna, A., Jain, S., Burgio, A., Bolshev, V., Panchenko, V., 2022. Blockchain-enabled supply chain platform for Indian dairy industry: Safety and traceability. Foods 11 (17), 2716. Available at https://doi.org/10.3390/FOODS11172716.

Knychala, M.M., Boing, L.A., Ienczak, J.L., Trichez, D., Stambuk, B.U., 2024. Precision fermentation as an alternative to animal protein: A review. Fermentation 10 (6), 315. Available at https://doi.org/10.3390/FERMENTATION10060315.

Kok Wah, J.N., 2025. AI-driven 3D and 4D food printing: Innovations for sustainability, personalization, and global applications. Food Reviews International 00 (00), 1–29. Available at https://doi.org/10.1080/87559129.2025.2502438.

Kokane, S.B., Anjaly, P.R., Thangalakshmi, S., Arora, V.K., 2024. Current trends in additive manufacturing based 4D food printing technology: A review. Future Foods 10 (March), 100450. Available at https://doi.org/10.1016/j.fufo.2024.100450.

Konstantinidis, F.K., Balaska, V., Symeonidis, S., Psarommatis, F., Psomoulis, A., Giakos, G., et al. (2023a). Achieving zero defected products in Diary 4.0 using digital twin and machine vision. *ACM International Conference Proceeding Series* (pp. 528–534). https://doi.org/10.1145/3594806.3596554.

Konstantinidis, F.K., Balaska, V., Symeonidis, S., Tsilis, D., Mouroutsos, S.G., Bampis, L., et al., 2023b. Automating dairy production lines with the yoghurt cups recognition and detection process in the Industry 4.0 era. Procedia Computer Science 217 (2022), 918–927. Available at https://doi.org/10.1016/j.procs.2022.12.289.

Kumar, U., Shankar, R., 2024. Smart dairy: Unleashing emerging ICT-enabled lean dairy supply chains through data-driven decision-making. International Journal of Information

Management Data Insights 4 (2), 100297. Available at https://doi.org/10.1016/J.JJIMEI.2024.100297.

Lavrijsen-Kromwijk, L., Demba, S., Müller, U., Rose, S., 2024. Impact of automation level of dairy farms in Northern and Central Germany on dairy cattle welfare. Animals 14 (24), 3699. Available at https://doi.org/10.3390/ani14243699.

Lee, Y.G., Han, K., Chung, C., Ji, I., 2024. Effects of smart farming on the productivity of Korean Dairy farms: A case study of robotic milking systems. Sustainability 16 (22), 9991. Available at https://doi.org/10.3390/SU16229991.

Li, C., Zhang, F., Leong Tan, M., Pan Chun, K., Neethirajan, S., 2024. Net zero dairy farming—Advancing Climate Goals with big data and artificial intelligence. Climate 12 (2), 15. Available at https://doi.org/10.3390/CLI12020015.

Liu, E., Yang, H., Sharma, S., van Leerdam, M.B., Niu, P., VandeHaar, M.J., et al., 2025. Agents are all you need: Pioneering the use of agentic artificial intelligence to embrace large language models into dairy science. Journal of Dairy Science 108 (12), 14038–14049. Available at https://doi.org/10.3168/jds.2025-26775.

Malik, M., Gahlawat, V.K., Mor, R.S., Hosseinian-Far, A., 2024. Towards white revolution 2.0: Challenges and opportunities for the industry 4.0 technologies in Indian dairy industry. Operations Management Research 17, 811–832. Available at https://doi.org/10.1007/s12063-024-00482-4.

Malik, M., Mor, R.S., Gahlawat, V.K., Hassoun, A., Jagtap, S., 2025. Drivers of Industry 5.0 technologies in dairy industry: An exploratory study. Sustainable Food Technology. Available at https://doi.org/10.1039/d5fb00156k.

Malik, M., Mor, R.S., Gahlawat, V.K., Kumar, V., 2026. Unlocking the potential: Hybrid blockchain and AI-enabled traceability model development and implementation in the dairy industry—Proof-of-concept. Transportation Research Part E: Logistics and Transportation Review 205 (May 2025), 104552. Available at https://doi.org/10.1016/j.tre.2025.104552.

Mhapsekar, R., Kilbane, D., Davy, S., Abraham, L., Fenelon, M., O'Shea, N., 2025. A systematic review of the internet of things and artificial intelligence applications in milk quality monitoring and analysis. International Journal of Dairy Technology 78 (3), e70049. Available at https://doi.org/10.1111/1471-0307.70049.

Michelena, Á., Fontenla-Romero, Ó., Luis Calvo-Rolle, J., 2025. A review and future trends of precision livestock over dairy and beef cow cattle with artificial intelligence. Logic Journal of the IGPL 33 (4). Available at https://doi.org/10.1093/jigpal/jzae111.

Mkumbukiy, A., Loghmani-Khouzani, T., Madani, K., Guenther, E., 2025. Agrifood systems' resilience for sustainable food security amid geopolitical tensions: a systematic literature review. Frontiers in Sustainable Food Systems 9, 1546851. Available at https://doi.org/10.3389/fsufs.2025.1546851.

Mourtzis, D., 2023. The metaverse in Industry 5.0: A Human-centric approach towards personalized value creation. Encyclopedia 3 (3), 1105–1120. Available at https://doi.org/10.3390/encyclopedia3030080.

Murphy, S.C., Martin, N.H., Barbano, D.M., Wiedmann, M., 2016. Influence of raw milk quality on processed dairy products: How do raw milk quality test results relate to product quality and yield? Journal of Dairy Science 99 (12), 10128–10149. Available at https://doi.org/10.3168/jds.2016-11172.

Naqvi, S.M.Z.A., Saleem, S.R., Tahir, M.N., Li, S., Hussain, S., Haq, S.I.U., et al., 2022. Role of 5G and 6G technology in precision agriculture. Environmental Sciences Proceedings 23 (1), 3. Available at https://doi.org/10.3390/ENVIRONSCIPROC2022023003.

Neculai-Văleanu, A.-S., Sanduleanu, C., Creangă, D.E., Brumă, I.S., Hoha, G.V., 2025. Digital Transformation in the dairy sector: A multidisciplinary approach to rural economic growth. In: Neculai-Văleanu, A.-S., Sanduleanu, C., Creangă, D.E., Brumă, I.S., Hoha, G.V. (Eds.), Tech-economic dynamics and regional growth strategies. IGI Global Scientific Publishing, pp. 139–168. Available at https://doi.org/10.4018/979-8-3373-2898-0.CH005.

Neethirajan, S., 2023. Artificial intelligence and sensor technologies in dairy livestock export: Charting a digital transformation. Sensors 23 (16), 7045. Available at https://doi.org/10.3390/S23167045.

Neethirajan, S., 2025. Safeguarding digital livestock farming—A comprehensive cybersecurity roadmap for dairy and poultry industries. Frontiers in Big Data 8, 1556157. Available at https://doi.org/10.3389/fdata.2025.1556157.

Nettle, R., Ingram, J., 2025. Smart farming technologies and changes to farm work: New insights into on-farm experiences. Technological Forecasting and Social Change 218, 124227. Available at https://doi.org/10.1016/j.techfore.2025.124227.

Neves, A., Alves, A.S.F., Azamfirei Ionita, V., Matias, J.C.O., Teixeira, S., Alcácer, V., et al., 2025. Key drivers and barriers of industry 4.0 for sustainable practices in small and medium-sized enterprises. Sustainable Futures 10, 101464. Available at https://doi.org/10.1016/J.SFTR.2025.101464.

Nguyen, A., Francis, M., Windfeld, E., Lhermie, G., Kim, K., 2024. Developing an immersive virtual farm simulation for engaging and effective public education about the dairy industry. Computers and Graphics (Pergamon) 118 (December 2023), 173–183. Available at https://doi.org/10.1016/j.cag.2023.12.011.

Nikolova-Alexieva, V., Valeva, K., Terziyska, M., Shakev, N., 2025. Collaborative robots as an engineering tool for the transition of the food industry to Industry 5.0. Engineering Proceedings 100 (1), 57. Available at https://doi.org/10.3390/ENGPROC2025100057.

Nozari, H., Szmelter-Jarosz, A., Ghahremani-Nahr, J., 2021. The ideas of sustainable and green marketing based on the internet of everything—The case of the dairy industry. Future Internet 13 (10), 266. Available at https://doi.org/10.3390/FI13100266.

Nsabiyeze, A., Zhang, M., Li, J., Zhao, Q., Zhang, X., 2025. Precision livestock farming for climate-resilient livestock management: A review of real-time monitoring and decision support systems. Journal of Cleaner Production 524 (August), 146454. Available at https://doi.org/10.1016/j.jclepro.2025.146454.

Oztuna Taner, O., Çolak, A.B., 2024. Dairy factory milk product processing and sustainable of the shelf-life extension with artificial intelligence: A model study. Frontiers in Sustainable Food Systems 8 (June), 1–13. Available at https://doi.org/10.3389/fsufs.2024.1344370.

Pacheco, H.A., Hernandez, R.O., Chen, S.Y., Neave, H.W., Pempek, J.A., Brito, L.F., 2025. Invited review: Phenotyping strategies and genetic background of dairy cattle behavior in intensive production systems—From trait definition to genomic selection. Journal of Dairy Science 108 (1), 6–32. Available at https://doi.org/10.3168/jds.2024-24953.

Palma, O., Plà-Aragonés, L.M., Mac Cawley, A., Albornoz, V.M., 2025. AI and data analytics in the dairy farms: A scoping review. Animals 15 (9), 1291. Available at https://doi.org/10.3390/ANI15091291.

Pan, Y., Liu, J., Wang, J., Gao, Y., Ma, N., 2024. Invited review: Application of biosensors and biomimetic sensors in dairy product testing. Journal of Dairy Science 107 (10), 7533–7548. Available at https://doi.org/10.3168/jds.2024-24666.

Pavithra, T., Rawat, S., Sunil, C.K., 2025. Recent trends in blockchain traceability of food products: A review. Food and Humanity 5 (May), 100666. Available at https://doi.org/10.1016/j.foohum.2025.100666.

Peres, F.A.P., Bondarczuk, B.A., Gomes, L.de C., Jardim, L.de C., Corrêa, R.G.de F., Baierle, I.C., 2025. Advances in food quality management driven by Industry 4.0: A systematic review-based framework. Foods 14 (14), 2429. Available at https://doi.org/10.3390/FOODS14142429.

Piccarozzi, M., Silvestri, L., Silvestri, C., Ruggieri, A., 2024. Roadmap to Industry 5.0: Enabling technologies, challenges, and opportunities towards a holistic definition in management studies. Technological Forecasting and Social Change 205 (July 2023), 123467. Available at https://doi.org/10.1016/j.techfore.2024.123467.

Piccialli, F., Chiaro, D., Sarwar, S., Cerciello, D., Qi, P., Mele, V., 2025. AgentAI: A comprehensive survey on autonomous agents in distributed AI for industry 4.0. Expert Systems with Applications 291, 128404. Available at https://doi.org/10.1016/J.ESWA.2025.128404.

Priyashantha, H., 2025. World dairy system sustainability: A milk quality perspective. Frontiers in Sustainable Resource Management 4, 1572962. Available at https://doi.org/10.3389/FSRMA.2025.1572962.

Rajput, D.V., More, P.R., Adhikari, P.A., Arya, S.S., 2025. Blockchain technology in the food supply chain: A way towards circular economy and sustainability. Sustainable Food Technology. Available at https://doi.org/10.1039/D5FB00065C.

Rao, S., Neethirajan, S., 2025. Computational architectures for precision dairy nutrition digital twins: A technical review and implementation framework. Sensors 25 (16), 4899. Available at https://doi.org/10.3390/S25164899.

Ray, A., Sinha, C., Sharma, A.K., Khamrui, K., Hussain, S.A., Dabas, J.K., et al., 2026. AI-driven real-time monitoring and predictive control system for yoghurt fermentation. Food Control 182 (June 2025), 111857. Available at https://doi.org/10.1016/j.foodcont.2025.111857.

Rehman, W.U., Koondhar, M.A., Afridi, S.K., Albasha, L., Smaili, I.H., Touti, E., et al., 2025. The role of 5G network in revolutionizing agriculture for sustainable development: A comprehensive review. Energy Nexus 17 (December 2024), 100368. Available at https://doi.org/10.1016/j.nexus.2025.100368.

Richter, S., Stolz, H., Martinez-Cruz, A.L., Kachi, A., 2025. Animal welfare has priority: Swiss consumers' preferences for animal welfare, greenhouse gas reductions and other sustainability improvements in dairy products. Food Quality and Preference 123, 105350. Available at https://doi.org/10.1016/J.FOODQUAL.2024.105350.

Rojek, I., Mikołajewski, D., Piszcz, A., Małolepsza, O., Kozielski, M., 2025. Role of generative AI in AI-based digital twins in Industry 5.0 and evolution to Industry 6.0. Applied Sciences (Switzerland) 15 (18), 10102. Available at https://doi.org/10.3390/app151810102.

Ruan, Y., Unachukwu, P., Benjakul, S., Zhang, Y., Fu, Y., 2026. The emerging role of blockchain technology in ensuring food safety: A review. Trends in Food Science and Technology 168 (October 2025), 105489. Available at https://doi.org/10.1016/j.tifs.2025.105489.

Sala, D., Richert, M., 2025. Perspectives of additive manufacturing in 5.0 Industry. Materials 18 (2), 429. Available at https://doi.org/10.3390/MA18020429.

Sangode, P.B., 2025. Dairy 4.0: Embracing digital technology for sustainable farming: A review. Asian Journal of Dairy and Food Research 44 (2), 169–178. Available at https://doi.org/10.18805/ajdfr.DR-2242.

Serrano-Torres, G.J., López-Naranjo, A.L., Larrea-Cuadrado, P.L., Mazón-Fierro, G., 2025. Transformation of the dairy supply chain through artificial intelligence: A systematic review. Sustainability (Switzerland) 17 (3), 982. Available at https://doi.org/10.3390/su17030982.

Sharifmousavi, M., Kayvanfar, V., Baldacci, R., 2024. Distributed artificial intelligence application in agri-food supply chains 4.0. Procedia Computer Science 232 (2023), 211–220. Available at https://doi.org/10.1016/j.procs.2024.01.021.

Sharma, D., Singh, G., Kaur, A., 2025. Blockchain-enabled traceability frameworks and prospects for dairy supply chain management. In: In: Hassanien, P., Anand, A.E., Jaiswal, S., Kumar, A. (Eds.), Innovative Computing and Communications Proceedings of ICICC 2024 Vol. 3. Springer Nature, Singapore, pp. 565–576. Available at https://doi.org/10.1007/978-981-97-4152-6_42.

Sheikh, R.A., Ahmed, I., Faqihi, A.Y.A., Shehawy, Y.M., 2024. Global perspectives on navigating Industry 5.0 Knowledge: Achieving resilience, sustainability, and human-centric innovation in manufacturing. Journal of the Knowledge Economy 16 (5), 15997–16032. Available at https://doi.org/10.1007/S13132-024-02498-4.

Shu, L., Pearson, S., Liu, Y., Ferrag, M.A., Maglaras, L., Zhang, F., et al., 2022. 6G-enabled smart agriculture: A review and prospect. Electronics 11 (18), 2845. Available at https://doi.org/10.3390/ELECTRONICS11182845.

Shweta, Mohammed, A., Mohammadi, M., Sharma, S., Sharma, U., Rana, V.S., et al., 2025. Assessment of advancements and applications of robotics, artificial intelligence, and automated technology in the modern food sector. Applied Food Research 5 (2), 101261. Available at https://doi.org/10.1016/j.afres.2025.101261.

Singh, K.A., Patra, F., Ghosh, T., Mahnot, N.K., Dutta, H., Duary, R.K., 2025. Advancing food systems with Industry 5.0: A systematic review of smart technologies, sustainability, and resource optimization. Sustainable Futures 9, 100694. Available at https://doi.org/10.1016/J.SFTR.2025.100694.

Stokes, J.E., Rowe, E., Mullan, S., Pritchard, J.C., Horler, R., Haskell, M.J., et al., 2022. A "Good Life" for dairy cattle: Developing and piloting a framework for assessing positive welfare opportunities based on scientific evidence and farmer expertise. Animals 12 (19), 2540. Available at https://doi.org/10.3390/ani12192540.

Suresh Kumar, K., Balakrishnan, S., Janet, J., 2021. A cloud-based prototype for the monitoring and predicting of data in precision agriculture based on internet of everything. Journal of Ambient Intelligence and Humanized Computing 12 (9), 8719–8730. Available at https://doi.org/10.1007/s12652-020-02632-5.

Tangorra, F.M., Buoio, E., Calcante, A., Bassi, A., Costa, A., 2024. Internet of Things (IoT): Sensors application in dairy cattle farming. Animals 14 (21), 3071. Available at https://doi.org/10.3390/ANI14213071.

Uribe-Alvarez, R., Crofton, E., Kilcawley, K., Skibinska, I., Coleman-Vaughan, C., Murphy, C.P., et al., 2025. Sensory characterisation and volatile analysis of 3D-printed dairy protein-based snack structures. International Journal of Food Properties 28 (1), 2573182. Available at https://doi.org/10.1080/10942912.2025.2573182.

Vargas-Bello-Pérez, E., Espinoza Sandoval, O.R., Rodríguez-Piñeros, S., Ghavipanje, N., 2025. Blockchain and its future in the supply and value chains of the dairy industry. Journal of Dairy Research 92 (2), 127–134. Available at https://doi.org/10.1017/S0022029925101076.

Vasileiou, M., Leonidas, Kyrgiakos, S., Kleisiari, C., Lappas, P.Z., Tsinopoulos, C., et al., 2025. Digital transformation of food supply chain management using blockchain: A systematic literature review towards food safety and traceability. Business & Information Systems Engineering 2025 1–28. Available at https://doi.org/10.1007/S12599-025-00948-0.

Woo, S., Kim, Y., Kim, S., 2025. Converging extended reality and robotics for innovation in the food industry. AgriEngineering 7 (10), 322. Available at https://doi.org/10.3390/AGRIENGINEERING7100322.

Wu, Y., Wang, X., Hu, B., 2025. Wearable sensor-based analysis of human biomechanics in manual and cobot-assisted agricultural transplanting. Electronics 14 (10), 2043. Available at https://doi.org/10.3390/ELECTRONICS14102043.

Xing, R., Li, B., Dora, S., Whittaker, M., Mathie, J., 2025. Engineering-oriented explainable machine learning and digital twin framework for sustainable dairy production and environmental impact optimisation. Algorithms 18 (10), 670. Available at https://doi.org/10.3390/A18100670.

Zahid, A., Ferraro, A., Petrillo, A., De Felice, F., 2025. Exploring the role of digital twin and industrial metaverse technologies in enhancing occupational health and safety in manufacturing. Applied Sciences 15 (15), 8268. Available at https://doi.org/10.3390/APP15158268.

Zhang, J., Zhang, R., Yang, Q., Hu, T., Guo, K., & Hong, T., 2021. Research on application technology of 5G Internet of Things and big data in dairy farm. In: 2021 International Wireless Communications and Mobile Computing, IWCMC 2021, June 2021 (pp. 138–140). https://doi.org/10.1109/IWCMC51323.2021.9498643.

Závodná, L.S., Činčalová, S., Kozáková, P., 2025. AI-driven dairy farming: A case study of technological adoption and organizational adaptation in a czech SME. Economics of Agriculture 72 (4), 1437–1452. Available at https://doi.org/10.59267/ekoPolj25041437S.

Index

Note: Page numbers followed by "*f*" and "*t*" refer to figures and tables, respectively.

K

L

M

N

O

P

Q

R

S

U

V

W

Y

Printed and bound by CPI Group (UK) Ltd, Croydon, CR0 4YY
14/09/2026
02216408-0008